SUPERSTRONG FIELDS IN PLASMAS

Related Titles from AIP Conference Proceedings

To learn more about these titles, or the AIP Conference Proceedings Series, please visit
the webpage **http://proceedings.aip.org**

SUPERSTRONG FIELDS IN PLASMAS

Second International Conference on
Superstrong Fields in Plasmas

Varenna, Italy   27 August–1 September 2001

EDITORS
Maurizio Lontano
Istituto di Fisica del Plasma, C.N.R., Milan, Italy

Gérard Mourou
*NSF Center for Ultrafast Optical Science
Ann Arbor, Michigan*

Orazio Svelto
Politecnico di Milano, Milan, Italy

Toshiki Tajima
*Lawrence Livermore National Laboratory
Livermore, California*

SPONSORING ORGANIZATION
Istituto di Fisica del Plasma "P. Caldirola", C.N.R.

Melville, New York, 2002
AIP CONFERENCE PROCEEDINGS ■ VOLUME 611

Editors:

Maurizio Lontano
Istituto di Fisica del Plasma "P. Caldirola"
Consiglio Nazionale delle Ricerche
Via R. Cozzi, 53
20125 Milano
ITALY

E-mail: lontano@ifp.mi.cnr.it

Gérard Mourou
NSF Center for Ultrafast Optical Science
University of Michigan
CUOS, 2200 Bonisteel Boulevard
Ann Arbor, MI 48109
USA

E-mail: mourou@eecs.umich.edu

Orazio Svelto
CEQSE-CNR
Dipartimento di Fisica del Politecnico
Piazza L. da Vinci, 32
20133 Milano
ITALY

E-mail: orazio.svelto@fisi.polimi.it

Toshiki Tajima
Lawrence Livermore National Lab.
7000 East Avenue, L-395
Livermore, CA 94551
USA

E-mail: tajima1@llnl.gov

Image on cover and title page: Snapshot of the electron probability distribution in a circular polarized laser field taken after 2.5 cycles of a low-frequency field (see page 28).

The article on pp. 63-74 is modified from Zeitschrift f. Phys. Chem. **215** (2001) 1563, with permission from Oldenbourg Wissenschaftsverlag GmbH.

L.C. Catalog Card No. 2002102567
ISBN 0-7354-0057-1
ISSN 0094-243X
Printed in the United States of America

CONTENTS

1. FUNDAMENTAL ATOMIC AND PLASMA PROCESSES AND NONLINEAR PHENOMENA IN THE PRESENCE OF ULTRA-INTENSE FIELDS

2. RELATIVISTIC NONLINEAR OPTICS OF ULTRA-SHORT LASER PULSES IN PLASMAS

4. APPLICATIONS OF ULTRA-INTENSE FIELDS: PARTICLE ACCELERATION, INTENSE X-RAY SOURCES, INERTIAL CONFINEMENT FUSION, AND LABORATORY ASTROPHYSICS

Preface

Over the past few years, laser intensities have dramatically increased. Today, relatively compact laser systems can produce on target intensities in the 10^{21} W/cm^2, that is four to five orders of magnitude greater than was previously possible. In the next few years it is predicted that the available intensities will be increased by two to three orders of magnitude. This revolution in laser intensities gives access to new physical parameter regimes of the electromagnetic field impossible to access before in a terrestrial laboratory.

Entirely new nonlinear optical phenomena are expected, with applications in a broad area of science and technology including ultrafast x-ray and gamma-ray as well as electron and ion sources, time-resolved diffraction and spectroscopy of ultrafast chemical, biological and physical reactions, novel fusion concepts and plasma astrophysics.

This second edition of the Conference, the first one held in 1997, intended to pursue a highly specialized venue of discussion of the most recent theoretical and experimental results on the interaction of ultrastrong fields with ionized matter. It also represented an opportunity to present new ideas in this quickly evolving research field and to gather scientists with different backgrounds in order to effectively tackle the interdisciplinary aspects of the matter.

The volume is divided into the following sections:

1. Fundamental atomic and plasma processes and nonlinear phenomena in the presence of ultra-intense fields.
2. Relativistic nonlinear optics of ultra-short laser pulses in plasmas.
3. Solid density plasmas, cluster plasmas, fast ions and nuclear physics with intense lasers.
4. Applications of ultra-intense fields: particle acceleration, intense x-ray sources, inertial confinement fusion, and laboratory astrophysics.
5. Lasers for ultrahigh intensity physics.
6. Applications of superstrong pulses to high energy physics.

In the course of the Conference a Forum on Strong Lasers in High Energy Physics has been organized. It consisted of few short topical presentations focused on superstrong lasers and their enabling novel applications to the frontiers of high energy physics and high energy accelerator physics. Sect.6 contains the highlights of this special session.

Maurizio Lontano
Gerard Mourou
Orazio Svelto
Toshiki Tajima

Conference Staff

Conference Chairmen

G. Mourou   O. Svelto   T. Tajima

International Scientific Committee

F. Amiranoff   C. Barty   T. Brabec
M. Campbell   V.M. Gordienko   M.H.R. Hutchinson
Y. Kato   M. Lontano (Secretary) G. Mourou
A. Mysyrowicz   F. Pegoraro   B. Remington
I.N. Ross   R. Sauerbrey   A.M. Sergeev
O. Svelto   T. Tajima   H. Takuma
D. von der Linde

Local Organizing Committee

D. Farina   M. Lontano (Chairman)   C. Allocchio

Istituto di Fisica del Plasma, Consiglio Nazionale delle Ricerche, Milan, Italy

Conference Secretariat

D. Pifferetti

Acknowledgments

The conference was fully supported by the Istituto di Fisica del Plasma "P. Caldirola", Consiglio Nazionale delle Ricerche, Milano, Italy.

1. FUNDAMENTAL ATOMIC AND PLASMA PROCESSES AND NONLINEAR PHENOMENA IN THE PRESENCE OF ULTRA-INTENSE FIELDS

Dynamics of Molecules and Clusters in Intense Laser-light Fields

Kaoru Yamanouchi

Department of Chemistry, School of Science, The University of Tokyo
7-3-1 Hongo, Bunkyo-ku, Tokyo 113-0033, Japan

Abstract. The investigation of the dynamical behavior of molecules in intense laser fields has afforded us invaluable opportunities to understand fundamentals of the interaction between molecules and light fields as well as to manipulate molecules using characteristics of laser light fields. In the present article, new directions of this rapidly growing interdisciplinary research fields are discussed by referring to our recent studies.

INTRODUCTION

It has been known that molecules behave in a characteristic way in intense laser fields (1-4). When the intensity of light fields is increased from a perturbative regime, the direction of a molecular beam can be deflected (5-6) and the axes of molecules are oriented along the laser polarization direction (7-10). In more intense fields as high as 10^{13} W/cm^2, light-dressed states (11) are formed through the coupling between electronic states of a molecule and a light field, and the nuclear dynamics of molecules can be described by the motion of the wave packet on the adiabatic light-dressed PESs. It was found for a number of molecules that the geometrical structure of a molecule is deformed within the ultrashort laser pulse duration, and this phenomenon was ascribed to the formation of such light dressed states in the intense laser fields (12-15).

When the light field intensity surpasses 10^{14} W/cm^2, a multiple ionization process tends to dominate, and multiply charged molecular ions begin to be formed. On the other hand, enhancement of the ionization process in a specific range of the bond length occurs, which is called an enhanced ionization (16-19). If the light field intensity increases more, it reaches a classical tunneling regime, where electrons escape from a distorted Coulombic potential of a molecule. This multiple ionization results in the formation of multiply charged molecules, which then break into atomic fragment ions having a large kinetic energy originating from the Coulombic repulsion force within a molecule. This type of fragmentation processes is called a Coulomb explosion.

In the present article, I introduce our recent studies on molecules and clusters in intense laser fields by newly developed methods such as pulsed gas electron diffraction, Coulomb explosion imaging, and tandem-mass spectroscopy, and discuss new research directions for the next stage of the interdisciplinary field called molecular science in intense laser fields.

CP611, *Superstrong Fields in Plasmas:* Second Int'l. Conf., edited by M. Lontano et al.
© 2002 American Institute of Physics 0-7354-0057-1/02/$19.00

PROBING MOLECULAR DYNAMICS BY GAS ELECTRON DIFFRACTION

In intense linearly polarized laser fields, it has been known that one of the principal axis of a molecule can be aligned along the laser polarization direction. This is due to the torque imposed on the molecule which is generated through the interaction between non-resonant light fields and an induced dipole moment of a molecule as interpreted by Friedrich and Hershbach [7].

One of the promising approaches to probe directly the extent of the alignment of molecules in intense laser fields may be the measurement of their electron diffraction pattern. From long ago, a gas electron diffraction method has been known as that determines geometrical structure of molecules with high precision. If a molecular axis is aligned along the laser polarization directions, such a process is expected to be reflected sensitively on the electron diffraction pattern.

Recently, we recorded a gas electron diffraction pattern of CS_2 molecules whose molecular axes are aligned along the polarization direction of the intense laser fields [20]. In this apparatus, a pulsed electron beam is generated by irradiating a surface of a photo-cathode with a UV laser pulse through the photoelectric effect. The electron packet is then accelerated to interact with a sample gas introduced into a diffraction chamber through a pulsed nozzle. Because the pulse width of the electron packet is determined by that of the UV laser, it is possible to generate an ultrashort pulsed electron packet when adopting an ultrashort UV laser pulse. A simulation of the electron trajectory showed that the temporal packet of the electron beam could become as short as 1 ps, indicating that real time ultrashort dynamics of molecules could be probed as a series of snap shots of the gas electron diffraction patterns.

When a diffraction pattern is observed for molecules aligned along the laser polarization direction, it could be significantly different from that observed for molecules randomly oriented in space. For example, in the case of CS_2, when the distribution of the molecular axis is isotropic, the diffraction halo pattern becomes concentric. On the other hand, if the molecular axis is aligned along the laser polarization direction, the resultant diffraction pattern becomes that compressed along the laser polarization direction. In our recent measurements performed under the laser field intensity of $\sim 10^{12}$ W/cm^2 showed that an effect of the alignment is clearly reflected in the observed diffraction pattern. Because a spacing between neighboring halos is sensitively dependent on the geometrical structure of molecules, this approach is also promising to probe the structural deformation process of molecules in intense laser fields.

On the theoretical side, Fujimura and coworkers proposed a new approach to design an optimal laser pulse shape for controlling orientation as well as alignment of molecules [21]. The term "orientation" is defined for polar molecules in which a specific direction is defined along one of the molecular axes. For example, in the case of CO molecules, if they are aligned along the vertically polarized laser fields so that O atom is placed above C atom, it can be said that the orientation is achieved. This type of control of molecular orientation would become an important factor to control chemical reaction on the solid state surface under the intense laser fields.

COULOMB EXPLOSION IMAGING OF MOLECULES

In the case of polyatomic molecules, there is a process of structural deformation which could compete with the alignment process. It has not been clear so far whether these two processes occur rather sequentially or almost simultaneously. Recently we investigated the alignment and structural deformation processes of CS_2 molecules by aligning them using intense nano-second laser fields ($\sim 10^{12}$ W/cm^2) and by ionizing them using circularly polarized ultrashort intense laser fields through the Coulomb explosion, and successfully probed the alignment and deformation processes in real time within a nano-second laser pulse (10). It was found that the structural deformation from linear to bent occurs as the extent of the alignment proceeds until the laser pulse reaches the maximum, and after passing the maximum, the recovery of the geometrical structure form bent to linear occurs as the laser field intensity decreases, simultaneously with the decrease in the extent of the alignment. The structural deformation of CS_2 identified in this relatively weak laser field suggests that a dressed state is efficiently formed with an electronic state having the bent equilibrium geometry.

When molecules are exposed to the strong short pulsed laser light whose intensity exceeds $\sim 10^{14}$ W/cm^2, a Coulomb explosion process becomes evident. In the case of diatomic and triatomic molecules, momentum vector distributions of atomic fragment ions are sensitively dependent on the geometrical structure of molecules just before the Coulomb explosion. In our recent studies, such ultrafast structural deformation has been investigated for N_2 (12,22,23), NO (12,30), SO_2 (22), CO_2 (13), NO_2 (14), H_2O (15) by the method called mass-resolved momentum imaging (MRMI) (22-24), in which momentum distributions of atomic and molecular fragment ions are plotted in the form of two dimensional or three dimensional plot.

From the analysis of the observed MRMI maps of diatomic and triatomic molecules, it was identified that the critical bond length at which the Coulomb explosion occurs increases commonly from ~ 1.4 r_e to ~ 2.7 r_e as the charge number of the parent ion increases, where r_e denotes the equilibrium internuclear distance. Furthermore, it was also identified that the increase in the critical bond length exhibits an even-odd alternation upon the increase of the total charge number of the parent molecular ions. The more striking finding was the existence of an ultrafast deformation process of molecular skeletal geometry of triatomic molecules.

In the case of CO_2, in intense laser fields, its linear structure is deformed towards bent structure, resulting in a largely spread bond angle distribution centering at the linear configuration with the mean amplitude of as large as $40°$. Similarly, in the case of NO_2 whose equilibrium structure is bent, it was found that the probability of taking linear configuration increases significantly, and in the case of H_2O its structure changes into linear with a broad mean amplitude of $\sim 60°$. As demonstrated in our series of studies, the MRMI method is a powerful technique to elucidate such ultrafast structural deformation of molecules in intense laser fields.

The geometrical deformation may be interpreted by considering a set of dressed state potentials which are formed through the strong mixing of molecular electronic states with intense laser fields. As investigated for H_2^+(25), a prepared wave packet would evolve on such a deformed PES, resulting in the deformation of skeletal geometry. For example, an electronic state whose equilibrium structure is linear may be coupled through the light fields with an electronic state whose equilibrium structure is bent, and a wave packet located first at the linear configuration would tend to spread towards along the bent direction as the laser field intensity increases, resulting in an ultrafast structural deformation.

It would be also important to determine momentum vectors of all the fragment ions ejected by a single event of the Coulomb explosion of polyatomic molecules in order to derive information regarding the shape of multi-dimensional light dressed potential energy surfaces, on which a nuclear wavepacket evolves until the instance of the Coulomb explosion. Recently, we conducted the measurements of double and triple coincidence imaging maps of mutiply charged CS_2 formed in intense laser fields using a position sensitive detector, and derived the specific geometrical structure of individual molecules just before the Coulomb explosion (26).

IDENTIFICATION OF FATE-DETERMING STAGE BY TANDEM-MASS SPECTROSCOPY

By measuring simply the momentum vector distributions of fragment ions, it would be difficult to identify whether the structural deformation occurs at the neutral stage, singly charged stage, or multiply charged stage, since the ionization process competes with the formation of the dressed states. In order to identify the ion stage at which crucial dynamics occurs through the major coupling among its electronic states induced by the light fields, it would be worthwhile to prepare a specific charge state of molecules by a mass separation technique and irradiate them with strong laser light. This type of study was performed recently by our group (27) using a tandem type time-of-flight (TOF) mass spectrometer. By the first stage of the TOF mass separation, mass and charge selected parent ions are prepared at the interaction region, where they are irradiated with ultrashort pulsed strong laser light, and the momentum distribution of fragment ions are measured by the second stage of the TOF mass separation.

The tandem TOF measurements were performed first for benzene cations. By comparing the ionization and fragmentation processes of neutral benzene with those of singly charged benzene cations, it was identified for the first time that the most dominant coupling between the electronic states are induced at the singly charged stage in ultrashort intense laser fields (~50 fs) at ~400 nm. On the other hand, when the wavelength was ~800 nm, the dominant products were found to be doubly charged parent molecules, and no evidence was found for a specific coupling among its electronic states. This remarkable difference in the response of molecules in the laser light fields at the different wavelengths suggests a possibility of controlling of molecular dynamics in intense laser light field by treating wavelength as one of the major control parameters.

On the theoretical side, in order to interpret the experimental findings of the ultrafast structural deformation of molecules in intense laser light fields, Kono, Koseki, and Fujimura (32) introduced a new idea in which molecular orbital calculations of PESs is combined with an electrostatic model of the charge distribution within a molecule, and investigated the ultrafast dynamics of CO_2 in intense laser fields. They were able to identify that the most crucial stage of the dynamics occurs at the doubly charged CO_2^{2+} stage, and that the geometrical structure of CO_2 just before the Coulomb explosion determined from our experiment could be interpreted mostly by the coupling between the molecular states and the light fields occurring at the doubly charged stage.

It should be noted that the tandem type mass spectroscopy enables us to investigate the responses of mass selected cluster ions to intense laser fields. In molecular clusters, molecules are bound with each other through weak and anisotropic intermolcular forces originating from electrostatic interactions such as charge-dipole, dipole-dipole, and dispersion interactions as well as from hydrogen bonding. In intense laser fields, the charge distribution within such clusters is expected to be strongly perturbed and a new class of intra-cluster reactions may proceed. Recently, we investigated the decomposition reaction of aniline and its complex with ammonia molecules, and identified that intra-cluster reactions exhibit a characteristic dependence on the number of ammonia molecules bound to the aniline moiety (29).

NEW FRONTIERS FOR MOLECULAR SPECTROSCOPY

As described above, the systematic investigation of the behavior of molecules in intense laser fields has been launched only very recently, and a wide range of progress of this new interdisciplinary research field is expected in the near future. For example, in our recent study of O_2^+, more than three electronic states were found to be coupled to form a sequence of dressed states (30). This study is the first example of the successful interpretation of the nuclear dynamics of multi-electron molecules in intense laser fields on the basis of the information from molecular spectroscopy.

Another important aspect of intense laser fields is its capability of generating short-pulsed short wavelength light such as vacuum ultraviolet (VUV) (31) and extreme ultraviolet (XUV) (32-34) light. It is expected that a phenomenon like dissociative ionization, which is characteristic of such a short wavelength region, may become a new target of molecular spectroscopy in the VUV and XUV wavelength regions. In our group, by focusing strong ultrashort laser pulses into a rare gas medium, short-pulsed high odd-order harmonics in the range from the 7th order to the 15th order were generated. In order to investigate the dissociative ionization process of molecules through the electronically highly excited states (31), they were irradiated with the harmonics with the specific order separated by a monochromator. The photon energy corresponding with this XUV wavelength region between 40~100 nm is in the same range of typical dissociation energy of chemical bonds in a molecule. Therefore, a laser based pump-and-probe experiment in this wavelength range would be promising to gain new insight into photochemistry initiated by the braking of a specific chemical bond.

ACKNOWLEDGEMENTS

The results of our studies introduced in the present article were obtained mainly through the CREST project supported by JST (Japan Science and Technology

Corporation). I would like to thank my colleagues, Dr. K. Someda, Dr. A. Hishikawa, Dr. K. Hoshina, Dr. A.Iwamae, Dr. M. Kono, Dr. S. Liu, Dr. T. Sako, Dr. Y. Fukuda, Dr. R. Itakura, and Dr. A. Iwasaki, Mr. H. Hasegawa, Mr. I. Maruyama, Ms. J. Watanabe for their efforts in making the project fruitful. Finally, I thank Dr. H. Todokoro, Mr. T. Ohshima, and Mr. Y. Oze for their support in the construction of the gas electron diffraction apparatus.

REFERENCES

1. *Molecules in Laser Fields*; Bandrauk, A. D., Ed.; M.Dekker, New York (1993).
2. Sheehy, B.; Dimauro, L. F. *Ann. Rev. Phys. Chem.* **1996**, 47, 463.
3. Codling, K.; Frasinski, L. J. *J. Phys. B: At. Mol. Opt.* **1993**, 26, 783.
4. Normand, D.; Lompre, L. A.; Cornaggia, C.; *J. Phys. B.: At. Mol. Opt.* **1992**, 25, 1497.
5. Sakai, H.; Tarasevitch, A.; Danilov, J.; Stapelfeldt, H.; Yip, R.W.; Ellert, C.; Constant, E.; Corkum, P. B.; *Phys. Rev. A* **1998**, 57, 2794.
6. Seideman, T.; *J. Chem. Phys.* **1997**, 106, 2881.
7. Friedrich, B.; Herschbach, D.; *Phys. Rev. Lett.* **1995**, 74, 4623.
8. Sakai, H.; Safvan, C. P.; Larsen, J. J.; Hilligsøe, K. M.; Hald, K.; Stapelfeldt, H.; *J. Chem. Phys.* **1999**, 110, 10235.
9. Larsen, J. J.; Sakai, H.; Safvan, C. P.; Larsen, I. W.; Stapelfeldt, H. *J. Chem. Phys.* **1999**, 111, 7774.
10. Iwasaki, A.; Hishikawa, A.; Yamanouchi, K. *Chem. Phys. Lett. in press.*
11. Giusti-Suzor, A.; Mies, F. H.; DiMauro, L. F.; Charron, E.; Yang, B. *J. Phys. B: At. Mol. Opt.* **1995**, 28, 309.
12. Yamanouchi, K.; Hishikawa, A.; Iwamae, A.; Liu, S. *Phys. Elec. At. Col., AIP Conference Proceedings* **2000**, 500, 182.
13. Hishikawa, A.; Iwamae, A.; Yamanouchi, K. *Phys. Rev. Lett.* **1999**, 83, 1127.
14. Hishikawa, A.; Iwamae, A.; Yamanouchi, K. *J. Chem. Phys.* **1999**, 111, 8871.
15. Liu, S.; Hishikawa, A.; Iwamae, A.; Yamanouchi, K.; *Advances in Multiphoton Processes and Spectroscopy* 13, Y. Fujimura and R.J.Gordon Eds.,World Scientific, 2000, p.189.
16. Codling, K.; Frasinski, L. J. *J. Phys. B: At. Mol. Opt.* **1993**, 26, 783.
17. Posthumus, J. H.; Frasinski, L. J.; Giles, A. J.; Codling, K. *J. Phys. B: At. Mol. Opt.* **1995**, 28, L349.
18. T.Seideman, T; Ivanov, M. Y.; Corkum, P. B. *Phys. Rev. Lett.* **1995**, 75, 2819.
19. Villeneuve, D. M.; Ivanov, M. Y.; Corkum, P. B. *Phys. Rev. A* **1996**, 54, 736.
20. Hoshina, K.; Yamanouchi, K.; Ohshima, T.; Ose, Y.; Todokoro, H. *in preparation.*
21. Hoki, K.; Fujimura, Y. *Chem. Phys.* **2001**, 267, 187.
22. Hishikawa, A.; Iwamae, A.; Hoshina, K.; Kono, M.; Yamanouchi, K. *Chem. Phys. Lett.* **1998**, 282, 283.
23. Hishikawa, A.; Iwamae, A.; Hoshina, K.; Kono, M.; Yamanouchi, K. *Chem. Phys.* **1998**, 231, 315.
24. Iwamae, A.; Hishikawa, A.; Yamanouchi, K. *J. Phys. B: At. Mol. Opt.* **2000**, 33, 223.
25. Maruyama, I; Sako, T.; Yamanouchi, K. *in preparation.*
26. Hasegawa, I; Hishikawa, A; and Yamanouchi, K. *Chem. Phys. Lett. in press.*
27. Itakura, R.; Watanabe, J.; Hishikawa, A.; Yamanouchi, K. *J. Chem. Phys.* **2001**, 114, 5598.
28. Kono, H.; Koseki, S.; Shirota, M.; Fujimura, Y. *J. Phys. Chem. A* **2001**, 105, 5627.
29. Watanabe, J.; Itakura, R.; Hishikawa, A.; Yamanouchi, K. *in preparation.*
30. Hishikawa, A.; Liu, S.; Iwasaki, A.; Yamanouchi, K. *J. Chem. Phys.* **2001**, 114, 9856.
31. Fukuda, Y.; Iwamae, A.; Hosaka, K.; Hoshina, K.; Hishikawa, A.; Yamanouchi, K. *Proceedings of the Second Symposium on Advanced Photon Research, Japan Atomic Energy Research Institute* (2001), p.288.
32. Sekikawa, T.; Ohno, T.; Yamazaki, T.; Nabekawa, Y.; Watanabe, S. *Phys. Rev. Lett.* **1999**, 83, 2564.
33. Tamaki, Y.; Itatani, J.; Nagata, Y.; Obata, M.; Midorikawa, K. *Phys. Rev. Lett.* **1999**, 82, 1422.
34. Nakano, H; Goto, Y; Lu, P; Nishiakwa, T; Uesugi, N. *Appl. Phys. Lett.* **1999**, 75, 2350.

Non-Dipole and Relativistic Effects in Laser-Atom Interactions

C. J. Joachain*, N. J. Kylstra† and R. M. Potvliege†

*Physique Théorique, Université Libre de Bruxelles, C.P. 227, Boulevard du Triomphe,
B-1050 Bruxelles, Belgium
†Department of Physics, University of Durham, Durham DH1 3LE, United Kingdom

Abstract. Descriptions of multiphoton phenomena based on the non-relativistic Schrödinger equation and the dipole approximation can break down at high laser intensities currently available. We first discuss how non-dipole effects, of order $1/c$, due to the magnetic field component of a laser pulse, influence harmonic generation and atomic stabilization in the non-relativistic regime. Particular attention is devoted to photon emission by ions interacting with ultra-short, intense laser pulses. We then review progress in the theoretical study of relativistic effects arising when atoms interact with ultra-strong laser fields.

INTRODUCTION

Nearly all theoretical studies of multiphoton processes occurring in laser-atom interactions have assumed that the laser field can be described in the dipole approximation by a spatially homogeneous vector potential $\mathbf{A}(t)$. The electric field component of the laser field is then given by $\mathcal{E}(t) = -d\mathbf{A}(t)/dt$, and the magnetic field component vanishes, since $\mathcal{B} = \nabla \times \mathbf{A}(t) = 0$. In addition, relativistic effects have usually been neglected.

Within this non relativistic framework and the dipole approximation, strong field phenomena at *low* frequencies can be understood qualitatively in terms of a semi-classical recollision model developed by Kuchiev [1], van Linden van den Heuvell and Muller [2], Corkum [3] and Kulander, Schafer and Krause [4]. This model, often called the "simple man's model", is based on the idea that strong field ionization and harmonic generation proceed via several steps at low frequencies. In the first (bound - free) step, the active electron is detached from its parent ion (atom) core by tunneling or over-the-barrier ionization. In the second (free-free) step, the unbound electron interacts mainly with the laser field, so that its dynamics is essentially that of a free electron in the field, and can be treated to good approximation by using classical mechanics. As the phase of the field reverses, the electron can be accelerated back towards the parent core. If the electron does not return to the core, single ionization will occur. If it does return, then, in a third step, scattering or radiative recombination with the core may occur, leading to single or multiple ionization or to harmonic generation.

Strong field phenomena at *high* frequencies can be understood within the high-frequency Floquet theory of Gavrila and co-workers [5, 6]. The theory predicts that in this regime atoms can become stable against ionization with increasing laser intensity. Termed "atomic stabilization", this phenomenon has been extensively studied theoreti-

CP611, *Superstrong Fields in Plasmas:* Second Int'l. Conf., edited by M. Lontano et al.

cally in the dipole approximation, within the high-frequency Floquet theory as well as by using the Sturmian-Floquet method [7], the R-matrix-Floquet theory [8] and by direct numerical integration of the time-dependent Schrödinger equation [9].

At high laser intensities $(I > 10^{16} \text{ W cm}^{-2})$ currently available, these relatively simple descriptions of harmonic generation and stabilization can break down since non-dipole effects (e.g. magnetic field effects) and relativistic effects may become important [10, 11, 12, 13]. In the next two sections, we shall give a survey of classical and quantum mechanical methods which have been used to analyze non-dipole and relativistic effects in laser-atom interactions. Atomic units (a.u.) are used throughout the paper, unless otherwise indicated.

CLASSICAL CALCULATIONS

The Classical Equations of Motion

Consider a laser pulse linearly polarized along $\hat{\mathbf{x}}$ and propagating along $\hat{\mathbf{z}}$ with wave vector $\mathbf{k} = k\hat{\mathbf{z}}$, where $k = \omega/c$ and ω is the laser angular frequency. This pulse is described by the vector potential $\mathbf{A}(\eta) = A(\eta)\hat{\mathbf{x}}$, where $\eta = \omega t - \mathbf{k}\cdot\mathbf{r} = \omega(t - z/c)$. The corresponding electric and magnetic fields are $\boldsymbol{\mathcal{E}}(\eta) = -\partial\mathbf{A}(\eta)/\partial t = \mathcal{E}(\eta)\hat{\mathbf{x}}$ and $\boldsymbol{\mathcal{B}}(\eta) = \boldsymbol{\nabla} \times \mathbf{A}(\eta) = \mathcal{B}(\eta)\hat{\mathbf{y}}$, with $\mathcal{B}/\mathcal{E} = 1/c$. The Lorentz equation for an electron in this laser pulse is

$$\frac{\mathrm{d}}{\mathrm{d}t}\mathbf{p} = -\left[\boldsymbol{\mathcal{E}}(\eta) + \mathbf{v} \times \boldsymbol{\mathcal{B}}(\eta)\right] \tag{1}$$

where $\mathbf{p} = \gamma\mathbf{v}$, $\gamma = (1 - \beta^2)^{-1/2}$, $\beta = v/c$ and v is the magnitude of the electron velocity $\mathbf{v}$.

Neglecting all terms of order $1/c$, one finds that the electron motion is one-dimensional along the polarization axis $\hat{\mathbf{x}}$, with

$$\frac{\mathrm{d}}{\mathrm{d}t}v_x = -\mathcal{E}(\omega t) \tag{2}$$

Keeping terms to leading order in $1/c$, it is found that the electron motion is two-dimensional in the (xz) plane, with

$$\frac{\mathrm{d}}{\mathrm{d}t}v_x = -\mathcal{E}(\omega t) \quad , \quad \frac{\mathrm{d}}{\mathrm{d}t}v_z = -v_x\mathcal{B}(\omega t). \tag{3}$$

The electron motion along the polarization axis $\hat{\mathbf{x}}$ is unaffected by the magnetic field.

Relativistic Effects

As is well known, relativistic effects must be taken into account when describing target atoms or ions with high atomic number Z. Relativistic effects also occur when atoms interact with super-strong laser fields and are expected to be important when the electron quiver, or ponderomotive, energy $U_p = \mathcal{E}_0^2/(4\omega^2)$ is of the order of its rest mass energy so that the quantity

$$q = \frac{U_p}{c^2} = \frac{\mathcal{E}_0^2}{4\omega^2 c^2} \approx 1.33 \times 10^{-5} \left(\frac{\mathcal{E}_0}{\omega}\right)^2 \tag{4}$$

is of the order of unity. Thus, if $\omega = 0.043$ a.u., corresponding to a Nd:YAG laser, we see that $q = 1$ when $\mathcal{E}_0 = 11.8$ a.u., that is, when the intensity $I = 4.9 \times 10^{18}$ W cm^{-2}. For a Ti:Sapphire laser, $q = 1$ when $\mathcal{E}_0 = 15.6$ a.u., that is, when the intensity $I = 8.6 \times 10^{18}$ W cm^{-2}. For a laser of higher angular frequency $\omega = 1$ a.u., we have $q = 1$ when $\mathcal{E}_0 = 274$ a.u., corresponding to the very large intensity $I = 2.6 \times 10^{21}$ W cm^{-2}.

An important relativistic effect is the relativistic mass shift or dressing of the electron mass. This effect has been studied extensively [14, 15], and it was found that a relativistic electron propagates in an electromagnetic field like a particle with the variable ("dressed") mass $m^* = m(1 + 2q)^{1/2}$ which increases with the intensity of the field.

Magnetic field effects

To investigate the effects due to the magnetic component of the laser field, we start from the equations (3), obtained through leading order in $1/c$. For an electron initially at rest at the origin, the displacement along the polarization ($\hat{x}$) direction and the propagation ($\hat{z}$) direction are given respectively by

$$x(t) = \alpha(t) = \int^t A(\omega t')\, dt', \qquad z(t) = \frac{1}{2c} \int^t A^2(\omega t')\, dt'. \tag{5}$$

Since the electron velocity in the propagation direction, $v_z(t) = v_x^2(t)/(2c)$, is never negative, the drift in the propagation direction increases monotonically during the pulse. In particular, for a step pulse of electric field amplitude $\mathcal{E}_0$, we have

$$x(t) = \alpha(t) = \alpha_0 \sin \omega t, \qquad z(t) = \frac{\mathcal{E}_0^2}{2c\omega^2}\left[\frac{t}{2} + \frac{\sin(2\omega t)}{4\omega}\right], \tag{6}$$

where $\alpha_0 = \mathcal{E}_0/\omega^2$ is the (non-relativistic) excursion amplitude of a classical electron. The drift per cycle in the propagation direction is given by

$$\chi_0 \approx \mathcal{E}_0^2/(c\omega^3) = 4q/k. \tag{7}$$

The electron dynamics is essentially non-relativistic and the displacement of the electron in the propagation direction is comparable to, or larger than, the width of the electron wave packet, a, when $q \ll 1 \lesssim q/(ka)$. These inequalities, in turn, imply that $ka \ll 1$, which is the usual requirement for the applicability of the dipole approximation. However, for intense laser pulses the criterion leads to a long wavelength approximation in which both the electric and magnetic fields of the laser are spatially homogeneous.

Taking the width a of the electron wave packet to be one Bohr radius, and choosing $\omega = 0.057$ a.u. (corresponding to a Ti:Sapphire laser), it is found that $\chi_0 \approx a$ when the intensity $I \approx 1 \times 10^{15}$ W cm^{-2}, while for $\omega = 1$ a.u., intensities of approximately $I \approx 5 \times 10^{18}$ W cm^{-2} are required. The magnetic field component of a laser pulse should therefore be expected to start playing a significant role if the peak intensity exceeds these estimates. As shown below, this is indeed the case for high frequency stabilization. However, in harmonic generation by ions driven by a Ti:Sapphire field the spreading of the electron wave packet over a cycle is considerable, which means that higher intensities are required (roughly 10^{17} W cm^{-2}) for non dipole effects to become important. Relativistic effects (for which $q \approx 1$) require still higher intensities, as we have seen above, and as a consequence there is an intermediate regime of intensity where the dipole approximation is not valid but the dynamics is non relativistic.

Classical Monte-Carlo Calculations

Relativistic classical Monte-Carlo calculations for models of atomic hydrogen have been performed by Kyrala [16]. More recently, Keitel, Knight and Burnett [17] have studied the relativistic high-harmonic generation from a classical, highly excited, one-dimensional model hydrogen atom, for an intensity of 3.6×10^{20} W cm^{-2} and an angular frequency $\omega = 0.1$ a.u. They found that relativistic effects are significant.

Monte-Carlo calculations of ionization and harmonic generation have been performed in the high-frequency, high-intensity regime by Keitel and Knight [18], for a three-dimensional hydrogen atom. They found that stabilization is inhibited when the magnetic field is effective in inducing a motion of the electron in the propagation direction. This result is in agreement with comments by Grochmalicki et al. [19] and Katsouleas and Mori [20], who indicated the possibility of a breakdown of stabilization due to the failure of the dipole approximation.

QUANTUM TREATMENTS

The Non-Relativistic, Non-Dipole Schrödinger Equation

The non-relativistic, non-dipole time-dependent Schrödinger equation (TDSE) for a one-electron atom (ion) in a laser pulse linearly polarized along $\hat{\mathbf{x}}$ and propagating along $\hat{\mathbf{z}}$ reads (in a.u.)

$$i\frac{\partial}{\partial t}\Psi(\mathbf{r},t) = \left\{\frac{1}{2}[-i\nabla + \mathbf{A}(\eta)]^2 + V(r)\right\}\Psi(\mathbf{r},t), \qquad (8)$$

where $V(r)$ is the potential binding the electron and we recall that $\mathbf{A}(\eta) = A(\eta)\hat{\mathbf{x}}$ with $\eta = \omega(t - z/c)$. To lowest order in $1/c$, we have $\mathbf{A}(\eta) = \mathbf{A}(\omega t) + (z/c)\boldsymbol{\mathcal{E}}(\omega t)$. Transforming to the length gauge by writing $\Psi^L(\mathbf{r},t) = \exp[i\mathbf{A}(\omega t)\cdot\mathbf{r}]\Psi(\mathbf{r},t)$, we obtain for $\Psi^L(\mathbf{r}, t)$ the non-dipole TDSE

$$i\frac{\partial}{\partial t}\Psi^L(\mathbf{r}, t) = \left\{-\frac{1}{2}\nabla^2 + \left[\mathbf{r} - i\frac{z}{c}\nabla\right]\cdot\boldsymbol{\mathcal{E}}(\omega t) + V(r)\right\}\Psi^L(\mathbf{r}, t). \tag{9}$$

An exact solution of equation (9) can be obtained when $V(r) = 0$. This non-relativistic, non-dipole Volkov wave function is given by [21]

$$\Phi_{\mathbf{p}}(\mathbf{r}, t) = (2\pi)^{-\frac{3}{2}}\exp\left(i\mathbf{\Pi}(\mathbf{p}, t)\cdot\mathbf{r} - \frac{i}{2}\int^t\left[\mathbf{\Pi}(\mathbf{p}, t')\right]^2 dt'\right), \tag{10}$$

with $\mathbf{\Pi}(\mathbf{p}, t) = \mathbf{p} + \mathbf{A}(\omega t) + \frac{1}{c}\left[\mathbf{p}\cdot\mathbf{A}(\omega t) + \frac{1}{2}A^2(\omega t)\right]\hat{\mathbf{z}}$. This wave function describes the dynamics of a free electron interacting with a laser pulse in the intermediate intensity regime that lies between the non-relativistic regime where non-dipole effects can be neglected and the fully relativistic regime. It reduces to the usual non-relativistic dipole Volkov wave function when $1/c \to 0$.

Harmonic Generation Beyond the Dipole Approximation

A fully quantum mechanical description of harmonic generation that incorporates the basic ideas of the semi-classical "recollision model" in the non-relativistic limit and in the dipole approximation has been developed by Lewenstein *et al.* [22, 23]. This approach uses the strong field approximation (SFA) which consists of neglecting the interaction of the active electron with the core after it is detached by the field and before it scatters or recombines as it returns near the nucleus. In this approach, the unbound electron can thus be described by a non-relativistic Volkov wave packet. Using the non-dipole Volkov wave function introduced above makes it possible to calculate the dipole moment of the atom, $\mathbf{d}(t)$, taking into account the acceleration imparted to the electron by the magnetic field component of the pulse [21]. The spectrum of the emitted photons is then obtained by calculating $|\hat{\mathbf{x}}\cdot\mathbf{a}(\Omega)|^2$, for emission polarized parallel to the polarization direction of the incident pulse, and $|\hat{\mathbf{z}}\cdot\mathbf{a}(\Omega)|^2$, for emission polarized along the direction of propagation of the incident pulse; in these expressions, Ω denotes the angular frequency of the emitted photon and $\mathbf{a}(\Omega)$ the Fourier transform of $\ddot{\mathbf{d}}(t)$. It is worth noting that emission polarized in the z direction is forbidden by the dipole selection rules and therefore would not occur if the magnetic field component of the pulse was neglected.

As an illustration, we shall consider the photon emission by ions interacting with intense, ultra-short laser pulses. This phenomenon has attracted considerable interest in recent years, especially with respect to the possibility of providing a source of coherent, tunable, attosecond X-ray pulses. Research in this domain has been spurred by the development of laser systems capable of delivering ultra-short pulses consisting of only a few optical cycles, with peak intensities well above 10^{15} W cm^{-2} [24]. Indeed, atoms or ions exposed to such pulses can experience much stronger laser fields before ionizing that would be possible in longer pulses, and this in turns permits the generation of photons of much higher energies. Moreover, positive ions can survive

13

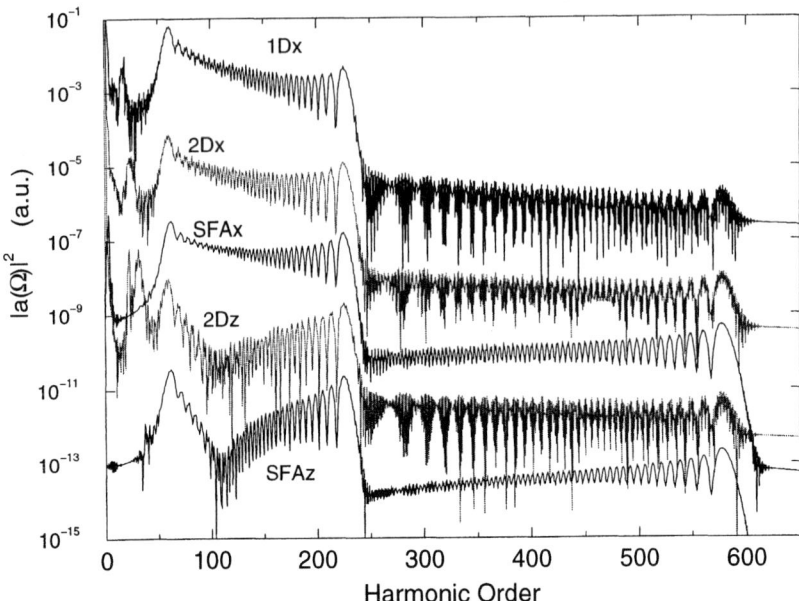

FIGURE 1. The magnitude squared of the Fourier transform of the dipole acceleration of He$^+$, in a.u., as a function of the photon energy (in units of $\hbar\omega$). As indicated, spectra were obtained for a 1D model, a 2D model and using the SFA (see text). Spectra for the emission of photons polarized along the laser polarization direction (1Dx, 2Dx, SFAx) as well as for emission with polarization parallel to the laser propagation direction (2Dz, SFAz) are shown. A two-cycle pulse with $\mathcal{E}_0 = 0.4$ a.u. and a carrier wavelength of 800 nm was used. From Kylstra *et al.* [21] and Potvliege *et al.* [27].

higher laser intensities than neutral atoms and therefore can generate more energetic photons [25, 26, 27].

In Fig. 1, we show the results obtained by Kylstra, Potvliege and Joachain [21] for photon emission by He$^+$ interacting with a two-cycle pulse with carrier wavelength $\lambda = 800$ nm (corresponding to a Ti-Sapphire laser). The peak intensity is 5.6×10^{15} W cm^{-2}. The results obtained by using the SFA including non-dipole effects, for the 3D ion, are compared with those obtained by Potvliege, Kylstra and Joachain [27] by integrating numerically the non-dipole time-dependent Schrödinger equation in two dimensions, with $\mathbf{r} \equiv (x, z)$, for a model He$^+$ ion. Also shown are results calculated in the dipole approximation for a one-dimensional model of the ion. The harmonic order is defined as Ω/ω. All the important features of the spectra are the same in the three calculations, despite their different dimensionality. They can be understood within the simple man's model of harmonic generation. For example, the two plateaus clearly visible in the figure arise from the radiative recombination of electrons ionized within different half-cycles of the laser pulse (each half-cycle contributes differently to the spectrum because of the rapid variation of the electric field amplitude during the pulse). We stress in particular that the non-dipole SFA reproduces all the main features of the spectra obtained by solving the time dependent Schrödinger equation for the 2D model. Differences in the shapes of the plateaus between the two calculations are expected in

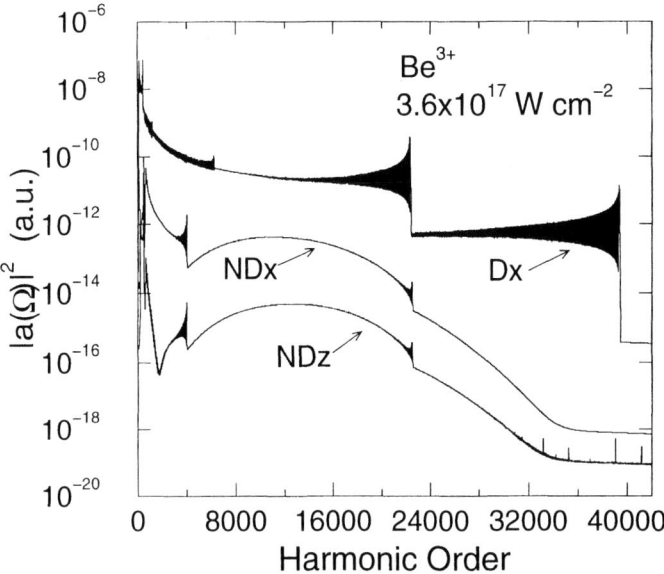

FIGURE 2. The magnitude squared of the Fourier transform of the SFA dipole acceleration of Be^{3+}, in a.u., as a function of the photon energy (in units of $\hbar\omega$). Dipole results (Dx) are displayed, as well as non-dipole results for emitted photons polarized along the laser polarization direction (NDx) and for emitted photons polarized along the laser propagation direction (NDz). A three-cycle pulse with $\mathcal{E}_0 = 3.2$ a.u. and a carrier wavelength of 800 nm was used.

view of the dependence of the dipole moments on the dimensionality of the model. The SFA results are lower than the 2D ones, and the 2D results lower than the 1D ones, but the ratio $|\hat{\mathbf{z}} \cdot \mathbf{a}(\Omega)|^2 / |\hat{\mathbf{x}} \cdot \mathbf{a}(\Omega)|^2$ is the same. There is no appreciable difference in the values of $|\hat{\mathbf{x}} \cdot \mathbf{a}(\Omega)|^2$ obtained with and without making the dipole approximation. We note in this respect that the displacement per half cycle due to the magnetic drift, $\chi_0/2$, which is equal to 5 a.u. at the intensity of 5.6×10^{15} W cm^{-2}, is much smaller than the width of the returning electron wave packet.

In Fig. 2, we show the photon emission by Be^{3+} interacting with a 3-cycle Ti:Sapphire pulse at the much higher peak intensity of 3.6×10^{17} W cm^{-2}. The pulse being longer than in Fig. 1, more plateaus are visible in the spectra. The magnetic field component of the pulse now affects photon emission markedly, as can be seen by comparing dipole and non dipole results: it causes a 'bending over' of the plateaus and reduces the strength of photon emission by two to four orders of magnitude. The change in the shape of the plateaus is also found in SFA calculations based on the Klein-Gordon equation [28]. The decrease in the strength of emission should be expected; the magnetic drift, 317 a.u. per half cycle at 3.6×10^{17} W cm^{-2}, is indeed sufficiently large for considerably reducing the overlap of the returning wave packet with the nucleus (if the electron is emitted in the direction of polarization of the field, as is assumed in the simple man's model). Because of the growing importance of the magnetic drift, photon emission saturates above 10^{17} W cm^{-2} peak intensity at the Ti:Sapphire wavelength [21, 29].

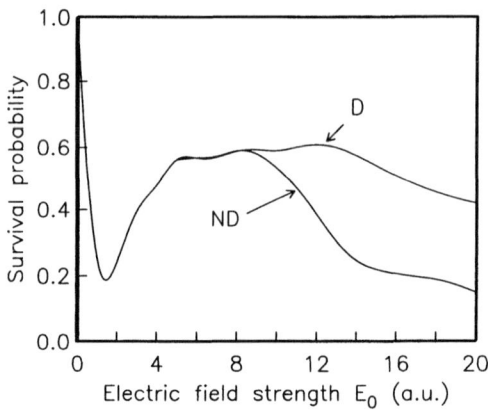

FIGURE 3. Survival probability as a function of the maximum electric field strength, $\mathcal{E}_0$, of the laser pulse. A 12 cycle trapezoidal pulse (3 cycle turn-on and turn-off, 6 cycle constant intensity) was used and $\omega = 1$ a.u. Dipole (D) and non-dipole (ND) results are shown, with the non-dipole results being obtained using the Hamiltonian containing the fully spatially-dependent vector potential. From Vazquez de Aldana *et al.* [33].

High-Frequency Stabilization Beyond the Dipole Approximation

We shall now discuss the influence of non-dipole effects on atomic stabilization in the high-frequency high-intensity regime. The validity of the dipole approximation in this regime has been examined for atomic hydrogen by Bugacov, Pont and Shake-shaft [30] who included multipole terms beyond the dipole approximation, and by Latinne, Joachain and Dörr [9], who solved numerically the non-dipole time-dependent Schrödinger equation for an angular frequency of $\omega = 2$ a.u. Latinne *et al.* found only small modifications in the ionization probability of the hydrogen atom ground state, up to an intensity $I = 2.4 \times 10^{19}$ W cm^{-2}.

More recently, Vasquez de Aldana and Roso [31], Kylstra *et al.* [32] and Vasquez de Aldana *et al.* [33] have studied a two-dimensional model hydrogen atom, described by a soft-core potential, interacting with a short laser pulse. The two-dimensional TDSE with a laser-atom interaction Hamiltonian containing the fully spatially dependent vector potential was numerically integrated on a uniform grid. The probability that the atom does not ionize during the laser pulse was determined by projecting the wave function at the end of the laser pulse onto the bound field-free states of the atom.

In Fig. 3, the ionization probability is shown as a function of the peak electric field strength of the laser, $\mathcal{E}_0$, for an angular frequency of $\omega = 1$ a.u. Results obtained in the dipole approximation (D) and non-dipole results are given. A trapezoidal pulse with three cycles turn-on (turn-off) and six cycles constant intensity was used. From the results, it is seen that the dipole approximation breaks down when $\mathcal{E}_0 > 10$ a.u. For this field strength, the displacement per cycle in the laser propagation direction is about 1 a.u. This displacement is comparable to the width of the electron wave packet along the z direction and is sufficient to reduce the influence to the atomic potential, thereby disrupting the electron dynamics required for stabilization. This leads to a pronounced decrease in the survival probability.

Relativistic Calculations

A quantum mechanical treatment of relativistic effects in the interaction of atoms with super-intense laser pulses requires the numerical solution of the time-dependent Dirac equation, which is a formidable task. It is therefore important to develop approximation methods, particularly in the low frequency (tunneling or over-the-barrier ionization) regime. In this case, relativistic effects essentially manifest themselves in the dynamics of the ejected electron in the laser field. The quasi-static tunneling theories of Keldysh [34] and Ammosov, Delone and Krainov [35] have been generalized to the relativistic regime by Krainov and Shokri [36] for linear polarization and by Krainov [37] for circular polarization. Similarly, a relativistic version of the strong field (KFR) approximation of Keldysh [34], Faisal [38] and Reiss [39] has been introduced by Crawford and Reiss [40].

The situation is different for the case of a high-frequency laser, i.e. in the stabilization regime. A generalization of the high-frequency Floquet theory [5, 6] to the relativistic high-frequency domain has been proposed by Kaminski [41] and Krstić and Mittleman [42]. Ermolaev [43] has used a one-dimensional version of the high-frequency approximation of Krstić and Mittleman to study the effect of the relativistic electron mass shift on the bound states of one-electron atoms in super-intense laser fields.

We now turn to ab-initio quantum mechanical treatments of relativistic effects in laser-atom interactions at high frequencies, in which time-dependent relativistic wave equations are solved numerically. The problem of solving the time-dependent Dirac equation for an atom in a super-intense laser field scales as $\mathcal{E}_0^3/\omega^4$, which favors the high-frequency regime. However, even in this case the calculations are extremely demanding numerically, and for this reason have been restricted to lower dimensionality. In addition, several of the calculations have been performed using the relativistic Schrödinger equation instead of the Dirac equation in order to reduce the computational effort. We shall discuss these calculations first. Protopapas *et al.* [44] have solved numerically the time-dependent relativistic Schrödinger equation in the Kramers frame for a model, one-dimensional atom in the high-frequency, high-intensity regime. Of course, retardation and magnetic field effects are not included in one-dimensional model calculations. On the other hand, relativistic effects due to the dressing of the electron mass by the laser field can be investigated. Protopapas *et al.* found differences between the relativistic and non-relativistic treatments, the stabilization of the model atom being increased when solving the relativistic Schrödinger equation rather than its non-relativistic counterpart. They attributed this difference to the relativistic mass shift of the electron in the field.

Taïeb *et al.* [45] have solved numerically the time-dependent relativistic Schrödinger equation for a one-dimensional atom in a two-color field. A high-frequency ($\omega_H = 10$ a.u.), ultra-intense ($I_H = 5 \times 10^{21}$ W cm^{-2}) laser pulse is used to dress the atom in the stabilization regime. A low frequency ($\omega_L = 0.2$ a.u.), less intense ($I_L = 10^{15}$ W cm^{-2}) laser pulse is then used to probe the wave function. Taïeb *et al.* found that the peaks in

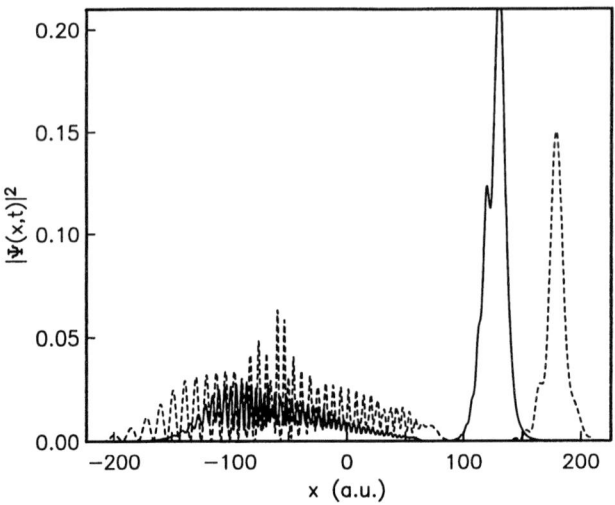

FIGURE 4. Probability densities of the Dirac (solid line) and Schrödinger (dashed line) wave packets at the end of the ninth cycle for a laser pulse with a four cycle $\sin^2$ turn-on, an angular frequency of $\omega = 1$ a.u., and a peak electric field of strength $\mathcal{E}_0 = 175$ a.u. From Kylstra *et al.* [46], with permission from IOP Publishing Limited.

the electron spectrum ("above threshold ionization" peaks) are shifted with respect to the corresponding non-relativistic calculation.

Let us now consider the time-dependent Dirac equation. Kylstra *et al.* [46] and Lenz *et al.* [47] have solved this equation numerically for a model one-dimensional atom. In one dimension, the four-component Dirac equation reduces to two uncoupled, two-component equations, corresponding to the two possible orientations of the electron spin. The calculations of Kylstra *et al.* were performed in momentum space, and results were obtained in the high-frequency, high-intensity regime. The wave function in momentum space was expanded on a basis of field-free (positive and negative energy) eigenstates, obtained by diagonalizing the field-free Dirac Hamiltonian using a basis of B-spline functions. The resulting differential equations were then propagated in time. It should be noted that for an angular frequency of $\omega = 1$ a.u., relativistic effects become important when, in a.u., $\mathcal{E}_0/c \simeq 2$. The magnitude of the laser-interaction term in the Dirac Hamiltonian is then of order c^2, which is comparable to the mass term. As a result, propagation in time requires time steps such that $\Delta t \ll 1/c^2$ a.u. Kylstra *et al.* found that for a peak electric field strength $\mathcal{E}_0 = 175$ a.u. and an angular frequency $\omega = 1$ a.u. (such that $q = 0.4$), relativistic effects become apparent. However, even under these extreme conditions, the Dirac wave function remains essentially localized in a superposition of field-free bound states and low-energy continuum states. We show in Fig. 4 the results of Kylstra *et al.* for the Dirac and Schrödinger probability densities, at the end of the 9th laser cycle, when the electric field is maximum. The peak in the Dirac probability density corresponds to the relativistic classical excursion amplitude, 124 a.u. The peak in the Schrödinger probability density occurs at $\alpha_0 = 175$ a.u., which is the value of the non-relativistic classical excursion amplitude. Kylstra *et al.* also found that the coupling, by the laser field, of the field-free positive and negative energy states is responsible for the relativistic corrections to the quiver motion. At the end of the laser pulse, the ionization probability is 0.52 for the Dirac wave function and 0.58 for the Schrödinger wave function, showing that the Dirac wave function is more stable against ionization. The energy distribution of the ionized electrons, obtained from the Dirac wave function at the end of the pulse, shows that most of the electrons are emitted at low energies.

Rathe *et al.* [48] have solved numerically the Dirac equation for a model two-dimensional atom in a super-intense, high-frequency laser pulse. They demonstrated the effects of the magnetic field on the wave packet and spin dynamics of the system. In particular, the electron was found to be pushed in the laser pulse propagation direction and moved with a superimposed figure-of-eight motion towards the absorbing boundaries of the box. The spin-flip probability was found to be substantial.

CONCLUSIONS

Important progress has been made recently in the study of non-dipole and relativistic effects in laser-atom interactions at high intensities. In particular, a non-dipole strong field approximation has been developed and applied to study photon emission by ions interacting with intense, ultra-short pulses of carrier wavelength $\lambda = 800$ nm, and it was found that the magnetic field component of the pulse is significant beyond a peak intensity of 10^{17} W cm^{-2} at this wavelength. It has also been shown that non-dipole effects can strongly modify the laser-atom interaction in the high-frequency, high-intensity (stabilization) regime.

Relativistic effects have been studied, especially by performing three-dimensional Monte Carlo simulations and quantum treatments of model one- and two-dimensional atoms using the time-dependent relativistic Schrödinger equation, Klein-Gordon and Dirac equations, in the high-frequency high-intensity regime. Dressed mass effects, magnetic field effects and spin effects have been investigated, but our present understanding of relativistic phenomena at ultra-high intensities is still incomplete. However, this domain of research is evolving rapidly, and a number of new research areas are being explored. Examples include the investigation of spin effects in intense laser fields [49] and atoms subjected to an intense laser pulse and a strong, static magnetic field [50].

REFERENCES

1. Kuchiev, M. Y., *JETP Lett.*, **45**, 404 (1987).
2. van Linden van den Heuvell, H. B., and Muller, H. G., "Limiting Cases of Excess-Photon Ionization", in *Multiphoton Processes*, edited by S. J. Smith and P. L. Knight, Cambridge Univ. Press, 1988, p. 25.
3. Corkum, P. B., *Phys. Rev. Lett.*, **71**, 1994 (1993).
4. Kulander, K. C., Schafer, K. J., and Krause, J. L., "Dynamics of Short-Pulse Excitation, Ionization and Harmonic Conversion", in *Super-Intense Laser-Atom Physics*, edited by B. Piraux, A. L'Huillier, and K. Rzazewski, Plenum Press, New York, New York, 1993, p. 95.
5. Gavrila, M., *Adv. At. Mol. Opt. Phys. Suppl.*, **1**, 435 (1992).
6. Gavrila, M., "Stabilization of Atoms in Ultra-Strong Laser Fields: a Decade Later", in *Multiphoton Processes*, edited by L. DiMauro, R. R. Freeman, and K. C. Kulander, AIP Conf. Proc. No. 525,

2000, p. 103.

7. Dörr, M., Potvliege, R. M., Proulx, D., and Shakeshaft, R., *Phys. Rev. A*, **43**, 3729 (1991).
8. Dörr, M., Burke, P. G., Joachain, C. J., Noble, C. J., Purvis, J., and Terao-Dunseath, M., *J. Phys. B*, **26**, L 275 (1993).
9. Latinne, O., Joachain, C. J., and Dörr, M., *Europhys. Lett.*, **26**, 333 (1994).
10. Protopapas, M., Keitel, C. H., and Knight, P. L., *Rep. Progr. Phys.*, **60**, 389 (1997).
11. Joachain, C. J., Dörr, M., and Kylstra, N. J., *Adv. At. Mol. Opt. Phys.*, **42**, 225 (2000).
12. Joachain, C. J., and Kylstra, N. J., *Laser Phys.*, **11**, 212 (2001).
13. Kylstra, N. J., Potvliege, R. M., Worthington, R. A., Patel, A. S., Knight, P. L., Vazquez de Aldana, J. R., Roso, L., and Joachain, C. J., "Intense Laser-Atom Interactions: Beyond the Dipole Approximation", in *Super-Intense Laser-Atom Physics*, edited by B. Piraux and K. Rzazewski, Kluwer Academic, 2001, p. 345.
14. Brown, L. S., and Kibble, T. W. B., *Phys. Rev.*, **133A**, 705 (1964).
15. Eberly, J. H., and Sleeper, A., *Phys. Rev.*, **176**, 1570 (1968).
16. Kyrala, G. A., *J. Opt. Soc. Am. B*, **4**, 731 (1987).
17. Keitel, C. H., Knight, P. L., and Burnett, K., *Europhys. Lett.*, **24**, 539 (1993).
18. Keitel, C. H., and Knight, P. L., *Phys. Rev. A*, **51**, 1420 (1995).
19. Grochmalicki, J., Lewenstein, M., and Rzążewski, K., *Phys. Rev. Lett.*, **66**, 1038 (1991).
20. Katsouleas, T., and Mori, W. B., *Phys. Rev. Lett.*, **70**, 1561(C) (1993).
21. Kylstra, N. J., Potvliege, R. M., and Joachain, C. J., *J. Phys B*, **34**, L55 (2001).
22. Lewenstein, M., Balcou, P., Ivanov, M. Y., L'Huillier, A., and Corkum, P., *Phys. Rev. A*, **49**, 2117 (1994).
23. Salières, P., L'Huillier, A., Antoine, P., and Lewenstein, M., *Adv. At. Mol. Opt. Phys.*, **41**, 83 (1999).
24. Brabec, T., and Krausz, F., *Rev. Mod. Phys.*, **72**, 545 (2000).
25. Preston, S. G., Sanpera, A., Zepf, M., Blyth, W. J., Smith, C. G., Wark, J. S., Key, M. H., Burnett, K., Nakai, N., Neely, D., and Offenberger, A. A., *Phys. Rev. A*, **53**, R31 (1996).
26. Casu, M., Szymanowski, C., Hu, S., and Keitel, C. H., *J. Phys. B*, **33**, L411 (2000).
27. Potvliege, R. M., Kylstra, N. J., and Joachain, C. J., *J. Phys B*, **33**, L743 (2000).
28. Milosević, D. B., Hu, X. S., and Keitel, C. H., *Phys. Rev. A*, **63**, 011403(R) (2001).
29. Walser, M. W., Keitel, C. H., Scrinzi, A., and Brabec, T., *Phys. Rev. Lett.*, **85**, 5082 (2000).
30. Bugacov, A., Pont, M., and Shakeshaft, R., *Phys. Rev. A*, **48**, R4027 (1993).
31. Vazquez de Aldana, J. R., and Roso, L., *Opt. Express*, **5**, 144 (1999).
32. Kylstra, N. J., Worthington, R. A., Patel, A., Knight, P. L., Vázquez de Aldana, J. R., and Roso, L., *Phys. Rev. Lett.*, **85**, 1835 (2000).
33. Vazquez de Aldana, J. R., Kylstra, N. J., Roso, L., Knight, P. L., Patel, A. S., and Worthington, R. A., *Phys. Rev. A*, **64**, 013411 (2001).
34. Keldysh, L. V., *Sov. Phys.-JETP*, **20**, 1307 (1965).
35. Ammosov, M., Delone, N., and Krainov, V., *Sov. Phys.-JETP*, **64**, 1191 (1986).
36. Krainov, V. P., and Shokri, B., *J. Opt. Soc. Am. B*, **9**, 1231 (1992).
37. Krainov, V. P., *J. Phys. B*, **32**, 1607 (1999).
38. Faisal, F. H. M., *J. Phys. B*, **6**, L89 (1973).
39. Reiss, H. R., *Phys. Rev. A*, **22**, 1786 (1980).
40. Crawford, D. P., and Reiss, H. R., *Phys. Rev. A*, **50**, 1844 (1994).
41. Kaminski, J., *Z. Phys. D*, **16**, 153 (1990).
42. Krstić, P. S., and Mittleman, M. H., *Phys. Rev. A*, **42**, 4037 (1990).
43. Ermolaev, A. M., *J. Phys. B*, **31**, L65 (1998).
44. Protopapas, M., Keitel, C. H., and Knight, P. L., *J. Phys. B*, **29**, L591 (1996).
45. Taïeb, R., Véniard, V., and Maquet, A., *Phys. Rev. Lett.*, **81**, 2882 (1998).
46. Kylstra, N. J., Ermolaev, A. M., and Joachain, C. J., *J. Phys. B*, **30**, L449 (1997).
47. Lenz, E., Dörr, M., and Sandner, W., "A Picturebook of Relativistically Driven Wavepackets", in *Super-Intense Laser-Atom Physics*, edited by B. Piraux and K. Rzazewski, Kluwer Academic, 2001, p. 355.
48. Rathe, U. W., Keitel, C. H., Protopapas, M., and Knight, P. L., *J. Phys. B*, **30**, L531 (1997).
49. Hu, S. X., and Keitel, C. H., *Phys. Rev. Lett.*, **83**, 4709 (1999).
50. Wagner, R. E., Su, Q., and Grobe, R., *Phys. Rev. Lett.*, **84**, 3282 (2000).

X-Ray Pulse Train Generation by Long-Lived Atoms and Ions in Superintense Laser Field

Alexander M. Sergeev and Mikhail Yu. Ryabikin

Institute of Applied Physics, Russian Academy of Science
46 Ulyanov Str., Nizhny Novgorod 603950, Russia

Abstract. Coherent X-ray production via high-order harmonic generation in the stabilization regime is studied in detail in numerical experiments on two-dimensional model atoms and ions. The temporal behavior and the ellipticity dependence of harmonic intensities are shown to differ dramatically from those inherent to the well-studied low-frequency regime of harmonic generation. These peculiar properties can be used for X-ray short pulse generation and probing atomic stabilization.

INTRODUCTION

The progress in studies of strong-field phenomena has been promoted by the development of high-power laser systems that are able to produce coherent optical radiation with the electric field strengths comparable with or exceeding the Coulomb binding field in atoms. Laser-matter interactions in this intensity regime are extremely nonlinear [1] that is manifested most clearly by the production of very high harmonics of the incident field. High-order harmonic generation (HHG) is of great interest as the effective means for producing coherent X-ray radiation. Furthermore, it opens the ways to attosecond pulse production.

Among the most surprising results in the theory of strong-field laser-atom interactions has been the prediction of atomic stabilization against ionization in a very intense high-frequency laser field. In this paper we examine the HHG in the stabilization regime as a source for attosecond X-ray pulse train production.

LOW-FREQUENCY AND HIGH-FREQUENCY REGIMES OF HIGH-ORDER HARMONIC GENERATION

High-order harmonic generation has been observed in numerous experiments on various gases with different laser sources (KrF, Nd:YAG, Ti:sapphire, etc. [1]). In experiments on HHG carried out so far the *low-frequency* regime was implemented in which the laser photon energy $\hbar\omega_0$ is much smaller than the atomic ionization potential I_p. The most striking discovery in this frequency regime has been the very characteristic shape of the harmonic spectrum that is highly nonperturbative: a fast decline in the conversion efficiency for the few lowest-order harmonics is followed by

CP611, *Superstrong Fields in Plasmas:* Second Int'l. Conf., edited by M. Lontano et al.

a long plateau where harmonics have approximately the same intensity. This plateau ends up abruptly at photon energies exceeding the value that has been found to obey the universal cut-off formula [2]:

A physical understanding of the cut-off law (1) is provided by the semiclassical model suggested by Kulander et al. [3] and Corkum [4]. According to this model, HHG process includes three steps: (1) atomic electron tunneling through the barrier created by the superposition of the Coulomb potential of the ion and the instantaneous electric field of the laser pulse, (2) acceleration of the free electron in the oscillating electric field of the laser pulse, and (3) its recollision with the parent ion that results in the emission of high-energy photon due to the Bremsstrahlung mechanism. Simple analysis based on the classical Newton equation for an electron driven by a constant-amplitude linearly polarized laser field with neglect of an ionic attraction shows that $3.17\ U_p$ is just the maximum kinetic energy that the electron can gain prior to its recollision with the ion. The fully quantum-mechanical extension of this model has been developed later [5, 6] that is consistent with these classical arguments.

$$E_{max} \approx I_p + 3.17\ U_p \tag{1}$$

where $U_p = E_0^2/4\omega_0^2$ is the ponderomotive energy of a free electron in the laser field of constant amplitude E_0.

As follows from the semiclassical picture, the mechanism for HHG described above is very sensitive to the factors leading to distortions of the electron trajectories. The typical example is the laser ellipticity that strongly affects the efficiency of the underlying "recollision" mechanism [7]. The other factor crucial for HHG is the effect of the magnetic field of the laser pulse, as has been confirmed by both classical [8] and quantum-mechanical [9] simulations. Indeed, the Lorentz force exerted by the magnetic field of the pulse deflects electrons along the wave propagation direction thus diminishing their interaction with the parent ions. This effect can be significant for HHG in the few-optical-cycle regime, since electrons detached from atoms during the fast turn-on of the intense laser pulse can gain very high velocities prior to coming back to the parent ions.

The interest to HHG is caused mainly by the prospects of its use as a tunable coherent light source in extreme ultraviolet and soft X-ray region. Moreover, different ways to use HHG for producing attosecond pulses of X-ray radiation were considered [10-14]. As follows from the cut-off formula (1), higher photon energies can be reached with increasing intensity of the driving field. However, a fundamental limitation arises from the atomic bound-state depletion at the saturation intensity at which all atoms become ionized. Several approaches have been suggested in order to overcome this limitation. First, very short driving pulses can be used with which atoms can survive to higher field strengths and electrons can acquire substantially higher kinetic energies during free motion than with long pulses [12, 15]. This idea has been implemented in experiments on He atoms driven by 26-fs pulses from a Ti:sapphire laser, and the generation of ultrashort pulses of radiation at the wavelength as short as 2.7 nm has been obtained for the first time [16]. Second, one may use multiply charged ions that have much higher ionization potential than neutrals [17]. Third, atomic stabilization may be utilized that is an unusual regime in which an electron can remain localized in the vicinity of the parent ion up to very high laser intensities.

The possibility of atomic stabilization against ionization in a very intense *high-frequency* laser field with $\hbar\omega_0 > I_p$ has been discovered for the cases of both constant-amplitude (adiabatic stabilization) and pulsed (dynamic stabilization) laser field in theoretical studies and numerical experiments on both 1D and 3D model atoms [18-20]. Stabilization has been proved to be associated with the formation of localized electron wave packets. The details of localization dynamics and the structure of localized wave packets have been studied for different laser polarizations [21-25]. The effect of the magnetic field of the laser pulse on atomic stabilization was studied by many authors, and the existence of a rather wide "stabilization window" in the intensity dependence of the atom survival probability has been shown recently in numerical experiments on 2D models [25-27].

Atomic stabilization, on the one hand, is of interest as a rather counterintuitive effect of decreasing ionization probability with increasing laser intensity. On the other hand, it provides a long-lived highly nonlinear medium that possesses a number of uncommon features. The possibility of high-order harmonic generation at dynamic stabilization with both linearly [20, 28] and circularly [23, 25] polarized laser fields has been pointed out and explained from the structure arguments. It was shown that high-order harmonics in this regime are produced in the form of long trains of attosecond pulses with unusual time dependence [25]. Just as in the low-frequency regime [9, 29-31], the magnetic field of high-intense laser pulse suppresses harmonic production, but, as shown in Ref. [25], in a wide range of laser intensities coincident with the atomic "stabilization window" rather efficient HHG still persists.

Below we will examine in detail the peculiarities of harmonic generation process in the high-frequency regime in a range of laser parameters corresponding to the atomic dynamic stabilization. We will compare the temporal behavior as well as the ellipticity dependence of HHG in the stabilization regime with those inherent to the well-studied low-frequency regime of HHG [5, 6].

HARMONIC GENERATION IN THE TWO-DIMENSIONAL MODEL OF ATOM

In this section we present the results of numerical experiments on the 2D model for a hydrogen atom. Such model is known to be useful for studying polarization [32, 33] and magnetic-field [9, 34] effects in strong-field laser-atom interactions. We assumed the atomic electron to be initially in the ground state of a smoothed Coulomb potential

$$V(r) = -\left(a^2 + x^2 + y^2\right)^{-1/2} \tag{2}$$

with $a=0.8$, the ionization potential I_p being equal to 0.5 (in atomic units). The electron wave-packet behavior and the harmonic generation were studied for different ellipticities ε of the laser field taken as $\mathbf{E}(t) = f(t)(\mathbf{e}_x E_x \sin\omega_0 t + \mathbf{e}_y E_y \cos\omega_0 t)$ with $E_y = \varepsilon E_x$ and $E_x^2 + E_y^2 = E_0^2$. The pulse shape was taken trapezoidal with two-cycle ramps. For the sake of comparison, calculations were made for two values of the laser frequency, ω_0 =0.057 (λ=800 nm, low-frequency regime) and ω_0 =1 (λ=45.6 nm, high-frequency regime). The laser intensities for these two regimes were chosen so as to provide harmonic spectra of approximately the same width. At ω_0 =1 the range of laser field strengths corresponding to the "stabilization window" spans the interval between $E_0 \approx 1.5$ and $E_0 \approx 15$, the upper limit being determined by the magnetic field of the laser pulse [25-27]. As we deal with the values of E_0 below this upper limit, the dipole approximation is justified. The Schrödinger equation was integrated numerically using the split-operator fast Fourier transform technique [35] on a square grid with absorbing boundaries, the center of the grid being placed at the nucleus. The time-frequency analysis of the dipole acceleration has been performed by means of the wavelet transform [36] that proves to be the very convenient mathematical tools for probing the temporal behavior of HHG [37, 38].

Time-Frequency Analysis of HHG

Figures 1a and 1b show the HHG power spectra calculated for the case of linearly polarized driving field with ω_0 =0.057, E_0 =0.125 (10-cycle pulse) and ω_0 =1, E_0 =12.5 (34-cycle pulse), respectively. The harmonic spectrum in Fig. 1a exhibits a plateau followed by a cutoff that is the attribute of the low-frequency regime of HHG. The spectrum in Fig. 1b contains a rather monotonically decreasing series of very sharp harmonic peaks without clear cutoff that is peculiar to the dynamic stabilization regime [1].

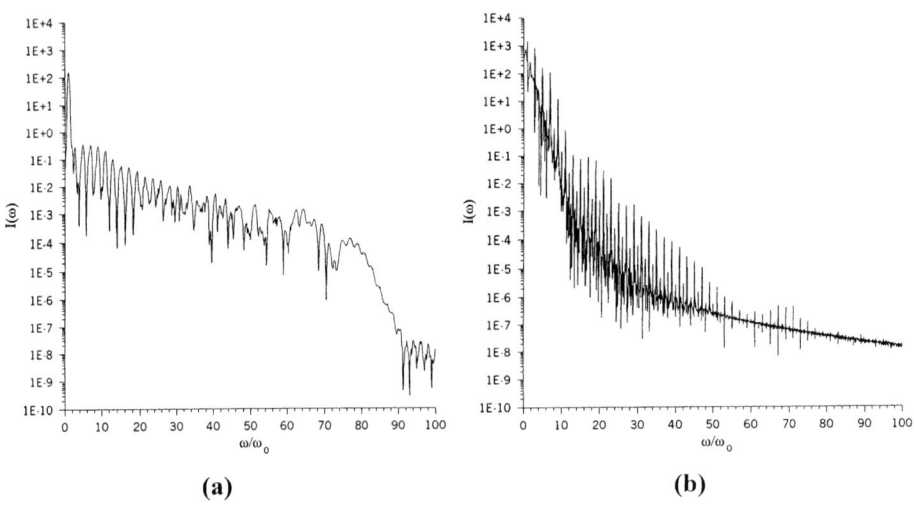

(a)   (b)

FIGURE 1. Harmonic spectrum for a 2D model atom in a linearly polarized laser pulse for a) 10-cycle pulse with ω_0 =0.057 and E_0 =0.125 and b) 34-cycle pulse with ω_0 =1 and E_0 =12.5.

The temporal behavior of high-order harmonics is shown in Figs. 2a and 2b (time dependence of the Morlet coefficient for 43rd harmonic is plotted). These figures also show the time dependence of the atom survival probability defined as the overall norm of the wave function remaining on a grid.

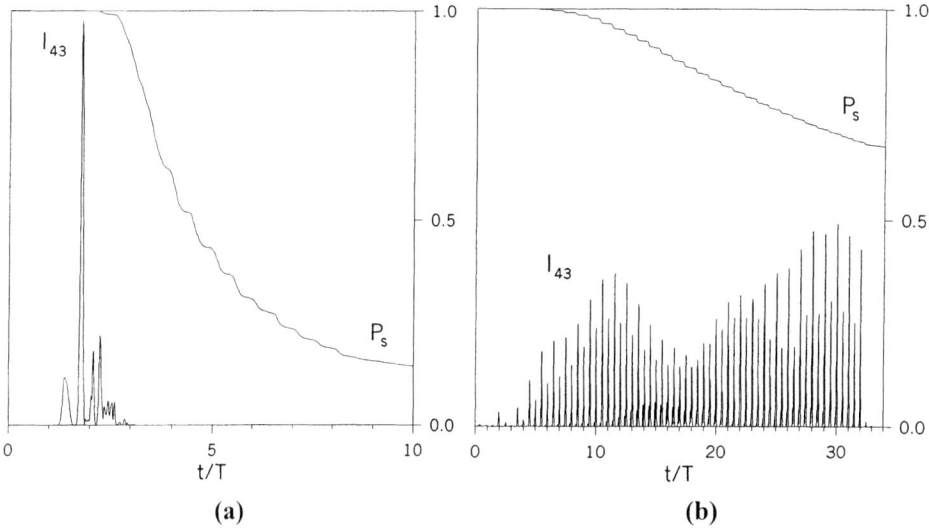

FIGURE 2. Time profile of the 43rd harmonic (I_{43}) and atom survival probability (P_s) for the same pulse as in a) Fig. 1a and b) Fig. 1b.

It is seen in Fig. 2a that, because of rapid ionization and subsequent free-electron wave packet spreading, high-order harmonics of a low-frequency laser field are produced during the short period of time (about several cycles of the field) close to the interval of atomic ionization. Indeed, as seen in Fig. 3a where the snapshot of the electron probability distribution taken after 2.5 cycles of the laser field is shown, the wave packet of ionized electrons soon after the pulse turn-on extends in longitudinal and transversal directions to dimensions greatly exceeding the classical electron excursion amplitude and the Bohr radius, respectively. This rapid electron delocalization leads to fast decay of polarization signal. Because of low periodicity of the process, harmonic peaks seen in Fig. 1a are not very sharp.

In contrast to that, at high laser frequency the wave packet has no time to spread noticeably during one cycle of the field. Due to ionic attraction experienced by one or another wing of the packet when its center approaches one of the turning points, its spreading in the course of time gives place to localization [22]. Such localized three-peak wave packet is seen in Fig. 3b where the snapshot of the electron probability distribution taken after 10 cycles of the linearly polarized laser field with $\omega_0 = 1$ and $E_0 = 12.5$ is shown. The side peaks correspond to the minima of Kramers-Henneberger potential while the central peak originates from the superposition of bound dressed states populated due to the fast turn-on of the pulse. These sharp features give rise to the harmonic production. The main contribution is due to the central part of the packet that encounters the ion twice per cycle at high speed, thus producing short bursts of the Bremsstrahlung radiation. High periodicity of the process leads to very high sharpness of harmonic peaks seen in Fig. 1b. As follows from Fig. 2b, this source of high-order harmonics is very long-lived. Over a long period of time harmonic peaks increase, actually, in spite of ionization.

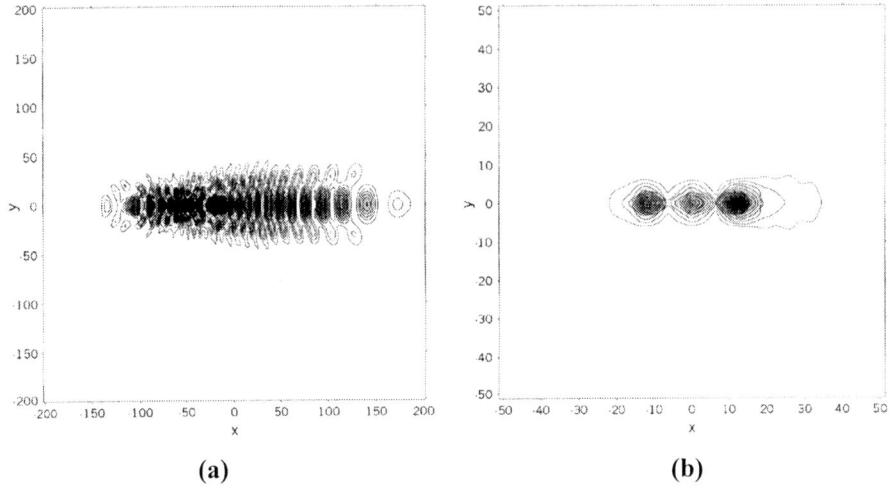

(a)   (b)

FIGURE 3. Snapshot of the electron probability distribution in a linearly polarized laser field taken after a) 2.5 cycles of a low-frequency field with the pulse parameters same as in Fig. 1a and b) 10 cycles of a high-frequency field with the pulse parameters same as in Fig. 1b.

It follows from a comparison between Figs. 1a and 1b that, although very high harmonics in the stabilization regime are produced much less efficiently than in the low-frequency regime [1], for not very large harmonic numbers (up to $N=23$) the conversion efficiency in the former case is comparable with or even higher than in the latter one. Note, by the way, that less than 20 harmonics of radiation with $\lambda=45.6$ nm are necessary to reach the "water window" between $\lambda=2.3$ and 4.4 nm instead of more than 180 harmonics needed when using the driving field with $\lambda=800$ nm.

A more detailed time-frequency analysis reveals two extra useful properties of harmonic generation process in the dynamic stabilization regime. Figs. 4a and 4b show the 3D time-frequency profiles of HHG for two regimes discussed above.

It is seen that in the low-frequency regime (Fig. 4a) different harmonics are produced efficiently at different times because of predominant contributions of different groups of electrons. As time increases, the frequency profile becomes more and more complicated that can be explained by increasing contributions of electron trajectories with multiple returns to the nucleus. These numerical observations are quite consistent with the analysis based on both the semiclassical model [3, 4] and the fully quantum-mechanical analytical theory [5]. Contrarily, the time-frequency profile in the stabilization regime (Fig. 4b) is very regular. All harmonics are produced synchronously with maximum conversion reached near the moments $t_n = nT/2$, i.e. when the center of the wave packet hits the ion. It enables to produce attosecond pulses without additional efforts aimed at the synchronization of different spectral components of the generated field. Moreover, the moments t_n are independent of the laser intensity that makes easier to realize phase matching conditions in the macroscopic medium.

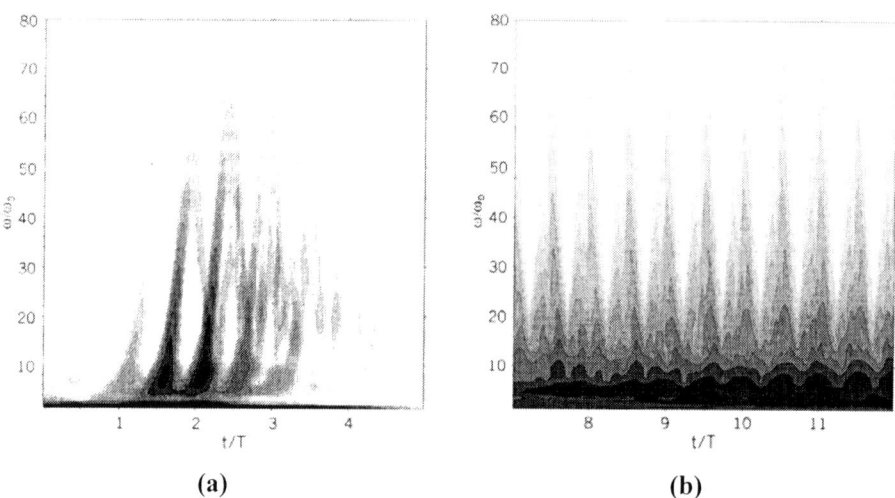

(a)   (b)

FIGURE 4. 3D time-frequency profile of HHG for a) 10-cycle laser pulse with $\omega_0 = 0.057$ and $E_0 = 0.1$ and b) 34-cycle laser pulse with $\omega_0 = 1$ and $E_0 = 12.5$.

Ellipticity Dependence of HHG

As mentioned above, the HHG process in the low-frequency regime is known to be very sensitive to the laser ellipticity [7]. In a classical picture, the y-component of the laser field deflects the electron from the parent ion, that's why the electron misses the ion and, consequently, doesn't participate in producing the Bremsstrahlung radiation. Quantum-mechanically, only due to the wave-packet spreading the harmonic emission, though weak, remains possible. The example of the wave packet at nonzero laser ellipticity is shown in Fig. 5a. The parameters are the same as in Fig. 3a except for the ellipticity equal here to 0.3. Two portions of electron probability distribution detached from the nucleus at two subsequent half-cycles are clearly seen that miss the origin almost entirely. The result is the strong suppression of harmonic emission. At circular laser polarization the free-electron wave packet has the appearance of a spiral unwinding from the origin (Fig. 6a). Electrons never revisit the vicinity of the nucleus and, hence, don't produce harmonics.

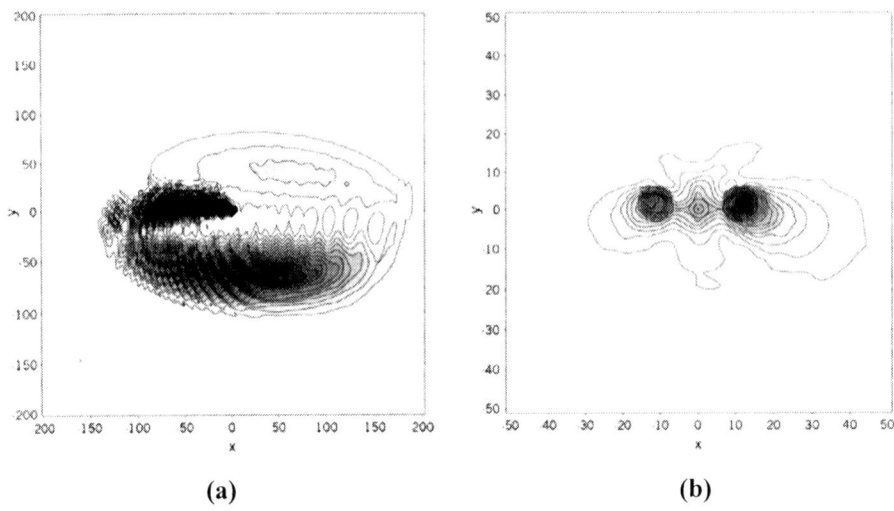

(a) **(b)**

FIGURE 5. Snapshot of the electron probability distribution in an elliptically polarized laser field with ε=0.3 taken after a) 2.5 cycles of a low-frequency field with the pulse parameters same as in Fig. 1a and b) 10 cycles of a high-frequency field with the pulse parameters same as in Fig. 1b.

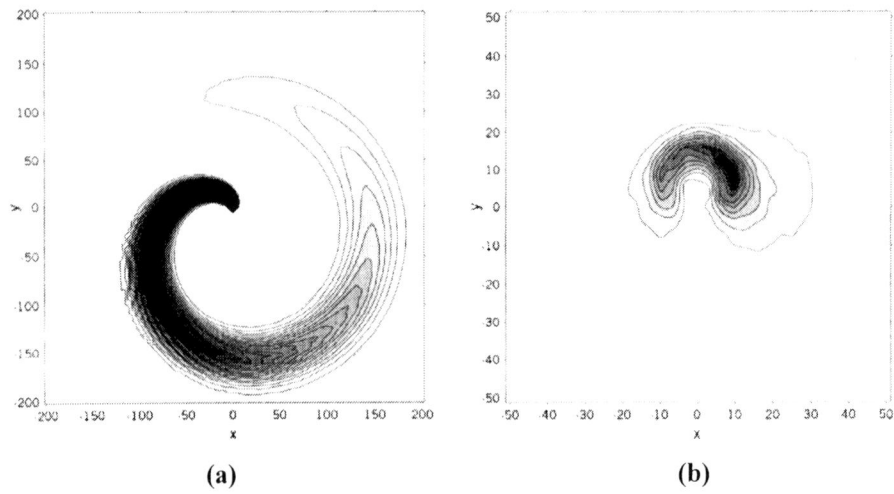

(a) **(b)**

FIGURE 6. Snapshot of the electron probability distribution in a circularly polarized laser field taken after a) 2.5 cycles of a low-frequency field with the pulse parameters same as in Fig. 1a and b) 10 cycles of a high-frequency field with the pulse parameters same as in Fig. 1b.

Figs. 7a and 7b show the ellipticity dependence of the efficiency of low-order (N=13) and high-order (N=43) harmonic production calculated for two regimes described above. It is seen that this dependence in the dynamic stabilization regime is much weaker than that in the low-frequency regime, especially for low-order harmonics.

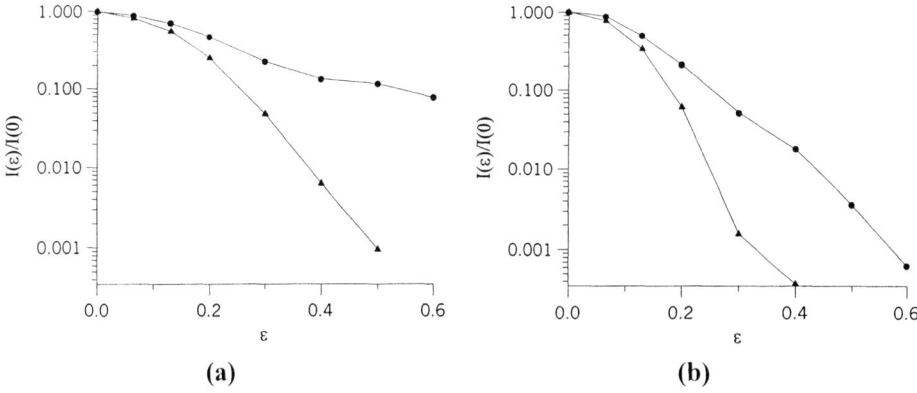

FIGURE 7. Ellipticity dependence of the intensity of a) 13-th harmonic and b) 43-th harmonic for a low-frequency (triangles) and high-frequency (circles) laser field. Parameters of the pulses are the same as in Fig. 1.

"STABILIZATION WINDOW" AND HARMONIC GENERATION WITH NEGATIVE IONS

The foregoing results relate to the ground-state stabilization of the hydrogen atom. The laser parameters required for stabilization in this case are inaccessible with up-to-date laser technique. The main limitation is due to the radiation frequency of modern high-intense laser sources that is usually much lower than the frequency of the electron motion in neutral atoms. Obviously, this limitation can be overcome by using quantum systems with lower electron binding energy, e.g. negative ions.

We have carried out the calculations for the hydrogen negative ion driven by the second harmonic radiation of the Ti:sapphire laser. We used a single-electron 2D model for H^- ion. The short-range potential binding an outer electron to the atomic core was chosen in the form similar to that used in 1D calculations in Ref. [22]:

$$V(x,z) = -V_0 \frac{\exp\left(-\sqrt{x^2 + z^2 + b^2}\right)}{\sqrt{x^2 + z^2 + a^2}} \qquad (3)$$

Here V_0=24.856, a=5.09, b=3.3; ionization potential I_p=0.0278 (0.75 eV), laser frequency ω_0=0.114 (λ=400 nm).

Using this 2D model, we explored the effect of the magnetic field of the laser pulse on the stabilization of the H^- ion. The numerical results shown in Fig. 8 are rather similar to that obtained earlier for the H atom [25-27]. The "stabilization window" is seen that extends from E_0 =0.035 to E_0 =0.4.

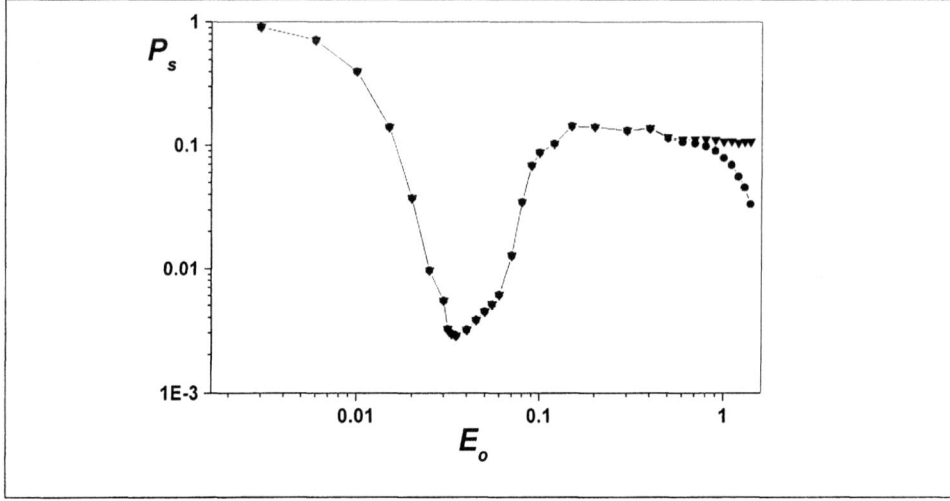

FIGURE 8. Survival probability for the the H$^-$ ion driven by 14-cycle pulse with ω_0 =0.114 as a function of the electric field strength with (triangles) and without (circles) the dipole approximation.

It's worth notice that the upper boundary of the "stabilization window" in Fig. 8 gets to the laser intensity domain where the single-electron approximation for H$^-$ ion becomes questionable [39]. In this area, electron correlation may have destabilizing effect [40], added to the effect of the magnetic field, but, nevertheless, the intensity range, though somewhat narrower, remains in which stabilization takes place. In this regime, H$^-$ ion exhibits the prolonged highly nonlinear response that is the source of the sequence of attosecond pulses.

CONCLUSIONS

The majority of results presented above relates to the case of the interaction of the laser field with an atom in its ground state. The laser parameters required for stabilization in this case are not yet accessible with up-to-date laser technique. The experiments in which the indication of stabilization has been obtained were performed on atoms prepared in highly excited circular states [41]. The mechanisms of stabilization in this case are not well established up to now. The other possibility to observe stabilization is to use negative ions that have much lower ionization potential than neutrals. Finally, great advances have been made recently in producing intense coherent radiation in the vacuum ultraviolet range with free electron lasers [42]. It opens the way to a number of new applications in atomic and high-field physics, and, in particular, interactions of atoms with a very intense electromagnetic field in the high-frequency regime, and, specifically, their stabilization, can be studied experimentally in the near future.

In conclusion, atomic electron interacting with a very intense high-frequency laser pulse provides a long-lived source of harmonic radiation. The peculiarities of the wave packet evolution in the dynamic stabilization regime cause a number of unusual properties of the harmonic generation process among which are its long-term persistence despite ionization and tolerance to ellipticity of the driving field. Specific time-frequency dynamics of the process favor the production of attosecond pulse trains in macroscopic medium. These properties are of practical interest from the viewpoints of X-ray pulse generation and probing this very special regime of strong-field laser-matter interactions.

ACKNOWLEDGMENTS

This work was supported by the Russian Foundation for Basic Research (grant 01–02–18006) and the Presidium of Russian Academy of Science (Fundamental Complex Program "Quantum Macrophysics").

REFERENCES

1. Protopapas, M., Keitel, C. H., and Knight P. L., *Rep. Prog. Phys.* **60**, 389-486 (1997).
1. Krause, J. L., Schafer, K. J., and Kulander, K. C., *Phys. Rev. Lett.* **68**, 3535-3538 (1992).
3. Kulander, K. C., Schafer, K. J., and Krause, J. L., in *Super-Intense Laser-Atom Physics* (NATO AST Series, Series B: Physics, Vol. 316), edited by B. Piraux et al., New York: Plenum Press, 1993, p. 95.
4. Corkum, P. B., *Phys. Rev. Lett.* **71**, 1994-1997 (1993).
5. Lewenstein, M., Balcou, Ph., Ivanov, M. Yu., L'Huillier, A., and Corkum, P. B., *Phys. Rev. A* **49**, 2117-2132 (1994).
6. Becker, W., Long, S., and McIver, J. K., *Phys. Rev. A* **50**, 1540-1560 (1994).
7. Dietrich, P., Burnett, N. H., Ivanov, M., and Corkum, P.B., *Phys. Rev. A* **50**, R3585- R3588 (1994).
8. Keitel, C. H., and Knight, P. L., *Phys. Rev. A* **51**, 1420-1430 (1995).
9. Kim, A. V., Ryabikin, M. Yu., and Sergeev, A. M., *Usp. Fiz. Nauk* **169**, 58-66 (1999) (*Phys. Usp.*, **38**, 54-61 (1999)).
10. Corkum, P. B., Burnett, N. H., and Ivanov, M. Y., *Opt. Lett.* **19**, 1870-1872 (1994).
11. Antoine, P., L'Huillier, A., and Lewenstein, M., *Phys. Rev. Lett.* **77**, 1234-1237 (1996).
12. Sergeev, A. M., Vanin, E. V., and Kim, A. V., *Proc. SPIE* **2701**, 235-246 (1996).
13. Schafer, K. J., and Kulander, K. C., *Phys. Rev. Lett.* **78**, 638-641 (1997).
14. Christov, I. P., Murnane M. M., and Kapteyn, H. C., *Phys. Rev. Lett.* **78**, 1251-1254 (1997).
15. Christov, I. P., Zhou, J., Peatross, J., et al., *Phys. Rev. Lett.* **77**, 1743-1746 (1996).
16. Chang, Z., Rundquist, A., Wang, H., et al., *Phys. Rev. Lett.* **79**, 2967-2970 (1997).
17. Hu, S. X., Milošević, D. B., Becker, W., and Sandner, W., *Phys. Rev. A* **64**, 013410 (2001).
18. Pont, M., Valet, N. R., Gavrila, M., and McCurdy, C. W., *Phys. Rev. Lett.* **61**, 939-942 (1988).
19. Su, Q., Eberly, J. H., and Javanainen, J., *Phys. Rev. Lett.* **64**, 862-865 (1990).
20. Krause, J. L., Schafer, K. J., and Kulander, K. C., *Phys. Rev. Lett.* **66**, 2601-2604 (1991).
21. Reed, V. C., Knight, P. L., and Burnett, K., *Phys. Rev. Lett.* **67**, 1415-1418 (1991).
22. Grobe, R., and Fedorov, M. V., *Phys. Rev. Lett.* **68**, 2592-2595 (1992).
23. Patel, A., Protopapas, M., Lappas, D. G., and Knight, P. L., *Phys. Rev. A* **58**, R2652-2655 (1998).
24. Patel, A., Kylstra, N. J., and Knight, P. L., *Opt. Express* **4**, 496-511 (1999).
25. Ryabikin, M. Yu., and Sergeev, A. M., *Opt. Express* **7**, 417-426 (2000).
26. Kylstra, N. J., Worthington, R. A., Patel, A., et al., *Phys. Rev. Lett.* **85**, 1835-1838 (2000).

High-Order Harmonic Generation in the Few-Optical-Cycle Regime

S. Stagira[a], E. Priori[a], G. Sansone[a], M. Nisoli[a], G. Cerullo[a],
S. De Silvestri[a], P. Villoresi[b], L. Poletto[b], G. Tondello[b], C. Altucci[c],
R. Bruzzese[d] and C. de Lisio[d]

[a]Istituto Nazionale per la Fisica della Materia, Dipartimento di Fisica, Politecnico di Milano, Piazza L. Da Vinci 32, 20133 Milano, Italy

[b]Istituto Nazionale per la Fisica della Materia, Laboratorio di Elettronica Quantistica – D.E.I., Università di Padova, Padova, Italy

[c]Istituto Nazionale per la Fisica della Materia, Dipartimento di Chimica, Università della Basilicata, Potenza (Italy)

[d]Istituto Nazionale per la Fisica della Materia, Dipartimento di Scienze Fisiche, Università di Napoli Federico II, Napoli (Italy)

Abstract. Production of high-order harmonics excited by high-intensity laser pulses is a promising tool for the generation of coherent, extreme ultraviolet radiation. Properties of such radiation source are strongly affected by temporal and spatial behavior of the driving laser pulses. In this work we describe the generation of high-order harmonics using multi-cycle and few-cycle laser pulses; the nonlinear process is characterized as a function of focusing geometry and driving pulse duration; numerical simulations of harmonic generation are compared to experimental results, allowing to single out the unique properties that arise using driving pulses in the few-optical-cycle regime.

INTRODUCTION

The availability of optical compression techniques for generation of high-intensity ultrashort laser pulses [1-3] has driven the nonlinear optics in the few-optical-cycle regime. New phenomena have been observed in this novel field, such as carrier-wave Rabi flopping [4] and coherent wave packet excitation [5]. For few-cycle driving pulses, well known nonlinear effects are also noticeably altered: owing to the very high peak intensity reached, the electromagnetic field becomes comparable to the atomic field; thus pure strong field phenomena take the place of multiphoton processes in laser-matter interaction [6]. Moreover the nonlinear phenomena driven in this regime

CP611, *Superstrong Fields in Plasmas:* Second Int'l. Conf., edited by M. Lontano et al.
© 2002 American Institute of Physics 0-7354-0057-1/02/$19.00

by ultrashort pulses can be affected by the carrier-envelope phase of the electric field [7,8].

High-order harmonic generation (HHG) is a typical signature of non-perturbative laser-matter interaction; it deserves a strong interest, because it represents an attractive tool for the production of ultrafast coherent pulses in the UV and soft X-ray regions of the spectrum using a tabletop laser system; moreover this processes can be extended to useful applications, such as atomic and molecular spectroscopy, solid state and plasma diagnostic and XUV nonlinear optics [9,10]. In a simplified picture, the HHG process can be described in the following way: when an intense, short laser pulse interacts with an atomic gas, tunneling ionization occurs followed by acceleration of the freed electron by the laser field and generation of high-order harmonics during the recombination process. The harmonic conversion efficiency is expected to increase upon decreasing the excitation pulse duration [11-13] and higher photon energies are also expected because the electrons are released into stronger laser fields. Coherent emission up to the "water window" (4.4-2.3 nm) was indeed demonstrated in helium by using sub-10-fs pulses [14] obtained by the hollow fiber compression technique.

In order to track the evolution of HHG upon changing the interaction parameters, we present in this work a characterization of harmonic emission as a function of some accessible experimental variables. Numerical simulations of harmonic generation will also be presented and compared to the experimental results. Our aim is to single out the properties of harmonic generation with compressed pulses and to provide some guidelines for the optimization of HHG process.

EXPERIMENTAL SETUP

HHG experiments were performed using both multi-cycle (25 fs) and few-cycle (sub-10-fs) laser pulses. 25-fs pulses, centered at 795 nm, are generated by a Ti:sapphire laser system, with energies up to 0.8 mJ, at 1-kHz repetition rate. Sub-10-fs pulses are generated by the hollow fiber compression technique, that will be described in more detail in the following. In order to generate high-order harmonics of the fundamental radiation, the excitation pulses are focused onto a noble gas jet by a 25-cm focal length silver mirror. The laser-gas interaction length is ~1 mm, with a gas pressure of ~50 torr. Harmonic emission is analyzed using a high-resolution flat-field soft X-ray spectrometer [15], based on a 1200 lines/mm gold-coated spherical variable-line-space grating optimized for the 5-40 nm spectral region. The spectrum is acquired by a single-stage microchannel-plate open intensifier with a magnesium-fluoride photocathode and an output on a phosphor screen, optically coupled by an objective to a high-resolution charge coupled device (CCD) camera with low read-out noise. The instrument is designed for the simultaneous acquisition of the spectrum and the far field pattern of the harmonic beam. The spectrometer operates without an entrance slit because of the limited size of the emitting source, which has been estimated to be of the order of 15 μm at full width at half maximum (FWHM). The spectrometer has been calibrated in order to provide an absolute measurement of the emitted photon flux.

Hollow fiber compression technique

Optical compression of high-energy pulses is achieved by hollow fiber technique. The 25-fs laser pulses are coupled into a 60-cm-long argon-filled hollow fiber (inner diameter of 500 µm). Owing to the high intensity reached, the pulses undergo spectral broadening by self-phase-modulation (SPM) during propagation in the gas. The linearly polarized output beam is almost diffraction limited. Broadening of the spectrum is essential to the generation of compressed pulses; nevertheless the pulses at the output of the fiber are chirped (i.e. the new optical frequencies generated by SPM are dispersed in time). Compression is obtained propagating such pulses in a high throughput dispersive delay line realized using chirped mirrors [16,17]. Typical pulse duration ranges from 5 to 7 fs; Figure 1 shows temporal and spectral behavior of the compressed pulses as measured by spectral phase interferometry for direct electric field reconstruction (SPIDER) [18]. The use of the hollow fiber presents the advantages of a guiding element with a large-diameter single mode and of a fast nonlinear medium with a high threshold for ionization. Wave propagation along hollow fibers can be thought of as occurring by grazing incidence reflections at the dielectric inner surface. Since the losses caused by these reflections greatly discriminate against higher-order modes, only the fundamental mode, with large and scalable size, will be transmitted through a sufficiently long fiber. For fused silica gas-filled hollow fibers the fundamental mode is the EH_{11} hybrid mode, whose intensity profile as a function of the radial coordinate r is given by $I_0(r) = I_0 J_0^2(2.405 r / a)$, where I_0 is the peak intensity, J_0 is the zero-order Bessel function and a is the capillary radius [19]. The good spatial properties of the beam at the output of the compression stage result in good conversion efficiency and spatial properties of the generated harmonics, as will be shown in the following. For this reason, and in order to keep the same spatial pattern of the exciting beam, evacuation of the hollow fiber and removal of chirped mirrors was performed in order

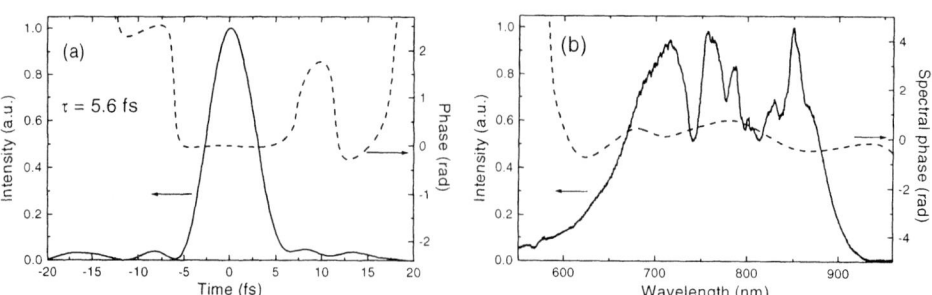

FIGURE 1. SPIDER characterization of the compressed pulses in temporal (a) and spectral (b) domain.

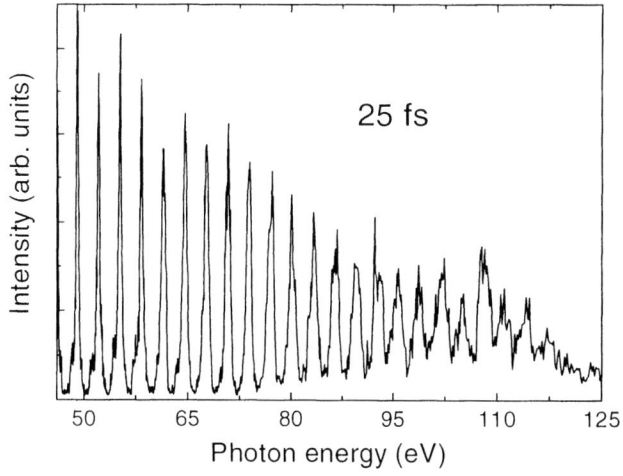

FIGURE 2. Harmonic spectrum obtained in neon for multi-cycle driving pulses.

to generate 25-fs pulses with a diffraction-limited beam.

HIGH ORDER HARMONIC GENERATION

Spectral characterization

Figure 2 shows a typical harmonic spectrum obtained using 25-fs pulses in neon, for a peak intensity of 7×10^{14} W/cm^2. In the transition region between the so called plateau

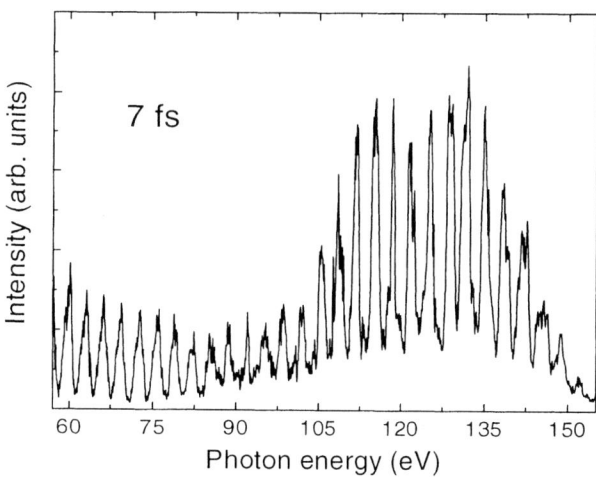

FIGURE 3. Harmonic spectrum obtained in neon for few-cycle driving pulses.

and the cutoff of the spectrum, the harmonic efficiency decreases with nonlinear order; well distinct peaks appear up to the 75th harmonic, but a continuos background is present in the cutoff. A striking behavior is the increase of harmonic linewidth with order, which is surprising at a first view. This result must be compared to the spectrum detected using 7-fs pulses in similar experimental conditions, which is shown in Figure 3. In this case harmonic efficiency slightly decreases approaching the cutoff region, but a recover of the spectrum intensity is observed above 100 eV; a continuos background is still present in the cutoff. It is worth pointing out that, for comparable peak intensities, few-optical cycle driving pulses allow to generate higher photon energies with respect to longer pulses, as can be seen from Fig. 2 and Fig. 3.

A strong difference between the two spectra concerns the dependence of harmonic linewidths on photon energy. Figure 4 shows the linewidths measured for multi-cycle (lower panel) and few-cycle (upper panel) driving pulses; as can be seen from the figure, for few-cycles driving pulses the linewidth is almost constant on a large energy range. For 25-fs driving pulses, the wavelength dependence of the linewidths [20] can be at least partially understood in terms of the different duration of the emission of the harmonics: those in the plateau are emitted over several optical cycles, while those approaching the cutoff are emitted only over a few cycles, close to the laser peak intensity. For 7-fs excitation pulses, the linewidth remain nearly constant since all the harmonics are emitted over a few cycles.

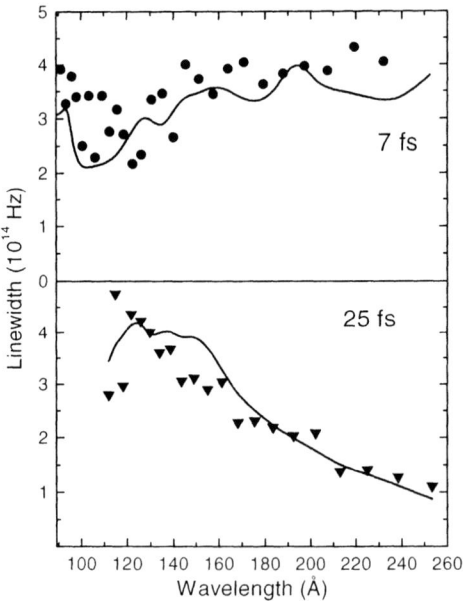

FIGURE 4. Harmonic linewidth vs. emission wavelength for 7-fs (upper panel) and 25-fs (lower panel) driving pulses. Solid lines: results from theoretical model (see text).

Spatial properties of harmonic emission

In order to explore in more detail the role of focusing conditions on the harmonic generation, we have measured the divergence of the harmonic beam as a function of gas jet position with respect to the laser focus and nonlinear order. It must be pointed out that the spectrometer makes the projection of the two-dimensional profile of the harmonic beam in the direction perpendicular to the spectral dispersion; harmonic divergence can be anyway extracted from the acquired spectra with data elaboration. Figure 5 shows the harmonic divergence as a function of nonlinear order, determined for 7-fs driving pulses, for different positions of the gas jet with respect to laser beam waist. Hereafter such relative displacement will be called z, and will be taken as positive when the laser beam is focused before the gas jet. The determination of beam waist position was obtained in two different and independent ways: using a CCD camera and looking for the maximum UV plasma emission from the gas jet. Both the two methods gave the same result. As can be seen from the figure, the divergence increases with harmonic order in the plateau region and a slight decrease is observed in the cutoff; nevertheless the low signal doesn't allow a complete characterization at high photon energy. The minimum divergence is measured for $z \cong -1$ mm, i.e. for the gas jet positioned slightly before the laser beam waist. It is worth noting that the intensity profile of the harmonic beam projected by the spectrometer appears regular, without any annular structure, for every nonlinear order and beam waist position. This is in contrast with both experimental and theoretical findings reported in literature for longer driving pulses [21] with a different spatial profile, where annular structure in harmonic generation was observed for positions of the gas jet before the laser beam waist. Moreover no appreciable harmonic signal was generally observed in such experiments, when the jet was positioned in the laser beam waist.

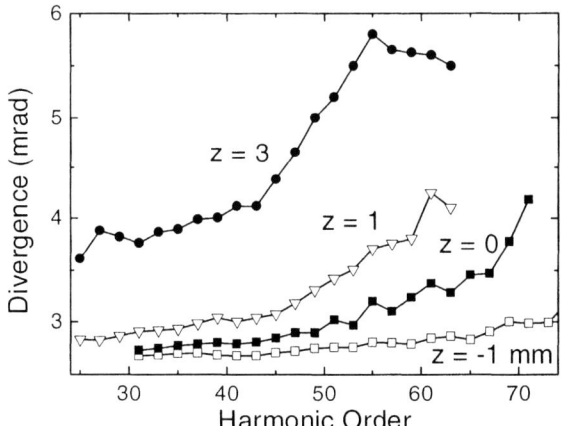

FIGURE 5. Harmonic divergence vs. nonlinear order for 7-fs driving pulses.

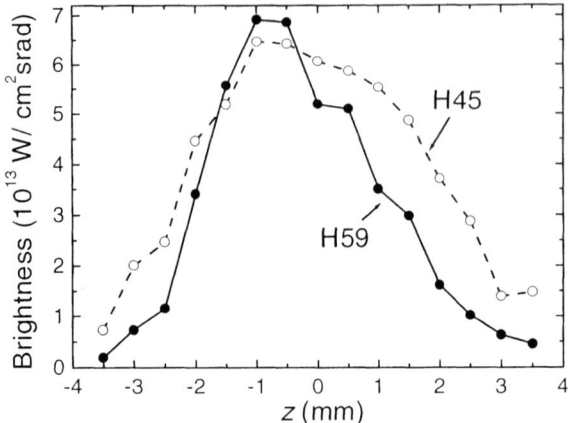

FIGURE 6. Brightness of 45th and 59th harmonics vs. beam waist position.

The harmonic brightness was also determined as a function of z. Figure 6 shows the brightness measured for the 45th and 59th harmonics for different values of z, using a 7-fs exciting pulse; a spot size of the emission area ranging from 13 to 20 μm was assumed for the different harmonic orders. As can be seen, the brightness is a smooth, quasi-symmetric function of z, showing a maximum for the gas jet position slightly before the laser beam waist. This behavior is observed across the whole spectrum, for both multi-cycle and few-cycle pulses; for this reason we relate it mainly to the focusing geometry of the driving laser beam, because only small changes are observed in the two different temporal regimes.

Theoretical model for HHG

On the basis of the previous observation, one can infer that both temporal and spatial behavior of the driving pulses must be taken into account in order to describe exhaustively the harmonic generation process. To this end we have performed numerical simulations of the HHG process, with the aim to distinguish between spatial and temporal effects. In our numerical model, the single-atom response is calculated using the Strong Field Approximation [22] which analytically solves the Schrödinger equation using the following assumptions: (i) all bound states in the atom are neglected, except the ground state; (ii) in the continuum, the electron is treated as a free particle moving in the laser electric field, with no influence of the atomic potential. In our model the SFA is used in a non-adiabatic form, so that the full electric field of the laser pulse is used to calculate the nonlinear dipole moment $d_{nl}(t)$ (using atomic units) as:

$$d_{nl}(t) = 2\,\mathrm{Re}\Big\{i \int_{-\infty}^{t} dt' \left(\frac{\pi}{\varepsilon + i(t-t')/2}\right)^{3/2} d^{*}\big[p_{st}(t',t) - A(t)\big] d\big[p_{st}(t',t) - A(t')\big]$$

$$\times \exp\big[-iS_{st}(t',t)\big] E_1(t')\Big\} \exp\left[-\int_{-\infty}^{t} w(t')\,dt'\right]$$

(1)

where: $E_l(t)$ is the electric field of the laser pulse (assumed to be linearly polarized); $A(t)$ is its associated vector potential; ε is a positive regularization constant; p_{st} and S_{st} are the stationary values of momentum and quasi-classical action; d is the dipole matrix element for bound-free transitions; $w(t)$ is the tunnel ionization rate from the ground state, calculated using the Ammosov, Delone and Krainov (ADK) model. The single atom response is inserted as a source term in the equations governing the evolution of the fundamental and the harmonic beams, which are solved in cylindrical coordinates and assuming radial symmetry. The propagation of the fundamental wavelength beam in the ionizing gas is described in paraxial approximation by the equation:

$$\nabla_{\perp}^{2} E_1(r,z',t') - \frac{2}{c}\frac{\partial^2 E_1(r,z',t')}{\partial z'\partial t'} = \frac{\omega_p^2(r,z',t')}{c^2} E_1(r,z',t')$$

(2)

where: c is the speed of light; ω_p is the plasma frequency; r is the transverse coordinate; t' and z' are the temporal and propagation coordinates in a moving frame, related to the laboratory frame by the transformations $z'=z$ and $t'=t-z/c$. This equation takes into account both temporal plasma-induced phase modulation and spatial plasma lensing effects on the fundamental beam, while it does not consider the linear gas dispersion and the depletion of the fundamental beam during the HHG process, which are negligible under our experimental conditions. For the propagation of the harmonic field $E_h(r,z,t)$, we solve the following equation:

$$\nabla_{\perp}^{2} E_h(r,z',t') - \frac{2}{c}\frac{\partial^2 E_h(r,z',t')}{\partial z'\partial t'} = \mu_0 \frac{\partial^2 P_{nl}(r,z',t')}{\partial t'^2}$$

(3)

where: $P_{nl}(r,z,t) = [n_0 - n_e(r,z,t)]d_{nl}(r,z,t)$ is the nonlinear polarization generated in the gas; n_0 and n_e the initial atomic density and the electron density respectively. In this equation the free electron dispersion is neglected because the plasma frequency is much lower than the frequencies of high order harmonics. The nonlinear dipole moment is calculated at each position inside the jet by using the corresponding fundamental field derived by solving Equation (2).

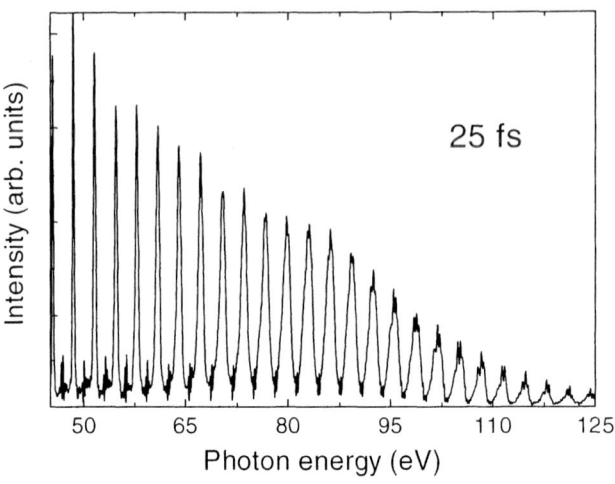

FIGURE 7. Calculated harmonic spectrum for 25-fs driving pulses.

The model based on relations (1-3) was used to simulate the experimental results previously reported. Figure 7 shows the harmonic spectrum calculated for 25-fs driving pulses in conditions similar to those met in our experiment. The simulation is in good agreement with the measured spectrum shown in Fig. 2; both the cutoff position and the decreasing in spectrum amplitude are well reproduced. Note that the peak linewidth increases with photon energy also in the calculated spectrum, as shown by the solid line reported in Fig. 4 (lower panel). Numerical simulations were also performed in the few-cycle regime. The calculated spectrum for 7-fs driving pulses is

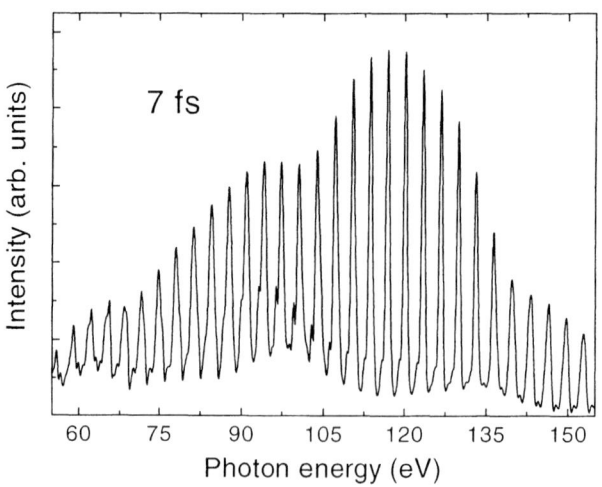

FIGURE 8. Calculated harmonic spectrum for 7-fs driving pulses.

shown in Figure 8; the agreement between simulated and experimental results is still good; the behavior of harmonic linewidths is also satisfactorily reproduced, as show by the solid line reported in the upper panel of Fig. 4.

The theoretical model can also reproduce the spatial properties of harmonic emission. Nevertheless, by using in the calculation a fundamental beam with Gaussian intensity profile, we were unable to reproduce the experimental data. In particular, in this case the results of the calculation predict a harmonic divergence decreasing upon moving the gas jet away from laser focus in the propagation direction. Moreover the spatial harmonic profile is narrow and regular only when the gas jet is placed well after focus ($z \geq 3$ mm). When the jet is moved near and before the focus the profile becomes annular, as shown by dashed line in Figure 9(a), which presents the radial intensity profiles of 45th harmonic for $z = 0$ in the case of Gaussian excitation beam with 7-fs pulses. We have then performed simulations considering the actual spatial profile of the fundamental beam at the output of the hollow fiber: a zeroth-order Bessel function truncated in correspondence of the first zero. Figure 10 shows the FWHM of the harmonic angular distribution as a function of the harmonic order, for two different positions of the gas jet. The angular distributions have been calculated by integrating the two-dimensional profile of the harmonic beam over one direction, in order to reproduce the operation of the spectrometer. In this case, the numerical simulation presents a qualitative agreement with experimental data; in particular, the divergence

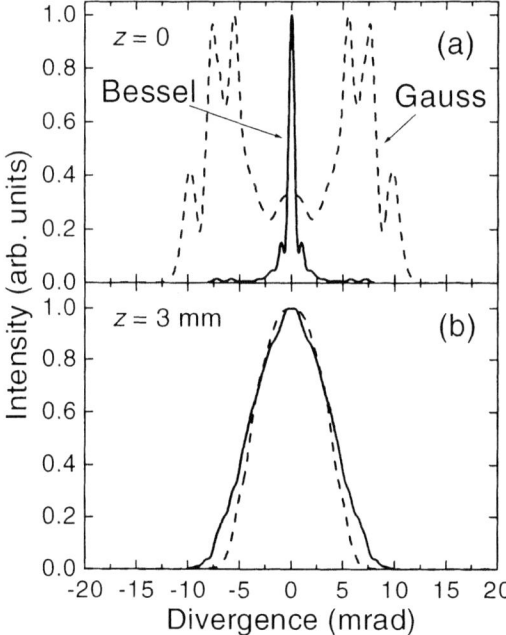

FIGURE 9. Calculated harmonic profiles generated by truncated-Bessel (solid line) and Gaussian (dashed line) fundamental beam for different positions of gas jet with respect to laser beam waist.

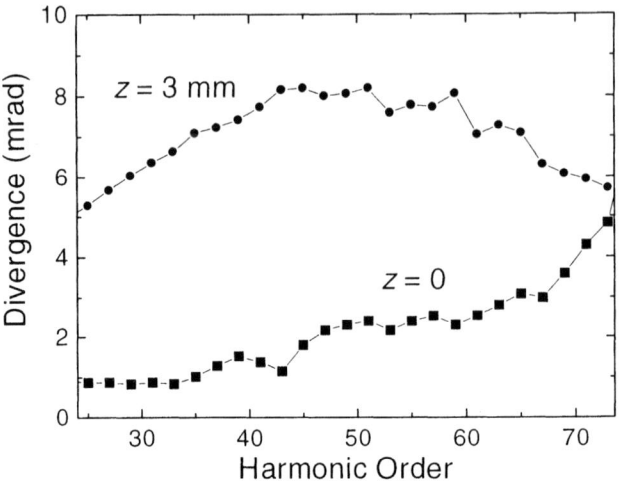

FIGURE 10. Calculated harmonic divergence vs. nonlinear order for 7-fs driving pulses.

correctly increases upon increasing the z value. The slight quantitative disagreement may be due to an unavoidable uncertainty of some experimental parameters (shape of the gas jet, gas pressure, etc.). Moreover, in agreement with the experimental results, the calculated harmonic beam displays regular spatial profiles even when the gas jet is moved near the laser focus, as shown by solid lines in Figs. 9(a)-(b). When the gas jet is located well beyond the laser focus ($z \geq 3$ mm, see Fig. 9(b)) the profiles are of bell shape, irrespectively of the order. On the other hand, when moving the jet around the laser focus (see Fig. 9(a)), the harmonic profile becomes very narrow, and does not display any annular structure. Upon increasing the nonlinear order, the profile presents narrow central peaks and long wings. It is important to note that the spectral characteristics of the harmonic beam are not significantly affected by the spatial profile of the fundamental beam.

CONCLUSIONS

High-order harmonic generation was investigated in both the multi-cycle and the few-cycle regime. Characterization of the XUV emission showed that harmonic beam with good spatial properties can be generated with 7-fs driving pulses from a hollow fiber compressor. Owing to the high brightness potentially achieved in this regime, such harmonic beam could be useful for nonlinear optics experiments in the XUV spectral range.

ACKNOWLEDGMENTS

This study was partially supported within the framework of the Istituto Nazionale per la Fisica della Materia under the project "Femtosecond soft-X-ray generation by high energy laser pulses".

REFERENCES

1. Nisoli, M., De Silvestri, S., and Svelto, O., *Appl. Phys. Lett.* **68**, 2793-2795 (1996).
2. Nisoli, M., De Silvestri, S., Svelto, O., Szipöcs, R., Ferencz, K., Spielmann, Ch., Sartania, S., and Krausz, F., *Opt. Lett.* **22**, 522-524 (1997).
3. Nisoli, M., Stagira, S., De Silvestri, S., Svelto, O., Sartania, S., Cheng, Z., Tempea, G., Spielmann, Ch., and Krausz, F., *IEEE J. Select. Top. Quantum Elect.* **4**, 414-420 (1998).
4. Mücke, O. D., Tritschler, T., Wegener, M., Morgner, U., and Kärtner, F. X., *Phys. Rev. Lett.* **87**, 057401, (2001).
5. Cerullo, G., Lanzani, G., Pallaro, L., and De Silvestri, S., *J. Mol. Struct.* **521**, 261-270 (2000)
6. Brabec, T., and Krausz, F., *Rev. Mod. Phys.* **72**, 545-591 (2000).
7. Dietrich, P., Krausz, F., and Corkum, P. B., *Opt. Lett.* **25**, 16-18 (2000).
8. Jones, D. J., Diddams, S. A., Ranka, J. K., Stentz, A., Windeler, R. S., Hall, J. L., and Cundiff, S. T., *Science* **288**, 635-639 (2000).
9. Salières, P., Le Déroff, L., Auguste, T., Monot, P., d'Oliveira, P., Campo, D., Hergott, J.-F., Merdji, H., and Carré, B., *Phys. Rev. Lett.* **83**, 5483-5486 (1999).
10. Descamps, D., Lynga, C., Norin, J., L'Huillier, A., Wahlstrom, C. G., Hergott, J. F., Merdji, H., Salieres, P., Bellini, M., and Hansch, T. W., *Optics Letters* **25**, 135-137 (2000).
11. Kan, C., Burnett, N. H., Capjack, and C. E., Rankin, R., *Phys. Rev. Lett.* **79**, 2971-2794 (1997).
12. Zhou, J., Peatross, J., Murnane, M. M., Kapteyn, H. C., and Christov, I. P., *Phys. Rev. Lett.* **76**, 752-755 (1996).
13. Schafer, K.J., and Kulander, K.C., *Phys. Rev. Lett.* **78**, 638-641 (1997).
14. Spielmann, C., Burnett, N. H., Sartania, S., Koppitsch, R., Schnuerer, M., Kan, C., Lenzner, M., Wobrauschek, P., and Krausz, F., *Science* **278**, 661-663 (1997).
15. Poletto, L., Tondello, G., and Villoresi, P., *Rev. Sci. Instrum.* **72**, 2868-2874 (2001).
16. Tempea, G., Krausz, F., Spielmann, C., and Ferencz, K., *IEEE J. Select. Top. Quantum Elect.* **4**, 193-196 (1998).
17. Matuschek, N, Kärtner, F. X., and Keller, U., *IEEE J. Select. Top. Quantum Elect.* **4**; 197-208 (1998).
18. Iaconis, C., and Walmsley, I. A., *IEEE J. Quantum Electron.* **35**, 501-509 (1999).
19. Marcatili, E. A. J., and Schmeltzer, R. A., *Bell Syst. Tech. J.* **43**, 1783-1809 (1964).
20. Villoresi, P., Ceccherini, P., Poletto, L., Tondello, G., Altucci, C., Bruzzese, R., de Lisio, C., Nisoli, M., Stagira, S., Cerullo, G., De Silvestri, and S., Svelto, O., *Phys. Rev. Lett.* **85**, 2494-2497 (2000).
21. Salières, P, L'Huillier, A., and Lewenstein, M, *Phys. Rev. Lett.* **74**, 3776-3779 (1995)
22. Priori, E., Cerullo, G., Nisoli, M., Stagira, S., De Silvestri, S., Villoresi, P., Poletto, L., Ceccherini, P., Altucci, C., and Bruzzese, R., *Phys. Rev. A* **61**, 063801 (2000).

Relativistic laser-particle interaction:
From single electrons to multi-particle systems

Guido R. Mocken*, Yousef I. Salamin[1]*, Karen Z. Hatsagortsyan[2]* and
Christoph H. Keitel[3]*

*Theoretische Quantendynamik, Fakultät für Physik, Universität Freiburg,
Hermann-Herder-Straße 3, D-79104 Freiburg, Germany

Abstract. The relativistic dynamics of single particles and thin crystals is investigated in very short, intense laser pulses. Single particle acceleration from 100 MeV to GeV energies is shown for electrons in Gaussian beams of few micron widths and crossed laser beams of 0.1 radians crossing angles. High harmonic generation in the X-ray regime is predicted from thin crystals due to their regularity in few-femtosecond pulses. Quantum features such as relativistic tunneling and spin effects are discussed.

INTRODUCTION

The single particle picture of laser matter interaction is often appropriate if the density of the material is low as in gases or if the pulse duration is so short that collective dynamics may not build up. The advantage of such a picture is high numerial accuracy and intuitive understanding even for complicated laser field geometries or complex atomic systems involving full relativistic and quantum features. We evaluate single particle dynamics in Gaussian and crossed beam geometries with emphasis on GeV particle acceleration and stability. The radiation spectrum is evaluated arising from crystals in very short, intense laser pulses. We predict high harmonic generation in the hard x-ray regime and point out quantum features.

The work presented here is split into four parts which all involve dynamics in ultra high intensity laser radiation. The first one deals with just a single particle, namely an electron, in a single laser beam. That beam, however, is described as accurately as possible, i.e., it is modelled as a true gaussian beam up to fifth order in the diffraction angle. The second part also deals with a single electron, but here it is subjected to the field generated by the superposition of two crossed laser beams, which in turn are modelled in a somewhat simpler way. In our third contribution, we use plane wave laser light, but the setup itself is much more complicated, involving the interaction of the laser field with a periodic atomic structure, containing both ions and electrons.

---

[1] Permanent address: Physics Department, Birzeit University, P.O. Box 14, Birzeit, West Bank, Palestine.

[2] Permanent address: Department of Theoretical Physics, Yerevan State University, A. Manoukian Street 1, 375049 Yerevan, Armenia.

[3] Email address: keitel@uni-freiburg.de

CP611, *Superstrong Fields in Plasmas:* Second Int'l. Conf., edited by M. Lontano et al.
© 2002 American Institute of Physics 0-7354-0057-1/02/$19.00

Whereas the first two examples illustrate the dynamics of otherwise free electrons inside one or two beams of intense laser light and especially the energy gain that can be achieved using these accelerator schemes, the third one is aimed to put forward x-ray amplification by laser controlled coherent bremsstrahlung. Finally we consider fully quantum relativistic pulsed laser-ion interactions based on the assumption that the various single-ion radiation sources are arranged to be phase-matched.

ELECTRON ACCELERATION

Gaussian laser beams

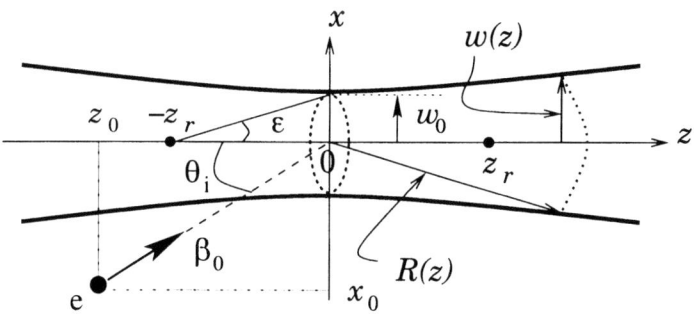

FIGURE 1. The beam geometry. Cross section through the focus is circular and has a radius w_0; a cross section through an arbitrary point z on axis is also circular with a radius given by $w(z) = w_0 \sqrt{1 + (z/z_r)^2}$. With $k = 2\pi/\lambda$, the Rayleigh length is $z_r = kw_0^2/2$ and the diffraction angle is $\epsilon = w_0/z_r$. (Reproduced from [1] with permission.)

The beam shown in Fig. 1 may be modelled by a vector potential linearly-polarized along $+x$. In order to describe beam widths down to few microns, the diffraction angle ϵ is included in our evaluations up to fifth order. The vector potential obeys a wave equation whose solution yields the following electric field components [2, 3, 4]

$$E_x = E \left\{ S_0 + \epsilon^2 \left[\xi^2 S_2 - \frac{\rho^4 S_3}{4} \right] + \epsilon^4 \left[\frac{S_2}{8} - \frac{\rho^2 S_3}{4} \right. \right.$$
$$\left. \left. - \frac{\rho^2 (\rho^2 - 16\xi^2) S_4}{16} - \frac{\rho^4 (\rho^2 + 2\xi^2) S_5}{8} + \frac{\rho^8 S_6}{32} \right] \right\}, \tag{1}$$

$$E_y = E\xi\upsilon \left\{ \epsilon^2 S_2 + \epsilon^4 \left[\rho^2 S_4 - \frac{\rho^4 S_5}{4} \right] \right\}, \tag{2}$$

$$E_z = E\xi \left\{ \epsilon C_1 + \epsilon^3 \left[-\frac{C_2}{2} + \rho^2 C_3 - \frac{\rho^4 C_4}{4} \right] + \epsilon^5 \left[-\frac{3C_3}{8} \right. \right.$$
$$\left. \left. - \frac{3\rho^2 C_4}{8} + \frac{17\rho^4 C_5}{16} - \frac{3\rho^6 C_6}{8} + \frac{\rho^8 C_7}{32} \right] \right\}. \tag{3}$$

Similarly, the magnetic field components are given by

$$B_x = 0, \tag{4}$$

$$B_y = E \left\{ S_0 + \epsilon^2 \left[\frac{\rho^2 S_2}{2} - \frac{\rho^4 S_3}{4} \right] \right.$$
$$\left. + \epsilon^4 \left[-\frac{S_2}{8} + \frac{\rho^2 S_3}{4} + \frac{5\rho^4 S_4}{16} - \frac{\rho^6 S_5}{4} + \frac{\rho^8 S_6}{32} \right] \right\}, \tag{5}$$

$$B_z = E\upsilon \left\{ \epsilon C_1 + \epsilon^3 \left[\frac{C_2}{2} + \frac{\rho^2 C_3}{2} - \frac{\rho^4 C_4}{4} \right] + \epsilon^5 \left[\frac{3C_3}{8} \right. \right.$$
$$\left. \left. + \frac{3\rho^2 C_4}{8} + \frac{3\rho^4 C_5}{16} - \frac{\rho^6 C_6}{4} + \frac{\rho^8 C_7}{32} \right] \right\}. \tag{6}$$

In Eqs. (1-6), we let $\xi = x/w_0$, $\upsilon = y/w_0$, $\zeta = z/z_r$ and

$$E = E_0 \frac{w_0}{w} \exp\left[-\frac{r^2}{w^2} \right]; \quad E_0 = \text{constant}, \tag{7}$$

$$S_n = \left(\frac{w_0}{w} \right)^n \sin(\psi + n\psi_G); \quad n = 0, 1, 2, \cdots \tag{8}$$

$$C_n = \left(\frac{w_0}{w} \right)^n \cos(\psi + n\psi_G). \tag{9}$$

Furthermore, $k = \omega/c$, $kA_0 = E_0$, $r^2 = x^2 + y^2$, and $\rho = r/w_0$. These equations were derived from a vector potential polarized along $+x$, which has an amplitude A_0, and a frequency ω. Also, $\psi = \psi_0 + \psi_P - \psi_R + \psi_G$, where ψ_0 is a constant phase, that is taken to be zero throughout this work, $\psi_P = \eta = \omega t - kz$ is the plane wave phase, $\psi_G = \tan^{-1} \zeta$ is the Guoy phase associated with the fact that a Gaussian beam undergoes a total phase change of π as z changes from $-\infty$ to $+\infty$, $\psi_R = kr^2/(2R)$ is the phase associated with the curvature of the wave fronts, and $R(z) = z + z_r^2/z$ is the radius of curvature of a wave-front intersecting the beam axis at the coordinate z. The fields given above satisfy Maxwell's equations $\vec{\nabla} \cdot \vec{E} = 0 = \vec{\nabla} \cdot \vec{B}$, plus terms of order ϵ^6. For a detailed derivation see [5].

Particle dynamics and energy gain

Figure 2a, which is similar to those shown in [1, 5], illustrates the dynamics of a single electron that is injected into the gaussian beam with an initial velocity vector $\vec{\beta}_0$ (which corresponds to a value of $\gamma_0 = 1/\sqrt{1 - \vec{\beta}_0^2}$) as shown in figure 1. The initial position of the particle at $t = t_i = 0$ is $\vec{r} = -(0.3\text{cm}, (s - 0.3\text{cm}) \tan\theta_i, 0\text{cm})$. The parameter s will later be used to shift the beginning of the trajectories in $\pm x$-direction while keeping the injection angle constant. Three cases are shown: The electron can be reflected off the laser beam, it can be transmitted right through, or captured within it. By the latter we mean that it travels within the indicated beam boundary for as long as our simulation goes, which in this case means $\omega t_f = \pi \times 10^6$, where ω denotes the laser frequency. At

46

$t = t_f$, the transmitted and reflected electrons come out with an energy gain of 89.6 keV and 24.3 keV, respectively, whereas the captured one still oscillates within the range 72.0 to 72.3 keV. These gains are possible due to the assymetric interaction with fractions of whole laser cycles. Intermediate gains reach the MeV range, and it is only in this illustrative example that the particle loses most of this again. The final gain can be much higher, as will be shown now.

In figure 2b, which is similar to ones in [1, 5], we now show that using reflected electrons, it is easily possible to reach gains in the MeV range. We lower the injection angle and the initial velocity, vary s/z_r from -1 to 4 and measure the energy gain $W(\eta) = mc^2(\gamma(\eta) - \gamma_0)$ after an interaction time t_f.

We see gains of over one hundred MeV for certain values of s and a quite modest intensity parameter of $q = 10$. As shown in [5], up to a few GeV are within reach for $q = 100$. A further reduction of the beam waist w_0 is also often favorable. We also note the importance of the higher order corrections in ϵ to the fields.

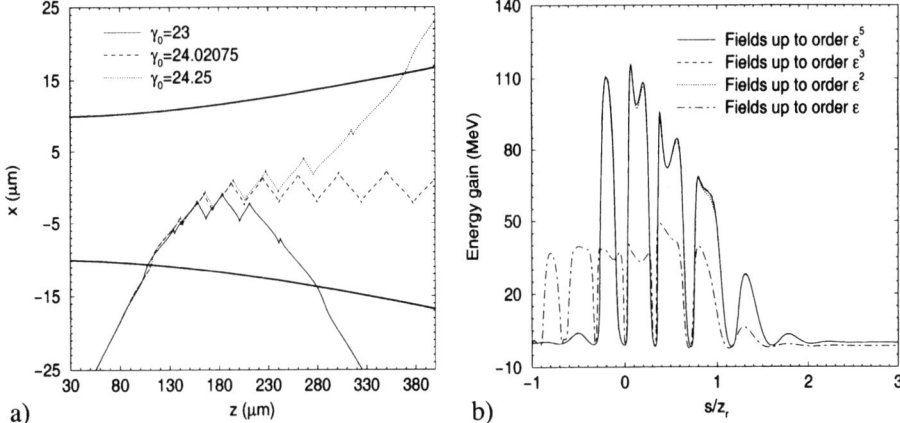

FIGURE 2. **a)** A reflected (solid line), a captured (dashed), and a transmitted (dotted) electron trajectory. The thick dark line marks the beam boundary (by which we mean the curve $\pm w(z)$ as given in the caption of figure 1). The laser wavelength used is $\lambda = 1.056\mu$m, the intensity parameter $q = \frac{eE_0}{mc\omega}$ is chosen as $q = 10$, the initial position corresponds to $s = 0.5z_r$, the injection angle is set to $\theta_i = 15^o$. The beam waist radius at focus w_0 is chosen to be 10μm. **b)** Variation of the energy gain with s when contributions to the fields up to different orders of ϵ are taken into account. Here $\theta_i = 5^o$, $\gamma_0 = 10$, and $\omega t_f = \pi \times 10^5$. (Similar to [1, 5].)

CROSSED LASER BEAMS

We investigate analytically and numerically the dynamics of a single electron, in vacuum, in the presence of a pair of laser beams [6] crossing at an arbitrary angle [7]. Figure 3 illustrates the spatial arrangement of the beams, which are denoted by their wave vectors $\vec{k}_1$ and $\vec{k}_2$, respectively. Note the interaction region where the two beams intersect. This is the region of interest. In a coordinate system such as in figure 3, the wave vectors

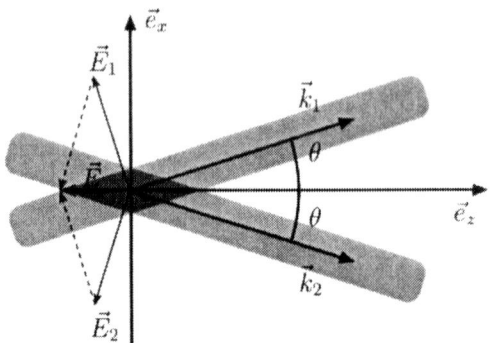

FIGURE 3. Schematic diagram of the electron interaction with the crossed laser beams.

can be expressed as

$$\vec{k}_{1,2} = \pm\frac{\omega}{c}\left(\vec{e}_x \sin\theta \pm \vec{e}_z \cos\theta\right). \tag{10}$$

Using the label η for the field phases $\eta_{1,2} = \omega t - \vec{k}_{1,2}\cdot\vec{r}$, the expressions for the electric and magnetic fields become:

$$\vec{E}_{1,2} = \pm E_0\, g(\eta_{1,2})\,\cos(\eta_{1,2})\left(\vec{e}_x\cos\theta \mp \vec{e}_z\sin\theta\right) \tag{11}$$

$$\vec{B}_{1,2} = \pm E_0\, g(\eta_{1,2})\,\cos(\eta_{1,2})\,\vec{e}_y \tag{12}$$

Herein $g(\eta)$ denotes a suitably chosen envelope function, e.g. $g(\eta_{1,2}) = \sin^2\left(\frac{\eta_{1,2}}{12}\right)$. For all points on the z axis this yields:

$$\eta_1, \eta_2 \to \eta = \omega\left(t - \frac{z}{c}\cos\theta\right) \tag{13}$$

$$\vec{B}(0,0,z) = \vec{0} \tag{14}$$

$$\vec{E}(0,0,z) = -2E_0\, g(\eta)\,\cos(\eta)\sin(\theta)\vec{e}_z \tag{15}$$

Our analysis evolves from exact solutions to the relativistic equations of motion:

$$\frac{d\vec{p}}{dt} = -e(\vec{E} + \vec{\beta}\times\vec{B}), \qquad \frac{d\mathcal{E}}{dt} = -ec\vec{\beta}\cdot\vec{E} \tag{16}$$

$(\frac{\mathcal{E}}{c}, \vec{p})$ is the particle's energy-momentum vector and $\vec{\beta}$ its velocity vector scaled by the speed of light, as usual. In our case this means

$$\frac{d(\gamma\beta_z)}{dt} = 2q\omega\sin\theta\, g(\eta)\,\cos(\eta) \tag{17}$$

$$\frac{d\gamma}{dt} = 2q\omega\sin\theta\, g(\eta)\beta_z\,\cos(\eta) \tag{18}$$

where $q = \frac{eE_0}{mc\omega}$. The above equations can be solved analytically for $\beta_z(\eta)$, $\gamma(\eta) = 1/\sqrt{1-\beta^2}$, $\frac{dz}{d\eta}(\eta)$ as shown in [8], and numerically for $z(\eta)$. For an electron that is injected into the interaction region right on the z-axis and with an initial velocity vector that is also pointing in this direction, the x- and y-component equations result in no motion due to the vanishing of the magnetic field on the z axis, which means that the electron will move on a line only along the z direction. Our analysis predicts average acceleration gradients of several TeV/m over short distances, using laser field intensities of up to 10^{22} Wcm^{-2}[8].

The radiation losses of the accelerated electron can be calculated using the Larmor formula:

$$P(t) = \frac{2}{3}\frac{e^2}{c}\gamma^6 \left\{ \left[\frac{d\vec{\beta}}{dt}\right]^2 - \left[\vec{\beta} \times \frac{d\vec{\beta}}{dt}\right]^2 \right\} \tag{19}$$

In our case, one obtains

$$P(\eta) = \frac{2}{3}\frac{e^2}{c}[2q\omega\sin\theta\, g(\eta)\cos(\eta)]^2 \tag{20}$$

$$\frac{d\mathcal{E}}{dt}(\eta) = mc^2[2q\omega\sin\theta\, g(\eta)\cos(\eta)]\beta(\eta) \tag{21}$$

which, when using $q = 10, \gamma_0 = 30, \theta = 0.1, g(\eta_{1,2}) = \sin^2\left(\frac{\eta_{1,2}}{12}\right)$, shows peak intensities of only a few μW for $P(\eta)$. The rate of change in energy (21), on the other hand, is of the order of a few hundred Watts.

Initial vertical displacement and angular distribution

For completeness, effects on the electron trajectories and electron energy gain due to departure from the ideal initial conditions, like the existence of a distribution of initial velocities, off-axis injection, and realistic beam representation, are studied. This is done purely numerically. The electronic motion becomes two dimensional in the first two cases, and as the magnetic field no longer vanishes when moving away from the z-axis, its influence becomes important. We point out the regime of stability for off-axis injection (see figure 4), the overall stability for injection at an angle (see figure 5), and the easy-to-recognize influence of the non-vanishing magnetic field on all these electron trajectories.

The main point learned from these studies is, that, although the electronic motion becomes much more complicated when a non-zero magnetic field comes into effect, it is in principal possible to keep the electron close to the z-axis for as long as desired, if the deviations from the ideal initial conditions are not too large.

Energy gain for a realistic beam representation

Whereas the electron will always lose any energy that it has intermediately gained during the interaction with a complete laser pulse, an asymmetry can be introduced by

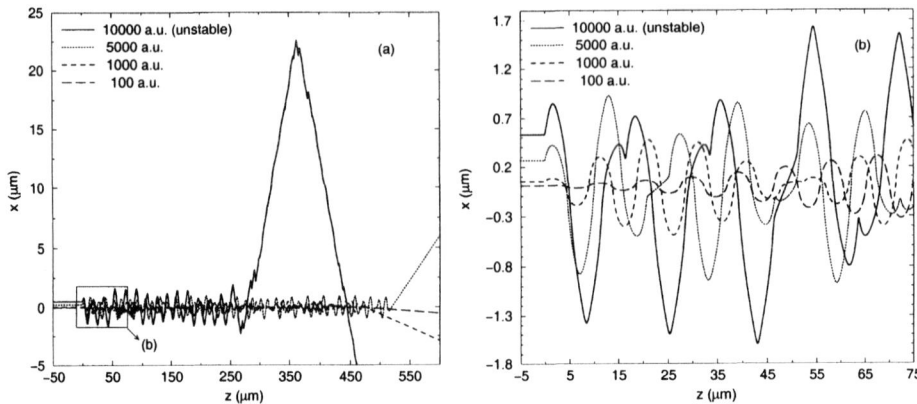

FIGURE 4. Influence of an initial vertical displacement on the trajectory of an electron subjected to the field of two crossed beams ($\theta = 0.1$ rad) modeled by plane-wave patterns with $\lambda = 1\mu m$. At $\gamma_0 = 1.5$ and $q = 1$, displacements of 100 to 5000 a.u. (0.005 to 0.265μm) result in trajectories that oscillate about the z-axis for the whole duration of the laser pulse (100 cycles), whereas an initial offset of 10000 a.u. (0.529μm) makes the electron leave the vicinity of the z-axis after some time. (b) is a magnified portion of (a) at the very beginning of the trajectories. Reproduced from [8] with permission.

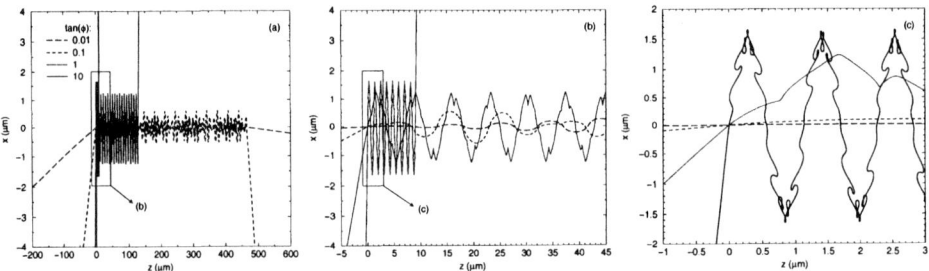

FIGURE 5. Influence of an angular distribution of initial velocities on an electron subjected to the field of two crossed beams ($\theta = 0.1$ rad) modeled by plane-wave patterns with $\lambda = 1\mu m$. At $\gamma_0 = 1.5$ and $q = 1$, all angles $\phi = \sphericalangle(\vec{v}_{initial}, \vec{e}_z)$ with $\tan(\phi)$ between 0.01 and 10 result in trajectories that oscillate about the z-axis for the whole duration of the laser pulse (100 cycles). (b) and (c) are magnified portions of (a) and (b), respectively. Reproduced from [8] with permission.

using spatially confined laser beams instead of infinitely extended plane waves. Then, it is possible that the electron passes through the right spatial edge of the limited interaction region before the laser pulse is turned off. Using a suitable set of initial parameters, we find out that the electron leaves the interaction region with spatially confined laser beams with energy gain in the GeV range [8]. This is shown in figure 6. Longitudinal focusing is effected by an envelope $g(\eta)$ consisting of a 3-cycle $\sin^2$ turn-on, a 4-cycle flat top and a 3-cycle $\sin^2$ turn-off. Transversely, the beams are focused using three different intensity profile functions: 1) A rectangular profile, 2) a $\sin^2$ profile, 3) a gaussian profile - each of them normalized to yield the same power when integrated over the beam cross section.

As can be seen, the sharper the spatial confinement of the interaction, the more gain can be reached - up to almost 2 GeV for the rectangular profile, but only about 0.4 GeV for the more realistic gaussian one.

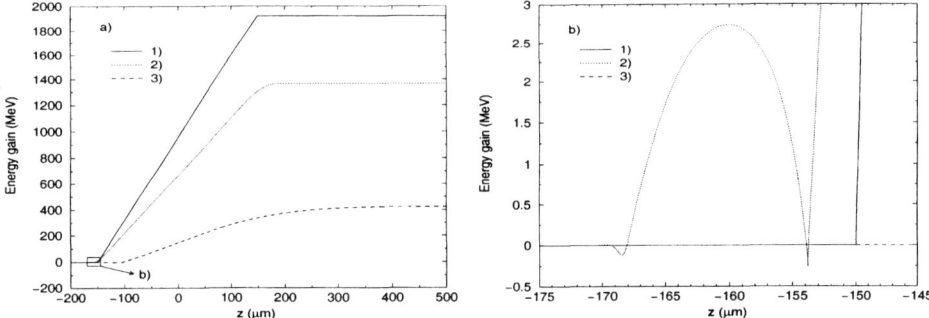

FIGURE 6. Electron energy gain vs the forward distance when the influence of the pulse shape and of the beam profiles are taken into account. The electron, whose initial energy corresponds to $\gamma_0 = 1.5$, is subjected to the field of two beams crossing at an angle $\theta = 0.01$ rad. The beam intensity corresponds to $q = 100$ at a wavelength $\lambda = 1\mu m$. The beam cross section is taken as circular with a maximum radius $1.5\mu m$. Figure (b) is a magnified portion of (a). See text for details. Reproduced from [8] with permission.

HIGH-FREQUENCY LIGHT GENERATION

X-Ray Amplification by Laser Controlled Coherent Bremsstrahlung

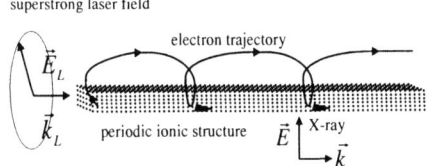

FIGURE 7. Scheme of interest: Electrons extracted from a thin solid layer propagate along and oscillate in a superstrong laser field ($\vec{k}_L$: wave vector and $\vec{E}_L$: electric field of the laser wave) and interact with a periodic ionic structure. The generation of coherent X-rays ($\vec{k}$: wave vector and $\vec{E}$: electric field of the X-rays) is depicted at the areas of interaction. Reproduced from [9] with permission.

We have proposed an alternative approach towards intense coherent X rays deviating clearly from the known main streams - X-FEL (X-ray free electron laser) and HHG (high harmonic generation) - however equally involving ideas of both of them [9]. We consider a thin solid layer (or some other thin periodic ionic structure) to interact simultaneously with a short superintense laser pulse and a weak high-frequency field as probe (see figure 7). Under the influence of the high-power laser pulse the electrons are extracted almost instantaneously from the layer. The periodic brief interactions with the far slower moving and thus still quasiregular ionic structure then lead to high-frequency coherent bremsstrahlung as shown here by the amplification of the probe field. Thus as in an

FEL, we consider stimulated radiation from essentially free electrons which however are extracted from the laser-driven solid layer rather than being injected into the system. With respect to conventional HHG schemes, the coherent X-ray emission arises also from harmonic generation due to laser-induced periodic nonlinear interactions of the electrons with ions. However, in our situation the electron possesses highly relativistic energies and multiphoton scattering dominates over recombinations to bound states. From a more technical point of view we would like to emphasize that we consider the more efficient *stimulated* X-ray emission in our scheme while conventional HHG is known to operate well spontaneously.

The main problem facing the realization of the proposed scheme is the destruction of the periodic ionic structure in the superstrong laser field. It is shown in [9] that the destruction process can be slow enough as compared to the time necessary for the X-ray radiation. That has been confirmed as well by PIC simulations [10], where no substantial ion dynamics was notable during the short interaction with the laser pulse.

We calculate the small signal gain G of X-ray radiation for our scheme of interest in the framework of classical radiation theory. The high power pump laser field is considered exactly while the scattering potential as well as the probe field of X-ray radiation are treated perturbatively to second order.

It was shown that the constructive interference among the various scattering events secures high coherence of the harmonics and an enhancement of the gain as compared with bremsstrahlung in a plasma [11].

There are various distinct regimes of amplification depending on the characteristics of the width of the resonance of the process. The most favorable regime for the probe amplification is what we call the exponential instability (EI) regime. This is defined by the situation when the gain G is large enough, such that the spectral width of the probe wave, due to its fast increase in intensity, prevails the resonance width due to limited interaction time, as well as that imposed by the momentum spread Δ of the electron beam

$$G \succeq \max\{1/L,\ N\Delta/mc\lambda_L\}, \tag{22}$$

where L is interaction length, Δ is the electron momentum spread, λ_L is the laser wavelenght, and $N = \frac{d}{a}$ is the ratio of the longitudinal period of the electron trajectory d and the ionic separation length a prior the laser pulse. In this regime the gain is found on the basis of a self-consistent set of Maxwell equations for the probe wave and the equation of motion for the electrons. The scattering potential of each ion is specified as a screened Coulomb potential. The ions are assumed to be centered along a regular lattice with a Gaussian distribution for the random variation from the crystal knot centers. In the case of a strong laser field ($\xi = \frac{eE_L}{mc\omega_L} \gg 1$, where E_L and ω_L are the electric field and frequency of the laser field) and concentrating on the high harmonic regime ($n \gg 1$), the X-ray small signal gain is then given by the expression

$$
\begin{aligned}
G &= \sqrt{3}\left(\frac{\pi}{4}\right)^{1/3}\exp\left\{-\frac{4\pi^2 u^2}{3a^2}\right\} \\
&\times \left[\frac{Z^2 n_e n_i r_0^3 \xi^6 \lambda_L^2 \gamma_o^3 m^4 c^4}{n^2 a^4 \Lambda^4 R_\perp^2}L_\perp^4(1+\beta_o)^4 I_{n-N}(\zeta)K_{n-N}(\zeta)\right]^{1/3}
\end{aligned}
\tag{23}
$$

52

where u is the mean deviation of electrons from the crystal knot points, r_0 is the classical radius of the electron; n_e, n_i are the electron and ion densities, respectively; γ_o the initial Lorenz-factor of the electron, $\beta_o = v_o/c$, v_o the electron initial velocity; $\overline{\varepsilon_o} = \varepsilon_o - c(mc\xi)^2/2\Lambda$ the renormalized energy of the electron in the laser field, $n = \omega/\omega_L$ is the harmonic number, $R_\perp$ is the transversal size of the ion structure (with respect to the laser propagation direction), $\zeta = \lambda_L \xi \gamma_o (1+\beta_o)/a$ and $I_n(\zeta), K_n(\zeta)$ are the modified Bessel functions. $\Lambda = p_{ox} - \epsilon_o/c$, where p_{ox} and ϵ_o are the electron's initial energy and its momentum in the laser wave propagation direction.

The gain dependence of the generated radiation frequency is shown in Fig 8.

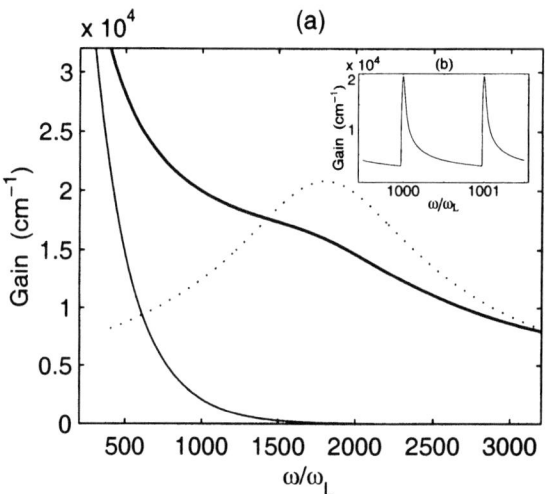

FIGURE 8. a) The envelope function of the gain G according to Eq. (23) as a function of the probe frequency in the EI regime (thick line). The electron and ion densities are $n_e = 10n_i = 10^{23}\text{cm}^{-3}$, the laser wavelength is $\lambda_L = 3.5 \cdot 10^{-6}\text{cm}$, the laser intensity parameter is $\xi = 6$, $R_\perp = 2.1 \cdot 10^{-5}\text{cm}$, the ion charge $Z = 10$, the ion structure period $a = 3.5 \cdot 10^{-8}\text{cm}$, the initial velocity $v_0 = 0$, the transversal size of the ionic structure is $L_\perp = 7 \cdot 10^{-6}\text{cm}$, $u/a = l_i^{\parallel}/a = 0.087$, and the momentum spread of the electron beam is $\Delta/mc = 10^{-4}$. The thin line describes the corresponding gain for the multiphoton Compton radiation amplification in a plasma [12, 13] for the same parameters. The dotted line depicts the function $I_{n-N}(\zeta) \cdot K_{n-N}(\zeta)$ multiplied by $2.5 \cdot 10^7$ which peaks at the multiphoton parameter $\bar{n} = \lambda_L \xi^2/2a$. b) A window of the fine structure of the gain corresponding to the thick line in a). Reproduced from [9] with permission.

Employing the KrF laser wavelength with $\lambda_L = 248\text{nm}$ and an intensity of $I \simeq 10^{21}\text{W/cm}^2$ ($\xi = 6$) and a Pb crystal with $n_e = 10n_i = 10^{23}\text{cm}^{-3}$, $a = 4.95 \cdot 10^{-8}\text{cm}$, $L_\perp = 5.2 \cdot 10^{-6}$ cm, $R_\perp = 2.5 \cdot 10^{-5}\text{cm}$, $L = 4 \cdot 10^{-4}$ cm and further $Z = 10$, $\Delta/mc = 3 \cdot 10^{-5}$, then the EI regime operates yielding very high gain, e.g. for the $n = 1000$ harmonic $G \simeq 1.8 \cdot 10^4\text{cm}^{-1}$. Compared to this at the same parameters the multiphoton Compton effect gives rise to merely $\tilde{n} \simeq 200$ harmonics [12, 13]. In this scheme, even for the very short interaction time of $\tau_{int} = 10\text{fs}$, the enhancement of radiation is substantial as the gain-length product, characterizing the lasing possibility at $\lambda = \lambda_L/n = 2.5\text{Å}$, is $GL = 7$. This gain-length product is somewhat less, but competitive with the gain of leading present day and near future X-ray lasers.

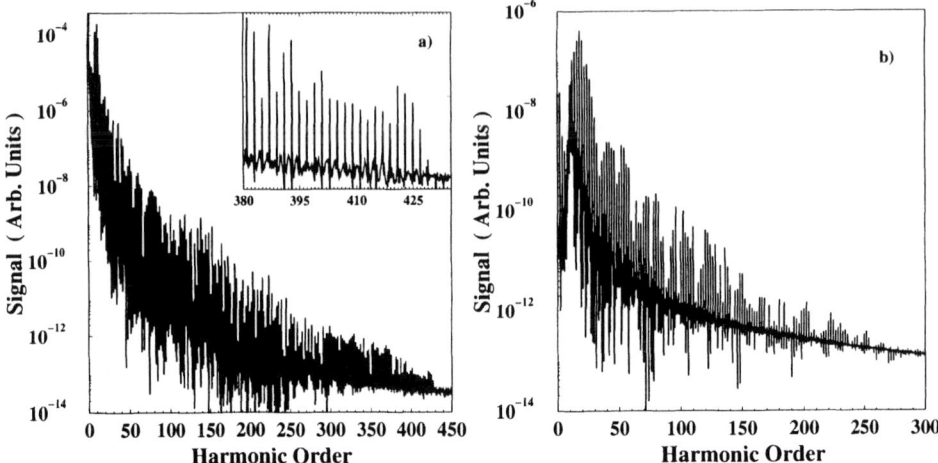

FIGURE 9. The quantum relativistic harmonic spectrum emitted from Z=4 ions. The laser intensity is $10^{17}W/cm^2$, and the KrF laser wavelength with 248nm is applied. The laser pulse has a linear 10-cycle turn-on and is followed by a 10-cycle duration with constant amplitude. (a) displays the emitted harmonic radiation polarized along the laser polarisation direction and (b) the radiation polarised along the laser propagation durection. The 427th harmonic is visible in the the cut-off regime. From [14] with permission.

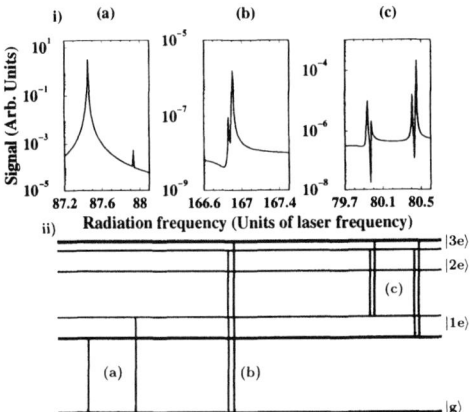

FIGURE 10. i). Segments of the radiation spectrum of the laser driven ion in the relativistic multiphoton regime involving spin-orbit coupling. The fully quantum mechanical approach includes spin-orbit coupling, the relativistic mass shift and Zitterbewegung. Figures (a), (b) and (c) (from left to right) are associated, respectively, with transitions from the first excited $|1e\rangle$ to the ground state $|g\rangle$, the third excited $|3e\rangle$ to ground state $|g\rangle$ and third excited $|3e\rangle$ to the first excited $|1e\rangle$. The spectral lines split into doublets ((a) and (b)) and a four-line structure (c) due to the spin-orbit interaction. The laser parameters involve a wavelength of 527nm, an intensity of $7 \times 10^{16}W/cm^2$, a 5.25-cycle linear turn-on and a 100-cycle duration with constant amplitude. The ion has an effective charge of Z=12. ii). The schematic diagram of state-splitting as induced by the enhanced spin-orbit interaction due to the intense laser field. We note that asymmetric states split as opposed to symmetric states. Transitions (a), (b) and (c) are associated with the corresponding spectral lines in i). From [14] with permission.

Relativistic Quantum Effects

Quantum effects are often of relevance in relativistic laser-atom interaction, especially with regard to tunneling dynamics, quantum interferences and spin-orbit coupling. This shall not be focused in this article, however for a recent discussion on this see [14]. We will merely concentrate here on high frequency light generation based on relativistic quantum dynamics. Rather than periodic scattering as described before, multiple tunneling and recollisions are a further origin of high harmonic generation. Relativistic tunneling was shown to be directed both in the laser polarisation and propagation directions [15]. In Fig. 9 we find relativistic high harmonics above 2 keV based on relativistic tunneling and recollision dynamics involving e.g. the relativistic mass shift and spin-orbit coupling. The radiation polarized along the laser propagation direction is an indication that the wave packet velocity is not negligible to the speed of light. Classical treatments would not even reproduce this figure approximately because without tunneling the electrons could not leave the ion vicinity and could not be accelerated significantly.

We add for completion that spin effects are also notable in the dynamics and spectra of high harmonic radiation [16, 17, 18, 19, 20]. Most pronounced however they are in the relativistic multiphoton regime. Multiply charged ions in intense laser fields clearly show spin-orbit splitting in the spectral response, as visible in Fig. 10 [21, 14].

ACKNOWLEDGMENTS

This work was supported by Deutsche Forschungsgemeinschaft (Nachwuchsgruppe within SFB 276) and DAAD (Gastdozenten and Short Visits Program).

REFERENCES

1. Salamin, Y. I., and Keitel, C. H., *submitted* (2001).
2. Davis, L. W., *Phys. Rev. A*, **19**, 1177 (1979).
3. Barton, J. P., and Alexander, D. R., *J. Appl. Phys.*, **66**, 2800 (1989).
4. Scully, M. O., and Zubairy, M. S., *Phys. Rev. A*, **44**, 2656 (1991).
5. Salamin, Y. I., Mocken, G. R., and Keitel, C. H., *to be submitted* (2001).
6. Esarey, E., Sprangle, P., and Krall, J., *Phys. Rev. E*, **52**, 5443 (1995).
7. Salamin, Y. I., and Keitel, C. H., *Appl. Phys. Lett.*, **77**, 1082 (2000).
8. Salamin, Y. I., Mocken, G. R., and Keitel, C. H., *submitted* (2001).
9. Hatsagortsyan, K. Z., and Keitel, C. H., *Phys. Rev. Lett.*, **86**, 2277 (2001).
10. Verboncoeur, J. P., et al., *Comput. Phys. Commun.*, **87**, 199 (1995).
11. Fedorov, M. V., *Sov. Phys. JETP*, **51**, 795 (1966).
12. Sprangle, P., and Esarey, E., *Phys. Rev. Lett.*, **67**, 2021 (1991).
13. Esarey, E., and Sprangle, P., *Phys. Rev. A*, **45**, 5872 (1992).
14. Hu, S. X., and Keitel, C. H., *Phys. Rev. A*, **63**, 053402 (2001).
15. McNaught, S. J., et al., *Phys. Rev. Lett.*, **78**, 626 (1997).
16. Latinne, O., Joachain, C. J., and Dörr, M., *Europhys. Lett.*, **26**, 333 (1994).
17. Rathe, U. W., Keitel, C. H., Protopapas, M., and Knight, P. L., *J. Phys. B*, **30**, L531 (1997).
18. Walser, M. W., and Keitel, C. H., *J. Phys. B: At. Mol. Opt. Phys.*, **33**, L221 (2000).
19. de Aldana, J. R. V., and Roso, L., *J. Phys. B: At. Mol. Opt. Phys.*, **33**, 3701 (2000).
20. Walser, M. W., et al., *to be submitted* (2001).
21. Hu, S. X., and Keitel, C. H., *Phys. Rev. Lett.*, **83**, 4709 (1999).

Toward Attosecond Metrology

Michael Hentschel[1], Reinhard Kienberger[1], Markus Drescher[2],
Paul Corkum[3] and Ferenc Krausz[1]

[1]Institut für Photonik, Technische Universität Wien, Gusshausstr. 27/387, A-1040 Wien, Austria
[2]Fakultät fütr Physik, Universität Bielefeld, D-33615 Bielefeld, Germany
[3]National Research Council of Canada, Ottawa, Ontario, Canada K1 A 0R6

Abstract. Ultrashort-pulse lasers constitute the fastest probes available for tracing transitions between different states of matter. They allow measuring time intervals on a femtosecond time scale and provide access to fundamental physical, chemical and biological processes on a microscopic scale in the time domain. The fastest speed at which events can be followed is limited by the laser pulse duration. Sub-10 femtosecond pulses are available in the visible and near infrared spectral range. They comprise only a few optical cycles, hence the phase of the carrier with respect to the pulse envelope becomes a significant issue. A method for controlling this phase is described in this paper.

Several fundamental atomic processes such as inner-shell electronic relaxation or ionization by optical tunneling, however, take place within a fraction of the oscillation period of visible or near-infrared radiation and require very short probes for being investigated. Isolated bursts of radiation of the order of 1 femtosecond or shorter are required to reliably trace these dynamics never before accessed in the time domain. These single bursts need to be carried at substantially shorter wavelengths, preferably in the extreme ultraviolet (XUV) or X-ray regime, if bound atomic electrons are to be accessed in ultrafast spectroscopy.

High harmonic generation using few-cycle pulses is ideal for this purpose.

OPTICAL WAVEFORM SYNTHESIS

Pulses delivered from a KLM laser are very stable in respect to the pulse to pulse variations of the envelope and the carrier frequency. They show an excellent energy noise behavior and timing jitter, the deviation from the repetition frequency.

When pulses getting shorter and reaches the two-cycle limit it is necessary to include also the offset phase between the envelope and the carrier in the description of the electric field E(t) from laser pulses:

$$E_n(t) = A(t) * e^{-i\Omega_0 t - i\varphi_{ceo}} + c.c. \tag{1}$$

A(t) denotes the complex pulse envelope, which can be measured by some techniques (FROG, SPIDER), Ω_0 is the carrier frequency (which can be a function of time t, when a frequency chirp is included). The new parameter to be measured an controlled is the **carrier-envelope offset phase** φ_{ceo} included in Eq. 1. For longer pulses including more than 4 optical cycles or so within the half width of the intensity, φ_{ceo} is not important to consider because the peak of the pulses carrying many

CP611, *Superstrong Fields in Plasmas:* Second Int'l. Conf., edited by M. Lontano et al.

oscillations is not well defined. However, interactions governed by the electric field [1] become strongly dependent on φ_{ceo} in the few-cycle regime. Very recently the access to φ_{ceo} experienced an increased attention using different methods [2]-[7].

The pulse-train shown in the upper part of Fig. 1 characterizes the state of pulse generation up to some month ago, when we had no access to the carrier-envelope offset phase φ_{ceo} at all. For two successive pulses picked out of the train, φ_{ceo} is not constant and can be arbitrary between 0 and 2π. The phase of two successive pulses changes with nearly the same value, we named **round-trip phase shift** $\Delta\varphi$, which we could measure some years ago [8].

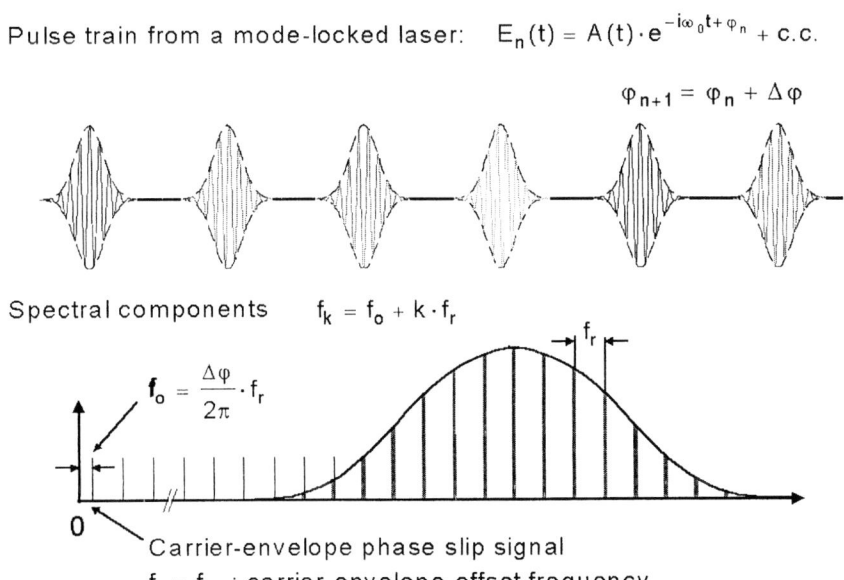

Pulse train from a mode-locked laser: $E_n(t) = A(t) \cdot e^{-i\omega_0 t + \varphi_n} + \text{c.c.}$

$$\varphi_{n+1} = \varphi_n + \Delta\varphi$$

Spectral components $\quad f_k = f_o + k \cdot f_r$

$$f_o = \frac{\Delta\varphi}{2\pi} \cdot f_r$$

Carrier-envelope phase slip signal

$f_o \equiv f_{ceo}$: carrier-envelope-offset frequency

FIGURE 1. Unstabilized pulse train from a modelocked laser.

In the upper part the electric field of a pulse-train is plotted in the time domain. The carrier-envelope phase φ_n of successive pulses have different values. The plotted round-trip phase-shift $\Delta\varphi = \pi/2$ would lead to every fourth pulse having the same ceo-phase.

The lower part depicts the same situation in the frequency domain, where the periodicity of the pulse-train implies an equidistant frequency comb. The virtual extension of the frequency comb to lower frequency ends at a distinct frequency f_0 which would only be zero when all pulses were exact replica of each other, i.e. the carrier-envelope phase φ_n of all pulses would be the same. Otherwise it is exactly the phase-slip frequency f_{ceo}.

Accessing the Phase Slip Frequency f_{ceo}

A way to get access to the carrier-envelope offset phase φ_{ceo} is heterodyne mixing with frequency converted radiation. Fig. 2 shows the basic principle.

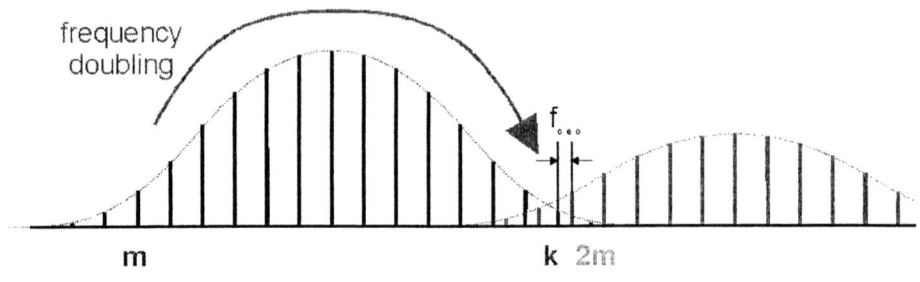

Beating of the fundamental $\qquad f_k = f_0 + k \cdot f_r$

$$\text{and SH} \quad 2 \cdot f_m = 2 \cdot f_0 + 2 \cdot m \cdot f_r$$

$$\text{for k=2m:} \quad 2 \cdot f_m - f_k = 2 \cdot f_0 - f_0 + (2m - k) \cdot f_r = f_0 \equiv f_{ceo}$$

FIGURE 2. Principle of determining f_{ceo}.

Due to a simple algebraic relation one can deduce f_{ceo} by mixing the high frequency part of the fundamental spectrum with the low frequencies of the doubled one. The beating signal is exactly the carrier-envelope offset frequency. Obviously this technique requires a spectral width spanning an optical octave. We met this prerequisite by frequency broadening in a single mode optical fiber in combination with a chirped mirror compressor.

Fig. 3 shows the experimental setup for the stabilization of f_{ceo} [10]. The upper and lower end of the spectrum are obtained from the light leaking through a chirped mirror (CP3) in the external pulse compressor and are separated by a dichroic mirror (DM). In this manner we do not consume any additional power from the laser for the phase locking procedure. Frequency doubling is performed in a BBO-crystal. The two "green" beams have different polarization and hence have to be recombined with a polarizer. Finally the beam is spatially dispersed with a grating to select the beating signal with a slit. A locking electronics provides the signal for controlling the spectral position of the beating signal, hence f_{ceo}. In this case the fine tuning was achieved via adjusting the pump power. A similar approach was presented in [9].

As a result, we managed to fix the carrier-phase slip to an arbitrary frequency between zero and the repetition rate, e.g. 1 MHz. Locking it to zero was not possible because of strong noise in the low frequency range. Picking pulses at an integer fraction of the locking frequency provides a train of identical pulses, e.g. for seeding an amplifier chain. However, ensuring phase locked pulses after amplification requires an additional feedback from the amplifier exit because of nonlinear phase distortions.

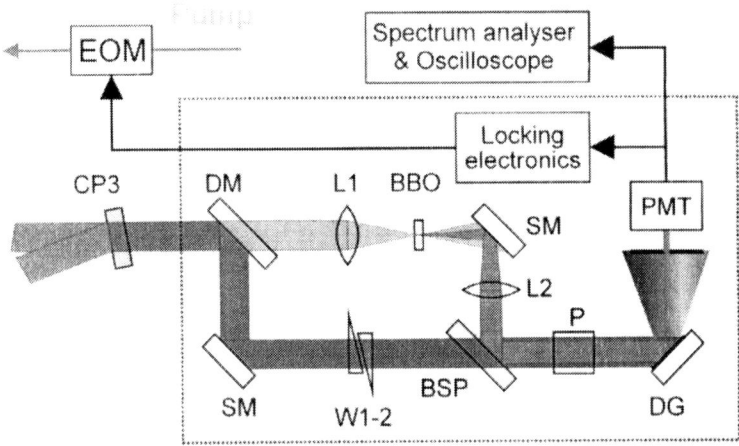

FIGURE 3. Experimental setup for phase stabilization.

GENERATION & MEASUREMENT OF NEAR-FS PULSES

The well known process of high-order harmonic generation results in XUV-radiation at odd multiples of the laser frequency. A typical spectrum is shown in Fig. 4 down to 7 nm wavelength [11].

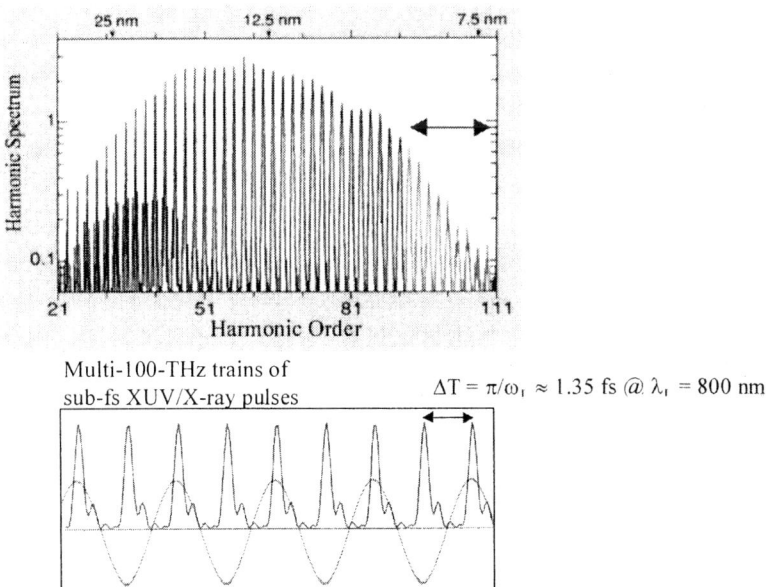

FIGURE 4. Typical harmonic spectrum and corresponding temporal envelope of a selected frequency band.

By selecting a number of harmonics (arrow in Fig. 4, upper part) one can expect radiation bursts of sub-laser cycle duration at a repetition rate of twice the laser frequency (Fig. 4, lower part) [13]-[17]. The high repetition frequency, however, limits their usability for spectroscopic applications.

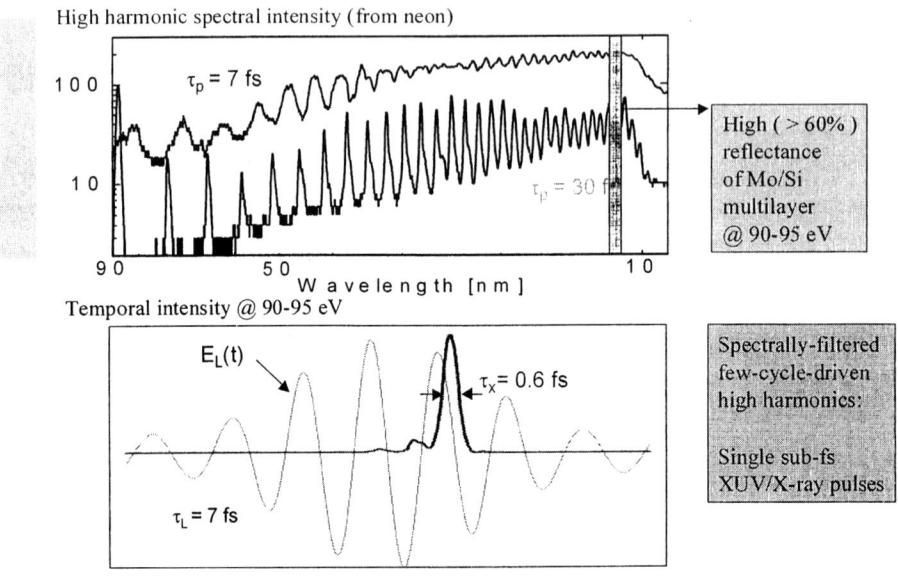

FIGURE 5. Schematic for generating single, sub-fs x-ray bursts.

Using few-cycle pulses as a driver, high-order harmonics can be emitted in a single, sub-fs duration pulse of XUV radiation. Fig. 5 shows two exemplary spectra generated with a multi-cycle and a few-cycle laser pulse, 7 fs and 30 fs respectively. While the lower curve exhibits discrete lines up to the cut-off harmonics, the upper spectrum merges into a quasi-continuum around 13 nm. A filtered band from this region corresponds to a short XUV pulse shown in the lower part of Fig. 5.

Cross-Correlation for Pulse Duration Measurement

For determining the duration of the XUV pulses generated in our source we applied a technique involving a cross-correlation of the x-ray pulse and the driving laser pulse. Atoms are being photoionized by the XUV pulse in presence of the laser field. The kinetic energy of the photoelectrons is characterized with an electron spectrometer. As demonstrated in previous work [18]-[21] the laser pulse causes a reduction of the electron energy by its ponderomotive potential and the generation of side bands due to interaction with laser photons, depending on the delay between XUV and laser pulse. However, this method quits the job when the XUV-pulses get shorter than the laser period as the side bands broaden and start to merge. At this point neither the amplitude of the individual side bands nor the ponderomotive down shift can be determined reliably.

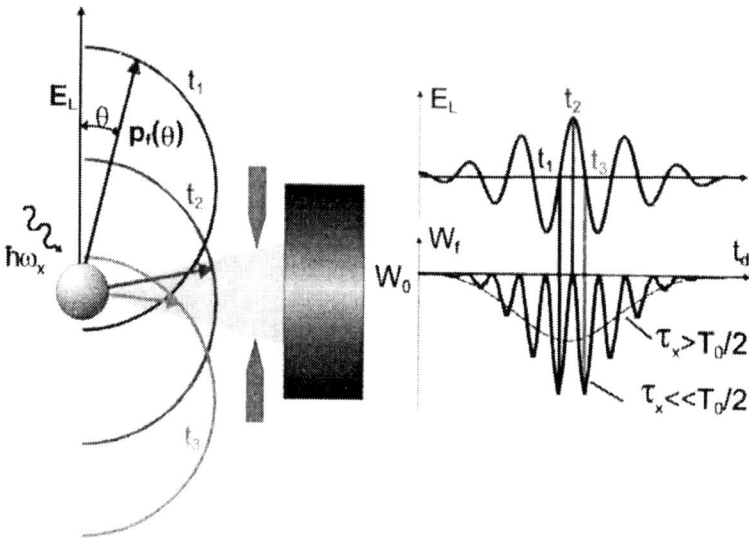

FIGURE 6. Emission characteristics of photoelectrons under influence of the laser field.

In a modified detection geometry we measure the kinetic energy of the photoelectrons emitted perpendicular to the electric field of the laser in a limited detection angel only [22]. In this way the generation of side bands is completely suppressed and the pure ponderomotive down shift can be observed. A classical picture of the process is shown in Fig. 6.

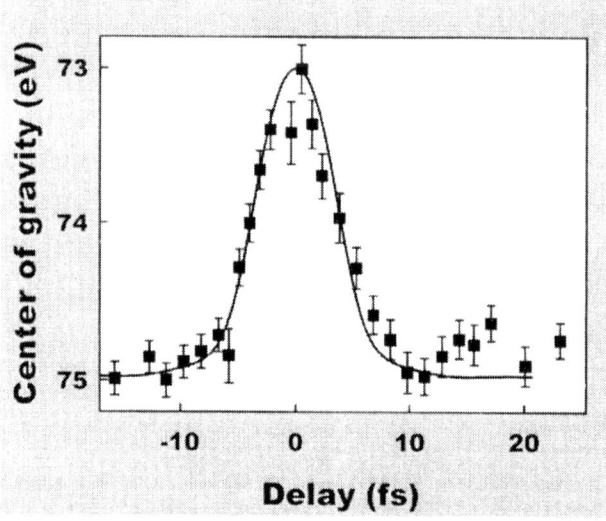

FIGURE 7. Cross-correlation of XUV and laser pulse.

The laser field transfers a transversal momentum to the photoelectrons resulting in a modified angular emission distribution. By confining the detection angel this translates into a net reduction of the kinetic energy, as depicted in the right hand part of Fig. 6. For x-ray pulses shorter than half the laser period the energy shift correlates directly with the square of the electric field, whereas for longer pulses the oscillatory behavior is smeared out.

Fig. 7 shows the position of the center of gravity of the photoelectron spectrum in dependence of the pulse delay. The deconvolution reveals a maximum XUV pulse duration of 2 fs.

ACKNOWLEDGMENTS

This work was supported by the Austrian Science Fund (grants Y44-PHY and F016) and the European ATTO network.

REFERENCES

1. Th. Brabec and F. Krausz, Rev. Mod. Phys. **72**, 545-592 (2000)
2. H.R. Telle, G. Steinmeyer, A. E. Dunlop, J. Stenger, D. H. Sutter, and U. Keller, Appl. Phys. B **69**, 327-332 (1999)
3. G. Steinmeyer, D. H. Sutter, U. Keller, J. Stenger, and H. R. Telle in Conference on Ultrafast Phenomena, OSA Technical Digest Series (Optical Society of America, Washington, D.C., 2000) Paper MF2, pp. 99-101
4. M. Mehendale, St. Mitchell, D. Velleneuve, and P. Corkum, in Conference on Ultrafast Phenomena, OSA Technical Digest Series (Optical Society of America, Washington, D.C., 2000) Paper MF8, pp. 115-117
5. R. J. Jones and J.-C. Diels, in Conference on Ultrafast Phenomena, OSA Technical Digest Series (Optical Society of America, Washington, D.C., 2000) Paper MF31, pp. 182-184
6. D. J. Jones, S. A. Diddams, J. L. Hall, S. T. Cundiff, J. K. Ranka, R. S. Windeler, and A. J. Stentz, in Conference on Ultrafast Phenomena, OSA Technical Digest Series (Optical Society of America, Washington, D.C., 2000) Paper TuC1, pp. 290-293
7. A. Poppe, A. Apolonski, G. Tempea, Ch. Spielmann, F. Krausz, T. Udem, R. Holtzwarth and T. W. Hänsch, in Conference on Ultrafast Phenomena, OSA Technical Digest Series (Optical Society of America, Washington, D.C., 2000) Paper WE3, pp. 608-610
8. L. Xu, Ch. Spielmann, A. Poppe, T. Brabec, F. Krausz, and T. W. Hänsch, **21** 008-2010 (1996)
9. D. J. Jones, S. A. Diddams, J. K. Ranka, A. Stentz, R. S. Windeler, J. L. Hall, S. T. Cundiff, Science **288**, 635-639 (2000)
10. A. Apolonski, A. Poppe, G. Tempea, Chr. Spielmann, Th. Udem, R. Holtzwarth, T. W. Hänsch, and F. Krausz, Phys. Rev. Lett. **85**, 740-744 (2000)
11. J. Macklin *et al.*, Phys. Rev. Lett. **70**, 766 (1993)
12. A. L'Huillier, P. Balcou, Phys. Rev. Lett **70**, 774 (1993)
13. P. Antoine *et al.*, Phys. Rev. Lett. **77**, 1234 (1996)
14. K. Schäfer, K. Kulander, Phys. Rev. Lett. **78**, 638 (1997)
15. I. Christov *et al.*, Phys. Rev. Lett. **78**, 1251 (1997)
16. N. Papadogiannis *et al.*, Phys. Rev. Lett. **83**, 4289 (1999)
17. P. M. Paul *et al.*, Science **292**, 1689 (2001)
18. J. Schins *et al.*, JOSA B **13**, 197 (1996)
19. T. Glover *et al.*, Phys. Rev. Lett. **76**, 2468 (1996)
20. A. Bouhal *et al.*, JOSA B **14**, 950 (1997)
21. E. Toma *et al.*, Phys. Rev. A **62**, 061801(R) (2000)
22. M. Drescher *et al.*, Science **291**, 1923 (2001), published online 15 February 2001, 10.1126/science.1058561

Femtosecond Time Resolution
in X-ray Spectroscopy*

D. von der Linde, K. Sokolowski-Tinten, Ch. Blome, C. Dietrich,
A. Tarasevitch, A. Cavalleri[1] and J. A. Squier[2]

Institut für Laser-und Plasmaphysik, Universität Essen, D-45117 Essen, Germany
[1]*Materials Science Div., Lawrence Berkeley National Laboratory, Berkeley, CA 94720, USA*
[2]*University of California at San Diego, La Jolla, CA 92093-0339, USA*

Abstract. For a long time the capability to perform measurements with femtosecond time-resolution belonged exclusively to the domain of optics. However, in the last few years laser-driven X-ray sources have been developed which enable femtosecond time-resolution to be extended to the X-ray regime.

1. INTRODUCTION

Ever since Wilhelm Conrad Roentgen discovered X-rays more than a hundred years ago, "Roentgen-Strahlung" has played a very important role in science. One of the greatest achievements of modern science, our extensive knowledge of the atomic structure of matter, has been made possible to a large extent through the use of X-rays.

During the last few years an exciting new perspective has come forth in the X-ray field: the possibility of performing X-ray measurements with femtosecond time resolution. New types of radiation sources are being developed that provide X-rays pulses of unprecedented short width, roughly speaking 10^{-13} to 10^{-14} s. As far as X-ray structural analysis is concerned, until recently one could only retrieve, in essence, a time-averaged static picture of the structure. In the near future, femtosecond X-ray pulses will enable the observation of the instantaneous configuration of the atoms and the recording of ultrafast changes in the atomic structure.

Over the past two decades, femtosecond time-resolved optical spectroscopy has enabled the study of a wide variety of ultrafast phenomena. The award of the Nobel Prize in chemistry to Ahmed Zewail in 1999 for his pioneering work in femtochemistry highlights the success of ultrafast optical science [1]. The perspective opening up now, with femtosecond X-ray pulses at our disposal, is the combination of atomic scale spatial and temporal resolution which will reveal both the structure and the evolution of matter.

2. LASER-DRIVEN SOURCES OF X-RAYS PULSES

The success of lasers in the production of X-rays is basically due to the strong electromagnetic field associated with intense laser pulses. During the interaction with the laser

*Modified version of article in Zeitschrift f. Phys. Chem. **215** (2001) 1563, with permission of Oldenbourg Wissenschaftsverlag

CP611, *Superstrong Fields in Plasmas:* Second Int'l. Conf., edited by M. Lontano et al.
2002 American Institute of Physics 0-7354-0057-1

63

field, electrons can pick up a kinetic energy of many tens of thousands of electron volt or even more, depending on the field strength. The electronic energy can be converted into high-energy photons, that is short wavelength electromagnetic radiation, by a variety of electron-photon interaction processes.

A convenient way of generating keV X-rays is to focus an intense femtosecond laser onto the surface of solid material in order to produce a short-lived microplasma of high density and high electronic kinetic energy. The microplasma emits a short burst of incoherent X-rays consisting of bremsstrahlung radiation and characteristic line emission, in particular K_α-lines [2,3]. The latter type of narrow line emission is particularly useful in the X-ray diffraction experiments described in this article. Typically, the spatial dimensions of the microplasma are several microns in diameter and a few tens of nanometers in depth. Thus, we have a nearly point-like source of X-ray pulses.

The duration of the X-ray bursts is limited by the duration of the laser pulses, and also to some extent by the duration of plasma expansion and electron cooling. Under suitable conditions, subpicosecond X-ray pulses can be expected [4]. However, measurement of the width of the X-ray pulses is quite difficult. With a few exceptions precise data are still lacking.

The widely accepted explanation of the origin of the K_α-emission [2,5] assumes that the energetic electrons escape from the primary hot, highly ionized plasma and penetrate into underlying colder material. Here they knock out electrons from inner electronic shells of the atoms, producing core holes. Recombination of the holes with electrons from outer shells leads to characteristic line emission, a mechanism very much like that in an ordinary X-ray tube.

A common implementation of a laser-plasma X-ray source uses laser pulses focused onto a thin metallic wire as a target [6]. To avoid the target erosion caused by the intense laser radiation, the wire is continuously pulled across the focal plain, thereby providing a fresh wire surface for each laser pulse. A photograph of our wire target is shown in Fig. 1. An advantage of this type of X-ray source is that it can be continuously operated over hours, simply by supplying a sufficiently large spool of wire.

Titanium is very suitable as target material for an X-ray source to be used in diffraction experiments. The photon energy of the Ti-K_α-line emission is 4.51 keV which corresponds to a wavelength of 0.274 nm. A low-resolution X-ray spectrum of the Ti-wire

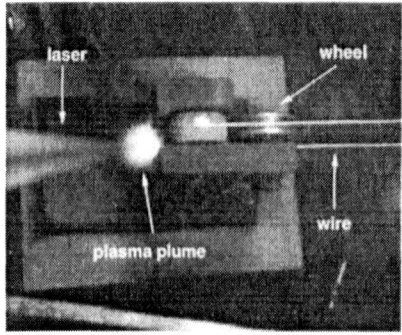

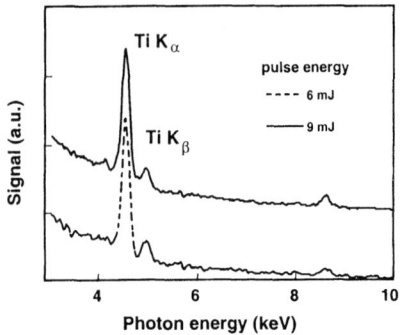

FIGURE 1. Photograph of the X-ray wire target. View from above.

FIGURE 2. X-ray spectrum of the laser-driven Titanium source.

source is shown in Fig. 2. The spectrum was obtained by analyzing the distribution of the electrical charge on the individual photosensitive elements (pixels) of a CCD chip after exposure to the X-rays from the titanium target. One can clearly distinguish the dominant Ti-K$_\alpha$-line at 4.5 keV and the much weaker K$_\beta$-line at 4.9 keV. The doublet structure of the K$_\alpha$-line due to spin-orbit splitting is not resolved.

3. FOCUSING OF MULTI-KILOVOLT X-RAYS

If one wishes to use such an incoherent, point-like X-ray source for experiments, an X-ray mirror with high reflectivity and a large acceptance angle for the efficient collection and focusing of X-ray photons is required. Bent crystals are excellent candidates as they allow one to make use of the strong Bragg reflection of X-rays from the lattice planes of crystals [7]. Figure 3 shows a typical arrangement of a bent crystal in a Rowland geometry. We use a toroidally bent silicon (311) crystal platelet designed for one-to-one imaging of a Ti-K$_\alpha$ point source at 4.5 keV. The horizontal and the vertical radius of curvature of the toroidal mirror must satisfy R$_v$/R$_h$=sin2Θ_B, where Θ_B is the Bragg angle.

The experiments described in this article were carried out with laser pulses from a standard terawatt CPA (chirped pulse amplification) titanium sapphire laser (λ = 800 nm). The laser system produced pulses of 120 fs duration with pulse energies up to 150 mJ at a repetition rate of 10 Hz.

Figure 4a depicts the X-ray intensity distribution in the focal plane of the toroidal Si (311) mirror, recorded with the help of an X-ray sensitive CCD camera. An approximately circular spot of about 80 microns in diameter was observed in the focal plane. The number of X-ray photons per pulse in the focal spot was approximately 10^4 for laser pulses of 30 mJ on the wire target.

For efficient collection of the X-rays emitted into the solid angle covered by the mirror, the Bragg condition must be satisfied at all points over the surface area of the

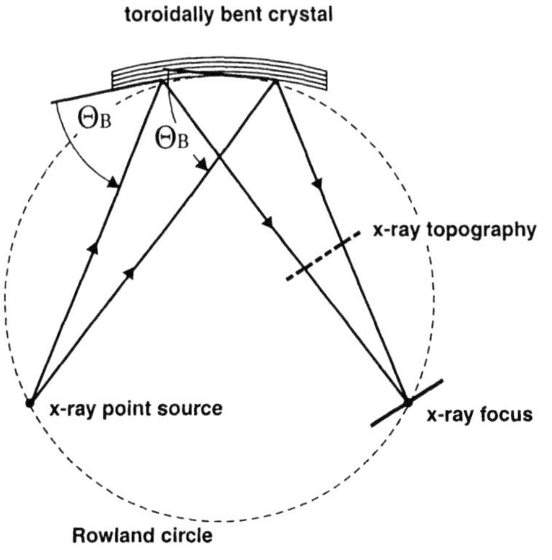

toroidally bent crystal

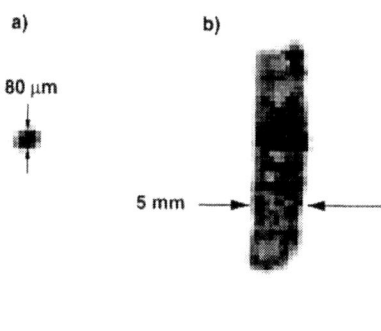

Θ_B Θ_B

x-ray topography

x-ray point source   x-ray focus

Rowland circle

FIGURE 3 Rowland-circle geometry for imaging of the X-ray point source.

a)   b)

80 μm

5 mm

FIGURE 4. a) X-ray focus obtained with the Si (311) mirror; b) X-ray topography of the mirror surface. The mirror has a rectangular shape. The width is 5 mm.

bent crystal. To demonstrate that this is indeed attainable, the intensity distribution of the reflected X-rays was measured in a plane closer to the mirror surface, as indicated by the dashed line in Fig. 3. The measured spatial distribution shown in Fig. 4b represents an X-ray topography of the bent crystal. Closer inspection indicates that with the exception of a relatively narrow border zone the Bragg condition is satisfied everywhere else on the surface so that the largest part of the mirror actually contributes to the reflection of the incident X-rays.

4. TIME-RESOLVED X-RAY DIFFRACTION

To perform ultrafast time-resolved measurements in the X-ray regime one can use suitable variants of pump-probe techniques that are well established in optical time-resolved spectroscopy. Figure 5 illustrates an optical pump/X-ray probe scheme for measuring ultrafast time-resolved X-ray diffraction, say from a single crystal. The basic laser pulses are divided into two beams with an adjustable time delay. The X-ray probe pulses are obtained from a microplasma generated by one of the beams, typically the more energetic one. A portion of the X-rays is collected and focused onto the specimen under study by means of a suitable X-ray mirror. The optical pump pulse and the X-ray probe pulse must spatially overlap on the sample surface. The incident X-ray probe beam typically covers an angular range much larger than the angular width of the reflection characteristic of the crystal ("rocking curve"). The diffracted X-rays are detected by a CCD-camera with a detector area large enough to enable complete rocking curves to be recorded.

Changes in the crystalline structure following the excitation by the pump pulse can be measured by recording the resulting changes of the rocking curve as a function of the delay time. The time resolution is determined by the duration of the X-ray pulses, which can be as short as a few hundred femtoseconds.

We shall briefly discuss two recent experiments: Observation of ultrafast, laser-induced solid-liquid phase transitions (Section 5), and picosecond acoustic transients developing after these transitions (Section 7).

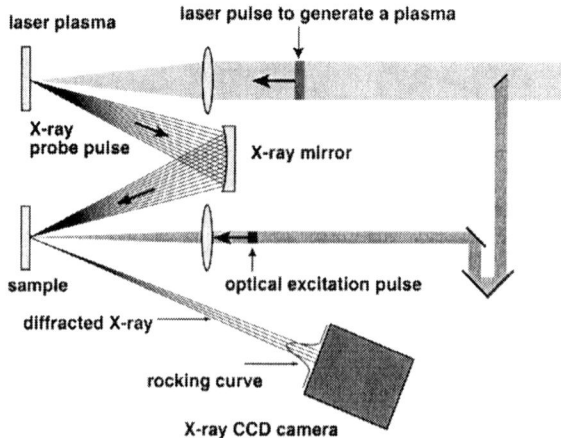

FIGURE 5. Optical pump/X-ray probe scheme for time-resolved X-ray diffraction experiments.

5. ULTRAFAST STRUCTURAL PHASE TRANSITIONS

There is extensive evidence to show that the transition from the solid to the liquid phase takes place on a time scale of a few hundred femtoseconds, when a semiconductor material such as Si or GaAs is very strongly photoexcited by an ultrashort laser pulse [8]. The process was discovered in the early eighties [9,10] and attracted a lot of attention because the speed of the apparent phase transition could not be explained by a normal thermal melting process. Theory indicates that such an ultrafast structural transformation occurs in covalent semiconductors when an electron-hole concentration of approximately 10^{22} cm^{-3} is achieved, that is, when more than ten percent of the covalent bonds are broken [11-13]. Until very recently, the only indication of such ultrafast solid-to-liquid transitions came from optical data [8,14]. However, because optical radiation cannot resolve the actual atomic configuration it was not possible to obtain direct evidence of a structural change.

As an example of the type of information available from optical measurements, time-resolved reflectivity spectra of Si are depicted in Fig. 6. These data show the evolution of the reflectivity of crystalline Si after photo-excitation by an intense femtosecond pulse. They indicate that a change from crystalline to metallic liquid reflectivity occurs within a few hundred femtoseconds. Obviously, the possibility of this type of extremely fast phase transition represents an ideal case to be studied by ultrafast X-ray diffraction.

Observation of the lattice structure by X-ray diffraction requires wavelengths shorter than the lattice constants, i. e. a few Angstroms or photon energies of several kilovolts. Generally speaking, X-rays of this wavelength penetrate deeply into the crystal, typically much farther than the thickness of the molten surface layer that is produced by the laser pulse. This thickness is typically less than 100 nm. Under these conditions X-ray diffraction would be strongly dominated by the bulk crystal. It would essentially ignore the very thin molten surface layer, unless special precautions were taken.

To overcome the problem we have used thin crystalline layers of germanium, which were grown by surfactant-mediated heteroepitaxy on standard Si (111) wafers [15]. The key point is that one obtains a heterostructure with a highly perfect crystal layer whose

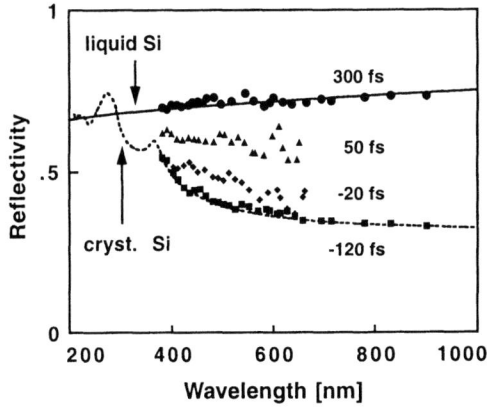

FIGURE 6. Spectra of the optical reflectivity of Si. Dotted line: Solid state. Full line: Liquid state. Data points: Time-resolved measurements during femtoseond laser-induced melting.

lattice constant does not match the lattice constant of the substrate. In fact, the lattice mismatch is quite large in the Ge/Si (111) system, and the Bragg reflections from bulk Si and from the Ge surface layer are clearly separated (difference of the diffraction angles $\Delta\Theta \approx 1°$). This is demonstrated by Fig. 7, which shows the CCD-camera read-out of the (111)-reflections from a 390 nm Ge/Si heterostructure. These data represent a one-minute recording with the X-ray pulses from our Ti-K_α-source. The strong sharp line is due to the Si substrate, whereas the broader and weaker peak represents the diffraction from the 390 nm Ge layer.

These Ge/Si heterostructures proved to be very useful in recent time-resolved X-ray diffraction experiments in which ultrafast melting and related lattice processes were studied. For example, this type of sample has been used in the time-resolved X-ray diffraction experiments by Siders et al. [16]. This work has provided, for the first time, direct structural evidence that intense femtosecond optical excitation of covalent semiconductors can indeed trigger a solid-to-liquid phase transition, as suggested by the earlier optical investigations. These results were confirmed and extended in more recent measurements for which we worked with a prepulse-optimized Ti-K_α-source and the Si (311) toroidal X-ray mirror described above [17].

In these experiments the basic scheme in Fig. 5 was employed. Briefly, the 4.51 keV probe beam from the Ti-K_α-source was focused on the laser-excited surface area of a Ge/Si heterostructure wafer. The Bragg angle for diffraction from the (111) lattice planes of Ge at 4.51 keV is about 25°, corresponding to an angle of incidence of 65° of the X-ray probe beam. The Ge film was photoexcited using a small fraction of the fundamental laser pulses at 800 nm. This wavelength corresponds to a secondary direct band gap of Ge. Thus, Ge is strongly absorbing at 800 nm, while the optical absorption of the Si-substrate is negligible. In the optical pump/X-ray probe experiment discussed here we measured the Bragg reflection from the Ge-film (thickness 170 nm) as a function of the time delay between the X-ray pulse and the 800 nm excitation pulse.

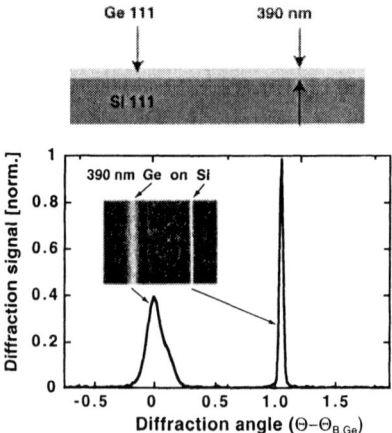

FIGURE 7. Bragg diffraction from (111) planes of a 390 nm Ge/Si heterostructure. Insert: CCD camera recording of the diffracted X-rays.

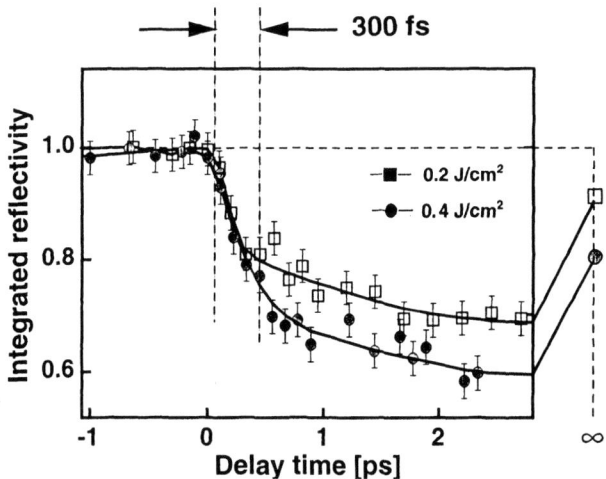

FIGURE 8. (111) X-ray diffraction from the crystalline Ge layer: Diffraction efficiency (integrated reflectivity) versus delay time. Infinity symbol: Measurement a few minutes after the pump pulse.

Results of the experiments are shown in Fig. 8. The measured X-ray diffraction signal, integrated with respect to the diffraction angle ("integrated reflectivity"), is plotted versus delay time for two different energy fluences of the excitation pulses. The remarkable feature of the data is the initial fast drop of the X-ray diffraction within a few hundred femtoseconds after the pump pulse, followed by a more gradual further decrease over several picoseconds. The two data points to the right of the time axis represent the diffraction signal from the laser-excited spot measured a few minutes later (delay time "infinity").

From the observed decrease of the X-ray diffraction we conclude that a portion of the Ge crystal underwent structural disordering within a few hundred femtoseconds. For example, the 25% decrease in the diffraction efficiency indicates a loss of crystalline order in a 40 nm layer in the first 300 fs, that is, the formation of a disordered or molten layer. A Debye-Waller effect due to lattice heating can be ruled out as an explanation because the temperatures that would be required to produce the observed effect are much too high. An angular shift resulting from a possible laser-induced lattice distortion can also be excluded because in the experiment a sufficiently large angular range was simultaneously recorded (see rocking curves in Fig. 11).

The observed structural change is much too fast to be explained by an ordinary thermal melting process, even if strongly superheated conditions were assumed. If we used the sound velocity for an estimate of the ultimate upper limit of the propagation speed of the melt-front, it would take at least 8 ps to melt a 40 nm layer.

On the other hand, it was observed that after the initial rapid drop, the diffraction efficiency continued to decrease over several tens of picoseconds. In fact, extensive measurements on the picosecond time scale (data not shown here) indicated that the electronically driven non-thermal ultrafast phase transition is followed by a rapid thermal melting process, presumably under highly superheated conditions. This subsequent thermal melting could be driven both by heat transfer from the hot liquid and heat directly deposited in the solid phase behind the molten layer. From the observed rate of the decrease in the X-ray diffraction efficiency one can estimate the propagation velocity of the melt front to be 850 m/s. This value is in good agreement with earlier measurements of the melt front velocity during picosecond laser melting experiments [18,19] .

The diffraction measurements on the laser-exposed spot at delay time „infinity" show that the molten layer has re-crystallized. However, we observed only a partial recovery of the diffraction. Examination of the final surface morphology indicated that weak laser ablation had taken place causing the removal of a 10 to 20 nm thick surface layer.

6. X-RAY PULSE DURATION

The time resolution in optical pump/X-ray probe experiments is ultimately limited by the duration of the X-ray pulses. At the present time, satisfactory methods for the measurement of ultrashort multi-keV X-ray pulses are not yet available. Great progress has been made in X-ray streak cameras but time-resolution significantly less than one picosecond is still difficult to achieve [20]. On the other hand, our X-ray diffraction experiments and those of others [21-23] have furnished clear evidence of an X-ray time response as short as a few hundred femtoseconds. The apparent time response is the result of a convolution of the X-ray pulse width and the actual material response (in our case the speed of the structural disordering of the semiconductor crystal).

An estimate of the X-ray pulse duration can be obtained by assuming an instantaneous, step-like material response in the X-ray diffraction experiments. Figure 9 shows fits of the measured X-ray diffraction data using calculated convolutions. A step-like material response and Gaussian-shaped X-ray pulses were assumed to calculate the convolution. It can be seen that good fits of the data are obtained with X-ray pulse widths between 250 and 350 fs. We conclude that the X-ray pulse duration of the Ti-K$_\alpha$-source must be approximately 300 fs with an uncertainty of ± 50 fs.

So far, no effort whatsoever has been made to minimize the X-ray pulse duration by optimizing the parameters under which the X-ray source is operated. We are quite confident that it will be possible to substantially reduce the X-ray pulse width in the future.

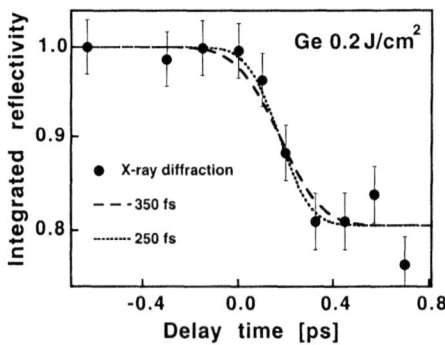

FIGURE 9. Fits of the X-ray diffraction data points assuming Gaussian-shaped X-ray probe pulses.

7. PICOSECOND ACOUSTIC PERTURBATIONS

The electronically induced ultrafast solid-to-liquid transition is completed within less than a picosecond, leaving behind a hot, pressurized layer of liquid Ge on top of some non-molten crystalline Ge (see Fig. 10a). The optical energy deposited in the material relaxes within a few picoseconds to some partial thermal equilibrium, and, as a result, thermally induced mechanical stress is set up in the Ge layer. Release of pressure and stress in the various strata of the material gives rise to interesting acoustic perturbations, which develop on a picosecond time-scale. These acoustic waves can be observed in X-ray diffraction through the accompanying lattice distortions, which lead to changes in the Bragg angle.

Figure 10b shows a qualitative picture of the physical situation after thermalization of the optical energy. The solid line shows the initial pressure/stress profiles, which are taken to be flat top, for sake of simplicity. The essential point is that there is a substantial pressure discontinuity at the liquid-solid interface. The initial stress in the Si substrate is negligible, because at 800 nm the optical absorption in Si is much less than in Ge.

Several acoustic perturbations start from the three interfaces: (1) A compression wave propagates from the liquid-solid interface into solid Ge and a rarefaction wave back into the liquid; (2) a rarefaction wave develops at the liquid-vacuum boundary and travels into the liquid; (3) compressive and expansive waves similar to, but weaker than (1), are launched from the Ge-Si interface. We believe that (1) is the dominant acoustic perturbation because pressure is expected to have a maximum in a 40 to 50 nm surface layer where the largest portion of the optical energy is deposited.

In addition to the acoustic perturbations, the liquid-solid interface (melt-front) moves forward into solid Ge. The liquid layer grows at the expense of the solid layer, resulting in a continuous overall reduction in the diffraction efficiency. As mentioned above, the

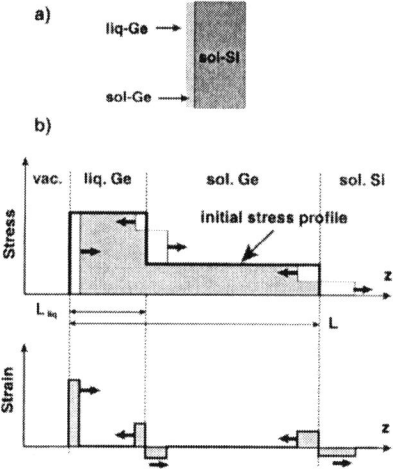

FIGURE 10. a) Ge/Si heterostructure with the molten Ge layer, the remaining crystalline layer and the Si substrate. b) Top: The initial stress profile (heavy solid line) and the developing stress waves. Bottom: The developing strain waves.

melt-front velocity can be estimated from the observed long term, picosecond time scale rate of decrease in the diffraction.

The different acoustic perturbations (1), (2) and (3) do not overlap for times shorter than the travel time between the interfaces. Later on, however, reflections occur at the various interfaces because of the mismatch of the acoustic impedance. The situation becomes quite complicated until the acoustic transients have finally damped out.

Figure 11 presents examples of measured time-dependent rocking curves, which illustrate some of the effects discussed in the foregoing paragraph. The dashed lines in Fig. 11 represent the rocking curves of the unperturbed Ge layer and Si substrate. The solid line in Fig.11a shows that at 0.8 ps after laser excitation the shape and the position of the rocking curve has not changed, but there is a reduction in the diffraction of Ge to somewhat less than 70 percent. This is the signature of the order-disorder phase transition as discussed in more detail in Section 5.

According to the discussion in the context of Fig. 10, the first acoustic effects to be noticed should be compression in Ge and, to a lesser extent, also in Si. The rocking curve for 13 ps, Fig. 11b, shows that this is indeed observed. There is a clear shift of the Ge line towards larger diffraction angles and a weak shoulder of the Si curve in the same direction.

The rarefaction wave approaching from the liquid surface is expected to effect the crystalline layer after times $t > L_{liq}/c_{liq} \approx 15$ ps, where $c_{liq} = 2660$ m/s is the sound velocity of liquid Ge. In fact, the Ge rocking curve at 27 ps, Fig. 11c, shows a new side band at lower diffraction angles and exhibits a doubly peaked structure. This shape indicates that at this time both compressed and expanded strata are present in Ge.

Finally, the data for 123 ps, Fig. 11d, reveal further points of interest. At this late stage the acoustic perturbations of the Ge layer have damped out leaving the relaxed, thermally expanded material, as indicated by a shift of the diffraction profile to smaller angles. The observed shift represents a thermal strain of $\varepsilon = 0.008$. This thermal expansion corresponds to a temperature of approximately 1100 to 1200 K. This value is close

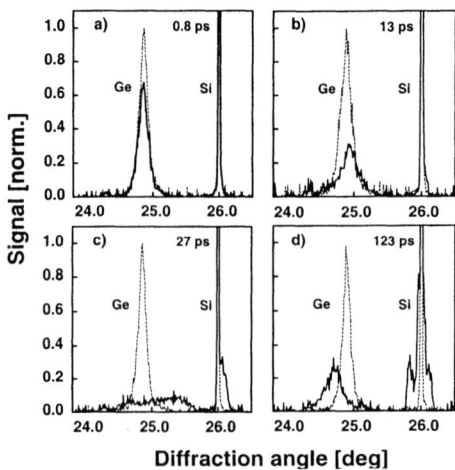

FIGURE 11. Examples of the measured rocking curves at different delay times.

to the melting temperature of Ge, suggesting that liquid and solid Ge have reached equilibrium near the melting point.

Concerning the Si rocking curve at 123 ps, the main feature is a pronounced side band at lower angles, presumably attributable to the thermal expansion of Si caused by heat transfer from the hot surface layers. However, the rocking curve of Si at 123 ps has a rather complex structure, and both expansive and compressive strain can be recognized.

8. CONCLUSION

Laser-driven short pulse X-ray sources have matured to the point where time-resolved X-ray spectroscopy with picosecond and femtosecond resolution is possible. We have shown that the new time-resolved X-ray techniques can reveal extremely fast structural changes and the dynamics of lattice distortions associated with acoustic perturbations.

An attractive feature of laser-driven X-ray sources is that they enable relatively small-scale, almost table-top type of experiments. On the other hand, large-scale facilities for the generation of very powerful ultrashort X-ray pulses will be available in the future, for example free electron lasers or other schemes that make use of high-energy electron beams.

At the present time, two highly developed disciplines of science are about to be tied together, namely X-ray science and ultrafast time-resolved spectroscopy. The combination of high time-resolution with atomic-scale spatial resolution should eventually enable researchers to record detailed, quasi-instantaneous pictures of complex atomic structures.

ACKNOWLEDGEMENTS

The authors are indebted to I. Uschmann and E. Förster for providing the X-ray mirror, to M. Horn-von Hoegen and M. Kammler for growing the heterostructure samples. We gratefully acknowledge financial support by the Deutsche Forschungsgemeinschaft, the German Academic Exchange Service and the XPOSE Network of the European Union. AC acknowledges travel funds from the National Science Foundation through grant INT-9981720.

REFERENCES

1. Zewail, A.H., *J. Phys. Chem.* **A 104**, 5660 (2000).
2. Rousse, A., Audebert, P., Geindre, J.P., Falliès, F., Gauthier, J.C., Mysyrowicz, A., Grillon, G., Antonetti, A., *Phys. Rev.* **E 50**, 2200 (1994)
3. Jiang, Z., Kieffer, J.C., Matte, J.P., Chaker, M., Peyrusse, O., Gilles, D., Korn, G., Maksimchuk, A., Coe, S., Mourou, G., *Phys. Plasmas* **2**, 1702 (1995)
4. Reich, Ch., Gibbon, P., Uschmann, I., Förster, E., *Phys. Rev. Lett.* **84**, 4846 (2000)
5. Bastiani, S., Rousse, A., Geindre, J.P., Audebert, P., Quoix, C., Harminaux, G., Antonetti, A., Gauthier, J.C., *Phys. Rev.* **E 56**, 7179 (1997)
6. Rose-Petruck, C., Jimeney, R., Guo, T., Cavalleri, A., Siders, C.W., Ràski, F., Squier, J.A., Walker, B.C., Wilson, K., Barty, C.P.J., *Nature (London)* **398**, 310 (1999)
7. Missalla, T., Uschmann, I., Förster, E., Jenke, G., von der Linde, D., *Rev. Sci. Instr.* **70**, 1288 (1999)

8. Sokolowski-Tinten, K., Bialkowski, J., von der Linde, D., *Phys. Rev. B* **51**, 14186 (1995)

9. Shank, C.V., Yen, R., Hirlimann, C., *Phys. Rev. Lett.* **50**, 454 (1983a)

10. Shank, C.V., Yen, R., Hirlimann, C., *Phys. Rev. Lett.* **51**, 900 (1983b)

11. Stampfli, P., Bennemann, K.H., *Phys. Rev. B* **49**, 7299 (1994)

12. Stampfli, P., Bennemann, K.H., *Phys. Rev. B* **46**, 10686 (1992)

13. Stampfli, P., Bennemann, K.H., *Phys. Rev. B* **42**, 7163 (1990)

14. Saeta, P., Wang, J.K., Siegal, Y.N., Bloembergen, N., Mazur, E., *Phys. Rev. Lett.* **67**, 1023 (1991)

15. Horn von Hoegen, M., *Appl. Phys. A* **59**, 503 (1994)

16. Siders, C.W., Cavalleri, A., Sokolowski-Tinten, K., Toth, C., Guo, T., Kammler, M., Horn von Hoegen, M., Wilson, K.R., von der Linde, D., Barty, C.P.J., *Science* **286**, 1340 (1999)

17. Sokolowski-Tinten, K., Blome, C. Dietrich, C., Tarasevitch, A., Horn von Hoegen, M., von der Linde, D., Cavalleri, A., Squier, J., Kammler, M., *Phys. Rev. Lett.*, in print

18. Danielzik, B., Harten, P., Sokolowski-Tinten, K., von der Linde, D., *Mat. Res. Soc. Symp. Proc.* **100**, 471 (1988)

19. von der Linde, D., In: *Resonances*, ed. by M. D. Levenson, E. Mazur, P. S. Pershan, and Y. R. Shen, World Scientific, Singapore, 1990, 337

20. Gallant, P., Forget, P., Dorchies, F., Jiang, Z., Kieffer, J.C., Jaanimagi, P.A., Rebuffie, J.C., Goulmy, C., Pelletier, J.F., Sutton, M., *Rev. Sci. Instr.* **71**, 3627 (2000)

21. Rischel, C., Rousse, A., Uschmann, I., Albouy, P.A., Geindre, J.P., Audebert, P., Gauthier, J.C., Förster, E., Martin, J.L., Antonetti, A., *Nature* **390**, 490 (1997)

22. Feurer, T., Morak, A., Uschmann, I., Ziener, Ch., Schwoerer, H., Förster, E., Sauerbrey, R., *Appl. Phys. B* **72**, 15–20 (2001)

23. Rousse, A., Rischel, C., Fourmaux, S., Uschmann, I., Sebban, S., Grillon, G., Balcou, Ph., Förster, E., Geindre, J.P., Audebert, P., Gauthier, J.C., Hulin, D., *Nature* **410**, 65 (2001)

X-ray Spectroscopy on Energy Transport and Deposition in Ultra-intensity Laser Produced Plasmas

H.Nishimura, T.Kawamura, Y.Ochi, R.Matsui, Y.Miao[a], S.Okihara, S.Sakabe, R.Kodama, K.A.Tanaka, Y.Kitagawa, Y.Sentoku, F.Koike[b], I.Uschmann[c], E.Förster[c], and K.Mima

Institute of Laser Engineering, Osaka University, Suita, Osaka, 565-0871 Japan
[a] Southwest Inst. Nuclear Phys. Chem., Mianyang 621900 Sichuan, 919-216, P.R.China
[b] Medical Department, Kitazato U., Sagamihara, Kanagawa 228-8555 Japan
[c] Institute of Optics and Quantum Electronics, U of Jena, Max-Wien-Platz Jena, 07743 Germany

Abstract. Observation of line radiation emanating from inner shell transitions in partially ionized high Z material is suggested as a diagnostic method for heating of dense plasma with an ultra-short high-intensity laser pulse. A new atomic-kinetics code specified for the analysis of the ionization-shift Kα lines was developed to derive electron temperature of the bulk plasma heated by hot electrons. Chlorine Kα to Cl^{15+} Heα lines from a solid planar target were observed with a 1 TW laser system at ILE Osaka. Comparison of the experimental results with the code prediction infers a ~100 eV temperature plasma on the target surface with a depth of about 0.6 μm at 1.5×10^{17} W/cm^2. Dependence of the temperature on laser intensity was found to be very weak.

INTRODUCTION

The physics of ultra-short laser plasma interactions with dense plasma is of wide interest as its close relevance to challenging researches on fs-x-ray radiation probing [1], energetic particle accelerations [2], and inertial confinement fusion (ICF) [3]. In the fast ignition approach to ICF, efficient energy transfer from ultra-intense laser to dense fusion plasma is one of the critical issues. In this study, a new spectroscopic method providing time- and space-resolved information has been expected for more quantitative understanding of the energy deposition rather than those provided by nuclear particle measurement [4]. Because of relatively small increase in temperature with currently available lasers and high ρR exceeding 0.1 g/cm^2 of the compressed fuel, conventional spectroscopic method based on K-shell lines from near-fully ionized plasma is not suitable. To have such a high ionization state in low temperature plasma, one can not avoid using low-Z material as a tracer but opacity for these K-shell lines is too high to escape from the deep, dense plasma.

CP611, *Superstrong Fields in Plasmas:* Second Int'l. Conf., edited by M. Lontano et al.
© 2002 American Institute of Physics 0-7354-0057-1/02/$19.00

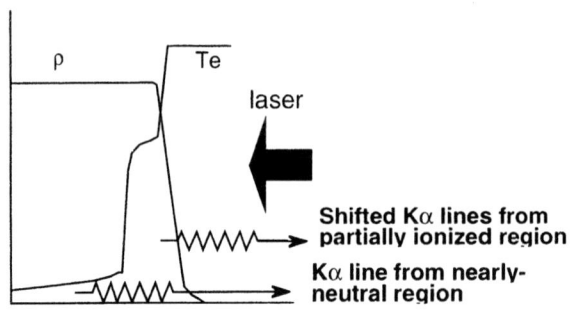

Fig. 1. A schematic of density and temperature profiles and corresponding Kα line emission regions.

Numerous works have been done to derive the fast electron spectrum and energy transfer efficiency by observing Kα lines from multi-layer target [5]. But, it has never been proven, by spectroscopic way in such experiments, that the region emitting Kα is substantially heated by hot electrons because of poor spectral resolution (~200 eV) when a charge-coupled-device (CCD) detector is used. As is predicted by fs-hydro-code simulations [6], thickness of the hot region could be much less than the hot electron penetration range so that the observed Kα line will be a signature of cold, near-neutral region deep in the target (see Fig. 1). In other experiments, x-ray spectrum ranging from Kα line to K-shell resonance lines have been observed by using a high-resolution x-ray crystal spectrograph [7]. The observed spectra consist of radiation arising from inner-shell transitions of partially ionized plasma. They decay critically with increase in over-layer thickness, showing also that the depth of hot region is much less than the range [7]. The inhibition of heat transport via hot electrons has been attributed to the space-charge effect, which is strongly dependent on electrical conductivity of target materials [8]. Therefore, in addition to the conventional K-shell line spectroscopy, we suggest observation and analysis of line radiation from inner shell transitions in partially ionized high Z material seeded in the target to improve understanding of energy deposition mechanism relevant to the fast ignitor experiments.

A new spectrum analysis code specified for the ionization-shift Kα lines has been developed to derive electron temperature of bulk plasma heated by hot electrons from laser-plasma interaction regions. This code treats a bi-Maxwellian plasma (*i.e.,* cold and hot electrons). Atomic processes are solved under collisional radiative equilibrium (CRE) state by balancing collisional-excitation and ionization with radiative decay (namely, Kα emissions), Auger, and other atomic processes. As a proof-of-principle experiment, x-ray spectra ranging from chlorine Kα to Cl^{15+} Heα lines emitted from a chlorinated plastic target were observed by using a toroidally-curved crystal spectrograph. Intensity ratios of the shifted Kα components have shown a very weak dependence on laser intensity. Comparison of the experimental results with the code prediction is discussed.

Atomic Rate-Equation Solver for Shifted Kα Lines

We developed a population kinetics code treating the atomic processes associated with 1s-inner-shell ionization and radiative transition from L-shell to K-shell orbits of chlorine atom to find a steady state solution for competition between the radiative and non-radiative decays [9].

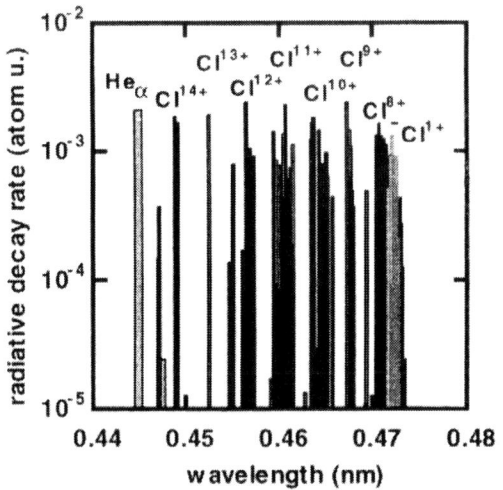

Fig. 2. Radiative decay rate vs. wavelength of chlorine Kα lines calculated by GRANT code.

Energy levels and radiative decay rates associated with the Kα emission were calculated by using a multi-configuration Dirac-Fock (MCDF) code GRASP [10]. Another atomic parameters were obtained using reliable formulae [9]. Figure 2 shows the radiative decay rates of Cl Kα lines from various ionization states. For Cl^+ to Cl^{13+} ions, we considered the inner-shell ionization supposing outer-most electrons is not excited (namely, $Cl^+:1s2s^22p^63s^23p^5,...,Cl^{13+}:1s2s^2 2p$). For Cl^{14+} ion, only $1s2s2p$ state is considered. Similar to other lines, primary Kα (Cl^+) transitions range between 0.4718 to 0.4729 nm due to sub-shell transitions. Energy differences from the primary Kα can be seen according to the charge states. Kα lines from O-like (represented by Cl^{9+}) to He-like Cl atoms are well separated from the primary line, and those from S-like (Cl^+) to Ne-like (Cl^{7+}) lines merge and are hard to distinguish each other. The F-like (Cl^{8+}) Kα line makes a wing of the primary lines in the blue side. To estimate net intensity of the Kα line emissions, Auger transition is an important decay process. The Auger rates in the code were calculated with an atomic code developed by Fritzsche *et al.* [11].

The free electron distribution $f(\varepsilon)_{all}$ was assumed to be a bi-Maxwellian distribution; one temperature is for the cold bulk component $f(\varepsilon)_{bulk}$ and the other is for the hot component $f(\varepsilon)_{hot}$; $f(\varepsilon)_{all}=(1-a)f(\varepsilon)_{bulk}+af(\varepsilon)_{hot}$ where a represents the fractional fast electrons in the total free-electrons. In the same manner, effective rate coefficients, individually calculated for two electron-components, are determined in proportion to the fraction a except for the radiative-, Auger-decays, and three-body recombination rates [9]. It will noteworthy that the fraction of hot electrons is estimated to be a few % of all free-electrons from particle-in cell (PIC) simulation [12].

After solving the kinetics code for representative plasma parameters, dependence of individual ionic states on cold and hot temperatures was investigated. The cold temperature was varied from 10-1000 eV and the hot from 1-100 keV. It was found that the average Z value obtained from the ionic population is strongly dependent on the cold temperature but quite-weakly on the hot temperature in particular T_{hot}>10 keV. Useful range as for plasma diagnostic for chlorine trace is found to be 30 eV < T_{cold} <200 eV. Above 200 eV, the use of K-shell lines and their satellite lines may be more practical rather than that of the shifted-Kα lines.

Experiment and Analysis

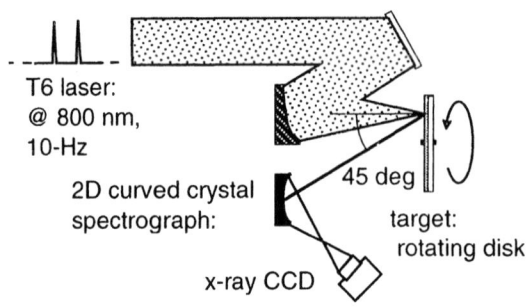

T6 laser:
@ 800 nm,
10-Hz

2D curved crystal
spectrograph:

45 deg

target:
rotating disk

x-ray CCD

Fig 3. Experimental set-up on T6 laser

Experiments were carried out by using Ti:sapphire laser system named T6 at ILE, Osaka. This laser provides energy of 50 mJ per pulse with the center wavelength of 800 nm at a 10-Hz repetition rate. In the present work, the pulse duration was varied from 132 to 442 fs for a constant energy to obtain dependence of electron temperatures. These pulses were focused an f/3 off-axis parabolic mirror onto a target at normal incidence with a focal spot diameter of 15 μm including 80% of incident energy, yielding intensities ranging from 7×10^{16} to 1.5×10^{17} W/cm^2. Figure 3 shows the experimental set-up with laser irradiation condition. The target was a 200 μm thick polyvinylchloride (C_2H_3Cl) sheet mounted on a rotating table to provide fresh surface for every shot. The density of the sheet was 1.4 g/cc. X-ray spectra from the target were observed with a 2-dimensionally-curved, crystal-spectrograph. The bent crystal was a quartz with meridional curvature radius of 200 mm and sagittal curvature radius of 178 mm. It was used in (10-1-1) reflection and set 1162 mm away from the target. The meridional direction includes the dispersion plane. Resultant image magnification was 1/8. X-ray spectra ranging from 0.43 to 0.48 nm in wavelength were recorded with a cooled, back-illumination CCD camera. Spectral resolution λ/dλ (here λ is the x-ray wavelength) was better than 10^4. The observation angle was 45 degrees from the target normal. In order to obtain clear signal on CCD nearly 10^4 shots was needed.

Figure 4 shows a typical spectrum obtained at 1.5×10^{17} W/cm^2. Between Cl Kα and Cl Kβ lines, inner shell transition lines corresponding to Kα Cl^{9+} (O-like) to Kα Cl^{14+} (Li-like), and He-like $1s^2$-$1s2p$ (Heα) line are seen. These were identified by GRANT-code used in the rate-equation solver. The Cl Kα line from Cl^{1+} atom appears to be wide partly because the signal is over-saturated in the CCD and partly because it involves Kα Cl^{1+} to Kα Cl^{8+} lines in its blue side. The other shifted Kα lines are also

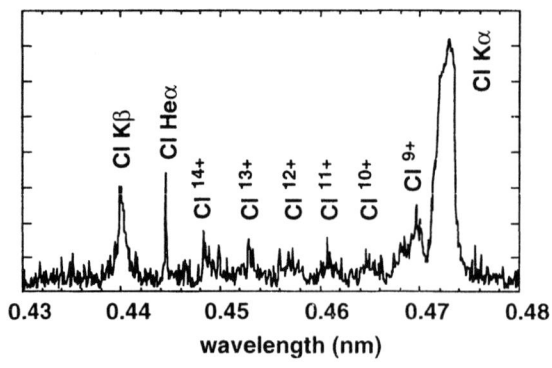

Fig. 4. Spectrum obtained for 132fs, 1.5×10^{17} W/cm^2.

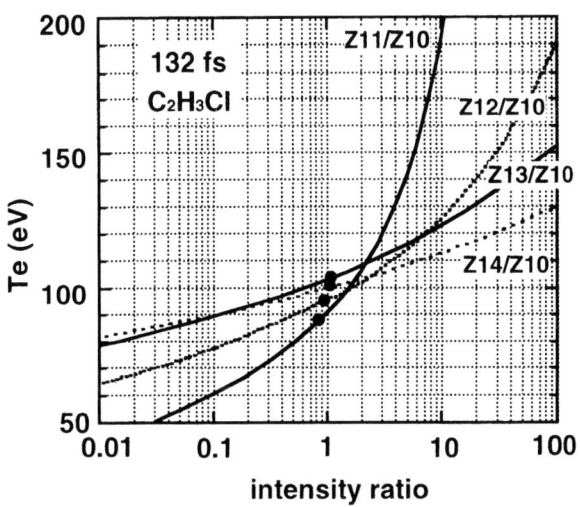

Fig. 5. Comparison of the experimental results with the code calculation. Legend of Z11/Z10, for example, represents the intensity ratio of Cl^{11+} to Cl^{10+} Kα lines. All data points infer electron temperature of 96 +/- 8 eV

wide but due to sub-level transitions for a given ionization state. It is noteworthy that the Heα ($1s^2:^1S_0$-$1s2p:^1P_1$) line, which consists of a single transition, is significantly narrow comparing to other lines. With decrease in laser intensity, the shifted-Kα lines from higher ionization states become relatively weaker than those from lower ionization states. This trend is consistent with the code prediction.

In separate experiment done under the same conditions, hot electrons expanding to the laser-forward side through a thin foil target were observed with an electron spectrometer [13]. The hot electron temperatures derived from the slope of electron spectra have shown a scaling of the temperature on laser intensity as $29I_{17}^{0.74}$ keV specified for 800 nm laser wavelength laser light. Here I_{17} is the laser intensity normalized to 1×10^{17} W/cm^2. Therefore the hot electron temperature T_{hot} in the present experiment at 1.5×10^{17} W/cm^2 is about 40 keV

The experimental results were compared with the calculations to derive electron temperature of bulk plasma. In this calculation, T_{hot} is assumed to follow the experimental scaling and a is 1%. Plasma component was C_2H_3Cl with the ion density of 1.5×10^{22} cm^{-3}. Typical result is show in Fig. 5 for 1.5×10^{17} W/cm^2 with the pulse duration of 132 fs. Due to difficulty in isolation of Cl^{9+} O-like Kα-line from $Cl^+ \sim Cl^{8+}$ lines, intensity ratios are taken with respect to the Cl^{10+} N-like Kα-line. All data points are in a temperature range of 96+/- 8 eV. With decrease in laser intensity, the temperature becomes slightly lower but dependence on the laser intensity was pretty weak. For example, the temperature for 7×10^{16} W/cm^2 at 442 fs duration is 89 +/- 6 eV.

Depth of the hot plasma on the target surface can be roughly estimated by using absolute value of the shifted Kα photons detected on the CCD and assuming Lambertian distribution of x rays. For example, the number of photons of Cl^{13+} Kα line was 5.1×10^5 per shot in the hemi-sphere of laser incident side. Population density of $1s$-hole Cl^{13+} ion contributing to the emission was calculated from the code to be 3.7×10^{15} /cc. Then the depth of hot plasma was estimated to be 0.6 µm assuming a plasma of pancake-like shape on the target surface, and its surface area corresponds to that of the laser spot. Many authors have already noted such depth much shorter than the hot electron range [6, 7].

The hot plasma possesses only a few % of incident laser energy so that strong inhibition in hot electron conduction has been also found in the present spectroscopic measurements.

Conclusions

For use in plasma diagnostic with ultra-short high-intensity lasers, a new rate-equation solver for the shifted-$K\alpha$ lines has been developed. To validate the code, chlorine $K\alpha$ to Cl^{15+} $He\alpha$ lines from the solid planar target were observed with a 1 TW Ti:sapphire laser system. Comparison of the experimental results with the atomic code prediction infers 100 eV temperature plasma of 0.6 μm depth at 1.5×10^{17} W/cm^2.

This work is in the start-up phase. Further experiments will be made to find a profile of heat front created by energy deposition from hot electrons by using multi-layered targets. Laser intensity will be made much higher to provide an experimental database in the relativistic energy regime. These results will be utilized to improve our understanding of physics in energy transport and deposition in direct-relevance to the fast ignitor approach to ICF.

REFERENCES

1. Rischel, C., Rousse, A., Uschmann, I., Albouy, P., Geindre, J., Audebert, P., Gauthier, J., Foerster, E., Martin, J., and Antonetti, A., *Nature* **390**, 490-491 (1997).
2. Mackinnon, A.J., Broghesi, M., Hatchett, S., Key, M.H., Patel, P.K., Campbell, H., Schiavi, A., Snavely, R., Wilks, S.C., and Willi, O. *Phys. Rev. Lett.* **86**, 1769-1772 (2001)
3. Tabak, M., Hammer, J., Glinsky, M.E., Kruer, W.L., Wilks, S.C., Woodworth, J., Campbell, E.M., Perry, M.D., and Mason, R.J., *Phys. Plasmas* **1** 1626-1634 (1994).
4. Disdier, L., Garconnet, J-P., Malka, G., and Miquel, J-L., Phys. Rev. Lett. **82**, 1454-1458 (1999).
5. Wharton, K.B., Hatchett, S.P. Wilks, S.C., Key, M.H., Moody, J.D., Yanovsky, V., Offenberger, A.A., Hammel, B.A.. Perry, M.D., and Josi, C. *Phys. Rev. Lett.* **81**, 822-825 (1998).
6. Guethlein, G., Food, M.E., and Price, D., *Phys. Rev. Lett.* 77 1055-1058 (1996).
7. Rousse, A., Audebert, P., Geindre, J.P., Fallies. F., Gauthier, J.C., Mysyrowicz, A., Grillon, G., and Antonetti, A., *Phys. Rev.* **E 50**, 2200-2207 (1994).
8. Bell A.R., Davies, J.R. Guerin, S., and Ruhl, H., *Plasma Phys. Control. Fusion* **39**, 653-659 (1997).
9. Kawamura, T., Nishimura, H., *et al.*, to be submitted.
10. Dyall K.G., Grant, I.P., Johnson, C.T., Parpia, F.A., and Plummer, E.P., *Comput. Phys. Commun.* **55**, 425-456 (1989).
11. Fritzsche, S., and Fricke, B., *Phys. Scr.* **T41**, 45-50 (1992).
12. Sentoku, Y., Mima, Taguchi, T., Miyamoto, S., and Kishimoto, Y., *Phys. Plasmas* **5**, 4366-4372 (2000).
13. Okihara, S., Sakabe, S., *et al.*, to be submitted.

Reconnection processes in a laser plasma in the presence of counterstreaming electrons

F. Pegoraro*, F. Califano*, N. Attico*, G. Bertin† and S.V. Bulanov**

*Phys. Dept., University of Pisa and Istituto Nazionale Fisica della Materia, Pisa, Italy
†Phys. Dept., University of Milan, Milano, Italy
**General Physics Institute RAS, Moscow, Russia

Abstract. The magnetic field generated by the Electro-Magnetic Current Filamentation Instability (EMCFI) is shown to develop magnetic islands due to the onset of a fast reconnection instability that occurs on the electron time scale. The main source of free energy for both instabilities is essentially the initial anisotropy of the two counterstreaming electron populations. Numerical results of a 2D-3V Vlasov code are presented and interpreted in terms of the theory of collisionless reconnection in the whistler regime in an anisotropic plasma. These results are relevant to the process of magnetic channel coalescence in relativistic laser plasma interactions and confirm the dynamical role of the magnetic field generated in the wake of the laser pulse first remarked in Ref. [1].

INTRODUCTION

Magnetic field generation at the expense of the thermal energy of an anisotropic electron population and the conversion of magnetic energy into thermal electron energy in the presence of inhomogeneous currents represent two complementary features of the dynamics of a magnetic field in a plasma. In relativistic plasmas interacting with ultrashort ultraintense laser pulses, the first process has been related to the electromagnetic current filamentation instability (EMCFI) [2, 3] which is analogous to the well known Weibel instability [4] and can occur in the presence of two counterstreaming (cold) electron populations. In this case the magnetic field generation can be interpreted as resulting from the separation in space of the two populations due to the repulsion of oppositely directed currents. The second process is best known in the case of the magnetic field line reconnection instabilities that have been studied extensively both in the context of laboratory and of astrophysical plasmas. Recently it has been shown [5] that a similar "high frequency" reconnection process can affect the three-dimensional (3D) spatial distribution and the time evolution of the magnetic field generated by the EMCFI in a laser generated plasma. In particular magnetic reconnection allows magnetic channels to coalesce [1], as seen recently e.g. in 3-D Particle in Cell simulations in overdense plasmas in Ref.[6]. This fast reconnection occurs on the electron dynamical time scales [7] and, as in the case of the more standard Magnetohydrodynamic (MHD) reconnection, leads to the formation of magnetic islands. Its onset is allowed by the effect of electron inertia and, in kinetic regimes where phase space effects are important, by the contribution of "resonant electrons" [8]. We recall that the magnitude of the magnetic field generated by the EMCFI driven by the fast electron current and by the associated counterstream-

CP611, *Superstrong Fields in Plasmas:* Second Int'l. Conf., edited by M. Lontano et al.
© 2002 American Institute of Physics 0-7354-0057-1/02/$19.00

ing return current, produced under present experimental conditions by a relativistically intense laser pulse, is estimated to be extremely large, of the order of $100MG$ [9]. In dimensionless units, in terms of the associated cyclotron and plasma frequencies, we have $\Omega_e/\omega_{pe} \lesssim 1$, which indicates that in a relativistic laser plasma magnetic fields can affect both the single particle dynamics and the whole collective nonlinear plasma behaviour.

GEOMETRICAL CONFIGURATION AND RECONNECTION

In the investigations of the EMCFI in Refs. [3, 10] 2D Cartesian configurations were used. However, the quasistatic magnetic field generated by the EMCFI is spatially inhomogeneous and the current separation that is at the basis of the instability mechanism leads to strong current gradients that are prone to the development of reconnection-type processes. In order to study the combined development of these competing processes, a fully 3D description of the plasma is needed [11]. We will refer to a Cartesian configuration where the counterstreaming currents are directed along x and the magnetic field generated by the EMCFI lies along z and depends on y. Magnetic reconnection requires z dependent perturbations: the reconnection electric field is directed along z and leads to particle acceleration along z while the reconnection magnetic field has a y component and depends both on y and on z.

The "minimal" description of the combined magnetic field generation and reconnection is obtained in a 2D-3V configuration where all the vector fields have three components, but are independent of x. In this description, which is complementary to the one used in Refs. [3], the electrostatic plasma dynamics along the direction of the electron streams is suppressed. This configuration is sufficient to show that the "quasistatic" magnetic field generated by the EMCFI develops magnetic islands on a fast electron time scale because of the combined effect of the current gradients and of the anisotropy in the y-z plane induced by the deflection of the streams along y caused by the EMCFI magnetic field itself. This eventually leads to a 3-D isotropization of the electron distribution. The island formation occurs in the high frequency range that corresponds to whistler waves. In this frequency range, in a dissipationless, fluid and cold (or polytropic) plasma, the generalized vorticity $\mathbf{B}_e \equiv \mathbf{B} + (m_e c/e)\nabla \times \mathbf{u}_e$ is frozen in the electron fluid for processes occurring on time scales longer than the Langmuir period such that charge separation effects can be neglected [12]. Here $\mathbf{u}_e$ is the electron fluid velocity and we have adopted a nonrelativistic electron kinematics. In the case of spatial scales larger than the electron inertial skin depth $d_e \equiv c/\omega_{pe}$, the generalized vorticity reduces to the magnetic field $\mathbf{B}$, which is thus frozen in the electron fluid. In these conditions reconnection can occur [7] due to the formation of current sheets with characteristic size of the order or smaller than d_e. Under the conditions of interest for ultraintense laser plasma interaction, no true freezing constraint can be invoked, because of the intrinsically kinetic and anisotropic nature of the electron dynamics. Yet our numerical simulations show that in the case of two counterstreaming electron currents, the structure and the time evolution of the magnetic islands and of the flow patterns that are formed after the nonlinear phase of the EMCFI conform to those produced by magnetic field line reconnection instabilities on a fast electron time scale.

NUMERICAL SIMULATIONS AND RESULTS

The results presented in this paper are based on 2D-3V kinetic simulations of the nonlinear electron dynamics [5]. These results are obtained with a well tested, nonrelativistic Vlasov code. We integrate the selfconsistent Vlasov-Maxwell system of equations equations in the (y, z, v_x, v_y, v_z) phase space of electrons while ions are taken as a fixed neutralizing background. We use normalized quantities such that the normalized electron skin depth equals unity, the magnetic field is given by the ratio between the cyclotron frequency and the plasma frequency and time is measured in electron times ω_{pe}^{-1}.

We consider a somewhat idealized, spatially homogeneous, initial state with zero electric and magnetic fields and with two symmetric counterstreaming electron populations. These are modeled as the sum of two Maxwellians, with densities $n_1 = n_2 = 1/2$ and equal thermal velocity $v_{th}/v_1 = 1.7 \ 10^{-1}$, where $v_1 = -v_2$ are the velocities associated with the two counterstreaming currents. This symmetrical case, where the densities of the fast and of the slow electrons are equal and their velocities equal and opposite, does not fully represent the laser plasma conditions and is chosen on the basis of numerical convenience, so as to maximize the growth rate of the EMCFI. We recall that these simulations are extremely costly in terms of memory requirements and computational time. Periodic boundary conditions are used both in the y and in the z directions with box size 2π. Different resolutions corresponding to $32^2 \times 61^3$ and to $64^2 \times 61^3$ grid points respectively were used to verify that reconnection is not caused by numerical dissipation in the code. The ratio of the box size $L = 2\pi$ to the electron skin depth is such that it allows the fastest EMCFI (corresponding to $k_y d_e \approx 1$) to grow. In such a configuration, in terms of the standard Δ' parameter of reconnection theory and in the absence of anisotropy effects, we would expect the magnetic field produced by the EMCFI to be stable against reconnection. We recall that Δ' is a measure of the fraction of magnetic energy arising from the current spatial inhomogeneity that is available to reconnection.

At $t = 0$ a perturbed magnetic field of the form $\mathbf{B} = B_o \sin(y) \, \mathbf{e_z} + \delta\mathbf{B}(y,z)$ is added to the system. Here $B_o = 10^{-3}$, and $\delta\mathbf{B}(y,z)$ is a divergence-free random noise perturbation 10^{-2} smaller. This initialization allows the system to develop at first a coherent y-dependent magnetic field along z due to the EMCFI seeded by B_o. The random noise perturbation seeds the subsequent development of the reconnection process. The EMCFI develops with a growth rate $\gamma_{CF} \approx 0.28$ leading to the formation of a dipolar magnetic field $B_z(y)$ with amplitude $B_{max} \approx 0.27$ at saturation. Its field lines are shown in Fig. 1a at $t = 40$. A local decrease of the electron density $\delta n_e \approx -0.4$ occurs where $B_z^2(y)$ is largest. The nonlinear development of the EMCFI leads to the magnetic deflection of the electron streams in the y direction. In terms of an effective stress tensor, measuring the spatially averaged kinetic energy T_j in the jth direction, we find [5] that in this phase $T_x \approx T_y \equiv T_\perp \gg T_z \equiv T_\parallel$. These conditions lead to the onset of the reconnection process characterized by the exponential growth of the $k_z = 1$ Fourier component of the magnetic field $\delta\mathbf{B}(y,z)$ with growth rate $\gamma_R \approx 0.11$.

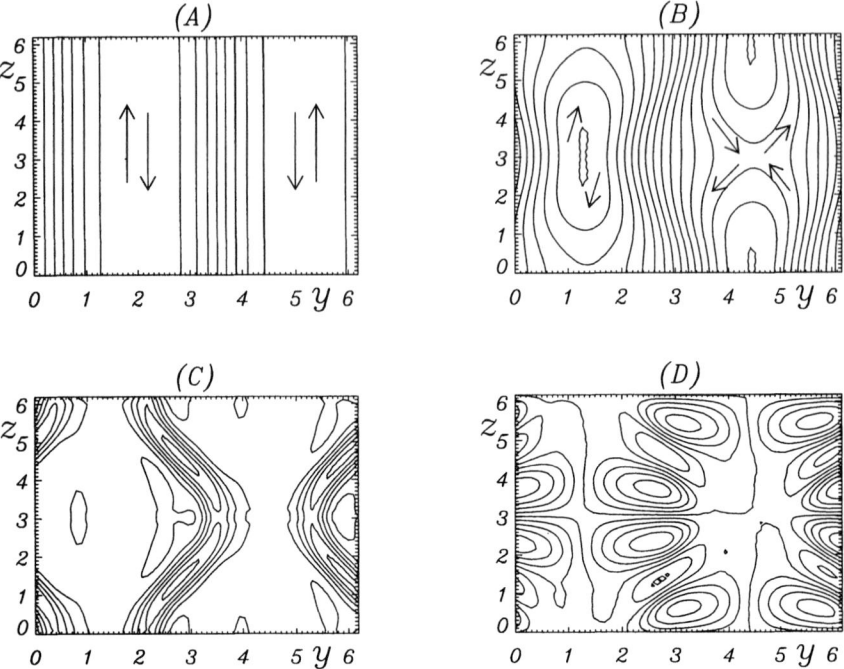

FIGURE 1. Development of the reconnection process. The magnetic field lines in the y-z plane are shown at the saturation of the EMCFI before the onset of reconnection in frame (a) and after the development of reconnection in frame (b). Frames (c) and (d) show the corresponding electron density and B_x distributions of frame (b).

MAGNETIC ISLANDS AND PRESSURE ISOTROPIZATION

The role of plasma anisotropy on the onset and growth rate of the reconnection instability, in a neutral sheet magnetic configuration analogous to the one produced by the EMCFI, has been studied long ago in the framework of the MHD equations and it has been shown that its growth rate is strongly enhanced if $(T_\perp/T_\parallel \gg 1)$ [13]. This effect is due to the coupling of the reconnecting mode with the (secondary) Weibel instability induced, in the case under consideration here, by the anisotropy in the y-z plane caused by the y deflection of the counterstreaming currents due to the EMCFI. The growth rate is purely determined by the dynamics of the electrons and is given by

$$\gamma_R = (k_z v_{thez}/\sqrt{\pi})\,(T_\parallel/T_\perp)\,(d_e^2/\delta_e)\,(\Delta' + \Delta_0')\,, \qquad (1)$$

where $\delta_e \sim L^{1/2}T_\perp^{1/4}/(2^{1/2}B_{max})^{1/2}$ is the scale of the so-called Parker orbits around the zero plane of the magnetic field generated by the EMCFI and $\Delta_0' = (\delta_e/d_e^2)\,[(T_\perp/T_\parallel) - 1]$. From the y-profile of the magnetic field and from the temperature ratios in the saturated phase of the EMCFI we estimate $\Delta' \simeq -1.3$ and $\Delta_0' \simeq 9.0$. Thus the instability

84

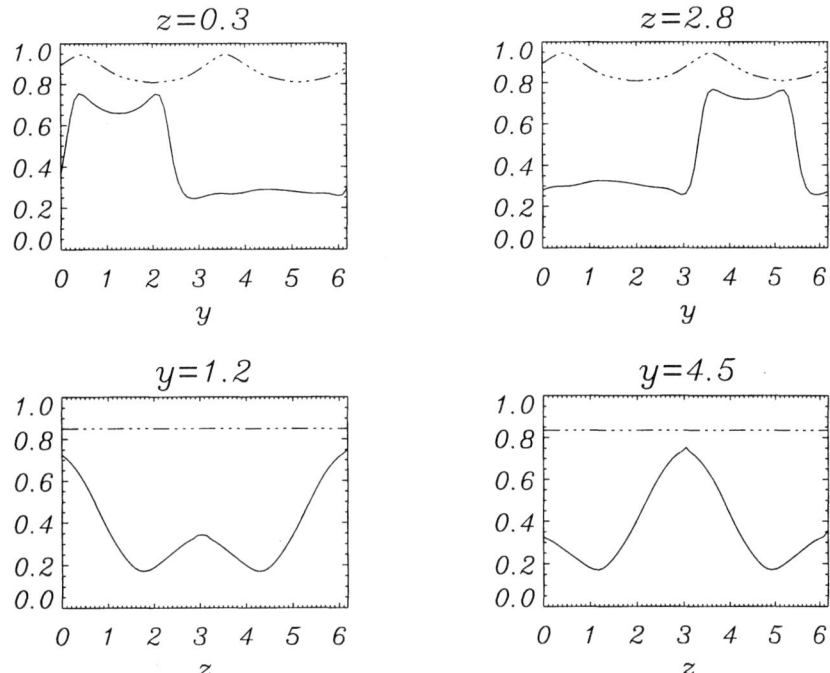

FIGURE 2. Cuts of the anisotropy factor $(T_\perp - T_\parallel)/(T_\perp + T_\parallel)$ in the y-z plane before and after reconnection (dashed-dotted and continuous lines, respectively). Note the correlation between the island position and the regions of almost complete isotropization.

condition $\Delta' + \Delta_0' > 0$ is well satisfied and the numerical growth rate is also found to be consistent with the value given by Eq.(1).

The development of the reconnection instability leads to the formation and growth of magnetic islands (see the projection of the field lines on the y-z plane in Fig. 1b at $t = 165$). The island structure is reproduced in the electron density in the y-z plane (Fig. 1c), in the electric field distribution (not shown here) and in the isocontours of $B_x(y,z)$ in Fig. 1d which give the projection of the current density field lines on the y-z plane. In the nonlinear phase of the reconnection instability the parallel temperature $T_\parallel$ increases while $T_\perp$ decreases, thus indicating a tendency toward full isotropization of the electron distribution. This isotropization is almost complete inside the magnetic islands where reconnection has taken place, as shown in Fig. 2 by the plots of the anisotropy factor $(T_\perp - T_\parallel)/(T_\perp + T_\parallel)$ at the start and in the nonlinear phase of the reconnection instability versus y at fixed z and versus z at fixed y.

CONCLUSIONS

These results prove that the field lines of the magnetic field generated by the EMCFI in the interaction of an ultra intense ultra short laser pulse with a plasma reconnect and develop magnetic islands on the electron time scale. In the model configuration used in our simulations, the reconnection growth rate is smaller than that of the EMCFI only by a factor ≈ 2. The analysis of the instability condition for reconnection and of the time behaviour of the anisotropy during the reconnection process shows that plasma anisotropy plays the major role in both processes and that the free energy for both instabilities originates from the initial anisotropy of the two counterstreaming electron populations.

ACKNOWLEDGMENTS

This work is supported by the INFM Parallel Computing Initiative and by MURST.

REFERENCES

1. G.A. Askar'yan, *et. al.*, Comm. Plasma Physics Contr. Fusion **17**, 35, (1995); F. Pegoraro, *et. al.*, Plasma Phys. Contr. Fusion **39**, B261, (1997).
2. V.Yu. Bychenkov *et. al.*, Sov. Phys. JETP **71**, 79 (1990).
3. F. Califano, *et. al.*, Phys. Rev. E **56**, 963 (1997); Journal of Plasma Physics **60**, 331 (1998); Phys. Rev. E **57**, 7048 (1998); Phys. Rev. E **58**, 7837 (1998).
4. E.W. Weibel, Phys. Rev. Lett. **2**, 83, (1959).
5. F. Califano, *et. al.*, Phys. Rev. Lett. **86**, 5293 (2001).
6. T. Taguchi, T. Antonsen Jr., C. Liu, K. Mima, Phys. Rev. Lett. **86**, 5055 (2001); H. Ruhl, to be published.
7. A.V. Gordeev, Sov. J. Plasma Phys. **13**, 838 (1987); S.V. Bulanov, *et al.*, Phys. Fluids B **4**, 2499 (1992); N. Attico, *et al.*, Phys. Plasmas **7**, 2381 (2000); Phys. Plasmas **8**, 16 (2001).
8. A.A. Galeev, *Handbook of Plasma Physics* **II**, eds. A A. Galeev, R.N. Sudan, 305 (1984); F. C. Hoh, Phys. Fluids **9**, 277 (1966); G. Laval, *et al.*, *Plasma Physics and Contr. Nucl. Fusion Research*, IAEA, Vienna, Vol. II, 736 (1966); B. Coppi, *et al.*, Phys. Rev. Lett. **16**, 1207 (1966); M. Dobrowolny, Il Nuovo Cimento B **55**, 427 (1968).
9. G.A. Askar'yan, *et. al.*, JETP Letters **60**, 240 (1994); Plasma Physics Reports **21**, 835, (1995); F. Pegoraro, *et. al.*, Physica Scripta T **63**, 262 (1996); A. Pukhov, *et. al.*, Phys. Rev. Lett. **76**, 3975 (1996); S.V. Bulanov, *et. al.*, Phys. Rev. Letters., 76, 3562, (1996).
10. Y. Sentoku, *et al.*, Phys. Plasmas **7**, 659 (2000); M. Honda, Phys. Plasmas **7**, 1606 (2000).
11. Y. Kazimura, *et al.*, Plasma Physics Reports **27**, 1 (2001).
12. A.S. Kingsep, *et al.*, 1990, *Reviews of Plasma Physics*, ed. by B. Kadomtsev, (Consultants Bureau, New York, N.Y.) **16**, 243.
13. B. Coppi, *et al.*, Annals of Physics **119**, 370 (1979); G. Bertin, *et al.*, *Proc. Int. Conf. on Plasma Physics (ICPP)* (Nagoya, Japan, 1980), **9P-II-33**; J. Chen, *et al.*, Phys. Fluids **24**, 2208 (1981); Phys. Fluids **27**, 1198 (1984); J. Ambrosiano, J. Geophys. Res. **91**, 113 (1986); Y. Shi, *et al.*, J. Geophys. Res. **92**, 12171 (1987); S. I. Vainshtein, *et al.*, Plasma Phys. **24**, 965 (1982); K. V. Gamayunov,*et al.*, Plasma Phys. Contr. Fusion **40**, 1285 (1998).

Non linear ion flow induced by a large amplitude wave in a plasma.

A. Cardinali, C. Castaldo, R. Cesario

Associazione EURATOM-ENEA sulla Fusione, Centro Ricerche Frascati, C.P. 65, 00044 Frascati, Rome, Italy

Abstract. The effect of a large amplitude electromagnetic wave on the ion motion of a magnetized plasma is considered by solving analytically the ion motion equations. The solution is obtained by multiple time scale analysis of the set of motion equations and at the second order in the expansion of a parameter which accounts for the strength of the field. A large ion drive velocity is found perpendicularly to the direction of the wave incident electric field, and when the gyration motion of the ion around the axial magnetic field is resonating with the considered wave frequency. Full numerical calculations are also performed and compared to the analytical calculation.

INTRODUCTION

In a magnetized plasma has been experimentally demonstrated that a poloidal ion sheared flow can be produced by coupling large amplitude waves in the ion cyclotron frequency range. This flow is essentially localized near the resonant layer, and can strongly improve the plasma confinement [1, 2]. Theoretical models of this RF induced poloidal flow were developed in the framework of fluid approach, which relies on the solution of the balance momentum equation against losses which are represented by the neoclassical viscosity. Critics to this model which implies plasma incompressibility have been adressed in Ref. [3] where a fluid compressible plasma model has been assumed and compared with the previous results. On the same article a kinetic approach based on the second order solution of the Vlasov equation has been proposed to calculate exactly the divergence of the second order pressure tensor. Nevertheless, the fluid based models does not clearly explain, in our opinion, how the wave, whose electric field is essentially oriented along the radial direction, can produce poloidal flow if the poloidal momentum of the wave is zero. The local balance of the poloidal momentum, indeed, implies a poloidal momentum transfer to the plasma, as a result of the wave interaction with the resonant ions. This is done, reasonably, by the Lorentz force, due to the toroidal magnetic field $\mathbf{B}_\phi$, acting on the resonant ions when they take radial momentum from the wave. Following the fluid model, such a mechanism is quite embedded in the convective term $< \mathbf{v} \bullet \nabla v_\theta >$ of the poloidal momentum balance (here the brackets denote a mean over magnetic surfaces and over time intervals long compared with the wave period, $\mathbf{v}$ is the fluid velocity of the resonant ions, and v_θ indicates the velocity component in the poloidal direction).

In this work, we attempt to explain the wave induced poloidal flows in terms of single particle dynamics of resonant ions. This could give, as a main purpose, an insight

CP611, *Superstrong Fields in Plasmas:* Second Int'l. Conf., edited by M. Lontano et al.

of the RF driven poloidal flow, and a guide line for a kinetic approach, in order to improve the fluid model of this process. To this end the single particle dynamics in a RF electromagnetic field is outlined in the framework of a perturbation theory (up to the second order in the expansion of a parameter which accounts for the strength of the field, $\lambda = qkE/m\Omega^2$, where q and m are the charge and the mass of the ion species, Ω the cyclotron frequency, and E the amplitude of the RF electric field), and the solution is analytically obtained by multiple time scale analysis of the motion equations.

ION DYNAMICS

In this Section we study the single particle motion of ions with electric charge q and mass m of a magnetized plasma, interacting with the RF electric field, modeled as a plane wave of amplitude $\mathbf{E} = A\mathbf{k}$, frequency ω, and wave vector $\mathbf{k} = k_\perp \mathbf{e}_x + k_\parallel \mathbf{e}_z$. The unit vectors $\mathbf{e}_z$ and $\mathbf{e}_x$ are respectively parallel and perpendicular to the magnetic field $\mathbf{B}$, and $\mathbf{e}_x$ is directed towards the plasma centre. The ion dynamics, in the adimensional variables: $\xi = k_\perp x$, $\eta = k_\perp y$, $\zeta = k_\parallel z$, and $\tau = \Omega t$, is given by the second order ODE system in time:

$$\ddot{\xi} = \lambda_1 cos(\xi + \zeta - \alpha\tau + \phi) + \dot{\eta}$$

$$\ddot{\eta} = -\dot{\xi} \tag{1}$$

$$\ddot{\zeta} = \lambda_2 cos(\xi + \zeta - \alpha\tau + \phi)$$

where: $\lambda_1 = qk_\perp^2 A/m\Omega^2$, $\lambda_2 = qk_\parallel^2 A/m\Omega^2$, $\alpha = \omega/\Omega$ and ϕ is the phase of the electric field acting on the ion at $\tau = 0$. The choice of the axes entails the equality $\dot{\eta} = -\xi$ obtained by integrating the second equation of the above system Eqs. 1 with the initial condition $\xi(\tau = 0) = 0$, and $\dot{\eta}(\tau = 0) = 0$.

In order to treat the ion dynamics in the frame of multiple-time-scale perturbation expansions [4], we assume: $\lambda_2 << \lambda_1 < 1$ where $\lambda_2 = O(\lambda_1^2)$ which leads to the following ordering for the physical quantities ξ and ζ:

$$\xi = \xi^{(0)} + \lambda_1\xi^{(1)} + \lambda_1^2\xi^{(2)} \tag{2}$$

$$\zeta = \zeta^{(0)} + \lambda_1^2\zeta^{(1)} + \lambda_1^4\zeta^{(2)} \tag{3}$$

The relevant equation will result:

$$\ddot{\xi} + \xi = \lambda_1 cos(\xi + \dot{\zeta}_0^{(0)}\tau - \alpha\tau + \phi) \tag{4}$$

where the third equation of the system Eq. 1 has been integrated to give: $\zeta^{(0)}(\tau) = \dot{\zeta}_0^{(0)}\tau$ where $\zeta^{(0)}(\tau = 0) = 0$, and $\dot{\zeta}^{(0)}(\tau = 0) = \dot{\zeta}_0^{(0)} = k_\parallel v_{z0}/\Omega = k_\parallel v_{thi}/\Omega$, the initial position and the initial component of the ion velocity parallel to the magnetic field.

A multiple-time-scale perturbation analysis can be applied to the above equation. We expand $\xi(\tau)$ in the small parameter λ_1 (see Eq. 2). In addition, however, we make use of the fact that the characteristic time scale for the non-linear growth of the amplitude is much longer than the time scale for the oscillation ω^{-1}. This allows to extend the number of time variables from one to $\tau_0, \tau_1, \tau_2, \ldots$. It holds:

$$d\tau_0/d\tau = 1; \qquad\qquad d\tau_1/d\tau = \lambda_1; \qquad\qquad d\tau_2/d\tau = \lambda_1^2 \qquad (5)$$

Eq. 2 becomes:

$$\xi = \xi^{(0)}(\tau_0, \tau_1, \tau_2) + \lambda_1 \xi^{(1)}(\tau_0, \tau_1, \tau_2) + \lambda_1^2 \xi^{(2)}(\tau_0, \tau_1, \tau_2) \qquad (6)$$

and the operator $d/d\tau$ will be:

$$\frac{d}{d\tau} = \frac{\partial}{\partial \tau_0} + \lambda_1 \frac{\partial}{\partial \tau_1} + \lambda_1^2 \frac{\partial}{\partial \tau_2} \qquad (7)$$

Substituting in the first of Eqs. 1, Eqs. 6-7, after Taylor-expanding the cosinus term, and equating to zero the coefficients of successive power of λ_1, we find at the zeroth order:

$$\left[\frac{\partial^2}{\partial \tau_0^2} + 1 \right] \xi^{(0)}(\tau_0, \tau_1, \tau_2) = 0 \qquad (8)$$

at the first order

$$\left[\frac{\partial^2}{\partial \tau_0^2} + 1 \right] \xi^{(1)}(\tau_0, \tau_1, \tau_2) = -2 \left[\frac{\partial^2 \xi^{(0)}}{\partial \tau_1 \partial \tau_0} \right] + cos(\xi^{(0)} + \alpha\tau + \phi) \qquad (9)$$

where

$$\alpha = \dot{\zeta}_0^{(0)} - \alpha \qquad (10)$$

This is the anharmonic oscillator law which holds at the first order in the expansion parameter.

At the second order we have

$$\left[\frac{\partial^2}{\partial \tau_0^2} + 1 \right] \xi^{(2)}(\tau_0, \tau_1, \tau_2) = -2 \left[\frac{\partial^2 \xi^1}{\partial \tau_1 \partial \tau_0} \right] - \left[\frac{\partial^2 \xi^{(0)}}{\partial \tau_1^2} \right] - \xi^{(1)} sin(\xi^{(0)} + \alpha\tau + \phi) \qquad (11)$$

The solution at the zeroth order gives:

$$\xi^{(0)} = A(\tau_1, \tau_2)cos(\tau_0) + B(\tau_1, \tau_2)sin(\tau_0) \qquad (12)$$

Note that the arbitrary constant A, and B are constant on the fast time variable τ_0, but depend on the slow time scale τ_1 and τ_2. At the first order, after substituting Eq. 12 in Eq.

9 and using the Fourier-Bessel expansion for the cosinus term, we have the following equation to be solved:

$$\ddot{\xi}^{(1)} + \xi^{(1)} = \sum_{n=-\infty}^{+\infty} \sum_{m=-\infty}^{+\infty} (i)^m J_n(B) J_m(A) \exp i(\alpha_{n,m}\tau_0 + \phi) +$$

$$\sum_{\kappa=-\infty}^{+\infty} \sum_{l=-\infty}^{+\infty} (-i)^l J_\kappa(B) J_l(A) \exp -i(\alpha_{\kappa,l}\tau_0 + \phi) \quad (13)$$

where J_n, J_m, J_κ and J_l are the Bessel function of order n, m, κ, l, respectively, and $\alpha_{n,m} = n + m + \alpha$ and $\alpha_{\kappa,l} = \kappa - l + \alpha$. A straightforward integration of Eq. 13 gives:

$$\xi^{(1)} = \left[C_1(\tau_0, A, B, \phi) - \tau_0 \frac{\partial A}{\partial \tau_1} \right] \cos \tau_0 +$$

$$+ \left[C_2(\tau_0, A, B, \phi) + \frac{\partial B}{\partial \tau_1} \right] \sin \tau_0 + c_1 \cos \tau_0 + c_2 \sin \tau_0 \quad (14)$$

where:

$$C_1 = \frac{\exp i\phi}{4} \mathcal{B}^{n,m} \left[\frac{\exp i(\alpha_{n,m}+1)\tau_0 - 1}{\alpha_{n,m}+1} - \frac{\exp i(\alpha_{n,m}-1)\tau_0 - 1}{\alpha_{n,m}-1} \right] +$$

$$- \frac{\exp -i\phi}{4} \mathcal{B}^{\kappa,l} \left[\frac{\exp -i(\alpha_{\kappa,l}-1)\tau_0 - 1}{\alpha_{\kappa,l}-1} - \frac{\exp -i(\alpha_{\kappa,l}+1)\tau_0 - 1}{\alpha_{\kappa,l}+1} \right] \quad (15)$$

$$C_2 = -\frac{i\exp i\phi}{4} \mathcal{B}^{n,m} \left[\frac{\exp i(\alpha_{n,m}+1)\tau_0 - 1}{\alpha_{n,m}+1} + \frac{\exp i(\alpha_{n,m}-1)\tau_0 - 1}{\alpha_{n,m}-1} \right] +$$

$$+ \frac{i\exp -i\phi}{4} \mathcal{B}^{\kappa,l} \left[\frac{\exp -i(\alpha_{\kappa,l}-1)\tau_0 - 1}{\alpha_{\kappa,l}-1} + \frac{\exp -i(\alpha_{\kappa,l}+1)\tau_0 - 1}{\alpha_{\kappa,l}+1} \right] \quad (16)$$

The functions $\mathcal{B}^{n,m}$ and $\mathcal{B}^{\kappa,l}$ are defined as:

$$\mathcal{B}^{n,m} = \sum_{n=-\infty}^{+\infty} \sum_{m=-\infty}^{+\infty} (i)^m J_n(B) J_m(A); \qquad \mathcal{B}^{\kappa,l} = \sum_{\kappa=-\infty}^{+\infty} \sum_{l=-\infty}^{+\infty} (-i)^l J_\kappa(B) J_l(A) \quad (17)$$

while c_1 and c_2 are two arbitrary constant which depend on the initial conditions, in this case $c_1 = c_2 = 0$.

Extending the number of time variables we have added an additional freedom in the perturbation analysis. This freedom is useful to remove, order by order, the time secularities which may occur in the solution for ξ_0, ξ_1, ξ_2. This assure that the perturbation solution is uniformly valid order by order. The condition to suppress secular terms leads to two differential equations for the amplitude and phase: $A(\tau_1)$ and $B(\tau_1)$

$$\left[C_1^{secular}(\tau_0, A, B, \phi) - \tau_0 \frac{\partial A}{\partial \tau_1} \right] = 0 \quad (18)$$

$$\left[C_2^{secular}(\tau_0,A,B,\phi) - \tau_0\frac{\partial B}{\partial \tau_1}\right] = 0 \tag{19}$$

If we average both equations on the phase ϕ which accounts for the phase shift between the wave field and the particle motion, we remark from Eqs. 15-16 that $< C_1^{secular} >_\phi$, and $< C_2^{secular} >_\phi$ are identically zero, the consequence is that $< A >_\phi$, and $< B >_\phi$ are constants on the time scale τ_1. This signifies that the averaged arbitrary constants will depend only on the slower scale time τ_2. A wave-particle momentum transfer along the ξ axis cannot be identified at this stage of the perturbed orbit calculation. The next order of approximation shows the above quoted non-linear effects. The formal solution is the same as Eq. 14 but the coefficients are much more complicated functions of trigonometric expressions. Averaging the solution on the phase ϕ we have that:

$$< \xi^{(2)} >_\phi = \left[< C_3(\tau_0,A,B,\phi) >_\phi -\tau_0\frac{\partial A}{\partial \tau_2}\right]\cos\tau_0 +$$
$$+\left[< C_4(\tau_0,A,B,\phi) >_\phi -\tau_0\frac{\partial B}{\partial \tau_2}\right]\sin\tau_0 + c_3\cos\tau_0 + c_4\sin\tau_0 \tag{20}$$

where $< C_3 >_\phi$ and $< C_4 >_\phi$ are lengthy functions of τ_0, $\mathcal{B}^{\kappa,l}\mathcal{B}^{n',m'}$, and $\mathcal{B}^{n,m}\mathcal{B}^{\kappa',l'}$ which for brevity we avoid to report here. c_3 and c_4, as before are two arbitrary constant which depend on the initial conditions, in this case $c_3 = c_4 = 0$. This time, the differential equations which comes from the removal of the secularity in τ_2 give the evolution of the arbitrary constants on the slower time scale τ_2.

In trying the secular terms in the functions $< C_3 >_\phi$ and $< C_4 >_\phi$ we obtain some rules for the indexes of the Bessel functions which appear in the definition of the above functions. In general there are a great number of the secularities belonging to different indexes of the Bessel functions. We have identified all the secularities:

$$< C_3^{secular} >_\phi = \frac{\tau_0}{16}\left[-\mathcal{B}^{l'-\alpha-4,l'}\mathcal{B}^{l'-\alpha-4,l'} + \mathcal{B}^{l'-\alpha-4,l'}\mathcal{B}^{l'-\alpha-4,l'}\right] \tag{21}$$

$$< C_4^{secular} >_\phi = \frac{\tau_0}{16}\left[-\mathcal{B}^{l'-\alpha-4,l'}\mathcal{B}^{l'-\alpha-4,l'} + \mathcal{B}^{l'-\alpha-4,l'}\mathcal{B}^{l'-\alpha-4,l'}\right] \tag{22}$$

where α is defined above in Eq. 10.

The relevant equations to be integrated are finally:

$$\left[< C_3^{secular}(\tau_0,A,B,\phi) >_\phi -\tau_0\frac{\partial A}{\partial \tau_2}\right] = 0 \tag{23}$$

$$\left[< C_4^{secular}(\tau_0,A,B,\phi) >_\phi -\tau_0\frac{\partial B}{\partial \tau_2}\right] = 0 \tag{24}$$

This is a system of two coupled differential equations for the amplitude and the phase, its solution gives the behaviour of the arbitrary constants on the slower time-scale τ_2.

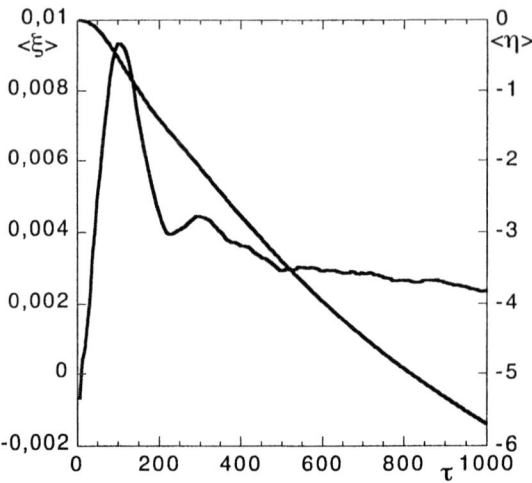

FIGURE 1. Plot of the centre guide drift η and its derivative −ξ vs time τ

The arbitrary constants vary on the slow time scale and show a saturation when $\tau \to \infty$. The numerical solution has been obtained by integrating the system Eqs. 1 along the ion gyration motion period τ for 100 different phases ϕ. A further numerical integration has been performed at each τ over all the phases. The obtained guiding centre drift, directed along the η-axis (see Fig. 1), and independent on the fast time scale τ, is in a good agreement with the shape predicted by the perturbation theory and shows a saturated behaviour for $\tau \to \infty$, as stated by the asympotic inspection of the solution of Eqs. 23-24. The poliodal velocity (the time derivative of η, Fig. 1 left scale) of the drift after an initial acceleration reaches a constant value.

COMMENTS AND CONCLUSIONS

We have demonstrated that a poloidal flow can be induced by a RF electromagnetic wave and it can be obtained on the basis of the single particle dynamics near an ion cyclotron resonance layer. An analytic expression of the poloidal velocity has been derived in the frame of a multiple-time scale perturbartion theory up to the second order in the small parameter λ_1 which account for the effects of the propagating field. Velocities of the order $10^5 - 10^6 cm/sec$ can be reached.

REFERENCES

1. LeBlanc B., et al., Phys. Rev. Letters, **82** 331 (1999).
2. Cesario R., et al. to be published on Phys. of Plasmas, October 2001.
3. L.A. Berry, E.F. Jaeger, D.B. Batchelor, Phys. Rev. Lett. **82** 1871 (1999).
4. R.C. Davidson, Methods in Nonlinear Plasma Theory, Academic Press, New York, 1972.

2. RELATIVISTIC NONLINEAR OPTICS OF ULTRA-SHORT LASER PULSES IN PLASMAS

Developments in Relativistic Nonlinear Optics

D. Umstadter, S. Banerjee, S. Chen, E. Dodd, K. Flippo, A. Maksimchuk,
N. Saleh, A. Valenzuela, and P. Zhang

Center for Ultrafast Optical Science, 1006 IST Bldg., University of Michigan, Ann Arbor
48109-2099

Abstract. We report recent results of experiments and simulations in the regime of peak laser intensities above 10^{19} W/cm^2, including the following topics: (1) electron and proton acceleration to energies in excess of 10 MeV in well collimated beams; (2) use of laser chirp to control the growth of plasma waves and acceleration of electrons by the Raman instability; (3) all optical injection and acceleration of electrons; (4) relativistic self-focusing by means of the mutual index of refraction of two overlapping laser pulses; (5) creation of a radioisotope by the reaction $^{10}B(d,n)^{11}C$; (6) high-order harmonic generation from relativistic free electrons in an underdense plasma.

INTRODUCTION

In the last decade, table-top lasers have undergone an orders-of-magnitude jump in peak power, with the invention of the technique of chirped pulse amplification [1]. They now have multi-terawatt peak powers and, when focused, can produce electromagnetic intensities well in excess of 10^{18} W/cm^2, which is high enough to cause nonlinearity with even unbound (ionized) electrons [2]. The nonlinearity arises, in this case, because the electrons oscillate at relativistic velocities in laser fields that exceed 10^9 V/cm, resulting in relativistic mass changes exceeding the electron rest mass and the light's magnetic field becomes important. The work done on an electron over the distance of a laser wavelength (λ) then approaches the electron rest mass energy ($m_e c^2$), where m_e is the electron rest mass and c is the speed of light. Thus, a new field of nonlinear optics, that of relativistic electrons, has been launched. Effects analogous to those studied with conventional nonlinear optics—self-focusing, self-modulation, harmonic generation, and so on—are all found, but based on this entirely different physical mechanism. Applications of these phenomena include compact and ultrashort pulse duration laser-based electron accelerators and x-ray sources.

In this paper, the results of our recent simulations and experiments are discussed in the following order: first, acceleration of electrons that are self-trapped; next, a technique to control the Raman instability by use of chirped pulses; then, some preliminary experiments in which one laser pulse injects electrons from the plasma into the plasma wave created by another. In the next section: proton acceleration, including the production of a short-lived radioisotope. Following that: high-order harmonic generation. Lastly: prospects for applications.

CP611, *Superstrong Fields in Plasmas:* Second Int'l. Conf., edited by M. Lontano et al.

ELECTRON ACCELERATION

For time periods that are short compared to an ion period, electrons are displaced from regions of high laser intensity, but ions, due to their much greater inertia, remain stationary. The resulting charge displacement provides an electrostatic restoring force that causes the plasma electrons to oscillate at the plasma frequency (ω_p) after the laser pulse passes by them, creating alternating regions of net positive and negative charge. The resulting electrostatic wakefield plasma wave propagates at a phase velocity nearly equal to the speed of light and thus can continuously accelerate properly phased electrons.

Acceleration of electrons by electron plasma waves is of current interest because the acceleration gradient (200 GeV/m) is much larger (four orders of magnitude larger) than in conventional rf linacs (< 20 MeV/m) [3, 4]. Several methods have been proposed for driving a large-amplitude high-phase-velocity plasma wave, such as the plasma wake-field accelerator, the plasma beat-wave accelerator, the laser wake-field accelerator (LWFA) and the self-modulated laser wake-field accelerator (SMLWFA). The LWFA and the SMLWFA have received considerable attention and shown rapid progress because of the development of table-top ultrashort-duration terawatt-peak-power lasers. In the SMLWFA, an electromagnetic wave (ω_o, k_o) decays into a plasma wave (ω_p, k_p) and another forward-propagating light wave ($\omega_o - \omega_p, k_o - k_p$) via the stimulated Raman forward scattering instability. In this case, the laser pulse duration is longer than an electron plasma period, $\tau >> \tau_p = 2\pi/\omega_p$.

We have recently studied theoretically the coherent control of the Raman-driven plasma waves by the use of a frequency chirp on the laser beam. Theoretical calculations show that a 12% bandwidth will eliminate Raman forward scattering for a plasma density that is 1% of the critical density. We find analytically that the amount of chirp needed to eliminate SRS,

$$b = -\frac{1}{4}\frac{\omega_0}{\omega_e}\frac{\gamma_0}{\tau_p}, \tag{1}$$

where ω_0 is the incident light's frequency, ω_e is the plasma wave frequency, τ_p is the full-width-half-maximum length of a pulse with a Gaussian profile, and γ_0 is the growth rate. The bandwidth for a chirped pulse can be written as $\Delta\omega_0 \simeq 2b\tau_p \simeq 1/2(\omega_0/\omega_e)\gamma_0$. The predicted changes to the growth rate are confirmed in two-dimensional particle-in-cell simulations. As shown in Fig. 1, a positive chirp (c) is found to increase the growth of Raman-driven plasma waves as compared with no chirp (a); a negative chirp (b) is found to decrease it [6]. In the LWFA, an electron plasma wave is driven resonantly by a short laser pulse ($\tau \sim \tau_p$) through the laser ponderomotive force. Self-guiding is possible when laser power exceeds the threshold for relativistic self-guiding, P_c. Whole beam self-focusing and, more recently, relativistic filamentation (a partial beam analog to the whole beam effect, or multiple filaments) have both been observed [5].

Several labs have observed the acceleration of MeV electrons by the SMLWFA, sometimes accompanied by self-guiding, but with large-electron energy spreads (most of the electrons have energies less than 5 MeV, with the number decaying exponentially with a temperature of ~ 1 MeV to just a few electrons at energies up to 100 MeV). The origin of the accelerated electrons is a subject of some debate. It has been attributed to

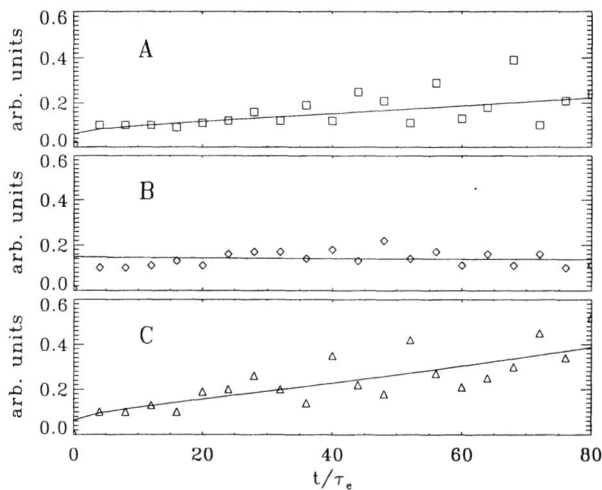

FIGURE 1. The amplitude of the density perturbations are plotted as a function of time. Plot A is the unchirped pulse, B, negatively, and C, the positively chirped pulses.

catastrophic wave-breaking of a relativistic Raman forward scattered plasma wave, and to wavebreaking of slower velocity Raman backscattered waves in both experiment and theory. A two-temperature distribution in the electron energy spectrum was observed [7] to accompany a multi-component spatial profile of the electron beam. In this case, electrons in the low energy range were observed to undergo an abrupt change in temperature, coinciding with the onset of extension of the laser channel due to self-guiding of the laser pulse, when the laser power or plasma density was varied. With the aid of a test-particle simulation, we have now determined that both of these effects appear to originate from the dynamics of trapping and detrapping as electrons oscillate in their orbits in the seperatrix from regions of acceleration to decelleration and from defocusing to focusing [7].

Laser acceleration of electrons is illustrated in Fig. 2. Here, an intense laser interacts with a gas jet located inside a vacuum chamber. The laser crosses the picture from left to right and is focussed by a parabolic mirror (right side of the picture). The supersonic nozzle (shown in the middle of the picture) is positioned with micron accuracy with a 3-axis micropositioner. The e-beam (up to 10^{10} electrons per shot) makes a small spot on a fluorescent (LANEX) screen (imaged with a CCD camera), shown in the upper left-hand corner of the picture. As shown in Fig. 2 [7], as the laser power increases, the divergence angle of the electron beam decreases. The lowest angle, 1°, obtained at the highest power, corresponds to a transverse geometrical emittance of $\varepsilon_\perp \leq 0.06\pi$ mm-mrad [7], which is an order of magnitude lower than that from the best conventional electron gun. This may be because a large acceleration gradient decreases the time over which space-charge can act to degrade the emittance.

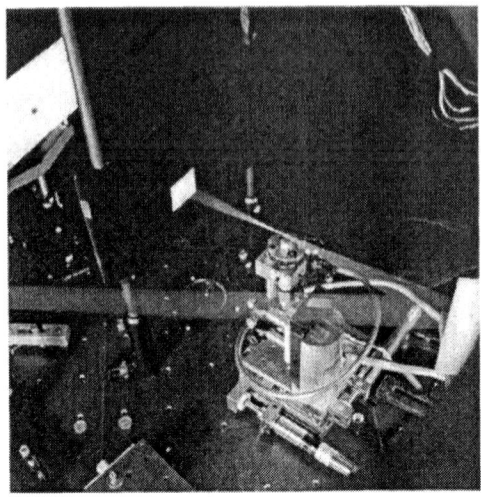

FIGURE 2. Photograph of the acceleration of an electron beam by a laser interacting with a gas jet inside a vacuum chamber. The laser (illustrated for the purpose of orientation) crosses the picture from left to right and is focused by a parabolic mirror (right side of the picture). The supersonic nozzle (shown in the middle of the picture) is position with micron accuracy with a 3-axis micropositioner. The e-beam makes a small spot on a white florescent (LANEX) screen, shown in the upper left-hand corner of the picture.

Optical Injection

The injection of electrons into plasma waves can occur uncontrollably by trapping of hot background electrons, which are preheated by other processes such as Raman backscattering and sidescattering instabilities [8], or by wave-breaking (longitudinal [9] or transverse [10]). Because the electrons in this case are injected into the plasma wave uniformly in phase space, large energy spreads result, as is typically observed in the SMLWFA regime. The injection can also be controlled by use of an external electron source (such as from an RF gun); however, because the pulse durations of the injected electron bunches in the experiments in which this method was tried were longer than the acceleration buckets, the energy spread was again large. It has been shown analytically and numerically that controlled injection might also be accomplished by means internal electrons, from the plasma itself, which are all put into the accelerating phase of the plasma wave by a separate laser pulse [11, 2]. Such a laser-driven plasma-cathode electron gun might eventually have (1) monoenergetic energy, (2) GeV/cm acceleration fields, (3) micron source size, (4) femtosecond pulse duration, (5) high brightness, (6) absolute synchronization between electrons and laser (for pump and probe experiments) and (7) compact size (university-lab scale).

As a first step towards the realization of monoenergetic beams by means of optical injection, we have studied experimentally, in the SMLWFA regime, a concept first discussed in the LWFA regime [11]. By crossing in a plasma two beams, each of duration 400 fs and wavelength 1 μm, one with high vacuum intensity ($a_0 = 1.0$) and the other lower intensity ($a_0 = 0.25$), we find that only when the beams are overlapped in space and time do we observe what appears to be diffraction of light from a standing wave (see Fig. 3) and the acceleration of electrons in the direction of the low power beam (see Fig. 4). Injection in this case is caused by the ponderomotive force of a standing

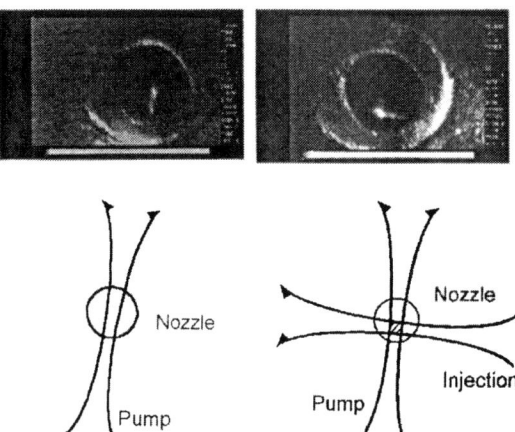

FIGURE 3. Photograph of the light Thomson scattered from the beam channels, without (left) and with (right) a crossed beam. Notice the creation of a standing wave at the bisector in the latter case.

wave created along the bisector of the two beams, which will stochastically heat the electrons to relativistic temperature and allow them to be caught in the plasma wave that was created by the low power beam. This process is similar to the "preheating" that occurs when the slow-phase-velocity plasma wave driven by Raman backscattering injects electrons into the faster plasma waves that are generated by Raman forward scattering [8]. However, in this case, the ponderomotive force of the standing wave is much higher because of its much shorter wavelength. It is also occasionally observed that the increase of the index of refraction resulting from the addition of the pulses can extend the distance over which the low-power beam remains focused. This is the first time that one laser beam has been used to trigger the acceleration of electrons by another laser beam, thus proving the LILAC principle.

PROTON ACCELERATION

Plasma ions can be accelerated to high energies by the formation of an electrostatic sheath due to charge displacement. The latter results from the initial preferential acceleration of electrons; the heavier ions are left behind due to inertia. Ions will also be accelerated by each other's unshielded charges in what has been termed a "Coulomb explosion." Among the many mechanisms that can accelerate the electrons are: thermal expansion, plasma waves, "$J \times B$ heating" or "vacuum heating." Recent simulations reveal a new mechanism [12] that occurs in cases where a significant preplasma exists, such as for pulses with typical laser-intensity contrast ratios. In this case, a standing wave is produced by the beating of the incident and reflected light, which can heat the

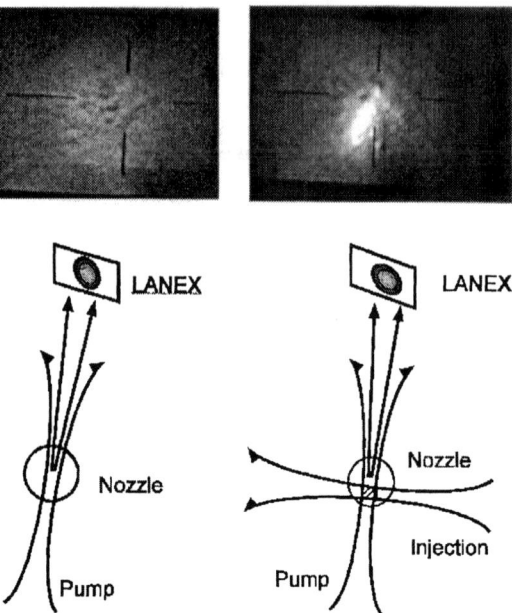

FIGURE 4. Photographs of the spatial profile of a laser accelerated electron beams without (left) and with (right) injection.

electrons to relativistic temperatures, as was described in the discussion of optical injection.

Energetic ions from underdense plasmas were accelerated by an electrostatic sheath, which was created by charge-displacement. Unlike earlier long-laser-pulse experiments, the displacement was due not to thermal expansion but to ponderomotive blow-out [13, 14]. When a helium-gas was used as the target, alpha particles were accelerated to several MeV in the direction orthogonal to the direction of laser propagation and along the direction of the maximum intensity gradient.

Several groups have reported the observation of ions originating from thin-film solid-density targets (or protons originating from monolayers of water on the target surface). Unlike previous long-pulse experiments, the protons were accelerated along the direction normal to the side of the target opposite to that upon which the laser was incident. The laser is focused with an off-axis parabola onto a thin-foil, held by a mesh that is positioned by a 3-axis micropositioner. A nuclear track detector, CR-39 , is used to measure the proton-produced pattern [15]. For instance, Fig. 5 shows the difference in the spatial profile of between a mylar target, an insulator, and aluminum, a conductor. Note that the proton beam from the latter is much smoother. In another experiment, protons were observed to be emitted in ring patterns, the radii of which depend on the proton energy, which was explained by self-generated magnetic fields [16]. Another recent result reported proton energies up to 60 MeV [17]. The results of these experiments indicate that a large number of protons (10^{13} p) can be accelerated, corresponding to current densi-

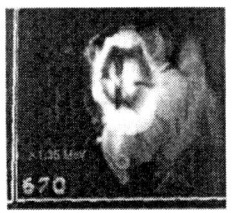 

FIGURE 5. Photograph of the spatial profile of a laser accelerated proton beams from a mylar target (left) and aluminum target (right). The proton beam is detected with CR-39, a nuclear track detector.

ties (10^8 A/cm^2) at the source that are nine orders-of-magnitude higher than produced by cyclotrons, but with comparable transverse emittances ($\varepsilon_\perp \leq 1.0\pi$ mm-mrad). The high end of the energy spectrum typically has a sharp cut-off, but, like the electrons is a continuum.

While the protons in several experiments originate from the front-side of the target [15, 16], in another [17], they originate from the back-side. Evidence for a back-side origin comes from results obtained when wedge-shaped targets were used. The proton beam was observed to point in the direction normal to the back side of the target, which was not perpendicular to the front surface. On the other hand, evidence for a front-side origin comes from an experiment in which deuterium was coated on a thin film of mylar and a boron target was placed behind it [18].

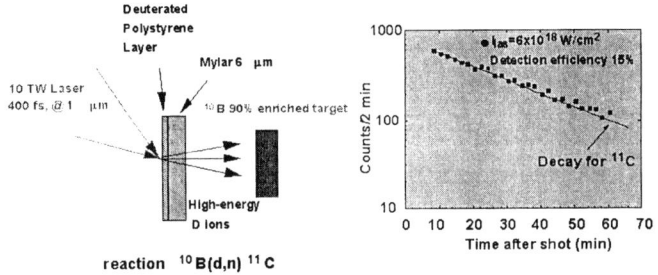

FIGURE 6. Experimentally recorded radioactive decay rate corresponds to that of ^{11}C, which decays by positron emission. The 511 keV gamma rays which were detected were created by annihilation of the electron-positron pairs.

High-order Harmonic Generation from Free Electrons

We report the first observation of high harmonics generated due to the scattering of relativistic electrons from high intensity laser light. The experiments were carried out with an Nd:Glass laser system with a peak intensity of 10^{19} W/cm^2 in underdense plasma. At high intensities, when the normalized electric field approaches unity, in addition to the conventional atomic harmonics from bound electrons, there is significant contribution to the harmonic spectrum from free electrons. The characteristic signatures of this are found to be the emission of even order harmonics (see Fig. 7), linear dependence on the electron density, significant amount of harmonics even with circular polarization and a much smaller spatial region over which these harmonics are produced (and emitted in a narrower cone) as compared to the atomic case. Imaging of the harmonic beam shows that it is emitted in a narrow cone with a divergence of 2-3 degrees.

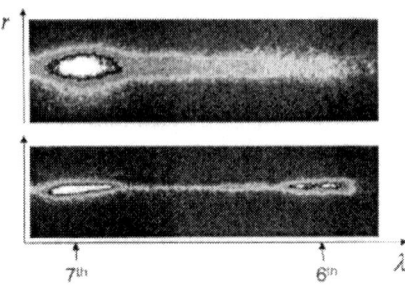

FIGURE 7. Spatial profile of the high order harmonics (a) $I = 5 \times 10^{16}$W/cm^2, $n = 10^{17}$ cm^{-3} and linear polarization (b) $I = 5 \times 10^{17}$W/cm^2, $n = 10^{18}$ cm^{-3} and circular polarization. Note that even orders appear only at high laser intensity and electron density.

PROSPECTS

Besides the beam of harmonics, discussed in the previous section, a promising avenue for the production of x-rays is based on the use of Compton scattering. Upon scattering with an electron beam, laser light can be upshifted due to a relativistic boost by a factor of $4\gamma^2$, which for electrons accelerated to 30 MeV corresponds to a factor of 10,000. Thus a 1 eV photon can be upshifted to 10 keV. Such coherent, ultrashort duration and energetic sources could enable ultrafast imaging on the atomic scale. If the source of the electron beam were a laser accelerator, the footprint of this synchrotron-like device would be small enough for it to fit in a university laboratory.

It was demonstrated that there are a sufficient number of electrons accelerated by laser-plasma accelerators to conduct pulsed radiolysis [19]. Time-resolved radio-biological studies with laser accelerated protons are also feasible.

It has been shown theoretically that a laser-induced burst of hot electrons or ions could be used as a spark to ignite a thermonuclear reaction with inertial confinement fusion in the so-called fast-ignitor laser fusion concept. In its original conception, a short but energetic laser pulse would drill through the under-dense plasma that surrounds the fusion core and a second shorter pulse would deposit energy in the core in the form of MeV electrons. In so doing, it would relax the otherwise stringent requirements on energy and symmetry of the long-pulse-duration heating and compression pulses. More recently, the use of a short pulse of ions for the ignitor has also been discussed.

The short-lived radioactive isotopes that were produced by laser acceleration of ions might be used to for cancer diagnostics such as in positron-emission tomography. Because they are created impulsively with a ion short pulse, they might also be used to study ultrashort duration isotope decay times in nuclear physics. If protons could be laser-accelerated to 70-160 MeV energies, they could be useful for proton therapy, which is now limited by the extraordinary expense of cyclotrons or synchrotrons and the large magnets required to transport the proton beams to the patient. Protons are superior to other forms of ionizing radiation for cancer treatment because of less straggling and their ability to deposit their energy over a narrower depth range.

ACKNOWLEDGMENTS

The authors acknowledge the support of the High Energy Physics Division, U.S. DOE, award DE-FG02-98ER41071 (electrons); Chemical Sciences, Geosciences and Biosciences Division, U.S. DOE (ions and x-rays); NSF (laser and plasma physics).

REFERENCES

1. Mourou, G A., Barty C P. J., and Perry, M. D., Physics Today, (Jan., 1998), and references therein.
2. Umstadter, D., Phys. Plasmas **8**, 1774 (2001), and references therein.
3. Tajima, T. and Dawson, J. M., Phys. Rev. Lett. **43**, 267, (1979).
4. Esarey, E., Sprangle, P., Krall, J., and Ting, A., IEEE Trans. Plasma Sci. **PS-24**, 252 (1996), and references therein.
5. Wang, X., Krishnan, M., Saleh, N., Wang H., and Umstadter, D., Phys. Rev. Lett. **84**, 5324 (2000).
6. Dodd, E., and Umstadter, D., Phys. Plasmas **8**, 3531 (2001).
7. Chen, S.-Y., Krishnan, M., Maksimchuk, A., Wagner, R. and Umstadter, D., Phys. of Plasmas **6**, 4739 (1999).
8. Bertrand, P., Ghizzo, A., Karttunen, S. J., Pattikangas, T. J. H., Salomaa, R. R. E., and Shoucri, M., Phys. Rev. E **49**, 5656 (1994).
9. Modena, A., Najmudin, Z., Dangor, A. E., Clayton, C. E., Marsh, K. A., Joshi, C., Malka, V., Darrow, C. B., Danson, C., Neely, D., and Walsh, F. N., Nature **377**, 606 (1995).
10. Bulanov, S. V., Pegoraro, F., Pukhov, A. M., and Sakharov, A. S., Phys. Rev. Lett. **78**, 4205 (1997).
11. Umstadter, D., Kim, J. K., and Dodd, E., Phys. Rev. Lett. **76**, 2073 (1996).
12. Sentoku, Y. et al., Appl. Phys. Letts. (submitted, 2001).
13. Sarkisov, G. S., Bychenkov, V. Yu., Novikov, V. N., Tikhonchuk, V. T., Maksimchuk, A., Chen, S. Y., Wagner, R., Mourou, G., and Umstadter, D., Phys. Rev. E **59** 7042 (1999).
14. Krushelnick, K., Clark, E. L., Najmudin, Z., *et al.*, Phys. Rev. Lett. **83**, 737 (1999).
15. Maksimchuk, A., Gu, S., Flippo, K., Umstadter, D., Bychenkov, V.Y., Phys. Rev. Lett. **84**, 4108 (2000).
16. Clark, E. L., Krushelnick, K., Davies, J. R., *et al.*, Phys. Rev. Lett. **84**, 670 (2000).
17. Snavely, R. A., Key, M. H., Hatchett, S. P., *et al.*, Phys. Rev. Lett. **85**, 2945 (2000).
18. Nemoto, K., Maksimchuk, A., Banerjee, S., Flippo, K., Mourou, G., Umstadter, D., Bychenkov, V. Yu., Appl. Phys. Lett. **78**, 595 (2001).
19. Saleh, N., Flippo, K., Nemoto, K., Umstadter, D., Crowell, R. A., Jonah, C. D., Trifunac, A. D., Rev. Sci. Instr., **71**, 2305 (2000).

Relativistic interaction of ultra-short laser pulses with plasmas

S.V.Bulanov[a,b], T.Zh.Esirkepov[b], D.Farina[c], F.F.Kamenets[b],
V.S.Khoroshkov[d], A.V.Kuznetsov[b], T.V.Liseikina[e], M.Lontano[c],
N.M.Naumova[a], K.Nishihara[f], F.Pegoraro[g], H.Ruhl[h], J.-I.Sakai[i],
D.V.Sokolov[b], I.V.Sokolov[a]

[a]General Physics Institute - RAS, Moscow, Russia
[b]Moscow Institute of Physics and Technology, Dolgoprudny, Russia
[c]Institute for Plasma Physics - CNR, Milan, Italy
[d]Institute of Theoretical and Experimental Physics, Moscow, Russia
[e]Institute of Computation Technologies – RAS, Novosibirsk, Russia
[f]Institute of Laser Engineering, Osaka University, Osaka, Japan
[g]University of Pisa and INFM, Pisa, Italy
[h]Max Born Institute, Berlin, Germany
[i]Toyama University, Toyama, Japan

Abstract. The review of the results obtained recently in the theory and in the computer simulations with 2D and 3D PIC codes of the interaction of high-intensity, ultra-short laser pulses with plasmas is presented. The laser interaction with underdense plasmas and its application to the problem of charged particle acceleration, of laser frequency upshifting, of relativistic self focusing, and of the generation of a quasistatic magnetic field is discussed. We present the results of the analytical description and of the PIC simulations of nonlinear coherent structures such as relativistic solitons and vortices and the production of high harmonics.

INTRODUCTION

When laser pulse interacts with plasmas, its shape changes due to various linear and non-linear effects as well as the light frequency changes with frequency up-shifting, down-shifting and high harmonic generation (see review in Ref. [1] and references therein). We see also formation of such the nonlinear coherent structures as self-focusing channels, relativistic solitons, vortices. Nonlinear coherent structures are characterized by local regions of ordered high amplitude electromagnetic field or/and of ordered motion. They have typical life-time that is much longer than a period of the electromagnetic field and a period of the particle motion inside them. Coherent structures at the present time are routinely seen in the 2D particle in cell simulations of the laser plasma interaction. The time life of these entities in collisionless plasmas is determined by their energy transformation into the energy of fast particles. It is substantially large to distinguish the coherent structures from strong plasma turbulence excited by the laser radiation. The pressure of the magnetic field associated with the electron vortices and the ponderomotive pressure of the trapped electromagnetic mode inside the solitons accelerate charged particles outwards. In an inhomogeneous plasma the solitons move toward the plasma-vacuum interface, where they convert trapped electromagnetic energy into the electromagnetic bursts, which make it is possible to

CP611, *Superstrong Fields in Plasmas:* Second Int'l. Conf., edited by M. Lontano et al.

detect them. On the other hand, the electron vortices move across the plasma density gradients. At the plasma vacuum interface the magnetic field, associated with the vortices, pushes the electrons inward, forming in such the way extend regions of non-compensated electric charge. In its turn, the charge separation electric field accelerates the ions [2]. When the laser radiation interacts with a thin slab of high density plasma, this leads to effective generation of fast ions at the rear side of the target. The ions can be focused by tailoring the target. In this case, the generated fast ions appear in the form of highly collimated beams. This opens up a way for broad range important applications in the laser fusion [3], for the particle injectors [4], and in the hadron therapy [5]. The process of the laser plasma interaction is accompanied by the high harmonic generation. The main mechanism of the high harmonic generation in the range of the relativistic intensity of the laser radiation is due to the reflection of the electromagnetic wave at the "oscillating mirror". The oscillating mirror appears due to the nonlinear motion of the electrons in the narrow region near the plasma boundary. The high harmonic generation efficiency becomes much higher when the laser pulse propagates inside a narrow channel. If the channel has thin walls, it makes possible to generate the high harmonics in the form of ultra short coherent wave packet.

It is well known that we must describe the laser-matter interaction within the framework of the relativistic theory when the dimensionless ratio, it is the normalized laser pulse amplitude, $a = eE_0 / m_e\omega_0 c$ is much larger than unity. Relativistic effects appear through the qualitative modification of the charged particle motion. In the electromagnetic wave the transverse component of the generalized momentum is constant, $\mathbf{p}_\perp - (e/c)\mathbf{A}_\perp = const$, and the kinetic energy $K = m_e c^2(\gamma - 1)$ and momentum $\mathbf{p}$ are given by the expressions: $K = m_e c^2 |\mathbf{a}_\perp|^2 / 2$, $p_\| = m_e c |\mathbf{a}_\perp|^2 / 2$, and $\mathbf{p}_\perp = m_e c \mathbf{a}_\perp$. The acceleration length, which the particle must pass to acquire the energy, is equal to $\xi_\perp = a_0\lambda$ in the transverse direction and it is $\xi_\| = a_0^2\lambda/2$ in the longitudinal direction. We see that in order to get this energy the particle must remain in the laser field in the region that has the size in the transverse direction not less than $a_0\lambda$ and the region size in the longitudinal direction must not be less than $a_0^2\lambda/2$.

A condition, $a_0 \gg 1$, when electrons become relativistic, corresponds to the focal spot size of the order of the laser wavelength and to the intensity equal to $I_e = 1.35\times10^{18}$ W/cm^2, for the 1 µm laser. The focal spot size in this case can be of the order of λ (see Ref. [6] discussed the laser pulse focusing into the one wavelength spot). We obtain for the laser power $P_e = 42$GW. The proton quiver energy becomes relativistic in the laser field with the normalized amplitude equal to $a_0 > m_p / m_e$. It corresponds to the laser radiation intensity $I_p = 4.5\times10^{24}$ W/cm^2 and the power $P_p = 1.4\times10^{17}$ W $= 140$PW. For the CO_2 lasers with $\lambda = 10$ µm we find $I_e = 1.35\times10^{16}$ W/cm^2, $P_e = 42$GW, $I_p = 4.5\times10^{20}$ W/cm^2, and $P_p = 1.4\times10^{15}$ W. As we see, the parameters, when the protons energy becomes relativistic, are out of reach for the nowadays lasers. However, as it shown in Refs. [1,7-9], the protons can gain the relativistic energy being accelerated in the electric field of the electric charge separation during interaction with plasmas of the laser pulse with much smaller amplitude. It is $a_0 > (m_p / m_e)^{1/2}$. In this case for the 1 µm laser we find its intensity to be $I_h = 2.5\times10^{21}$ W/cm^2 and its power is $P_p = 8\times10^{13}$ W $= 80$TW.

RELATIVISTIC SELF-FOCUSING OF LINEARLY
POLARIZED LASER PULSE

One of the most impressive nonlinear phenomenon in underdense plasmas is the self-focusing of the laser light, discovered by G. A. Askar'yan in Ref. [10]. The nonlinear evolution of an electromagnetic wave in an underdense plasma has been studied analytically in numerous publications under various simplifying assumptions, such as pulse circular polarization, quasistatic approximation and weak nonlinearity, or within the framework of the paraxial approximation. Linearly polarized pulses are more complex to study because the analytic simplifications, that follow in the case of circularly polarized pulses from their lack of harmonic content, do not apply. In addition the intensity of modern petawatt power laser pulses is so high that we cannot take advantage of the weak nonlinearity approximation. As is well known, in 3-D plasma configurations the role of nonlinearity becomes more important than in 1-D and 2-D cases because in 3-D configurations the phenomenon of wave collapse results in the development of a 3-D singularity as it was shown in Ref. [11]. In this case the role of three dimensional computer simulations can not be understated.

We illustrate the propagation of linearly-polarized, ultra intense laser pulses in underdense plasmas by using a three-dimensional PIC code REMP. The REMP code is massively parallelized and fully vectorized; it is described in Ref. [12]. We see new features of the laser light plasma interaction in three dimensional regimes. Some of these features were described in Refs. [13] and [14]. In Figs. 1 and 2 we present the results of 3D PIC simulations. The self-focusing of the linearly polarized laser pulse is anisotropic, and the light intensity distribution is different in the plane parallel to the direction of the light polarization and in the perpendicular plane (see also Ref. [15]). The self-generated magnetic field (Fig. 2 (a)) changes sign in the y=10 plane, as discussed in [16,17], and its structure corresponds to an electric current density concentrated in the plane z=10. Inside the self-focusing channel we see thin plasma layer, which corresponds to the plasma filament discussed in Refs. [1,8,9,18].

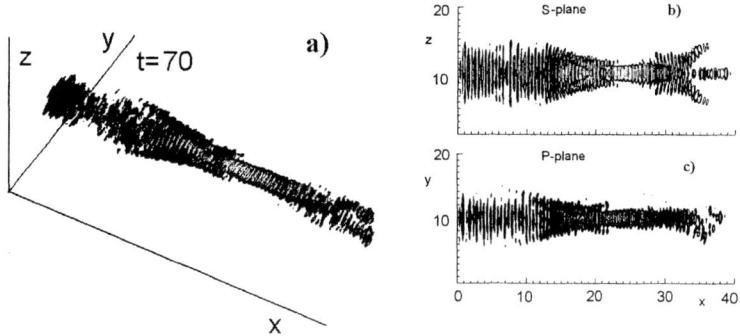

FIGURE 1. 3-D view of the isosurface of the electromagnetic energy density of a linearly polarized semi-infinite beam at t=70·2π/ω (a). 2-D projections the isosurface of the electromagnetic energy density on the x,z plane (b), on the x,y plane (c) at t=70·2π/ω.

In order to illustrate the properties of the ion motion inside the self-focusing channel,

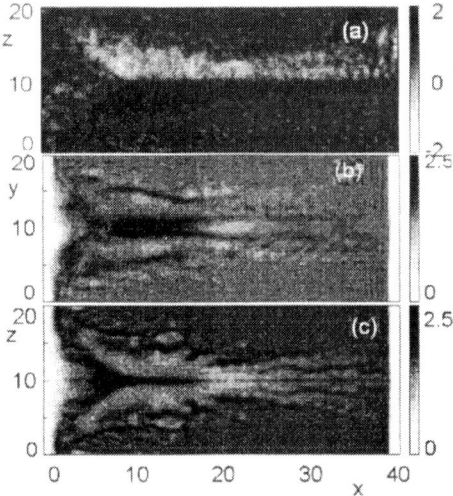

in Fig.3 (a) we present the ion trajectories calculated in 2D case, when an s-polarized laser pulse interacts with an overdense plasma. The dimensionless amplitude of the laser pulse is $a=3$. We see that the ions at the channel periphery are deflected outwards while the ions localized in the vicinity of the channel axis are not deflected. This results in the formation of the filament seen in the ion density distribution in Fig. 3 (b). In this figure we also see the channel shape and the bubbles inside and outside the channel. These bubbles correspond to the postsolitons discussed below and they were studied in details in Refs. [19-21].

FIGURE 2. Cross sections of the distribution of the z component of the magnetic field at y=10 (a), of the ion density at z=10 (b) and at y=10 (c).

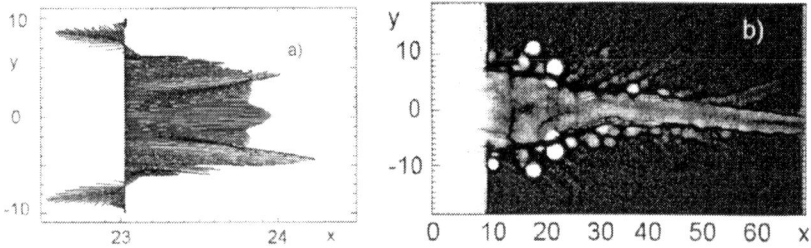

FIGURE 3. Ion trajectories inside the self-focusing channel (a) and the self-focusing channel and the postsolitons seen in the ion density distribution (b) at $t=60\times2\pi/\omega$.

RELATIVISTIC SOLITONS AND POSTSOLITONS

Recently in the interaction of a laser pulse with a plasma, a novel phenomenon has been identified [19-21] that directly involves the expansion of a cavity filled with electromagnetic energy, its interaction and coalescence with other cavities in the plasma. This phenomenon is related to the long time evolution of the relativistic subcycle electromagnetic solitons [22-27] that are produced when an ultrashort ultraintense laser pulse propagates in an underdense plasma. Relativistic solitons consist of electron density cavities inside which an electromagnetic field oscillates coherently with a frequency below the local plasma frequency and a spatial structure corresponding to half a cycle. Recently, they have been identified in 3D PIC simulations [21].

In inhomogeneous plasmas, in contrast to the electron vortices, which move across a density gradient, solitons move along the density gradient towards the lowest density level. When a soliton approaches the plasma-vacuum boundary some critical density level, it radiates its energy in the form of a low-frequency short electromagnetic burst [28]. On an ion timescale due to the ion acceleration caused by the time-averaged electrostatic field inside the soliton it evolves into the postsoliton, which is a slowly expanding bubble in the plasma. The typical energy of fast ions is of the order of $m_e c^2 a$, where a is the amplitude of the trapped electromagnetic field.

ELECTRON VORTICES

A laser pulse of finite length and width and very high intensity, propagating in underdense plasmas is subject to relativistic self-focusing and can propagate in the shape of a short, narrow ``bullet'' [29]. Such a laser pulse produces a quasistatic magnetic field wake in an initially unmagnetized plasma and the corresponding electron fluid motion takes the form of a vortex row [30] (see also [15]). The vortices seen in the magnetic field distribution are shown in Fig. 4. The distance between the vortices is comparable to, or in their final stage even larger than, the collisionless skin depth. The vortex row moves as a whole in the direction of the laser pulse propagation with a velocity much smaller than the pulse group velocity. The velocity of the vortex row decreases with increasing distance between the vortex chains that form the row.

The vortex system evolution is described by the Hasegawa-Mima equation [31]. In Ref. [30] it was shown that symmetric vortex raw is always unstable, but the antisymmetric raw can be stable similarly to the Von Karman vortex raw in Eulerian fluids [32].

In an inhomogeneous plasma the vortices move along the lines of constant density according to the Hasegawa - Mima equation. This is a consequence of the Eurthels theorem. When the laser pulse interacts with the finite length plasma slab, the magnetic field advected by the electron vortices appears at the rear side of the slab in the form of a dipole. Depending on the length scale of the plasma inhomogeneity the vortex pair either propagates in the longitudinal direction in a weakly inhomogeneous case or it breaks into two separated vortices in the case of a steep plasma boundary, and each of them moves in the opposite direction in the transverse direction.

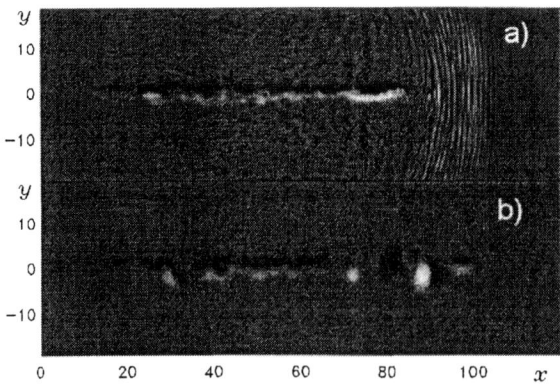

FIGURE 4. The z-component of the magnetic field at t = 120 $\times 2\pi/\omega$ (a), and at t=210$\times 2\pi/\omega$ (b).

In Fig. 5 we present the results of 2D PIC simulations of the laser pulse interaction with a slab of an underdense plasma. The laser pulse with the length 7.5λ is focused on the slab of a gradually inhomogeneous plasma (maximal density equals $0.5n_{cr}$) preceded by the low density preplasma with $n=0.1n_{cr}$. The pulse is focused on the left boundary of the plasma slab. At the focus the pulse amplitude is equal to $a=10$.

In the distribution of the ion density shown in Fig. 5a we see the self-focusing channel, the postsolitons inside the channel and aside it, and large void regions at the rare side of the slab. Large void regions here are made by the pressure of the magnetic field, that pushes plasma ions outward. Near the axis we see a plasma filament carrying the electric current in the layer between the regions with opposite sign magnetic field.

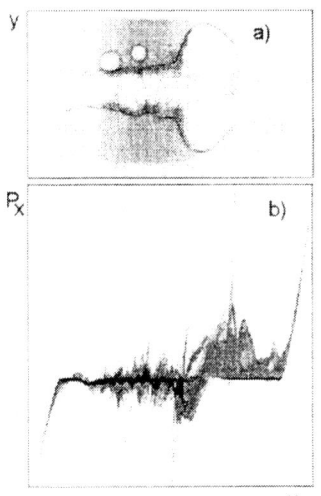

In the phase plane (Fig. 5 b) we see the ions accelerated in the backward direction, the ions accelerated in radial direction inside the postsolitons, the ions accelerated in the forward direction at the right hand side plasma vacuum interface. However, the ions accelerated in the electric current carrying filament acquire largest energy. It is about ≈ 20 MeV. The mechanism of the ion acceleration is connected with the disrupture of the filament at t ≈ 130×2π/ω.

FIGURE 5. The ion density distribution in the x,y plane (a) and the ion phase plane (p_x,x) (b) at t = 170×2π/ω.

ION ACCELERATION DURING INTERACTION OF THE LASER PULSE WITH THIN FOIL

As it has been mention above, effective ion acceleration requires lase radiation in the petawatt power range. In the simplest scheme of ion acceleration, the laser pulse interacts with a thin foil. When multiterawatt laser radiation interacts with a foil, matter is ionized in an interval shorter than a single optic oscillation period of the laser radiation, producing in such a way a collisionless plasma. Under the action of the laser radiation the electrons are expelled from a region on the foil with

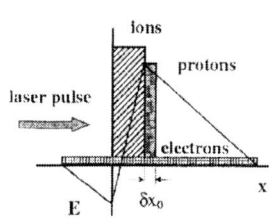

FIGURE 6. Scheme of a two-layer target. Density distribution of heavy ions, protons, and electrons and dependence of the electric field on the x coordinate

transverse size of the order of the diameter of the focal spot. For femtosecond long laser pulses with multiterawatt power, the typical time scale of the hydrodynamic expansion of a micron plasma slab is much longer than the laser pulse duration. Under these conditions ions remain at rest, which results in the formation of a positively charged layer of ions. However after a time interval equal to or longer than the inverse of the ion Langmuir frequency $\omega_{pi}^{-1}=(4\pi n_0 Z_i^2 e^2/m_i)^{-1/2}$ the ion layer explodes because of the Coulomb repulsion of electric charges with equal sign.

However if we compare the form of the energy spectra of the accelerated ions observed in the experiments or obtained in numerical simulations with the narrow energy spectrum required for various applications in controlled nuclear fusion, for the injectors, and in hadrontherapy, we see that the energy spectra of the laser accelerated ions are at present quite far from the required ones. It was found that the ion spectra below a maximum energy can be approximated by a quasi-thermal distribution with an effective temperature several times smaller than the maximal ion energy.

In order to improve the proton beam quality one can cut the beam into beamlets with a narrow spread in energy space. However, in this case the efficiency of the laser energy transformation into the energy of fast particles decreases significantly, and, what is more important, the number of fast particles decreases. A different approach which seems more promising, is connected with the use of multi-layer targets. In this scheme a thin foil is used as a target and its rear surface is coated with a thin hydrogen layer. When the ultra short laser pulse irradiates the target, the heavy atoms are partly ionized and the ionized electrons abandon the foil, generating an electric field due to charge separation. Because of their inertia the heavy ions remain at rest, while the lighter protons are accelerated.

A sketch of a double-layer target is shown in Fig.6. The first layer is made of heavy ions i.e., of ions with a sufficiently large value of μ/Z, where $\mu=m_i/m_p$, as required in order to respect the condition of their remain at rest during interaction with the laser pulse.

In addition, the target must be sufficiently thick so as to give a large enough electric field due to charge separation. This electric field has opposite sign on the two different sides of the target, has a zero inside the target and vanishes at a finite distance from it, as illustrated in Fig. 6. Since the number of protons is assumed to be sufficiently small not to produce any significant effect on the electric field, their behavior can be described within the framework of the test particle approximation. In addition, the transverse size of the proton layer must be smaller than the pulse waist. This is required in order to decrease the influence of the laser pulse inhomogeneity in the direction perpendicular to its direction of propagation. The effect of the finite waist of the laser pulse leads also to an undesirable defocusing of the fast ion beam. In order to compensate for this effect and to focus the ion beam we can use the properly deformed targets, as suggested in Refs.[8,28,33].

In Fig. 7 we show the results of numerical simulations of the proton acceleration during the interaction of a short, high power laser pulse with a two-layer target. Initially the target consists of two neutral plasma layers. The electron density in the heavy ion layer corresponds to the ratio $\omega_{pe}/\omega=3.0$ between the plasma and the laser frequencies. The second layer is localized at the rear side of the target and its electron density is smaller than the critical density and corresponds to $\omega_{pe}/\omega=0.7$. The size of the first layer is $0.5\lambda\times10\lambda$. The proton layer has size $0.1\lambda\times5\lambda$. A circularly polarized laser pulse with dimensionless amplitude $a=20$ a size $20\lambda\times15\lambda$. In Fig. 7a we present the energy spectrum of the protons at time $t=45\times2\pi/\omega$ and of the heavy ions at time t=97.5×2π/ω with the energy given per nucleon. In Fig. 7 b we show the electric charge distribution in the x,y plane at t=27×2π/ω. The proton layer moves in the x direction and that the distance between the two layers increases.

The electric field generated by the electric charge separation appears to be strong enough to accelerate the protons up. The energy per nucleon acquired by the heavy ion component is approximately ten times smaller than the proton energy. As seen in Fig. 7b, the heavy ion components have a wide energy spectrum while the protons form a quasi-mono-energetic bunch. The relative energy spread is approximately ΔE/E=0.1.

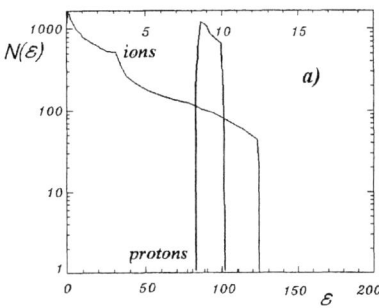
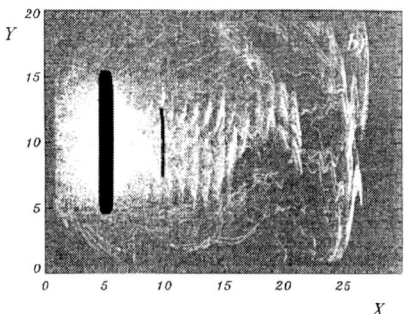

FIGURE 7. The proton and the heavy ion energy spectra (a); distribution of the electric charge density in the x,y plane

ION ACCELERATION DURING RECONNECTION OF MAGNETIC FIELD LINES IN SUCCESSIVE COALESCENCE OF FILAMENT CURRENTS

The interaction of plasma flows is accompanied by the generation of strong electric and magnetic fields. Even if both the plasma charge and the net plasma current vanish almost completely in the initial state, strong electric and magnetic fields arise due to the onset of electromagnetic instabilities. These instabilities are similar in character to the well-known Weibel instability [34] and are caused by the anisotropy of the particle distribution function in velocity [35,36]. This instability was incorporated in developing the theory of the generation of quasi-static magnetic fields in laser plasmas, whose anisotropy is related to the presence of high-energy electron flows.

In Refs. [37] it was investigated the collective electron beam stopping and ion heating in relativistic-electron-beam transport. They showed the nonlinear dynamics of coalescence of the electric current filaments, which resembles the coalescence of magnetic islands well known in the physics of magnetically confined plasmas [38]. In turn this process corresponds to the generic properties of the laboratory and space plasma dynamics in the magnetic field known as the reconnection of the magnetic field lines (see Refs. [38-40]). Here we notice that the importance of the magnetic reconnection for the laser plasma dynamics in the high intensity laser radiation regime has been forecasted in Ref. [41]. Recently the reconnection of the magnetic field lines

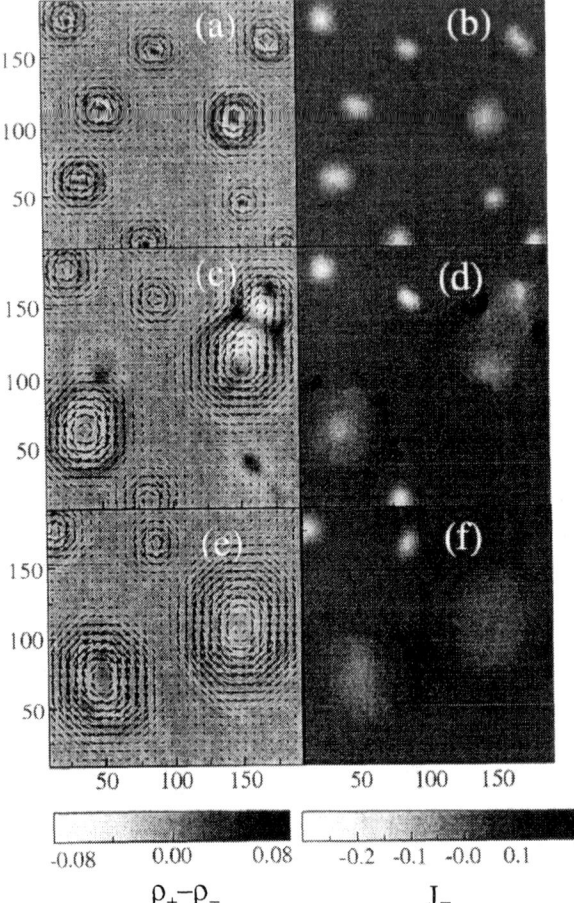

$\rho_+ - \rho_-$ J_z

in laser plasmas has been studied in Refs [42-46].

Here we present the results of 2D PIC simulations of the electromagnetic filamentation instability development in the configuration with two counter-streaming electron components moving initially along the z-axis in opposite direction with different velocities and densities. The velocities are $v_{e1}=0.9c$, $v_{e2}=-0.1c$ and the densities are $n_{e1}=0.1$ and $n_{e2}=0.9$. Thus net electric current in the system equals zero.

In the early stage of the electromagnetic filamentation instability there occur many small scale current filaments whose scale-length is of the order of the electron skin depth. Then the generated small scale current filaments start to merge and grow to large scale current filaments.

FIGURE 8. Distribution of the electric charge density and the magnetic field produced from current filaments (left hand side column) and distribution of the electric current density (right hand side column) at (I) $\omega_{pe}t=190$, (II) $\omega_{pe}t=885$, and (III) $\omega_{pe}t=1580$.

The time evolution of this process is shown in Fig. 8 with first row for $\omega_{pe}t=190$, second row for $\omega_{pe}t=885$, and third row for $\omega_{pe}t=1580$.

Here in the left hand side column the vector plots of magnetic fields B_x B_y produced by the current filaments with the distribution of the electric charge density (in grey colour) are presented. In the right hand side column of Fig. 8 we see the distribution of the z-component of the electric current density J_z. Each electric current filament has a core current (negative in the z-direction) and a shell return current (positive in the z-direction) surrounding the core current.

Fig. 9 shows the time history of (a) the electric field energy and (b) the ion kinetic energy in the system. In Fig. 9 (b) we see that the ion kinetic energy in the x,y plane increases through three successive steps. The first step of the ion acceleration continues till about $\omega_{pe}t=300$. During this first step the electromagnetic filament instability develops to the nonlinear stage where the small scale filaments merge and grow to large filaments shown in Fig.8. The ion acceleration occurs in a ring-shaped region corresponding to the electric charge separation of each filament.

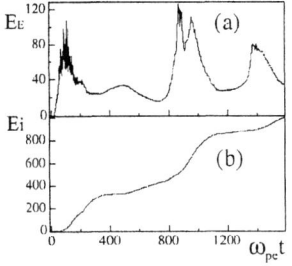

FIGURE 9 Time history of (a) electric field energy and (b) ion kinetic energy.

At the second step of the ion acceleration, it lasts from $\omega_{pe}t=900$ to $\omega_{pe}t=1000$, a strong electric field appears just after the coalescence of the large filaments. At the third step a strong charge separation appears near the magnetic reconnection region between two current filaments. In Fig.10(a) the ion kinetic energy in the x-y plane is shown.

The magnetic reconnection is triggered by the attraction force of two core electron currents, which merge into one core. During the magnetic reconnection the electrons in the current sheet are pushed away perpendicular to the direction of the core current coalescence, while the ions remain at rest during the process. Therefore strong charge separation is generated. This strong charge separation near the magnetic reconnection region results in the formation of strong electric fields mostly perpendicular to the local magnetic field. As seen in Fig. 10, ions are strongly accelerated with formation of a ring structure.

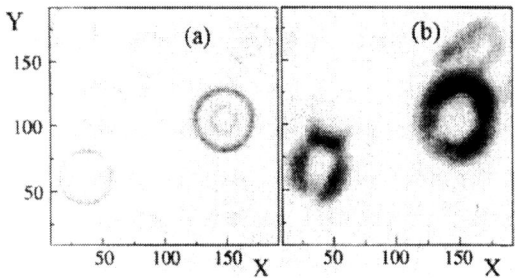

FIGURE 10. Ion kinetic energy normalized by the initial ion thermal energy in the x-y plane at (a) $\omega_{pe}t=250$ and (b) $\omega_{pe}t==1030$.

RELATIVISTIC WHISTLE: HIGH HARMONIC GENERATION BY ULTRA-INTENSE LASER PULSE INSIDE HOLLOW FIBER

Generation of high order harmonics of the electromagnetic radiation during interaction of high-intensity laser pulses with underdense and overdense plasmas presents a manifestation of one of the most basic nonlinear processes in physics. In the underdense plasmas the high garmonics are produced by the mechanism of the parametric excitation by the laser light of the electromagnetic and electrostatic waves with different frequencies. When the laser radiation interacts with the overdense plasmas it reflects back at the plasma-vacuum interface in the case of sharp plasma boundary or at the surface of critical density in the case of gradual density profile. The reflection layer of the plasma dragged by the electromagnetic wave back and forth as well as in the plane of the surface of the plasma-vacuum interface (in the plane of the critical surface) forms the oscillating mirror (see Refs. [47-52]). The spectrum of the reflected at the oscillating mirror light contains odd and even harmonics, which polarization and amplitude depend on the incidence angle of the pulse, its intensity and the pulse polarization.

High efficiency of the laser energy transformation into the energy of the high harmonic radiation can be achieved when the laser pulse propagates inside a hollow channel [53]. In this case for sufficiently long channel we have actually a multiple reflections of the laser pulse at the vacuum-plasma interface with a transformation of a portion of the laser energy into high harmonics at each reflection event.

If the laser pulse interacts with a fiber, i.e. when it propagates inside a hollow, thin wall channel, in addition, it becomes possible to extract the high frequency radiation from the fiber in the form of the wave, which propagates aside at some angle with respect to the fiber axis. Since only harmonics with the number above critical value, for which the fiber walls are transparent, propagate outwards, it provides an approach to make the high frequency radiation source with controlled properties, changing the fiber diameter and the wall thickness. We call the fiber in the regime of the laser energy conversion to the energy of high harmonic radiation "the relativistic whistle".

Within 2D approximation the fiber corresponds to the configuration made by two finite length thin foils parallel to the x-axis in the x,y plane. The wall thickness of the foil is l with a distance between the foils equal to L. We consider the foils comprised collisionless overcritical plasmas. assuming the ions to be at rest.

From the dispersion equation, $\omega^2 = k^2 c^2$, where ω is the wave frequency and $k^2 = k_x^2 + k_y^2$, we find the group velocity of the wave inside the channel to be equal to $v_{gr} = \partial\omega / \partial k_x = c(\omega^2 - \pi^2/L^2)^{1/2}/\omega$. This expression can be written in the form $v_{gr} = c \cos \theta$, where $\tan\theta = k_y/k_x = 1/(L^2/\lambda^2 - 1)^{1/2}$.

If we perform the Lorenz transformation into the reference frame moving with the group velocity along the x direction, the x component of the wave vector vanishes , $k'_x = 0$. We see that the high frequency waves excited by the fundamental mode at the plasma-boundary interface also have in this frame zero x component of the wave vector, i.e. propagate outwards in the normal direction to the fiber walls independently of the harmonic number. Performing the Lorenz transformation into the laboratory frame, we obtain that the waves radiated outwards propagate at the same angle θ with respect to the channel axis independently of the wave frequency. Their energy is localized inside of a slab co-moving along the x axis with the laser pulse. The selection rules of the harmonic generation at each wall (see Refs. [48-50]) show that, if the laser pulse forms inside the channel the s-polarized fundamental mode, the generated high harmonic radiation is s-polarized, if their frequency factor n is an odd number, and p-polarized for even n. The p-polarized fundamental mode generates both n-odd and n-even p-polarized high harmonic radiation.

The efficiency of the transformation of the fundamental mode into the n-th harmonic can be obtained within the framework of the thin foil approximation for the channel walls (see Ref. [50]). It is $(a_0 sin\theta/\varepsilon)^n$. Here $a_0 = eE/m_e c\omega$ is the dimensionless amplitude of the fundamental mode, $\varepsilon = 2\pi n_e e^2 l / m_e c\omega$ is the dimensionless parameter, which characterizes the wall thickness, and n_e is the electron density.

Since the problem of high harmonic generation requires substantially high spatial resolution, in PIC simulations we used the computation mesh with 50 cells per λ. The channel walls have a length $l_{ch}=25\lambda$ and width 2λ with a distance between the walls 0.7 λ. The electron density is equal to $8n_{cr}$. For the chosen parameters of the channel walls only high harmonics with the number higher than 4 can propagate through the walls. This condition follows from the boundary conditions for the electromagnetic wave at the plasma-vacuum interface for the oblique incidence at the angle θ. The s-polarized laser has been initialized in a vacuum region at the left hand side from the channel entry. The pulse aperture is f/1, the length equal 7.5λ and the dimensionless amplitude at the 1λ focus spot $a=3$, which corresponds to the intensity $I \approx 10^{19} Wcm^{-2}$. We notice here that the pulse focusing into the 1λ focus spot has been discussed in Ref. [6]. In our simulation model the entry of the channel has been smoothed to decrease the laser pulse reflection here as it is shown in Fig. 11. In this figure we present the distribution of the electron density in the channel walls and the distribution of the electromagnetic energy density of the laser pulse focused at the channel entry at $t=15\times2\pi/\omega$. We see that the smoothing of the channel entry provides almost reflection-less matching of the laser pulse and the channel.

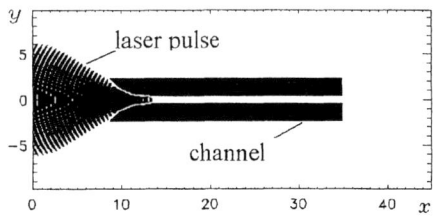

FIGURE 11. Distribution of the electron density in the channel walls and the distribution of the electromagnetic energy density of the laser pulse focused at the channel entry at $t=15 \times 2\pi/\omega$.

When the laser pulse propagates inside the channel a portion of its energy is transformed into the energy of high harmonic waves radiated outwards; the remain part of the laser pulse propagates through the channel toward the right hand side vacuum region, where it forms a diverging electromagnetic wave.

In Fig. 12 (a) we show the distribution of the z-component of the electric field in the x,y-plane at $t=45\times2\pi/\omega$. This figure presents the visualization of odd-number high harmonics, which, as it has been noticed above, have the s-polarization. We see the laser pulse inside the channel and the high harmonics aside the channel. The high harmonic region has a length along the x-axis equal to the laser pulse length and the length along the y-axis equal to $c\delta t \sin\theta$, where $\sin\theta \approx 1/2$ and $\delta t \approx 20\times2\pi/\omega$.

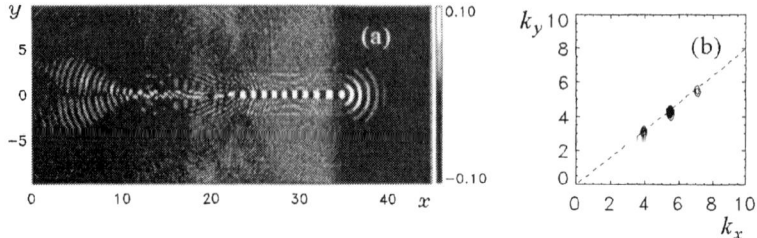

FIGURE 12. Distribution of the z-component of the electric field in the x,y-plane at t=45×2π/ω (a); spectrum of the s-polarized light at t=30×2π/ω (dashed line corresponds to k_y = 0.8 k_x) (b).

We performed the spatial Fourier transform of the z-component of the electric field inside a subdomain 18λ<x<28λ, 3λ<y<10λ at t=30×2π/ω. We show the spectrum of the s-polarized light inside this subdomain in Fig. 12 (b). We see the odd-number harmonics with the harmonic number n≥5. We notice that the discrete points in the k_x,k_y plane lie along the straight line given by expression k_y ≈ 0.8 k_x, i.e. close to the propagation angle.

In Fig. 13 (a) we present the distribution of the z-component of the magnetic field in the x,y-plane at t=45×2π/ω to show the even-number high harmonics. The even-number harmonics have the p-polarization. The fundamental mode of the laser pulse inside the channel does not contain the p-polarized component and we see only the high harmonics inside and aside the channel, as well as the low frequency surface mode in the vicinity of the outer vacuum-plasma interfaces on both sides of the channel. The even-number high harmonic outside the channel are localized in the same region as the odd-number harmonics.

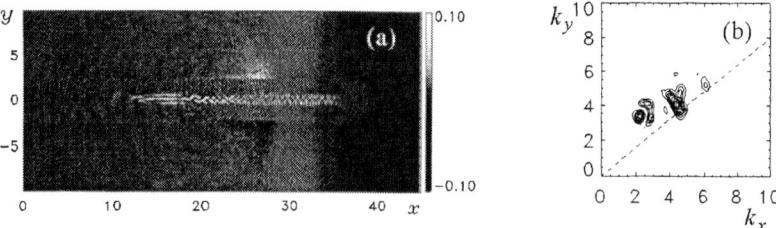

FIGURE 13. Distribution of the z-component of the magnetic field in the x,y-plane at t=45×2π/ω (a); spectrum of the s-polarized light at t=25×2π/ω (dashed line corresponds to k_y = 0.8 k_x) (b).

The spectrum of the even-number high harmonics calculated for the field inside a subdomain 16λ<x<26λ, 3λ<y<10λ at t=25×2π/ω is shown in Fig. 13 (b). We see the harmonics with even numbers n=4, n=6, and n=8. The discrete points in the k_x,k_y plane also lie along the straight line given by expression k_y ≈ 0.8 k_x.

REFERENCES

1. Bulanov,S.V., et al., "Relativistic Interaction of Laser Pulses with Plasmas", in *Reviews of Plasma Physics*, edited by V.D.Shafranov, Kluwer Acad./ Plenum Publ., New York, 2001, Vol. 22., p. 227

2. Kuznetsov,A.V., et al., *Plasma Phys. Rep.* **27**, 211 (2001)

3. Roth, M., et al., *Phys. Rev. Lett.* **86**, 436 (2001); Bychenkov,V.Yu., et al., *Plasma Phys. Rep.* **27**,(2001)

4. Krushelnik, K., et al., *IEEE Trans. Plasma Science* **28**, 1184 (2000)

5. Bulanov,S.V., Khoroshkov,V.S., *Plasma Phys. Rep.* **28**,(2002)

6. Mourou, G., et al., *Plasma Phys. Rep.* **28**,(2002)

7. Bulanov,S.V., et al., *Plasma Phys. Rep.* **25**, 701 (1999)

8. Bulanov,S.V., et al., *JETP Lett.* **71**, 407 (2000)

9. Sentoku,Y., et al., *Phys. Rev. E* **62**, 7271 (2000)

10. Askar'yan,G.A., *Sov. Phys. JETP* **15**, 8 (1962); *Sov. Phys. Usp.* **16**, 680 (1973)

11. Zakharov,V.E., *Sov. Phys. JETP* **35**, 908 (1972)

12. Esirkepov,T.Zh, *Comput. Phys. Comm.* **135**, 144 (2001)

13. Honda,T., et al., *J. Plasma Fusion Research* **75**, NO10-CD, 219 (1999)

14. Pukhov,A., Meyer-ter-Vehn,J. *Phys. Rev. Lett.* **76**, 3975 (1996)

15. Naumova,N.M., et al., *Phys. Plasmas* **8**, 4149 (2001)

16. Askar'yan,G.A., et al., *JETP Lett.* **60**, 251 (1994)

17. Borghesi,M., et al., *Phys. Rev. Lett.* **79**, 2686 (1997)

18. Esirkepov,T.Zh., et al., *JETP Lett.* **70**, 82 (1999)

19. Naumova,N.M., et al., *Phys. Rev. Lett.,* **87**, 185004 (2001)

20. Borghesi,M., et al., submitted to *Phys. Rev. Lett.* (2001)

21. Esirkepov,T.Zh., et al., in preparation.

22. Bulanov,S.V., et al., *Phys. Fluids B* **4**, 1935 (1992); Bulanov,S.V., et al., *Plasma Phys. Rep.* **21**, 600 (1995)

23. Esirkepov,T.Zh., et al., *JETP Lett.* **68**, 36 (1998)

24. Bulanov,S.V., et al., *Phys. Rev. Lett.* **82**, 3440 (1999)

25. Farina,D., et al., *Phys. Rev. E* **62**, 4146 (2000)

26. Farina,D., Bulanov,S.V., *Phys. Rev. Lett.* **86**, 5289 (2001); *Plasma Phys. Rep.* **27**, 680 (2001)

27. Lontano,M., et al., *Phys. Plasmas* **8**, 5113 (2001)

28. Sentoku,Y., et al., *Phys. Rev. Lett.* **83**, 3434 (1999)

29. Bulanov, S.V., et al., *Phys. Rev. Lett.* **74**, 710 (1995)

30. Bulanov,S.V., et al., *Phys. Rev. Lett.* **76**, 3562 (1996); Bulanov, S.V., et al., *Plasma Phys. Rep.* **23**, 660 (1997)

31. Hasegawa,A., Mima,K., *Phys. Rev. Lett.* **39**, 205 (1977)

32. Lamb,H., *Hydrodynamics* (Cambridge University Press, Cambridge, 1932)

33. Ruhl,H., et al., *Plasma Phys. Rep.* **27**, 411 (2001)

34. Weibel,E.W. *Phys. Rev. Lett.* **2**, 83 (1959)

35. Bychenkov,V.Yu., et al., *Sov. Phys. JETP* **71**, 709 (1990)

36. Califano,F., et al., *Phys. Rev. E* **57**, 7048 (1998); *Phys. Rev. E* **58**, 7837 (1998)

37. Honda,M., et al., *Phys. Rev. Lett.* **85**, 2128 (2000); *Phys. Plasmas* **7**, 1302 (2001)

38. Biskamp,D., *Nonlinear Magnetohydrodynamics* (Cambridge Univ. Press, Cambridge: 1993)

39. Berezinskij,V.S., Bulanov,S.V., Dogiel,V.A., Ginzburg,V.L., Ptushkin,V.S., *Astrophysics of Cosmic Rays* (North-Holland, Amsterdam, 1990)
40. Tajima,T., Shibata,K., *Plasma Astrophysics* (Addison-Wesley, Reading, 1997)
41. Askar'yan,G.A., et al., *Comments Plasma Phys. Controll. Fusion* **17**, 35 (1996)
42. Kazimura,Y., et al., *Plasma Phys. Rep.* **27**, 330 (2001)
43. Taguchi,T., et al., *Phys. Rev. Lett.* **86**, 5055 (2001)
44. Califano,F., et al., *Phys. Rev. Lett.* **86**, 5293 (2001)
45. Sakai,J.I., et al., in: Proceedings of ISSS-6 (2001)
46. Sakai,J.I, et al., in preparation
47. Bulanov,S.V., et al., *Phys. Plasmas* **1**, 745 (1994)
48. Lichters,R., et al., *Phys. Plasmas* **3**, 3425 (1996)
49. Von der Linde,D., AIP Conf. Proc. **426**, 221 (1997)
50. Vshivkov,V.A., et al., *Phys. Plasmas* **5**, 2752 (1998)
51. Zepf,M., et al., *Phys. Rev. E* **58**, 5253 (1998)
52. Tarasevitch,A., et al., *Phys. Rev. E* **62**, 023816-1 (2000)
53. Bulanov,S.V., et al., *Physics Letters A* **195**, 84 (1994)

Relativistic Channel Formation with Different Pulse Durations

A. Sjögren*, J. Mauritsson*, C. Delfin*, V. Lokhnygin*, C.-G. Wahlström*,
A. Pukhov† and G. D. Tsakiris†

*Department of Physics, Lund Institute of Technology, P. O. Box 118, S- 221 00 Lund, Sweden
†Max-Planck-Institut für Quantenoptik, Hans-Kopfermann-Strasse 1, 85748 Garching, Germany

Abstract. Relativistic channels are generated in an under-dense plasma by means of sub-picosecond laser pulses. The channel extension is characterized via the Thomson-scattered light and the impact of the laser pulse duration and chirp are studied. It is shown that, under certain conditions, the channel extends further when the pulse duration is increased. This is in contrast to many earlier reports, where the channel grows with increasing laser power. Some theoretical arguments are presented to support the experimental findings. The number of electrons, accelerated in the forward direction, and their energies are assessed. In some cases, it is found that the channel divides and the fragments extend in different directions. This raises the question if, and how, the channel formation can be externally manipulated.

INTRODUCTION

The possibility to form long and tight light channels in which the light intensity is very high ($> 10^{18}$ W/cm^2) appeals to scientists in many fields. The connection to various X-ray laser schemes and laser-driven particle acceleration is obvious [1, 2, 3].

When a gas with low atomic number is exposed to a short, high-intensity laser pulse the gas is completely ionized by the leading edge of the pulse, and the following interaction is to a large extent governed by the plasma frequency. At very high intensities, the velocities of the wiggling electrons can approach the speed of light. Hence, the mean electron mass becomes a function of the laser intensity. The electron mass couples the distribution of laser intensity to the refractive index distribution in the plasma. If the laser power is high enough, the effect of this nonlinear refractive index counteracts diffraction and a relativistic channel can form [4]. For a given electron density in the plasma there exist a critical peak laser power required for relativistic self-focusing. This critical power is given by $P_c \approx 17 \cdot (\omega_0/\omega_p)^2$ [GW], where ω_p is the plasma frequency of the ambient plasma and ω_0 is the laser frequency [5]. The ambient plasma frequency, in turn, is proportional to the square root of the electron density, $\omega_p \propto \sqrt{n_e}$. Relativistic channels have been observed in many experiments [4, 6, 7].

The rapid increase in peak power and compactness of table top laser systems based on the chirped-pulse amplification (CPA) technique allows a growing number of laboratories to realize experiments in the relativistic regime. Partly, the compactness of these high-power lasers comes about by a decrease in the laser pulse energy. The peak power is retained by a corresponding shortening of the pulse duration [8, 9]. However, as will

CP611, *Superstrong Fields in Plasmas:* Second Int'l. Conf., edited by M. Lontano et al.
© 2002 American Institute of Physics 0-7354-0057-1/02/$19.00

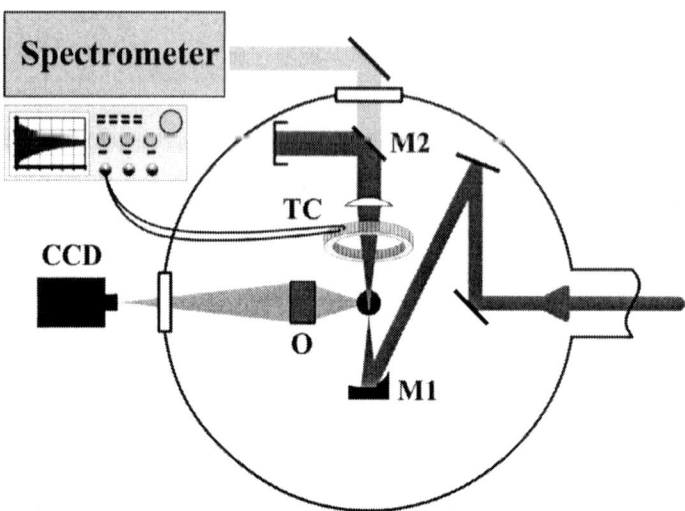

FIGURE 1. The experimental set-up with the side-scattering imaging system and the diagnostics for forward Raman scattered light. The laser enters from the right and is focused by an off-axis parabolic mirror, M1, into the gas-jet. The channels formed are imaged onto a CCD chip using an objective, O. After the gas jet, laser radiation at the fundamental laser wavelength is reflected by a dielectric mirror, M2, to a beam-dump, while forward Raman scattered light is transmitted through M2 to a spectrometer outside the vacuum chamber. The toroidal coil, TC, is used to measure the total charge of the electrons passing through.

be discussed below, if as long relativistic channels as possible are in demand, this is not necessarily the best way to go. We present here systematic experimental studies on how the laser pulse duration influences the extension of relativistic channels.

EXPERIMENTAL SET-UP AND METHODS

The experiments are realized through use of the 10 Hz multi-terawatt, femto-second laser at the Lund Laser Centre. It is a titanium-sapphire system based on the CPA technique and operates at 800 nm. The maximum energy after compression is up to 1 J.

By translating one of the gratings in the compressor, the pulse duration can be varied, 30 fs being the minimum. The pulse duration for each grating position is measured with a second-order auto correlator. The horizontally polarized laser beam is focused by a f/3, silver-coated, off-axis parabolic mirror. The laser focus is imaged and studied with an eight-bit dynamic-range beam profiler. The focal spot radius is roughly $r_0 = 3\,\mu$m at $1/e^2$ of the peak intensity, corresponding to a Rayleigh range of approximately 35 μm. The peak intensity is estimated to be a few times 10^{19} W/cm^2.

The experimental set-up is illustrated in Figure 1. The laser is focused 200 μm below

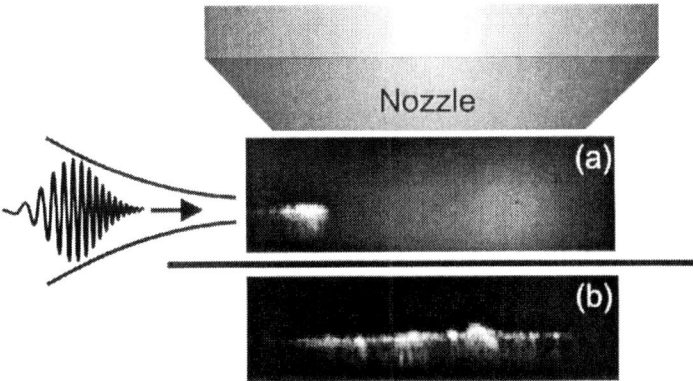

FIGURE 2. (a) The onset of relativistic channel formation. The electron density is $7 \cdot 10^{19}\,\mathrm{cm}^{-3}$, the pulse duration is 30 fs, and the energy is 50 mJ. (b) A relativistic channel that has grown through the gas medium. The length of the channel, 750 μm, equals 21 Rayleigh ranges. The electron density is $7 \cdot 10^{19}\,\mathrm{cm}^{-3}$, the pulse duration is 65 fs, and the energy is 120 mJ. The diameter of the gas orifice is in both cases 0.8 mm. For clarity the chirp of the pulse is strongly exaggerated.

the orifice of a pulsed, sonic gas jet. The gas orifice is interchangeable and in these experiments 0.5 mm or 0.8 mm opening diameters are used. The valve is backed with helium gas of up to 50 bar pressure for the smaller orifice and 70 bar for the larger. The opening time of the valve is a few milliseconds.

The electron density in the jet is assessed by measuring the frequency of the first Stokes component of the forward Raman-scattered light (generated in the laser-plasma interaction). Through a careful study of the data, the electron density is calibrated against the backing pressure. The electron density, ranging from $10^{19}\,\mathrm{cm}^{-3}$ to $10^{20}\,\mathrm{cm}^{-3}$, is found to be linearly proportional to the backing pressure. However, the larger orifice requires a higher backing pressure to reach a certain density.

The Thomson scattered light and the plasma recombination light is imaged at 90° to the propagation direction, using a 13.5 cm focal length objective. The resolution is estimated to 5 μm, so it is not possible to differentiate between whole beam self focusing and the potential formation of several parallell filaments. An interference filter, transmitting only close to 800 nm, was introduced to verify that most of the recorded light originates from the process of Thomson scattering. The measured length is therefore unaffected, to within a few per cent, by the presence of the plasma light. The width of the imaged channel, however, shrinks considerably with the interference filter in place. This is expected, because during the camera exposure, the plasma rapidly expands while it moves with a speed equal to that of the gas jet. The onset of channel formation, imaged without any filter, is shown in Figure 2(a). A typical image of a long channel is shown in Figure 2(b). Here neutral density filters are required to reduce the light intensity by several orders of magnitude.

A toroidally wound, air-filled coil with twenty turns is used to measure the total charge of the 'bunch' of electrons that are forwardly accelerated in the channel. The diameter of the coil is 15 cm and it is positioned 10 cm from the gas nozzle. The coil is connected to an oscilloscope via a coaxial cable. A bunch of electrons passing through the coil induces an exponentially decreasing resonance oscillation in the circuit. The amplitude of the oscillation is used to estimated the total charge of the electron bunch.

EXPERIMENTAL RESULTS AND DISCUSSION

In our experiments, which are partly presented elsewhere [10], it is found that the relativistic channels grow longer when increasing the ratio P/P_c. This is in agreement with previous reports [4, 6]. This is obtained if the pulse energy is increased (higher P) or if the electron density is increased (lower P_c). In the latter case, the channel extension saturates for high densities, which Gahn *et al.* attribute to the energy losses suffered by the pulse as it heats the electrons [6]. If, indeed, the energy losses are governed by electron heating, a few simple arguments can be brought forward. The losses are consequently proportional to the density of electrons, and to the mean energy of the heated electrons, which in this intensity regime is proportional to the square root of the intensity.

It is also observed in these experiments that the channel extends further when stretching the laser pulse in time, while adjusting the pulse energy to maintain a constant power. In effect, the ration P/P_c, is kept constant. With the energy loss mechanism above, this can be expected since, while keeping the laser peak power, and intensity, constant, more energy is available in the longer laser pulse. Therefore the pulse can propagate a longer distance before its energy is depleted.

The experiment is taken one step further, and all parameters are kept constant, except the pulse duration. The duration is increased, and it is found that the channel still grows, as illustrated in Figure 3. This result shows that, in fact, laser pulses of *lower* peak power can generate *longer* channels. The pulse lengthening corresponds to a decrease in the ratio P/P_c from 7.0 to 1.1. This behavior is studied for a wide parameter space. In particular, the growth is confirmed to be independent of the sign of the chirp of the stretched pulse. This result can also be understood in terms of energy losses due to electron heating. The reason is that as the intensity is decreased, less energy is deposited per unit length of the channel, and the laser pulse can penetrate a longer distance into the plasma. Therefore, if the peak power is much higher than the critical power, a stretching of the laser pulse will result in a longer channel. However, at every instant, the condition $P > P_c$ must be uphold, so if the laser pulse is stretched too much, channelling can not continue. For a given laser energy and electron density, and optimum pulse duration is therefore to be expected.

In the light of certain future applications, it is quite interesting to note that, in rare cases, during the experiments, the imaged Thomson scattered light reveals channel division. The division, of which examples are shown in Figure 4, seem to take place at random positions along the channel. This division phenomenon can be seen as an artefact in the experiments, and their origin might be the existence of sharp gas density gradients in the gas jet or an abnormal intensity or phase distribution in the cross section of the laser pulse. Presently, the process is neither fully understood, nor wanted. However, if the phenomenon can be understood and relativistic channels can be bent, split or combined

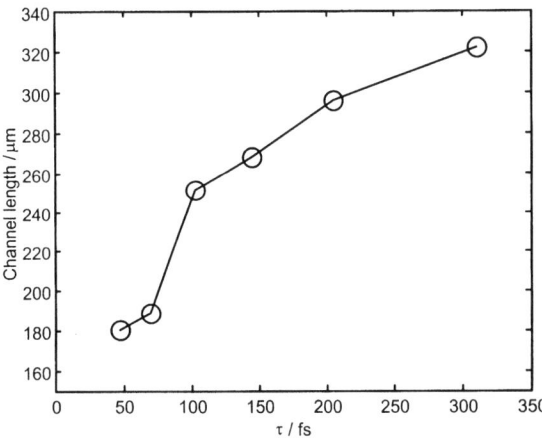

FIGURE 3. The length of the channel is increasing with increasing laser pulse duration. The electron density and energy are held constant, $n_e = 1 \cdot 10^{20}\,\text{cm}^{-3}$ and $E_{laser} = 0.1$ J, so the increase in pulse duration corresponds to a decrease in P/P_c from 7.0 to 1.1. The diameter of the orifice is 0.5 mm.

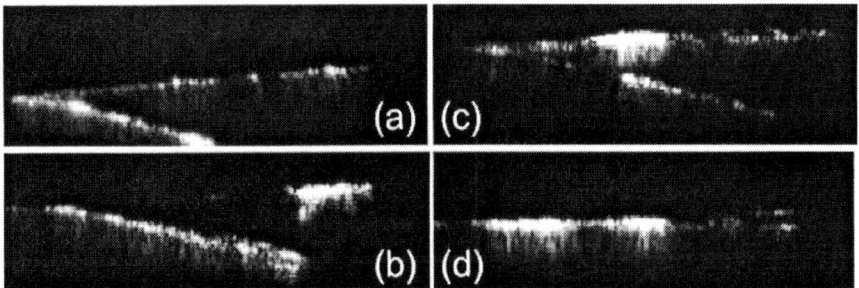

FIGURE 4. Channel division taking place at different positions along the propagation axis. The pulse duration is 90 fs (a), 65 fs (b,c), and 30 fs (d), the energy is 0.16 J (a), 0.20 J (b, c and d), and the electron density is $7 \cdot 10^{19}\,\text{cm}^{-3}$ in all the images. The diameter of the orifice is 0.8 mm.

at will, the implications can be of great importance to many kinds of activities in the area of relativistic self focusing.

A small fraction of the laser pulse energy is converted into electron kinetic energy in the forward direction. The acceleration is due to mechanisms such as direct laser acceleration (DLA) and self-modulated wake-field acceleration (SMWFA) [6, 11]. A simple electron spectrometer, consisting of an electron beam aperture, a permanent magnet and a scintillating plate, is employed to study these electrons. The measurements show that the electron kinetic energy spectrum extends well above 10 MeV. Because the channel is less than one millimeter long, the accelerating fields must be greater

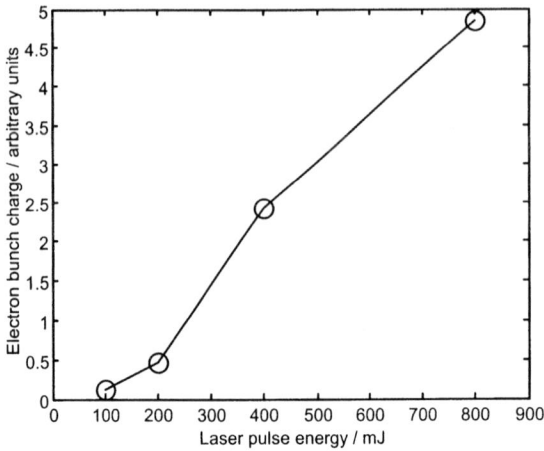

FIGURE 5. The number of electrons increase with laser energy. The electron density is $n_e = 5 \cdot 10^{19}\,\text{cm}^{-3}$ and the pulse duration is 50 fs. The orifice diameter is 0.8 mm.

than 100 MeV/cm. Letting the beam pass through the toroidally wound coil allows an estimation of the total charge of the electron bunch. The maximum bunch charge measured is greater than 10 nC. For increasing laser pulse energy the total charge increases, as illustrated in Figure 5, and so does the channel length. In Figure 6 the total charge is plotted for different laser pulse durations. It is found that longer pulses (longer channels) in this case accelerate less total bunch charge. Further experiments are planned in order to study the electron energy spectrum and bunch charge as a function of the channel length and the sign of the chirp.

Summarizing this discussion, the experiments reveal that laser pulses with reduced peak power, obtained by temporal stretching, can generate longer relativistic channels. Simple arguments are presented, which explain some of the qualitative features of the observed channel formation. The division of relativistic channels has been observed and forwardly accelerated electrons detected. These electrons are found to have a broad energy distribution that imply electric fields stronger than 100 MeV/cm. The total charge of the accelerated electrons is not directly correlated to the length of the relativistic channel.

ACKNOWLEDGEMENT

We acknowledge the support from the Swedish Research Council, and the Knut and Alice Wallenberg Foundation.

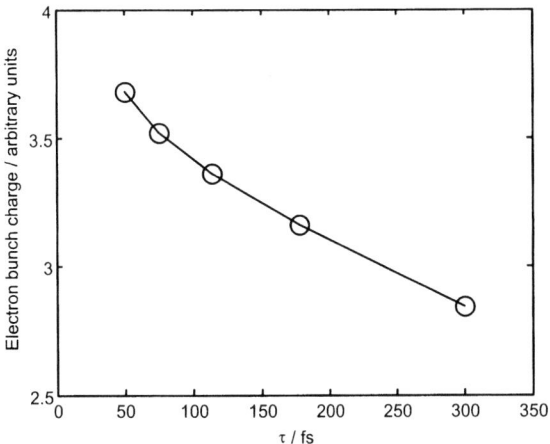

FIGURE 6. The number of electrons decrease with increasing laser pulse duration. The electron density is $n_e = 6 \cdot 10^{19}\,\mathrm{cm}^{-3}$ and the pulse energy is 0.4 J. The orifice diameter is 0.8 mm.

REFERENCES

1. Burnett, N. H., and Corkum, P. B., *J. Opt. Soc. Am. B*, **6**, 1195–1199 (1989).
2. Amendt, P., Eder, D. C., and Wilks, S. C., *Phys. Rev. Lett.*, **66**, 2589–2592 (1991).
3. Tajima, T., and Dawson, J. M., *Phys. Rev. Lett.*, **43**, 267 (1979).
4. Wagner, R., Chen, S.-Y., Maksimchuk, A., and Umstadter, D., *Phys. Rev. Lett.*, **78**, 3125–3128 (1997).
5. Esarey, E., Sprangle, P., Krall, J., and Ting, A., *IEEE J. Quantum Elect.*, **33**, 1879–1914 (1997).
6. Gahn, C., Tsakiris, G. D., Pukhov, A., Meyer-Ter-Vehn, J., Pretzler, G., Thirolf, P., Habs, D., and Witte, K. J., *Phys. Rev. Lett.*, **83**, 4772–4775 (1999).
7. Borisov, A. B., Borovskiy, A. V., Korobkin, V. V., Prokhorov, A. M., Shiryaev, O. B., Shi, X. M., Luk, T. S., McPherson, A., Solem, J. C., Boyer, K., and Rhodes, C. K., *Phys. Rev. Lett.*, **68**, 2309 (1992).
8. Albert, O., Wang, H., Liu, D., Chang, Z., and Mourou, G., *Opt. Lett.*, **25**, 1125–1127 (2000).
9. Wang, X. F., Krishnan, M., Saleh, N., Wang, H. W., and Umstadter, D., *Phys. Rev. Lett.*, **84**, 5324–5327 (2000).
10. Delfin, C., Lokhnygin, V., Mauritsson, J., Sjögren, A., Wahlström, C.-G., Pukhov, A., and Tsakiris, G. D., Influence of laser pulse duration on relativistic channels (2001), submitted.
11. Andreev, N. E., Gorbunov, L. M., Kirsanov, V. I., Pogosova, A. A., and Ramazashvili, R. R., *Jetp. Lett+*, **55**, 571–576 (1992).

Relativistic Self-focusing of Ultra-intense Laser Pulses and Ion Acceleration

James Koga*, Kazuhisa Nakajima† and Keisuke Nakagawa*

*Advanced Photon Research Center, JAERI, Kyoto-fu, 619-0215, Japan
†High Energy Accelerator Research Organization, Tsukuba, 305-0801,Japan

Abstract. Two dimensional particle-in-cell simulations are performed which show the formation of an extremely large electrostatic field near the front of a relativisitically self-focused laser pulse propagating in an underdense plasma. The size of the field is found to reach a maximum of ~ 6.5 TV/m for a 100TW laser pulse propagating over a distance of about 1mm in a plasma at about 3% of the critical density. Along with this large electrostatic field electrons of a maximum of 195MeV are also found to be produced. We propose using this field generated by relativistically self-focused ultraintense laser pulses to acclerate injected ions. Several factors contribute to the possible acceleration including the self-focused pulse group velocity, pulse depletion, and plasma density.

INTRODUCTION

Ultraintense laser interactions with matter can generate enormous numbers of highly energetic electrons, photons, and ions. Recent laser-matter interaction experiments have revealed acceleration of $\sim 10^{10}$ electrons with the maximum energy up to ~ 100 MeV and production of $\sim 10^{12}$ ions up to several tens of MeV[1].

In two-dimensional PIC simulations for propagation of ultraintense laser pulses of the order of 10^{20}W/cm^2 in underdense plasmas of $\sim 10^{20}$cm^{-3}, we have found that large amplitude of positive electrostatic fields of the order of a few TV/m are generated over a ~ 1 mm scale in the front of a relativistically self-focused laser pulse. These enormous fields imply the capablility of accelerating protons. The solitary structure of the electrostatic field with a spacial and temporal size of the order of μm can produce a femtosecond ion pulse. We propose a new acceleration mechanism for ions due to enormous accelerating fields generated by relativistically self-focused laser pulses in plasmas. This mechanism may open up a new regime of ultraintense laser-matter interactions and new fields of high energy particle physics.

STRONG ELECTROMAGNETIC FIELDS

The peak amplitude of the transverse electric field of a linearly polarized laser pulse is given by

$$E_L[\text{TV/m}] \simeq 2.7 \times 10^{-9} I^{1/2}[\text{W/cm}^2] \cong 3.2 a_0/\lambda_0[\mu\text{m}], \qquad (1)$$

CP611, *Superstrong Fields in Plasmas:* Second Int'l. Conf., edited by M. Lontano et al.
© 2002 American Institute of Physics 0-7354-0057-1/02/$19.00

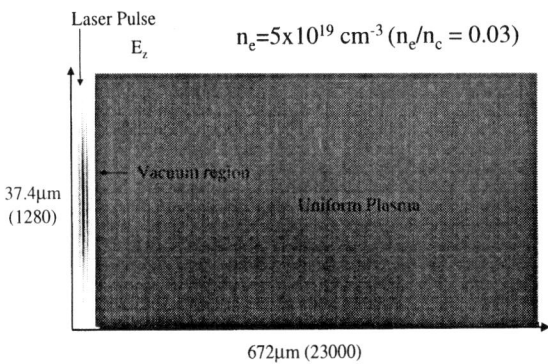

Laser Pulse
E_z

$n_e = 5 \times 10^{19}$ cm^{-3} ($n_e/n_c = 0.03$)

Vacuum region

Uniform Plasma

37.4μm
(1280)

672μm (23000)

FIGURE 1. Simulation setup

where I is the laser intensity, λ_0 is the laser wavelength, and a_0 is the laser strength parameter defined by $a_0 \equiv eA_0/m_e c^2$ in terms of the peak amplitude of the laser vetor potential A_0 and the electron rest energy $m_e c^2$. Using the laser peak intensity $I = cE_L^2/8\pi = ck^2A_0^2/8\pi$, the laser strength parameter is given by

$$a_0 = (2e^2\lambda_0^2 I/\pi m_e^2 c^5)^{1/2} \cong 0.85 \times 10^{-9}\lambda_0[\mu m]I^{1/2}[\text{W/cm}^2]. \quad (2)$$

Physically a_0 is equal to the normalized momentum of the electron quiver motion in the laser field. The corresponding magnetic field is given by $B_L[\text{T}] = E_L/c$ and the radiation pressure exerted by the laser intensity I is given by $P_L[\text{Bar}] = 0.1I/c[\text{J/cm}^3]$. Advances in laser technology provide us with ultraintense ultrashort lasers capable of generating high intensities more than 10^{20} W/cm^2 and short pulses less than 20 fs. Such laser pulses can generate an electric field of more than $E_L \sim 27$ TV/m. At these intensities, the magnetic field is $B_L \sim 10^5$ T and the radiation pressure exceeds $P_L \sim 30$ GBar.

SIMULATION PARAMETERS

To study the self-focusing of a high intensity short pulse laser in a plasma we use the code PCUBE (Progressive Parallel Plasma Code) which is a 2 dimensional fully relativistic particle-in-cell(PIC) code. The code has been parallelized to run on the Compaq ES40 227 node massively parallel computer. Figure 1 shows the initial configuration of the simulation. The simulation box is 672μm (23000 cells) by 37.4μm (1280 cells) in the x and y directions respectively. The boundary conditions are periodic in the y direction and outgoing in the x direction. There is a vacuum region at one end of the simulation box of length 21.9μm. The plasma density is chosen to be 5.3×10^{19}cm^{-3} which corresponds roughly to doubly ionized Helium gas at atmospheric pressure. There are 8 electrons and 8 ions in each simulation cell with an ion to electron mass ratio of 1836. The linearly s-polarized laser pulse (E_z, B_y) starts in the vacuum region on the left and

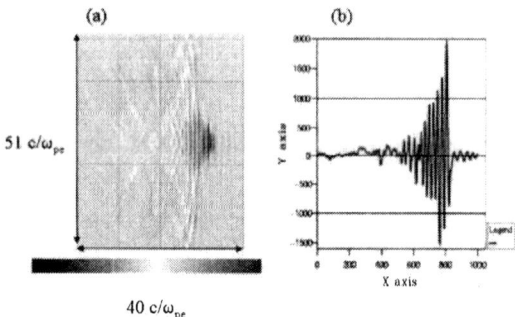

(a)

51 c/ω_{pe}

40 c/ω_{pe}

(b)

Y axis

X axis

FIGURE 2. (a) E_z field of the self-focusing laser pulse, (b) profile down the center of the pulse after the laser has propagated $340c/\omega_p$.

propagates to the right. The parameters of the laser are that of the 100 TW Ti:sapphire laser at the Japan Atomic Energy Research Institute [2]. The pulse length is 19 fs with a spot size of 10μm. The wavelength is 0.8μm. The corresponding unitless laser strength parameter $a_0 = 7.4$ where $a_0 = eE_0/m_0\omega_0c$, E_0 is the peak electric field, m_0 is the electron mass, and ω_0 is the laser frequency.

Using the formula for the critical power P_{cr} for the relativistic self-focusing of a Gaussian laser pulse [3]

$$P_{cr}[\text{GW}] = 17(\frac{\omega_0}{\omega_p})^2,\tag{3}$$

where ω_0 is the laser frequency and ω_p is the plasma frequency. With the current density $P_{cr} = 560\text{GW}$ so that $P/P_{cr} = 179$, where P is the laser power. Thus, the laser pulse should relativistically self-focus in the plasma. Also given the condition for length of the laser pulse L_t necessary for the optimum generation of a wake field [4]: $L_t = \pi c/\omega_p$, we find that $L_t = 2.3\mu$m whereas the laser pulse length is 4.9μm. Under these conditions a large wake field behind the pulse should not occur.

SIMULATION RESULTS

Figure 2(a) shows the laser pulse after it has propagated 256μm ($340c/\omega_p$). The laser pulse has relativistically self-focused and has filamented and the central portion has narrowed. From the line profile taken down the center of the pulse (Figure 2(b)) we can see that the front of the laser pulse has steepened. This is due to the fact that the front of the laser pulse has been depleted compared with the initial gaussian profile. The amplitude is approximately a factor of 2 higher than the initial laser pulse amplitude. Figure 3 (a) shows the electron density at the same propagation distance. In the central portion of the pulse the electrons have been completely ejected. An electron cavity from

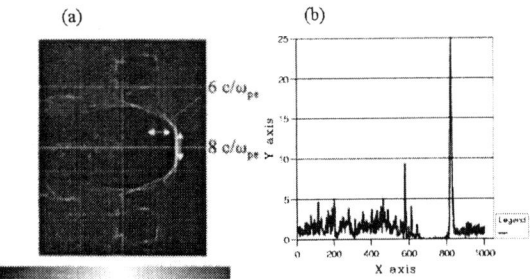

FIGURE 3. (a) Electron density of the background plasma after the laser has propagated $340c/\omega_p$. (b) Profile of the electron density down the center of the simulation box.

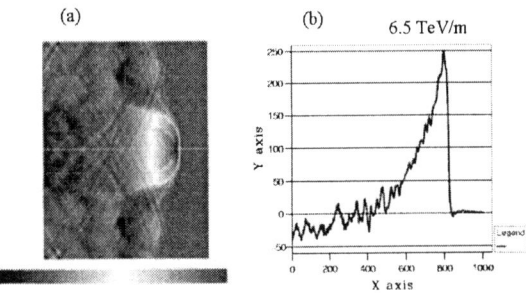

FIGURE 4. (a) The structure of the E_x electrostatic field after propagating $256\mu m(340c/\omega_p m)$. (b) Profile of the electron density down the center of the simulation box.

the main part of the pulse has formed which is $8c/\omega_p$ wide and $6c/\omega_p$ long. In the front of this cavity the electrons have built up. Figure 3 (b) shows a profile of the electron density down the center of the evacuated region. In the front of the laser pulse the density is 25 times the initial background plasma density or $1.3 \times 10^{21} \text{cm}^{-3}$. This is just below the critical density which is 32.9 times the background density or $1.74 \times 10^{21} \text{cm}^{-3}$.

Due to this large buildup of electrons at the front of the pulse, there is a large positive electrostatic field in the propagation direction of the laser pulse created there. Figure 4 shows the structure of the electric field after the same propagation distance. Figure 4(b) shows the line profile of the electric field down the center of the pulse. The electric field rises rapidly to a maximum of 6.5 TeV/m and gradually drops off until it becomes negative behind the laser pulse. We can compare this maximum field to the wave-breaking limit field [5]:

$$E_{WB} = \sqrt{2}(\gamma_p - 1)^{1/2} E_0 \tag{4}$$

$$E_0 = \frac{cm_e\omega_p}{e} = 96 n_p^{1/2}(\text{cm}^{-3})\text{V/m} \tag{5}$$

where $\gamma_p = (1 - v_p^2/c^2)^{-1/2}$. Using the linear group velocity for v_p

$$\beta_p = v_p/c = \sqrt{1 - \omega_p^2/\omega_0^2} \tag{6}$$

the wave-breaking limit field becomes $E_{WB} = 2.17$ TV/m. We see that the field generated in the front of the laser pulse is greater than the wave-breaking limit field.

The source of this large electrostatic field is the large density gradient created at the front of the laser pulse. The creation of this large density gradient can be understood from the equation of continuity:

$$\frac{\partial n_e}{\partial t} + \frac{\partial(n_e v_x)}{\partial x} = 0 \tag{7}$$

where n_e is the electron density, x is the propagation direction of the laser pulse, and v_x is the velocity of electrons in the propagation direction. In the frame comoving with the laser pulse we get[6]:

$$n_e = \frac{n_{e0}}{1 - \beta_x} \tag{8}$$

where n_{e0} is the initial electron density and β_x is v_x/c. We can see from this equation that where the background electron velocity is high the density is high. In Figure 5 is shown the electron density profile down the center of the pulse and the corresponding electron momentum in the propagation direction after the same propagation distance shown previously. In the figure we can see that the high density corresponds to the high electron momentum in agreement with Equation 8.

We can determine the peak of this electron density relative to the laser field by using the equations of motion of an electron in a plane wave. For a plane wave of the form

$$E(\phi) = a_0 \sin(\phi) \tag{9}$$

$$\phi = \omega[t - \frac{x(t)}{c}] \tag{10}$$

where a_0 is the normalized amplitude, ϕ is the phase, and ω is the frequency of the laser pulse, the velocity of the electron β_x is given by [7]:

$$\beta_x = 1 - \frac{2}{2 + a_0^2(\cos(\phi) - 1)^2}. \tag{11}$$

Using Equation 8 and 11 we can get an expression for the background electron density. In Figure 6 is plotted the amplitude of the plane wave and the background electron

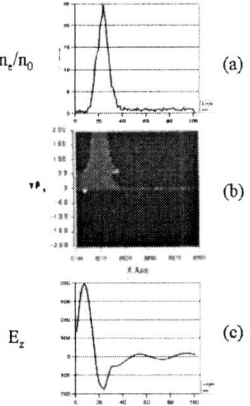

n_e/n_0 (a)

$\gamma\beta_x$ (b)

E_z (c)

FIGURE 5. (a) Close up of the electron density profile down the center of the simulation box after propagating $256\mu m(340c/\omega_p)$. (b) Profile of the electron momentum $\gamma\beta_x$ and (c) the laser field at the same time.

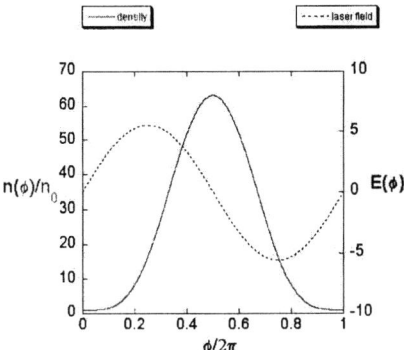

FIGURE 6. Background electron density normalized by the initial electron density (left axis, solid line) and laser amplitude (right axis, dotted line) as a function of the phase ϕ for a plane wave of amplitude $a_0 = 5$

density as a function of the phase ϕ for a plane wave of amplitude $a_0 = 5$. This value of a_0 corresponds to the field at the same time as Figure 5(c). From the plot we can see that the maximum in the electron density occurs at $\phi = \pi$. When we compare this to the simulation results in Figures 5(a) and (c), that this corresponds to the same position in the laser wave for the maximum density. The maximum value of the density disagrees with the theoretical prediction. This maybe attributed to the two dimensional nature of the simulation and the difference in the electron motion in the plasma from that of an electron in vacuum moving in a plane wave.

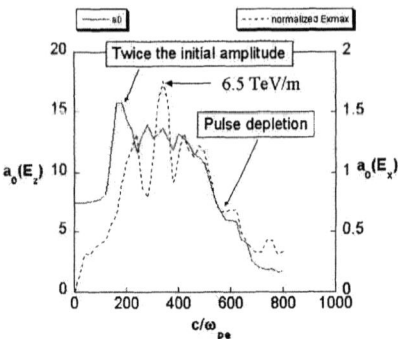

FIGURE 7. (Normalized E_z field of the self-focusing laser pulse (left axis, solid line) and electrostatic field E_x (right axis, dotted line) as a function of the laser pulse propagation distance.

Figure 7 shows the maximum normalized amplitude of the laser pulse E_z and the corresponding electrostatic field E_x generated by the pulse as a function of propagation distance of the laser pulse. The laser pulse relativistically self-focuses to a peak normalized amplitude of $a_0 = 15.7$ after propagating $160c/\omega_p$. This is more than twice the initial amplitude. After the initial peak in self-focusing amplitude, the pulse begins to deplete. This depletion is due to the absorption of the laser pulse by the plasma. The laser is propagating in a near vacuum at the speed of light while the electron cavity created by the laser pulse propagates at near the plasma group velocity. Theoretically in one dimension the depletion distance is [8]:

$$l_{pulse}\frac{\omega_0}{\omega_p} \approx 257\frac{c}{\omega_p} = 188\mu m \qquad (12)$$

This value is smaller than that seen from the simulation results. One factor may be the effect of the relativistic factor γ of the background electrons on the plasma frequency ω_p. Another factor may be attributed to two dimensional effects.

As shown in Figure 7 the electrostatic field peaks after the laser field peaks. The maximum electrostatic field is 6.5 TeV/m ($a_0 = 1.75$) which corresponds to the previous figures showing the structure of the laser pulse. It can be seen that throughout most of the laser propagation that the electrostatic field amplitude is about 10% of the laser field amplitude. Also, in conjunction with the laser pulse depletion the electrostatic field decreases.

HIGH ENERGY ELECTRONS

Even though the length of the laser pulse is greater than the optimal length to generate a large wakefield behind the laser pulse a large number of high energy electrons are

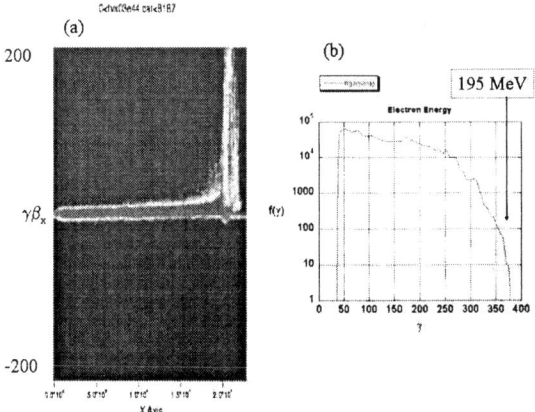

FIGURE 8. (a) $\gamma\beta_x - x$ phase space, (b) energy distribution of the background electrons after where electrons below $\gamma = 40$ are not included after the laser pulse has propagated $642\mu m(880c/\omega_p m)$.

observed behind the laser pulse. In Figure 8(a) is shown the $\gamma\beta_x - x$ phase space of background electrons after the laser pulse has propagated propagated $642\mu m$ ($880c/\omega_p$). It can be seen that there are a large number of high momentum electrons which are located behind the laser pulse which has depleted to less than $a_0 = 1.8$ as seen in Figure 7. For an electron in a plane wave with initial zero velocity [7]

$$\gamma\beta_x = a_0^2(\cos(\phi) - 1)^2/2 \tag{13}$$

which gives a maximum of $\gamma\beta_{x\text{max}} = 2a_0^2$. Using the maximum laser amplitude at the current propagation distance we get $\gamma\beta_{x\text{max}} \approx 6.5$ which is far below the measured maximum from the simulations. The simulation maximum is closer to the value for the maximum laser amplitude after just self-focusing $a_0 = 15.7$ giving $\gamma\beta_{x\text{max}} \approx 493$. Figure 8(b) shows the energy distribution $f(\gamma)$ of the background electrons behind the laser pulse. It can be seen that there is a broad range of energies (the distribution was cutoff for electrons below $\gamma = 40$). The maximum energy of the electrons corresponds to 195 MeV which is close to the maximum energy electrons of 252 MeV would have obtained propagating in a plane wave with $a_0 = 15.7$. Of course, in a plane wave there is no net gain in energy.

ION ACCELERATION SCHEME

We propose to use the large field created at the front of the pulse to accelerate injected protons to higher energies. In order to determine the minimum energy of injection of the protons we can use the theory developed for determining the minimum energy of injection for electrons in a wakefield [9]:

$$\gamma_{\text{max/min}} = \gamma_p(1 \pm \beta_p\sqrt{2\gamma_p\Delta\phi}) \tag{14}$$

133

$$\gamma_p \Delta\phi \ll 1 \qquad (15)$$

where $\gamma_{max/min}$ refers to the maximum and minimum energy of electrons which can be trapped in a potential well of a normalized potential difference $\Delta\phi$. The potential difference is:

$$\Delta\phi = \phi_{max} - \phi_{min} = a_0 \frac{m_e}{m_i} \qquad (16)$$

where a_0 is the normalized amplitude of the electrostatic field generated by the laser pulse. Using the maximum electrostatic field $a_0 = 1.75$ we get $\Delta\phi = 9.53 \times 10^{-4}$ and with $\gamma_p = 5.73$ resulting in $\gamma_p \Delta\phi = 5.46 \times 10^{-3}$ satisfying the condition for the validity of Equation 14. Using these values we get:

$$\gamma_{min} = 5.3 \qquad (17)$$
$$\gamma_{max} = 6.18 \qquad (18)$$

or correspondingly,

$$T_{min} = (\gamma_{min} - 1)m_i c^2 = 3.92 \text{GeV} \qquad (19)$$
$$T_{max} = (\gamma_{max} - 1)m_i c^2 = 4.72 \text{GeV} \qquad (20)$$

This implies that if we inject electrons at the minimum energy and can accelerate them to the maximum energy we can get an increase of up to 0.8 GeV. Due to the fact that the actual group velocity of the self-focused laser pulse differs from the one predicted by linear considerations we measured the veolcity of the pulse from the simulations. The values are:

$$\beta_{sim} = 0.952 \qquad (21)$$
$$\gamma_{sim} = 3.27 \qquad (22)$$

Using these values we get:

$$T_{sim} = (\gamma_{sim} - 1)m_i c^2 = 2.06 \text{GeV} \qquad (23)$$

which gives a lower injection velocity. In order to insure that some protons are injected in front of the electrostatic field as an initial condition a proton beam with an energy spread was input into the simulation. The beam was uniformly distributed in the y direction and placed in the same initial position as the laser pulse with a length of 29.2μm (1000Δ). The proton beam had a very low density so that the protons are only pushed by the electrostatic field generated by the laser pulse. Figure 9 shows the (a) initial and (b) final $\gamma - x$ phase space of the central portion of the injected proton beam at $630\Delta \leq y \leq 650\Delta$ where the center of the simulation is at $y = 640\Delta$. In Figure 9(a) is shown the initial proton beam with a spread in γ between 2.47 and 6.85. After the proton beam has propagated with the laser over 642μm (880c/ω_p) the proton beam has spread due to the initial spread in energies (Figure 9(b)). There is a gap between the front portion of the beam and the back part of the beam. This may be due to the electrostatic field generated

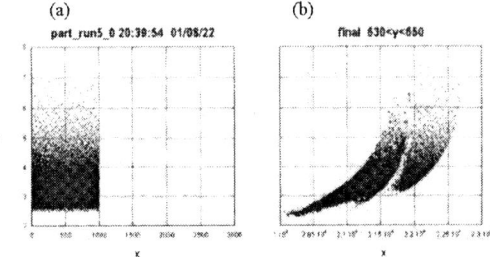

FIGURE 9. (a) Initial and (b) final $\gamma - x$ phase space of the central portion of the injected proton beam at $630\Delta \le y \le 650\Delta$ where the center of the simulation is at $y = 640\Delta$

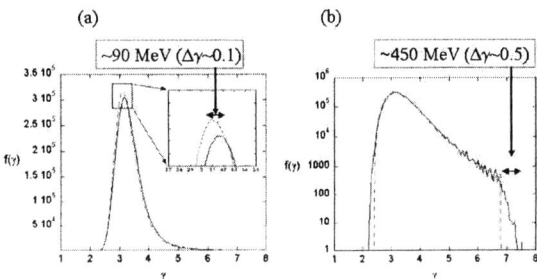

FIGURE 10. Initial (dotted lines) and final (solid lines) distribution $f(\gamma)$ of the injected proton beam. (a) linear scale, closeup of distribution (inset) and (b) log scale.

by the laser pulse. It can be seen in the plot that there are now protons which have energies greater than $\gamma = 7$. In Figure 10 is shown the energy distribution of the injected proton beam. In Figure 10(a) one can see that the final distribution (solid line) is shifted from the initial distribution (dotted line). In a closeup of the peak in Figure 10(a)(inset) the shift corresponds to $\Delta\gamma \approx 0.1$ or an equivalent acceleration of 90 MeV. In a log plot of the distribution in Figure 10(b) one can see that there is an increase in the high energy tail of the proton beam. One can see that this increase corresponds to $\Delta\gamma \approx 0.5$ or an equivalent acceleration of 450 MeV. The 90 MeV and 450 MeV increases in energy correspond to average acceleration gradients of 0.14 and 0.7 TeV/m, respectively. These gradients are reasonable in comparison to the maximum electrostatic field gradient of

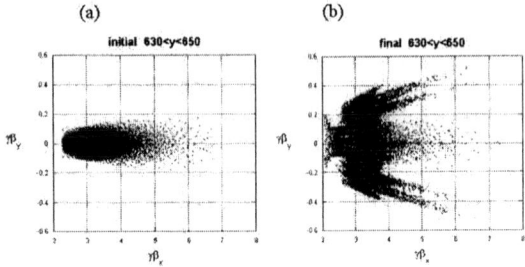

FIGURE 11. (a) Initial and (b) final $\gamma\beta_x - \gamma\beta_y$ phase space plots of the injected proton beam down the center at $630\Delta \le y \le 650\Delta$ where the center of the simulation is at $y = 640\Delta$

6.5 TeV/m.

In Figure 11 (a) and (b) are shown the initial and final $\gamma\beta_y - \gamma\beta_x$ phase space plots of the central part of the injected proton beam, respectively. The increase in the phase space area occupied by the beam is evident from Figure 11 (b) where the spread in the transverse momentum $\gamma\beta_y$ has increased by about a factor of 2 from the initial spread.

The optimal energy and position of the injected proton still needs to be determined. For this paper the main point was to show the possibility of acceleration of protons by the large electrostatic field generated at the front of the laser pulse.

There are several factors which affect or limit the acceleration of ions. One is the depletion of the laser pulse. In Figure 7 we showed the maximum electric field as a function of propagation distance. The maximum field was found to decrease in accordance with the depletion of the laser pulse. Thus with the depletion of the laser pulse the acceleration efficiency also drops. Since the depletion is slower at lower densities according to theory [8], it is better to accelerate protons using low density plasmas. At lower densities, however, the initial injection energy of the protons needs to be higher. Another factor affecting the acceleration is the "snaking" instability which causes the laser pulse to deviate from it's original propagation direction [10]. All these factors need to be considered for the optimal acceleration of protons.

CONCLUSION

In this paper we have proposed using the large electrostatic field created in the front of a relativistically self-focused laser pulse to accelerate ions. We have shown from 2 dimensional PIC simulations that an accelerating field of the order of 6.5 TeV/m can be excited by a 100 TW laser pulse propagating in a plasma at a density of 5.3×10^{19} cm^{-3}. The large field is found to be generated by the buildup of electrons at the front of the relativistically self-focused laser pulse. The buildup of electrons can be accounted for by using the continuity equation. Even though the laser pulse does not satisfy the optimum condition for the generation of a wake field, electrons from the background plasma are found to be accelerated up to 195 MeV. Protons injected in the front of such a self-focused pulse can be accelerated in coincidence with the electric field which moves at a velocity nearly equal to the laser group velocity in plasmas. The maximum kinetic energy gain for injected protons with an average energy of 2 GeV is found to be 450 MeV for a laser intensity of the order of 10^{20}cm^{-3} with a pulse duration of ~ 20 fs.. The acceleration will be limited by the depletion of the laser pulse. The next stage of this investigation will be the acceleration of stationary protons using an inhomogeneous plasma with a density gradient. These type of laser-matter interactions will open up a new regime in high energy beam science.

ACKNOWLEDGMENTS

J. K. wishes to thank M. Yamagiwa, Y. Ueshima, T. Tajima, B. Bulanov, and Y. Kato for useful discussions. Also, J. K. wishes to thank various COMPAQ computer personel for aid in running the simulations.

REFERENCES

1. Cowan, T. E., Roth, M., Johnson, J., Brown, C., Christl, M., Fountain, W., Hatchett, S., Henry, E. A., Hunt, A. W., Key, M. H., MacKinnon, A., Parnell, T., Pennington, D. M., Perry, M. D., Phillips, T. W., Sangster, T. C., Singh, M., Snavely, R., Stoyer, M., Takahashi, Y., Wilks, S. C., and Yasuike, K., *Nucl. Instr. Meth. Phys. Res. A*, **445**, 130–139 (2000).
2. Yamakawa, K., Aoyama, M., Matsuoka, S., Kase, T., Akahane, Y., and Takuma, H., *Opt. Lett.*, **23**, 1468–1470 (1998).
3. Sun, G. Z., Ott, E., Lee, Y. C., and Guzdar, P., *Phys. Fluids*, **30**, 526–532 (1987).
4. Tajima, T., and Dawson, J. M., *Phys. Rev. Lett.*, **43**, 267–270 (1979).
5. Akhiezer, A. I., and Polovin, R. V., *Sov. Phys. JETP*, **3**, 696–705 (1956).
6. Ueshima, Y., Kishimoto, Y., Sasaki, A., and Tajima, T., *Laser and Particle Beams*, **17**, 45–58 (1999).
7. Hartemann, F. V., Fochs, G. P., LeSage, G. P., Luhmann, N. C. J., Woodworth, J. G., Perry, M. D., Chen, Y. J., and Kerman, A. K., *Physical Review E*, **51**, 4833–4843 (1995).
8. Bulanov, S. V., Kirsanov, V. I., Naumova, N. M., Sakharov, A. S., and Shah, H. A., *Physica Scripta*, **47**, 209–213 (1993).
9. Esarey, E., and Pilloff, M., *Physics of Plasmas*, **2**, 1432–1436 (1995).
10. Naumova, N. M., Koga, J., Nakajima, K., Tajima, T., Esirkepov, T. Z., Bulanov, S. V., and Pegoraro, F., *Physics of Plasmas*, **8**, 4149–4155 (2001).

Nonlinear relativistic optics in the single cycle, single wavelength regime and kilohertz repetition rate

G. Mourou[1], Z. Chang[1], A. Maksimchuk[1], J. Nees[1], S. V. Bulanov[2], N. M. Naumova[2], V. Yu. Bychenkov[3], T. Zh. Esirkepov[4], F. Pegoraro[5] and H. Ruhl[6]

[1]*Center for Ultrafast Optical Science, University of Michigan, Ann Arbor, Michigan 48109-2099*
[2]*General Physics Institute of the Russian Academy of Sciences, Moscow, Russia 117942*
[3]*P. N. Lebedev Physics Institute of the Russian Academy of Sciences, Moscow, Russia 117924*
[4]*Moscow Institute for Physics and Technology, Moscow Region, Russia 141700*
[5]*Pisa University and Instituto Nazionale Fisica della Materia, Pisa, Italy*
[6]*Max Born Institute, Berlin, Germany*

Abstract. Pulses of few optical cycles, focused on one wavelength with relativistic intensities can be produced at a kilohertz repetition rate. By properly choosing the plasma and laser parameters, relativistic nonlinear effects, such as channeling and electron and ion acceleration to tens of megaelectronvolts are demonstrated.

INTRODUCTION

Laser intensity in the relativistic regime, i.e. greater than 10^{18} W/cm^2 for 1 μm light[1], has opened new frontiers in physics. At this intensity level electrons acquire a quiver energy greater than 0.5 MeV, corresponding to the rest mass of the electron. The relativistic character of electrons is dominated by a mass increase and a large ponderomotive force ($\vec{v} \times \vec{B}$) where $\vec{v}$ is the quiver velocity of the electrons and $\vec{B}$ the light's magnetic field. In interaction of a high-intensity light with matter a host of novel effects have been demonstrated: the production of high energy electron and ion beams[2,3], the generation of the directional γ–ray pulses[4], the demonstration of relativistic harmonics from solids[5], relativistic self-focusing[6] and nonlinear Thomson scattering[7], etc. The lasers involved in these studies although more compact than its predecessors are still very large and expensive with energy in the joule level, at repetition rate from 0.01 Hz to 10 Hz and with pulse duration greater than 100 fs. Thanks to the progresses in short pulse generation and the application of deformable mirrors for beam focusing, we have recently shown[8] that it is possible to produce relativistic intensities at a kilohertz repetition rate. The laser

CP611, *Superstrong Fields in Plasmas:* Second Int'l. Conf., edited by M. Lontano et al.
© 2002 American Institute of Physics 0-7354-0057-1/02/$19.00

pulse energy is in the millijoule range, with sub-ten femtosecond duration, i.e. in the single-cycle regime and a focused spot size of one wavelength dimension.

Because of their very short Rayleigh range of the order of 1 μm, these pulses would have only limited applications, such as harmonic generation from solid's interface. To extend their usage it is crucial to increase their interaction distance with the plasma. This condition is necessary, for instance, in particle acceleration (electron, positron, ion). In this paper we demonstrate numerically that in a situation, similar to single-mode graded index optics, if the laser numerical aperture, *NA*, is matched with the relativistic waveguide numerical aperture, set by the laser power and the plasma frequency, *single-mode propagation* of the relativistic pulse, over many Rayleigh ranges can be obtained.

This important result opens the door to a number of exciting prospects including: i) the possibility to produce few-femtosecond-duration, tens-of-MeV electron/positron/ion bursts, with a very compact system at a kHz repetition rate; ii) femtosecond injectors for electron/positron accelerators; iii) the production of para- and ortho-positronium, where the ~ 10 ps lifetime could be measured for the first time; iv) Thomson scattering on the accelerated electrons could provide a new source of incoherent or coherent x-ray pulses; v) Last but not least, the possibility to perform relativistic nonlinear optics on a millijoule system makes this nascent field accessible to a much wider scientific endeavor.

LASER-PLASMA MATCHING

It is convenient to express the laser field amplitude in terms of normalized vector potential $a = \dfrac{eA}{mc^2}$, where *A* is the laser field vector potential, *e* the charge of the electron, *m* the electron mass and *c* the speed of light. The value of *a* can be obtained from the expression $I\lambda^2 = 1.37 \cdot 10^{18}\,W/cm^2 \cdot a^2$, where *I* and λ are the laser intensity and wavelength. It is generally considered that the relativistic intensity threshold is reached for *a*=1, corresponding for λ= 800 nm to 2.10^{18} W/cm². We will also define the relativistic factor as $\gamma = (1 + a^2/2)$. The intensity or γ distribution across the beam will lead to a mass change and a plasma frequency radial distribution. These effects will produce a radial distribution in plasma index of refraction with a maximum on axis leading to self-focusing. The self-focusing will shrink the size of the laser beam to a single wavelength transverse dimension, increasing the laser intensity accordingly. The threshold for the self-focusing power is given by the expression[9]:

$$P_c = 16.2(\omega/\omega_{p0})^2 \quad \text{in GW,} \tag{1}$$

where $\omega_{p0} = (4\pi n_o e^2/m)^{1/2}$ is the plasma frequency at low intensity.

In order to avoid beam break up, we propose to match the numerical aperture of the input optics to the numerical aperture of the relativistic channel. Using the plasma index of refraction given by

$$n(r) = \sqrt{1 - \omega^2_p(r)/\omega^2} \;, \tag{2}$$

where $\omega_p = \left(4\pi n_o e^2/\gamma m\right)^{1/2}$ is the plasma frequency with relativistic correction.

The output of our laser[8] is approaching 1mJ in 10 fs or 100 GW. From equation (1) we find that a value of $\left(\omega/\omega_{po}\right)^2 \approx 5$, is necessary to reach relativistic self-focusing. A Taylor expansion of (2) gives

$$n(r) \approx 1 - \omega^2_p(r)/2\omega^2 \;. \tag{3}$$

Note that (3) is the expression of a simple quadratic graded index waveguide if ω_p is proportional to r. This condition is fulfilled, for a Gaussian beam described by $I(r) = I_0 \exp(2r^2/\sigma_o^2)$, where σ_0 is the laser beam size. A graded index channel waveguide, with the index profile

$$n(r) = 1 - \beta^2 r^2/2 \tag{4}$$

has a numerical aperture $NA = \beta b$ where b is the fiber aperture. By a proper identification between (3) and (4), we can derive the channel numerical aperture given by $NA \approx \omega_{p0}/\omega$. For $\left(\omega/\omega_{p0}\right)^2 \approx 5$, we find a $NA \sim 0.4$, corresponding to a waveguide diameter of about λ or 0.8 μm and $n_0 = 2.10^{20} \, cm^{-3}$. As we have shown this spot size could be obtained by using a deformable mirror and f/1 paraboloid. The channel NA will determine, for a given laser wavelength, the plasma density for the rest of the experiments.

To demonstrate the importance of optimum laser-plasma matching we performed simulations with the use of the 2D-3V PIC code[10]. The laser pulse with the wavelength $\lambda = 0.8$ μm, the intensity $I = 5.10^{19}$ W/cm^2 ($a = 4.8$), and the pulse-duration $\tau = 20$ fs, which is linearly polarized in the (x,y)-plane (p-polarized light) incidents on an underdense plasma slab placed at $10\lambda < x < 50\lambda$. The initial plasma density is $n_e = 0.5n_c$ and the ion to electron mass ratio is 1836. Figure 1 shows laser pulse propagation in a plasma when it is focused at the plasma-vacuum interface, $x = 10\lambda$ with a 1λ density gradient. This focusing corresponds to the optimum conditions for the laser-plasma matching. Just behind the focus the laser pulse becomes guided due to the relativistic self-focusing, as it is seen in Fig. 1 which shows the distribution of the electromagnetic energy density in the (x,y)-plane. At $t = 50(2\pi/\omega)$ the laser pulse loses almost all its energy. The ponderomotive pressure of the light forms the channel in plasma. The laser pulse accelerates electrons predominantly in the forward direction and ions in the transverse direction. The maximum electron energy is about 12 MeV. The maximum ion energy reaches $\varepsilon_{i,max} \sim 0.5$ MeV. This is close to the estimation that follows from the mechanism of "Coulomb explosion"[11] for the channel with diameter $d \sim \lambda$, $\varepsilon_{i,max} \approx \left(m_e/m_i\right)\left(\gamma c \tau/d\right)^2$ MeV ≈ 0.4

MeV. We observed, as well, that nonlinear ion wave breaking also contributes to the ion acceleration.

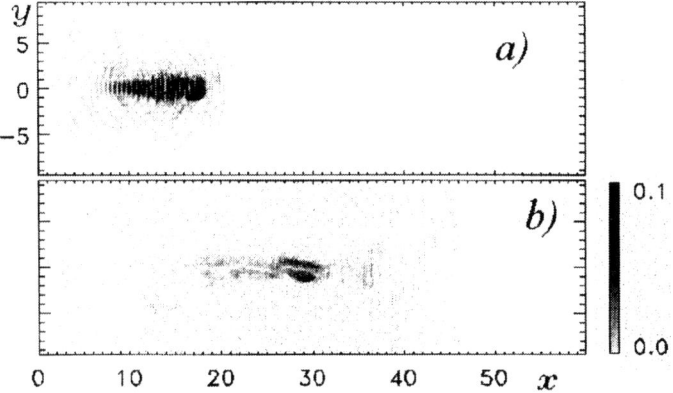

Figure 1. Electromagnetic energy density in the (x,y)-plane at t=30(2π /ω) (a) and 60(2π /ω) (b) for optimum laser-plasma matching

Figure 2 demonstrates the case of non-optimum laser-plasma matching when laser pulse is focused at the distance 5λ inside the plasma. Just behind the focus the break up of the laser pulse into several filaments appears. As a result the laser pulse energy depletion is much stronger than in the previous case. In addition, instead of well-formed channel several relatively short and wide channels appear.

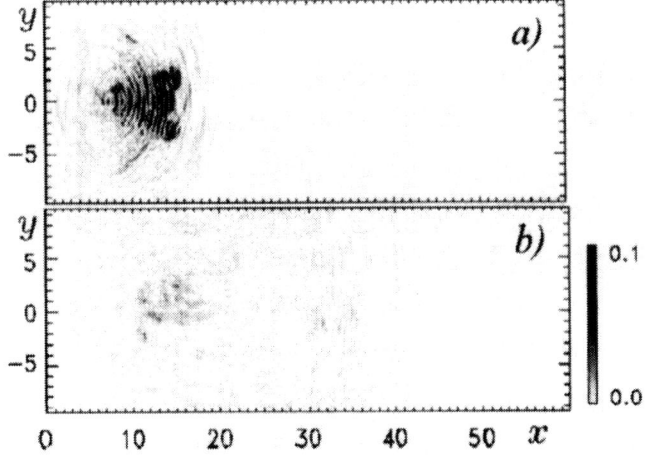

Figure 2. Electromagnetic energy density in the (x,y)-plane at t=30(2π /ω) (a) and 60(2π /ω) (b) for non-optimum laser-plasma matching

These features observed in PIC simulations correlate well with observations of laser channel produced in gas jet. This experiment was performed with a 400 fs laser pulses at $\lambda=1$ μm. Laser beam focused to a spot size of about 10 μm with a paraboloidal mirror f/3.3 interacted with He gas jet with density $n_e \sim 0.08$ n_c. The experimental setup was described in Ref. [11]. Figure 3 shows shadowgrams of a He plasma in the defocusing (a) and relativistic self-focusing and self-channeling dominated regimes (b) for the laser intensity of 6. 10^{18} W/cm^2 and different distances from the nozzle top. Increase in the distance from the nozzle top corresponds to a less sharp vacuum-gas interface, which leads to a breaking of laser-plasma matching conditions. The defocusing dominated regime is characterized by the formation of a short on-axis channel and off-axial laser filamentation. On other hand the regime of relativistic self-channeling characterized by the trapping of significant portion of the laser beam into a long axial plasma channel, which extends to the end of the gas jet.

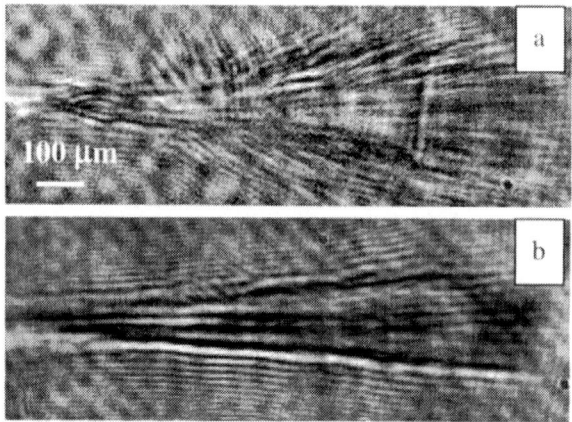

Figure 3. Shadowgrams of laser interaction with a He jet target for laser beam focused at a distance 1.5 mm (a) and 0.5 (b) from the top of supersonic gas nozzle. The probe beam is orthogonal to the pump beam and delayed by 10 ps. The high-intensity laser beam propagates from the left to the right. Relativistic channel is clearly visible on (b) as a bright on-axis line. An external plasma cone is formed due to He gas ionization by the spatial wings of the laser beam.

Our second example deals with laser-plasma matching at the front of solid target. Particularly, we discuss here ion acceleration in an overdense plasma in the interaction of the p-polarized laser pulse ($\lambda=0.8$ μm, I=10^{20} W/cm^2 (a=4.8), $\tau=20$ fs) with the aluminum foil. The foil of the thickness 0.8 μm and the density n=6.5n_c precedes with the low density plasma layer of 5.2 μm length, where the density rises up exponentially from zero to the critical density at the front side of the foil, x=10λ. Laser light is focused on the front side of the foil into the spot with diameter 0.8 μm. In Fig. 4 we show the distribution of the electron (a) and ion (b) densities in the *(x,y)*-plane, and the ion phase plane *(p$_y$,x)* in frame (c) at t=200(2π/ω). We can see that the maximum ion energy gain is about 48 MeV.

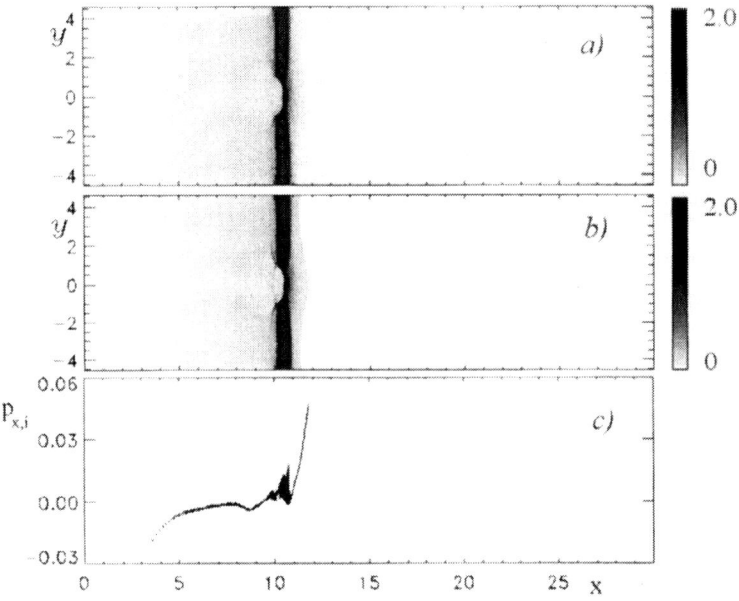

Figure 4. Results of the PIC simulations of the interaction of the 20 fs p-polarized laser pulse (a=6.8) with a thin slab of overderdense (n/n$_c$=6.5) plasma at t=200(2π /ω). The distribution in the *(x,y)*-plane of the electron (a) ion (b) density, and the ion phase plane *(p$_x$,x)* (c).

For the past ten years research on short-laser-pulse interaction with plasmas has basically striven for higher laser energy. Very little efforts have been devoted to the control of the laser-plasma matching. Here we show that this control allows to achieve the highest parameters for channeling and high-energy particle generation.

ACKNOWLEDGMENTS

This research was supported by the National Science Foundation through the Center for Ultrafast Optical Science, contract STC PHY8920108 and the Russian Foundation for Basic Research, grant No. 00-02-16063.

REFERENCES

1. G. A. Mourou, C. P. J. Barty, and M. D. Perry, *Physics Today*, January 1998, 22-28.
2. D. Umstadter, S. -Y. Chen, A. Maksimchuk, G. Mourou, and R. Wagner, *Science* **273**, 472-475 (1996).
3. A. Maksimchuk, S. Gu, K. Flippo, D. Umstadter, and V. Yu. Bychenkov, *Phys. Rev. Lett.* **84,** 4108-4111, (2000).
4. P. Norreys, M. Santala, E. Clark, M. Zepf, I. Watts, F. N. Beg, K. Krushelnick, M. Tatarakis, A. E. Dangor, X. Fang, P. Graham, T. McCanny, R. P. Singhal, K. W. D. Ledingham, A. Creswell, D. C. W. Sanderson, J. Magill, A. Machacek, J. S. Wark, R. Allott, B. Kennedy, and D. Neely, *Phys. Plasmas* **6**, 2150-2156 (1999).
5. D. von der Linde, T. Engers, and G. Jenke, P. Agostini, G. Grillon, E. Nibbering, A. Mysyrowicz, and A. Antonetti, *Phys. Rev. A* **52**, R25-7 (1995).
6. A. B. Borisov, A. V. Borovskiy, O. B. Shiryaev, V. V. Korobkin, A. M. Prokhorov, J. C. Solem, T. S. Luk, K. Boyer, and C. K. Rhodes, *Phys. Rev. A* **45**, 5830-5845 (1992).
7. S. -Y. Chen, A. Maksimchuk, and D. Umstadter, *Nature* **396**, 653-655 (1998).
8. O. Albert, H. Wang, D. Liu, Z. Chang, and G. Mourou, *Opt. Lett.* **25**, 1125-1127 (2000).
9. G.-Z. Sun, E. Ott, Y. C. Lee, and P. Guzdar, *Phys. Fluids* **30**, 526-532 (1987).
10. T. Zh. Esirkepov, *Comput. Phys. Comm.* **135,** 144-53 (2001).
11. G. S. Sarkisov, V. Yu. Bychenkov, V. N. Novikov, V. T. Tikhonchuk, A. Makismchuk, S. -Y. Chen, R. Wagner, G. Mourou, and D. Umstadter, *Phys. Rev. E* **59**, 7042-7054 (1999).

Relativistic Solitons
in Laser-Plasma Interaction

Lj. Hadžievski[a], M. S. Jovanović[b], M. M. Škorić[*a], K. Mima[b]

[a]Vinča Institute of Nuclear Sciences, Belgrade, Yugoslavia
[b]Institute of Laser Engineering, Osaka University, Osaka, Japan

Abstract. We discuss existence and stability of electromagnetic solitons in a relativistic interaction of a linearly polarized laser light with uniform underdense cold plasma. In a weakly relativistic NLS model with local and non-local cubic nonlinearities, an one-dimensional standing electromagnetic soliton trapped in a localized density well, is analytically shown to be stable. By fully relativistic fluid-Maxwell simulations existence of a family of large amplitude relativistic solitons is revealed. Simple analytical estimates give the relation between the maximum amplitude of the soliton vector potential and its eigen-frequency, in agreement with simulation data.

INTRODUCTION

Evolution of a relativistic laser pulse in a long-scale moderately underdense plasma was studied analytically and by computer simulation [1-2]. The circular polarization case was studied in much detail [2]; however, we treat a more complex linear polarization case. Linearly polarized laser light sets electrons into longitudinal motion by relativistic Lorentz force generating coupled longitudinal-transverse wave modes. In this situation, by relativistic fluid and particle simulations for long laser pulse we have recently analyzed nonlinear interplay between forward and backward stimulated Raman scattering and relativistic modulational instability [1]. Parametric down-cascade evolves into a weak turbulence, which saturates into a photon condensate at the bottom of the light spectrum. This phenomenon, similar to Langmuir condensate, corresponds to strong energy depletion and laser beam break-up, as observed in many simulations. In the final stage of saturation, behind the pulse front, the train of intense ultra short relativistic standing solitons is formed. It was estimated, for ultra-short laser pulses [2], that 30 to 40% of the energy can be trapped inside the low frequency electromagnetic solitons creating a significant channel of pulse energy conversion. In this paper, linearly polarized electromagnetic solitons are studied by two simple one-dimensional analytical models for weak and strong nonlinearity, respectively and compared with fully relativistic simulations.

---

* Invited speaker

CP611, *Superstrong Fields in Plasmas:* Second Int'l. Conf., edited by M. Lontano et al.
© 2002 American Institute of Physics 0-7354-0057-1/02/$19.00

WEAKLY RELATIVISTIC SOLITONS

The fully nonlinear relativistic one-dimensional wave equation, the continuity equation and the cold electron momentum equation, in the Coulomb gauge, read

$$\left(\frac{\partial^2}{\partial t^2} - c^2 \frac{\partial^2}{\partial x^2}\right)a = -\frac{\omega_p^2}{n_0}\frac{n}{\gamma}a,$$ (1)

$$\frac{\partial n}{\partial t} + \frac{\partial}{\partial x}\left(\frac{np}{m\gamma}\right) = 0,$$ (2)

$$\frac{\partial p}{\partial t} = -eE_{\parallel} - mc^2\frac{\partial \gamma}{\partial x},$$ (3)

where $a = eA/mc^2$ is the normalized vector potential in y direction, n is the electron density, p is the electron momentum in x direction, $\gamma = (1 + a^2 + p^2/m^2c^2)^{1/2}$, $E_{\parallel}$ is the longitudinal electric field, n_0 is the unperturbed electron density and $\omega_p = (4\pi e^2 n_0/m)^{1/2}$ is the background electron plasma frequency.

By expanding the right hand side of the (1) in the limit for $|a| \ll 1$ and $|\delta n| \ll 1$, introducing the normalized perturbed electron density $\delta n = (n - n_0)/n_0$ and dimensionless variables $x \to (c\omega_p^{-1})x$ and $t \to (\omega_p^{-1})t$ we obtain [3]

$$\left(\frac{\partial^2}{\partial t^2} - \frac{\partial^2}{\partial x^2}\right)a = -\left(1 + \delta n + \frac{a^2}{2}\right)a.$$ (4)

Further, combining the linearized equations of continuity (2) and the electron momentum (3) we get for the perturbed electron density [3]

$$\frac{\partial^2 \delta n}{\partial t^2} + n = \frac{1}{2}\frac{\partial^2}{\partial x^2}a^2.$$ (5)

In distinction to circular polarization [4], linearly polarized waves have odd harmonics of the vector potential and even harmonics of the electron density [3,5]; therefore, we can introduce the slow time varying complex envelopes in a form

$$a = \frac{1}{2}\left[Ae^{-it} + A^*e^{it}\right]; \qquad \delta n = N_0 + \frac{1}{2}\left(N_2 e^{-i2t} + N_2^* e^{i2t}.\right)$$ (6)

and find the envelopes N_0 and N_2 by substituting (6) into (5) and collecting the zero and second harmonic terms (e^{-i2t})

$$N_0 = \frac{1}{4}(|A|^2)_{xx}; \qquad N_2 = -\frac{1}{6}(A^2)_{xx}.$$ (7)

By substituting (6) and (7) into the wave equation (4) and collecting first harmonic terms (e^{-it}) we obtain the wave equation for the vector potential envelope A

$$i\frac{\partial A}{\partial t} + \frac{1}{2}A_{xx} + \frac{3}{16}|A|^2 A - \frac{1}{8}(|A|^2)_{xx}A + \frac{1}{24}(A^2)_{xx}A^* = 0.$$ (8)

The eq. (8) has a form of the nonlinear Schrödinger (NLS) equation [5] with two extra nonlocal (derivative) nonlinear terms. We can also derive the conserved quantities: photon number P

146

$$P = \int |A|^2 \, dx, \tag{9}$$

and Hamiltonian H

$$H = \frac{1}{2} \int \left\{ |A_{xx}|^2 - \frac{3}{16}|A|^4 - \frac{1}{8}[(|A|^2)_x]^2 - \frac{1}{6}|A|^2|A_x|^2 \right\} dx. \tag{10}$$

We further look for a stationary and localized solution of (8) in a form

$$A = \alpha(x)e^{i\lambda^2 t}, \tag{11}$$

with the boundary conditions

$$\alpha(\pm\infty) = 0, \quad \alpha(x) < \infty. \tag{12}$$

Under the assumptions (11) and (12) the first integration of (8) leads to

$$(\alpha_x)^2 = 2 \frac{\alpha^2 \lambda^2 \left(1 - \dfrac{3}{32}\dfrac{\alpha^2}{\lambda^2}\right)}{1 - \dfrac{\alpha^2}{3}}. \tag{13}$$

Additional integration of (13) gives an implicit localized- soliton solution

$$\pm \lambda x = \frac{1}{2\sqrt{2}} \ln \frac{\sqrt{1 - \dfrac{3}{32}\dfrac{\alpha^2}{\lambda^2}} + \sqrt{1 - \dfrac{\alpha^2}{3}}}{\left|\sqrt{1 - \dfrac{3}{32}\dfrac{\alpha^2}{\lambda^2}} - \sqrt{1 - \dfrac{\alpha^2}{3}}\right|} - \frac{4}{3}\lambda \ln \frac{\dfrac{1}{3}\sqrt{32\lambda^2 - 3\alpha^2} + \sqrt{1 - \dfrac{\alpha^2}{3}}}{\sqrt{\left|1 - \dfrac{32}{9}\lambda^2\right|}}, \tag{14}$$

with a soliton amplitude $\alpha_0 = \dfrac{4\sqrt{2}}{\sqrt{3}}\lambda$.

For the soliton strength λ above the critical value $\lambda \geq \lambda_c = 3/4\sqrt{2}$ ($\alpha_0 \geq \sqrt{3}$) the solution (11) has a form of a "cusp" soliton [5]; the centrally highly pointed waveform.

In the small amplitude limit $\lambda \ll \lambda_c$ one neglects the non-local (ponderomotive) terms and the solution (11) becomes the well-known secant hyperbolic (NLS) soliton.

To check the stability of the soliton (11) we use the stability criteria [6,7]

$$\frac{dP_o}{d\lambda^2} > 0, \tag{15}$$

where P_o is the soliton photon number defined by (9). The function $P_o(\lambda)$ is calculated in an analytical form (16) and also shown in Figure 1.

$$P_o(\lambda) = \frac{16}{3}\lambda + \frac{2}{\sqrt{2}}\left(1 - \frac{32}{9}\lambda^2\right)\ln \frac{1 + \dfrac{4\sqrt{2}}{3}}{\left|1 - \dfrac{4\sqrt{2}}{3}\right|}. \tag{16}$$

According to the condition (15) the soliton (14) turns to be stable in the region $\lambda < \lambda_s \approx 0.44$ ($\alpha_0 < \alpha_s \approx 1.44$) indicating that cusp solitons are also unstable ($\lambda_s < \lambda_c$). More generally, we can now conclude that small amplitude linearly polarized solitons ($\alpha_0 < 1$) within the weakly relativistic model (4) -(5) are stable.

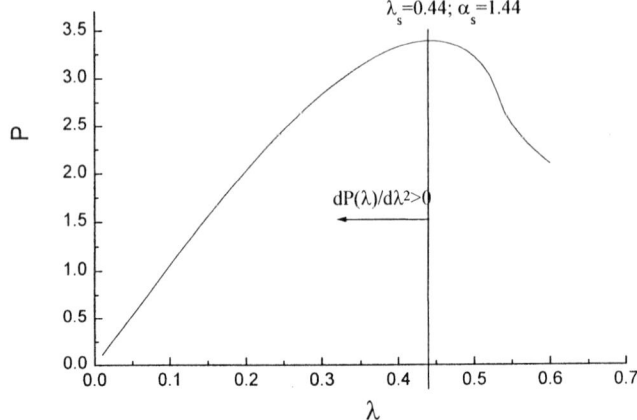

FIGURE 1. Photon number $P_o(\lambda)$ variation with an illustration of the stability criteria.

FULLY RELATIVISTIC MODEL

A direct numerical solution of the fully nonlinear relativistic one-dimensional fluid-Maxwell system (1)-(3) has been performed using the algorithm based on the second order accuracy Lax-Wendorff method and leap-frog central scheme [1].

Numerical data in Fig. 2. give spatio-temporal evolution of energy illustrating the generation of a train of intense ultra-short relativistic standing EM solitons inside the photon condensate. Large relativistic solitons $A_s(x)$, trapped inside their self-consistent density cavities (with $\omega_{sol} < \omega_p$), $\delta n_s(x)$ at several instants of time are shown in Fig. 3. Simulations clearly reveal that apart from stable small amplitude solitons (vide supra), also exists the family of large relativistic low-frequency solitons, which saturate at values close to $A_o \approx 3$.

As earlier analytical model fails for large amplitudes it is necessary to use a different approach. Namely, numerical data strongly support the following substitution

$$a(x,t) = A(x)\cos(\omega t); \quad p(x,t) = P(x)\sin(2\omega t);$$

$$\delta n(x,t) = N_0(x) + N_2(x)\cos(2\omega t) \tag{17}$$

Neglecting p^2 in γ, after a simple trigonometric transformation we get

$$\gamma = \sqrt{1+a^2} = \sqrt{1+A^2\cos^2(\omega t)} = \sqrt{(1+\frac{A^2}{2}) + \frac{A^2}{2}\cos(2\omega t)}. \tag{18}$$

The first term under the square root is larger than the absolute value of the second term, allowing the expansion

$$\gamma \approx \gamma_A + \frac{A^2}{4\gamma_A}\cos(2\omega t), \tag{19}$$

148

where $\gamma_A = \sqrt{1 + A^2/2}$. For instance, for $A = 2$, a maximum error in (19) is only 3%. Furthermore, under approximations (17-19) the following set of equations is obtained

$$A_{xx} = \left[\frac{5\gamma_A^2 - 1}{4\gamma_A} - \frac{\omega^2}{2}(3\gamma_A^2 - 1) + \frac{\gamma_A^2 + 2}{4\gamma_A^4} A_x^2 + \omega \frac{3\gamma_A^2 - 1}{2\gamma_A} P \right] A$$

$$N_0 = \frac{A_{xx}A}{2\gamma_A}\left(1 - \frac{A^2}{8\gamma_A^2}\right) + \frac{A_x^2}{2\gamma_A^3}\left(1 - \frac{3A^2}{8\gamma_A^2}\right)$$

$$N_1 = \frac{A_{xx}A}{2\gamma_A}\left(1 - \frac{A^2}{4\gamma_A^2}\right) + \frac{A_x^2}{2\gamma_A^3}\left(1 - \frac{3A^2}{4\gamma_A^2}\right) + 2\omega P$$

(20)

$$P = \frac{\omega(1 + \gamma_A^2)AA_x}{2\gamma_A^2\left(4\omega^2\gamma_A - N_0 + \frac{A^2}{16\gamma_A^2}N_1\right)}$$

Solving the system (20) under a condition that maximum soliton amplitudes saturate at $\delta n \geq -1$ gives $A_0 = 2.67$ and $\omega_0 = 0.72$ close to the values obtained by direct numerical simulation of the fully relativistic system ($A_0 \approx 2.9$ and $\omega_0 \approx 0.73$). We expect to address the stability problem for large solitons in a separate paper.

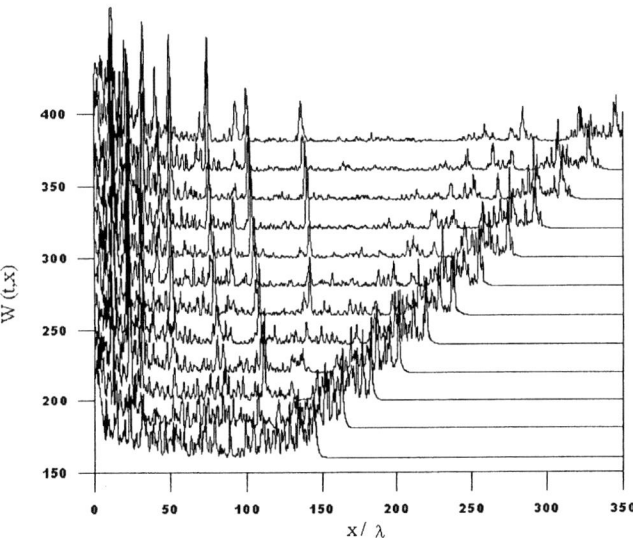

FIGURE 2. Spatial-temporal evolution of EM energy density inside a photon condensate, from 1D fluid-Maxwell simulation with a laser pump $a_0 = 0.5$ and $\omega_0 = 3.2\ \omega_p$ (after [1]).

In conclusion, we have analytically investigated existence and stability of relativistic electromagnetic solitons in cold underdense plasma. We analytically find 1D linearly polarized solitons in agreement with our relativistic fluid and some particle simulations [1,2]. Difference in linear and circular polarization is singled out, in particular, in a role of the 2nd harmonic term in the relativistic γ-factor present for linear polarization. The question of multi-dimensional effects, such as e.g. transverse stability of 1D solitons for symmetry breaking perturbation deserves future attention.

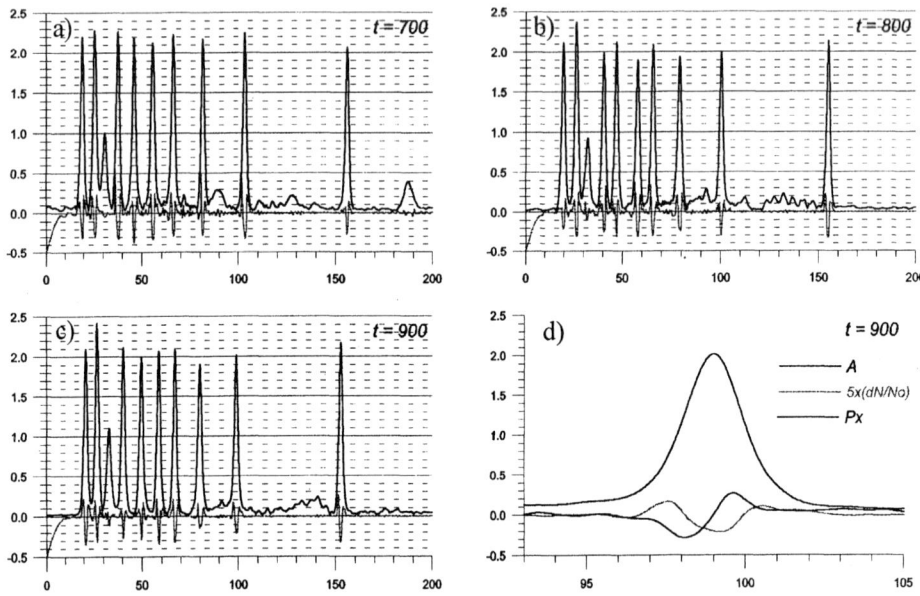

FIGURE 3. Ultra-relativistic standing solitons $A_s(x)$, $\delta n_s(x)$ at three instants of time (a, b, c) and enlarged wiew of the soliton located at $x_s \approx 100$ (d). A finite laser pulse with $a_0 = 0.6$ and $T = 500$ long, was injected into an underdense plasma ($n_0 = 0.1 n_{cr}$).

ACKNOWLEDGEMENTS

One of us M.S.J. acknowledges the research fellowship by Japan Society for Promotion of Science (JSPS). The work was supported in parts by Project 01E11 of the Ministry of Science and Technology of Republic of Serbia.

REFERENCES

1. Mima K., Jovanović M. S., Sentoku Y., Sheng Z.-M., Škorić M. M., Sato T., *Phys. Plasmas*, **8**, 2349-2356 (2001).
2. Bulanov S. V., Califano F., Esirkepov T. Zh., Mima K., Naumova N. M., Nishihara K., Pegoraro F., Sentoku Y., Vshivkov V. A., *Physica D,* **152-153**, 682-693 (2001*)* (and refereces therein)
3. Decker C. D., Mori W. B., Tzeng K.-C., Katsouleas T., *Phys. Plasmas* **3**, 2047-2056 (1996).
4. Kaw, P.K., Sen, A., Katsouleas, T., *Phys. Rev. Lett.* **68**, 3172-3175 (1992)
5. Akhiezer, A.I., Polovin, R.V., *Soviet Phys. JETP* **3**, 696-706 (1956).
6. Litvak. A. G., Sergeev A. M., *JETP Lett.* **27**, 517-520 (1978).
7. Vahitov N.G., Kolokolov A. A., *Izv. Vyssh. Uchebn. Zaved. Radiofizika* **16**, 1020-1028 (1973).

Bright and dark relativistic solitons in plasmas

D. Farina[*], S. V. Bulanov[†]

[*]Istituto di Fisica del Plasma, Consiglio Nazionale delle Ricerche, Milano, Italy,
[†]General Physics Institute of RAS, Moscow, Russia

Abstract. A set of nonlinear differential equations, that describes the moving relativistic solitons with the ion response taken into account, is investigated analytically and solved numerically. We study the influence of the ion motion on the soliton structure and show that depending on the propagation velocity we have either the bright, dark solitons or electromagnetic collisionless shock waves.

FORMULATION OF THE PROBLEM

Relativistic solitons have been seen in multi-dimensional particle in cell PIC simulations of laser pulse interaction with the underdense and the overdense plasmas [1]. The solitons are generated in the wake left behind the laser pulse and they propagate with the velocity well below the speed of light toward the plasma-vacuum interface. Here, they disappear suddenly radiating away their energy in the form of low frequency electromagnetic bursts [2]. Inside the soliton, dispersion effects due to the finite electron inertia are balanced by the nonlinearity of the media, due to the relativistic increase of the electron mass as well as to the plasma density redistribution under the ponderomotive force action, which pushes the particles (the electrons and at a longer time the ions) away from the region of the maximum electromagnetic field. The analytical theory of the relativistic electromagnetic solitons has been developed in Refs. [3–8]. In [5], it has been noticed that the ion contribution limits the velocity of the soliton propagation from below, and modifies the distribution of the electromagnetic field inside the soliton. In the case of relativistic but relatively low amplitude of the soliton, the ions can be assumed to be at rest during approximately $2\pi\omega_{pi}^{-1}$ periods of oscillations of the electromagnetic field inside the soliton. However, for a longer time interval, the ponderomotive pressure of the electromagnetic field inside the soliton starts to dig a hole in the ion density and the parameters of the soliton change [2,9]. Here, some new effects due to the influence on the soliton structure of the ion motion are discussed. In particular, the modifications of the bright soliton spectrum, and the occurence of new solutions is discussed, such as dark solitons and collisionless shock waves.

CP611, Superstrong Fields in Plasmas: Second Int'l. Conf., edited by M. Lontano et al.

We use the relativistic hydrodynamic approximation to describe both the electron and ion components, and assume the plasma to be cold with zero temperature. We consider the 1D (in which $\partial_y = \partial_z = 0$) Maxwell's equations for the wave vector and scalar potentials $\mathbf{A}$ and ϕ, and the hydrodynamic equations for the density and the kinetic momentum $\mathbf{p}_\alpha$ of electrons and ions ($\alpha = e, \imath$) in the Coulomb gauge ($\nabla \cdot \mathbf{A} = 0$). In this case, the relations $A_x = 0$, and $\mathbf{p}_{\perp\alpha} = -\rho_\alpha \mathbf{A}_\perp$ hold, where $\perp$ refers to the direction perpendicular to the x–direction, and $\rho_\alpha = (q_\alpha/q_e)(m_\alpha/m_e)$. We then look for a solution of the system of the form (circularly polarized wave) $A_y + iA_z = a(\xi)\exp[-i\omega t + ikx + i\theta(\xi)]$, with $\xi = (x - Vt)/(1 - V^2)^{1/2}$ (being V a velocity), while all the other quantities, ϕ, n_α, γ_α, and $p_{\|\alpha}$, are assumed to depend only on the variable ξ. Dimensionless quantities are used. Length, time, velocity, momentum, vector and scalar potential, and density are normalized over c/ω_{pe}, ω_{pe}, c, $m_\alpha c$, $m_e c^2/e$, and n_0, respectively, being $\omega_{pe} = (4\pi n_0 e^2/m_e)^{1/2}$ the electron plasma frequency, m_α the rest mass, and n_0 the unperturbed density. By imposing as boundary condition at the point $\xi = \xi_0$, (with, e.g., $\xi_0 = -\infty$) $a = \pm a_0$, $\phi = 0$, $n_\alpha = 1$, and $p_{x\alpha} = 0$ (plasma at rest), the following closed system of equations for the potentials is obtained, which describes coupled Langmuir and circularly polarized transverse electromagnetic waves

$$\frac{d^2\phi}{d\xi^2} = V\left(\frac{\psi_e}{R_e} - \frac{\psi_i}{R_i}\right), \tag{1}$$

$$\frac{d^2a}{d\xi^2} + a\left(\bar{\omega}^2 - \bar{k}^2\frac{a_0^4}{a^4}\right) = aV\left(\frac{1}{R_e} + \frac{\rho}{R_i}\right), \tag{2}$$

$$\frac{d\theta}{d\xi} = -\bar{k}\left(1 - \frac{a_0^2}{a^2}\right). \tag{3}$$

where $\rho \equiv |\rho_i| = m_e/m_i$, $\psi_\alpha = \Gamma_{0\alpha} + \rho_\alpha\phi$, and $R_\alpha = [\psi_\alpha^2 - (1 - V^2)(1 + \rho_\alpha^2 a^2)]^{1/2}$, with $\Gamma_{0\alpha} = (1 + \rho_\alpha^2 a_0^2)^{1/2}$. The longitudinal component of the kinetic momentum, the kinetic energy, and the density as a function of the potentials are given by $p_{x\alpha} = (V\psi_\alpha - R_\alpha)/(1 - V^2)$, $\gamma_\alpha = (\psi_\alpha - VR_\alpha)/(1 - V^2)$, $n_\alpha = V(\psi_\alpha/R_\alpha - V)/(1 - V^2)$. The system (1-2) can be put in Hamiltonian form, and has a first integral $H(a, a', \phi, \phi') = 1/2(1 - V^2)(a'^2 + \bar{\omega}^2 a^2 + \bar{k}^2 a_0^4/a^2) - 1/2\phi'^2 + V(R_e + R_i/\rho) = K$, where the value of the constant K is determined by the boundary condition, and the symbol $\prime$ denotes derivative with respect to ξ.

The above system of equations (1-3) is quite general, and admits different solitary solutions, depending on the value of the boundary condition for a. In the case of $a_0 = 0$, the system coincides with the one already analyzed in the literature [5,8,10], and the localized solutions are bright solitons with $\bar{k} = 0$. The phase is constant, $\theta = \theta_0$, so that the e.m. potential reads $a(\xi)\exp[-i\bar{\omega}\tau + i\theta_0]$, being $\tau = (t - Vx)/(1 - V^2)^{1/2}$ the proper time in the moving frame. In the case of $a_0 \neq 0$, dark solitons, gray solitons, and shock-wave solutions are found. In this case, the phase behavior is in general non trivial, and Eq. (3) has to be solved after Eqs.(1-2).

BRIGHT SOLITONS IN ELECTRON–ION PLASMAS

Assuming $a_0 = 0$, localized solutions are found for $\bar{\omega}^2 < 1 + \rho$ [10]. Since the equations are reversible under the transformation $\xi \to -\xi$, as $\phi \to \phi$, and $a \to \pm a$, we focus on the class of solutions with single humped ϕ profiles, and a profiles with number of zeroes p even or odd respectively. The eigenspectrum of the bright solitons has been determined, solving numerically Eqs. (1-2), as it was done, e.g, in Ref. [5].

The eigenspectra relevant to the case $p = 1, 2$ are shown in Fig. 1, together with the peak value of the potential inside the soliton, both for the case in which the ion dynamics is neglected, i.e., at $\rho = 0$ (dotted lines), and for the case when the ion dynamics is taken into account (solid lines), with the reference value $\rho = 1/1836$. If the ions are fixed, moving solitons are found for V larger than a (small) critical value V_c, ($V_c \approx 1.6 \times 10^{-3}$ for $p = 1$). At fixed V, the frequency spectrum is discrete, and $\bar{\omega}$ decreases with increasing p. The maximum value of ϕ is found at the minimum velocity, and increases with p. Quite different results are found when

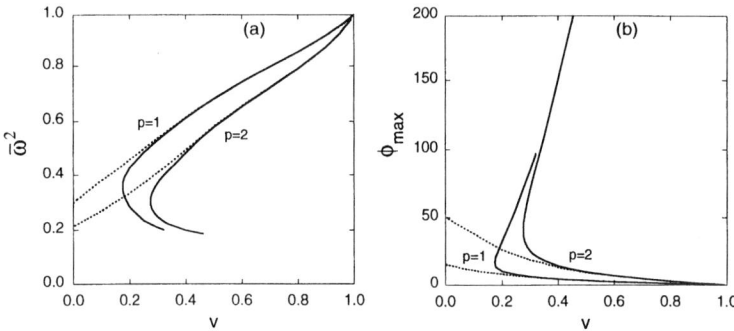

FIGURE 1. Behavior of the soliton frequency $\bar{\omega}^2$ versus the soliton velocity for $p = 1$ and $p = 2$, in the case of fixed (dotted lines), and movable ions (solid lines) (a); dependence of the peak value of the electrostatic potential ϕ inside the soliton (b).

the ion dynamics is taken into account (i.e., $\rho \neq 0$). For velocities smaller than a bifurcation value V_{bif}, no solution can be found. In the range $V_{bif} < V < V_{br}$, two solitary solutions are found, while for $V > V_{br}$ there is just one solution with a frequency value close to that of the $\rho = 0$ case. At high frequencies (and velocities), the ion dynamics is negligible, while it starts to play a role when the velocity diminishes. The ions tend to pile at the soliton center, while the electrons at its edges. In the range $V_{bif} < V < V_{br}$, the lowest frequency branch has features quite different from the highest branch: the potential amplitudes are larger, as it seen in Fig. 1(b), and the associated density profiles are much more peaked. At $V = V_{br}$, the branch ends since the soliton breaks. The peak ϕ and $|a|$ values are considerable larger on the lower frequency branch, corresponding to breaking.

The structure of the solution for $p = 1$ and a velocity value close to breaking is shown in Fig. 2, where the waveforms of the potential and of the densities, are plotted versus the variable ξ. Breaking occurs since the ion density diverges at $\xi = 0$, at $\phi_{br,0} = (1 - \sqrt{1 - V_{br}^2})/\rho$. The ions reach the velocity $V_{br} \approx 0.32$ in the center of the soliton, and the electrons at its edges (i.e., in these regions the particles move with the group velocity of the soliton). The density distributions (see Fig. 2(b)) are very peaked in the center for the ions, and at the edges for the electrons. The corresponding dependence of the ion velocity profile shows a cusp in the center. Such a cusp is the signature of the nonlinear wavebreaking. After the break, a portion of the ions will be injected into the acceleration phase. Moving together with the soliton, the ions are accelerated further. We can estimate the energy gained by the ions in this process (in units of $m_e c^2$) as $E_{ion} \approx \phi_{br,0}/(1 - V_{br}) \approx 70$ MeV. This shows that the soliton breaking can provide an additional mechanism for the generation of fast ions in laser irradiated plasmas.

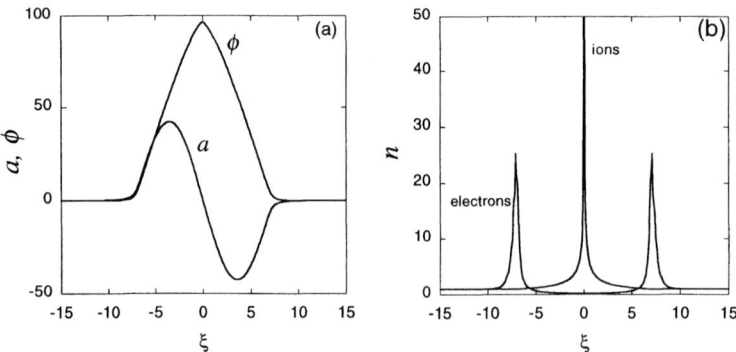

FIGURE 2. Electrostatic ϕ and vector potential a (a), electron and ion density (b) versus ξ for $p = 1$, $\omega^2 = 0.224$, and $V = 0.32$, close to the breaking velocity.

DARK SOLITONS IN ELECTRON–ION PLASMAS

We now consider the case with $a_0 \neq 0$, and are interested in the solutions of Eqs. (1,2) characterized by $da/d\xi = d\phi/d\xi = 0$ at $\xi = \pm\infty$. This condition corresponds to two sets of asymptotic solutions. The first one corresponds to the given boundary conditions $\phi = 0$, $a = 0$, when the following (nonlinear) dispersion relation is fulfilled

$$\Omega^2 \equiv \bar{\omega}^2 - \bar{k}^2 = \frac{1}{\Gamma_{0e}} + \frac{\rho}{\Gamma_{0i}}. \tag{4}$$

The other set reads $\phi = \phi_0 = (\Gamma_{0i} - \Gamma_{0e})/(1 + \rho)$, and $a = 0$, and is valid for $\bar{k} = 0$. In the following, we focus on the particular case $\bar{k} = 0$.

To determine the parameter range in the space (a_0, V) in which localized solutions can be found, Eqs. (1,2) are linearized first around the stationary solutions. Two characteristic velocities are identified, $V_s(a_0) < V_c(a_0)$, with $V_s^2(0) = V_c^2(0) = \rho/(1 - \rho + \rho^2)$. At given a_0, no solitary solution can be found at $V > V_c(a_0)$. Otherwise, three kind of solitary solutions can be found: i) at $V < V_s(a_0)$, $a \to \pm a_0$, $\phi \to 0$ at $\xi \to -\infty$, and $a \to \mp a_0$, $\phi \to 0$ at $\xi \to \infty$; ii) at $V_s(a_0) < V < V_c(a_0)$, $a \to \pm a_0$, $\phi \to 0$ at $\xi \to \pm\infty$; iii) at $V = V_s(a_0)$, $a \to \pm a_0$, $\phi \to 0$ at $\xi \to -\infty$, and $a \to 0$, $\phi \to \phi_0$ at $\xi \to \infty$. The first case corresponds to a dark solution (of kink type), the second to a gray soliton, while the third to a shock wave. Examples of such solutions are shown in Fig. 3.

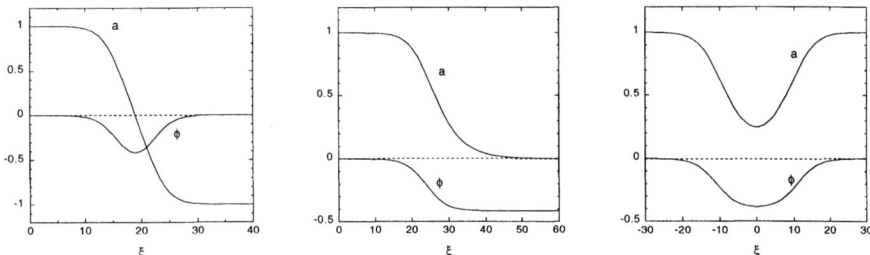

FIGURE 3. Behavior of ϕ and vector potential a versus ξ for $a_0 = 1$. Cases (a) to (c) refer to $V = 0.02$, $V = V_s$, and $V = 0.024$, i.e., to a 'dark soliton, a shock wave, and a gray soliton.

In the quasineutral approximation, which is valid in the longwavelength limit, $\phi'' \ll n_e - n_i$, the electron and ion density are assumed to be equal to each other, $n_e = n_i$, thus obtaining $\psi_e/R_e = \psi_i/R_i$. Then, the equation for the vector potential in the quasineutral approximation reads

$$\frac{d^2 a}{d\xi^2} + a\left(\bar{\omega}^2 - \bar{k}^2\frac{a_0^4}{a^4}\right) = a\frac{V\Gamma^2}{\Gamma_i\Gamma_e[\Gamma_0^2 - (1 - V^2)\Gamma^2]^{1/2}}. \tag{5}$$

The above equation has the following first integral $H_0(a, a') = 1/2\,(a')^2 + U(a) = K_0$, where K_0 is an integration constant, and $U(a) = (\bar{\omega}^2 a^2 + \bar{k}^2 a_0^4/a^2)/2 + V[\Gamma_0^2 - (1 - V^2)\Gamma^2]^{1/2}/[\rho(1 - V^2)]$. From the above equations, the problem is reduced to quadratures. In the weakly nonlinear limit (see [8]), explicit analytical expressions for the field a can be obtained.

The solutions of the full nonlinear system are obtained numerically similarly to the case of bright solitons. The obtained spectrum and the corresponding a_0 value are shown in Fig. 4 as a function of the velocity, for the case with $\bar{k} = 0$. At fixed V, the spectrum is continuous at low a_0, while it becomes discrete above a threshold value, which increases with V, and, finally, no solution can be found above a critical value. Solitary solutions are found up to $V \approx 0.051$, and $a_0 \approx 5.8$. Above these values, no solutions have been found. The discrete spectrum is made up by a very large number of different branches. The obtained solutions are characterized by the same kind of ϕ profile, two possible a profiles (corresponding to black or gray solitons at $V < V_s$ and $V > V_s$). Shock wave solutions are found at $V = V_s$. Breaking of the solutions occurs at inceasing a_0 in each branch, due to the divergence of the electron density.

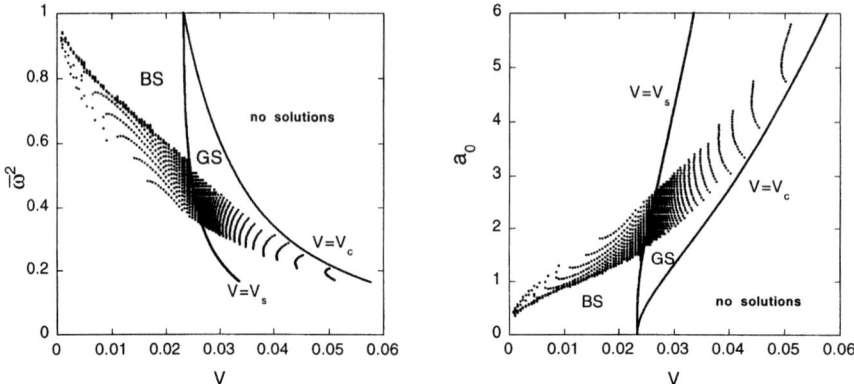

FIGURE 4. Plot of the Doppler shifted frequency $\bar{\omega}^2$ and of the maximum e.m. field a_0 as a function of V. The characteristic velocities V_c and V_s are also shown. The regions denoted by BS and GS correspond to the "continuous" spectrum for black and gray solitons, respectively.

Finally, we note that the electrostatic potential of a dark soliton gives rise to an electrostatic well (see Fig. 3). Thus dark (and gray) solitons can trap and advect the positively charged ions.

REFERENCES

1. Bulanov, S. V., *et al.*, *Physics of Fluids B* **4**, 1935 (1992); Bulanov, S. V., *et al.*, *Phys. Rev. Lett.* **82**, 3440 (1999); Bulanov, S. V., *et al.*, *J. Plasma Fusion Research* **75**, No. 5-CD, 506 (1999); Bulanov, S. V., *et al.*, *Physics of Plasmas* **1**, 745 (1994); Honda, T., *et al.*, *J. Plasma Fusion Research* **75**, No. 10 - CD, 219 (1999).
2. Sentoku, Y., *et al.*, *Phys. Rev. Lett.* **83**, 3434 (1999).
3. Gerstein, J. I., and Tzoar, N., *Phys. Rev. Lett.* **35**, 934 (1975).
4. Tsintsadze, N. L., and Tskhakaya, D. D., *Sov. Phys. JETP* **45**, 252 (1977); Tsintsadze, N. L., *et al.*, Phys. Rev. E **60**, 7435 (1999).
5. Kozlov, V. A., Litvak, A. G., and Suvorov, E. V., *Sov. Phys. JETP* **49**, 75 (1979).
6. P. K. Kaw, A. Sen, T. Katsouleas, *Phys. Rev. Lett.* **68**, 3172 (1992).
7. T. Zh. Esirkepov *et al.*, *JETP Lett.* **68**, 36, (1998).
8. Farina, D., and Bulanov, S. V., *Plasma Phys. Rep.*, **8**, 641 (2001)
9. Naumova, N., *et al.* Phys. Rev. Lett., **87** 185004 (2001).
10. Farina, D., and Bulanov, S. V., *Phys. Rev. Lett.* **86**, 5289 (2001).

Relativistic Electromagnetic Solitons in a High Temperature Plasma

Maurizio Lontano*, Sergei Bulanov¶, James Koga§

* Istituto di Fisica del Plasma, Consiglio Nazionale delle Ricerche, Milano, Italy
¶ General Physics Institute, Russian Academy of Sciences, Moscow, Russia
§ Advanced Photon Research Center, JAERI, Kizu, Japan

Abstract. The set of the relativistic hydrodynamic equations for a hot two-species plasma are derived and then reduced in order to study the existence of one-dimensional soliton-like distributions of the electromagnetic energy in an electron-positron plasma.
The investigation shows that *(i)* non-drifting bright solitons can exist in a hot electron-positron plasma, within a well defined range of plasma temperature; *(ii)* extremely high electromagnetic energy concentrations are possible in an ultrarelativistic plasma; *(iii)* the consistent plasma temperature develops strong spatial nonuniformities.

INTRODUCTION

The rapid development of the laser techinolgy occurred during the last decade has brought to a renewd interest the physics of relativistically intense solitons in plasmas. Particle-in-Cell numerical simulation of the interaction of an ultrashort ultraintense laser pulse with a preformed plasma show that a non negligible fraction of the laser energy is lost due to its trapping into quasi-stationary localized density depressions which are formed behind the pulse [1]. Due to plasma density gradients this energy is subsequently irradiated at the plasma boundary [2]. It is then of primary importance to investigate the nature of the formation of such peculiar electromagnetic energy distributions and try either to prevent it or to control in order to take advantage from it. On the other side, electromagnetic soliton-like structures are of concern for astrophysics and cosmology, as well. Among many reasons for this, it is believed that spatial density fluctuations in the early Universe (between 10^{-2} and 1 sec after the Big Bang) are at the origin of the galaxies and clusters of galaxies formation [3]. In addition, spatial temperature nonuniformities of the early hot plasma would cause the observed nonuniformities in the distribution of the cosmic microwave background radiation. In particular, it is conjectured that in the early epoch of the Universe evolution the matter was in the form of a mixture of electrons, positrons, and photons in thermal equilibrium at a temperature larger of $m_e c^2$ [3]. It is therefore of principal interest to study the physics of relativistic

CP611, *Superstrong Fields in Plasmas:* Second Int'l. Conf., edited by M. Lontano et al.
© 2002 American Institute of Physics 0-7354-0057-1/02/$19.00

electromagnetic soliton-like structure in hot plasmas, since the cold fluid model [4,5] is not always appropriate to investigate the problem in its generality.

THE HYDRODYNAMIC EQUATIONS FOR A HOT PLASMA

The set of relativistic hydrodynamic equations for a hot multi-component plasma can be derived from the *conservation of the particle number*

$$\frac{\partial N_s^\alpha}{\partial x^\alpha} = 0, \tag{1}$$

and from the *conservation of the energy and momentum* [6]

$$\frac{\partial T_s^{\alpha\beta}}{\partial x^\beta} = 0, \tag{2}$$

where the index s indicates the particle species (electrons, ions, others), and the Greek letters run from 0 up to 3, denoting the four components (0 for the time, and 1,2,3 for the space) of the four-vectors; the other definitions can be found in [6,7]. Eq.(1), and the time ($\alpha = 0$) and the three-vector components ($\alpha = i = 1,2,3$) of Eq.(2) we obtain the following set of hydrodynamic equations:

$$\frac{\partial N_s}{\partial t} + \nabla \cdot (N_s U_s) = 0 \tag{3}$$

$$\frac{\partial (R_s \mathbf{p}_s)}{\partial t} + \mathbf{U}_s \cdot \nabla (R_s \mathbf{p}_s) = -\frac{1}{N_s} \nabla P_s + Q_s (\mathbf{E} + \mathbf{U}_s \times \mathbf{B}) \tag{4}$$

$$d\left(\frac{P_s \gamma_s^{\Gamma_s}}{N_s^{\Gamma_s}}\right) = 0 \tag{5}$$

where

$$R_s = 1 + \alpha_s \frac{\gamma_s P_s}{N_s m_s} \tag{6}$$

with Q_s, $\alpha_s = \Gamma_s/(\Gamma_s - 1)$, Γ_s and m_s are the electric charge, the adiabatic index and the rest mass of the s-species, respectively. We take $\Gamma_s = 5/3$ for a cold, nonrelativistic gas, and $\Gamma_s = 4/3$ for a hot, ultrarelativistic plasma. The hydrodynamic variables which have been introduced for each s-species are: the particle density in the laboratory frame, $N_s = n_s \gamma_s$, connected to the proper particle density, n_s, through the relativistic factor $\gamma_s = (1 - v_s^2)^{-1/2}$; the fluid velocity, $\mathbf{U}_s$, the momentum, $\mathbf{p}_s$, the kinetic pressure, P_s.

$\mathbf{E} = -\nabla\phi - \dfrac{\partial \mathbf{A}}{\partial t}$ and $\mathbf{B} = \nabla \times \mathbf{A}$ are the electric and the magnetic field which satisfy the equations for the electrostatic and vector potentials

$$\nabla^2\phi = -\sigma \tag{7}$$

$$\nabla^2\mathbf{A} - \frac{\partial^2\mathbf{A}}{\partial t^2} - \frac{\partial(\nabla\phi)}{\partial t} = -\mathbf{J}, \tag{8}$$

where $\sigma = \sum_s Q_s N_s$ and $\mathbf{J} = \sum_s Q_s N_s \mathbf{U}_s$ are the particle and the current densities, respectively. Here and in the following we assume that the speed of light is unitary, except where otherwise specified. By introducing the temperature of the s-species through the relationship $P_s = n_s T_s$, where

$$T_s = T_{s0}\frac{n_s^{\Gamma_s - 1}}{n_{s0}^{\Gamma_s - 1}} \tag{9}$$

and $T_{s0} = P_{s0}/n_{s0}$; the quantities labelled with "0" refer to their unperturbed values, for example taken at $|\mathbf{r}| \to \infty$. Let us define the generalized momentum of the s-species, $\mathbf{P}_s = R_s \mathbf{p}_s + Q_s \mathbf{A}$; then Eq.(4) can be put in the following form

$$\frac{\partial \mathbf{P}_s}{\partial t} = \mathbf{U}_s \times \nabla \times \mathbf{P}_s - \nabla\left(Q_s\phi + m_s^{\text{eff}}\right) \tag{10}$$

where the *effective mass* of the particle of s-species, $m_s^{\text{eff}} = m_s \gamma_s R_s$ has been introduced. From Eq.(10) it is easy to verify that the generalized vorticity $\Omega_s = \nabla \times \mathbf{P}_s$ is preserved during the time evolution of the system. Moreover, the relativistic factor can be put in the implicit form

$$\gamma_s = \left(1 + \frac{p_{s\parallel}^2}{m_s^2} + \frac{Q_s^2|\mathbf{A}_\perp|^2}{m_s^2 R_s^2}\right)^{\frac{1}{2}}, \tag{11}$$

where R_s in general cannot be considered independent of γ_s.

THE ONE-DIMENSIONAL ELECTRON-POSITRON PLASMA

Let us consider a one-dimensional geometry, where all the physical quantities depend on x only. In order to have localized solutions (for example, soliton-like ones) $A_x = 0$. Moreover, by assuming that the radiation is circularly polarized, it is worth introducing the complex amplitudes $A_\perp(x,t) = A_y(x,t) + iA_z(x,t)$ and $P_{s\perp}(x,t) = P_{sy}(x,t) + iP_{sz}(x,t)$. The full set of one-dimensional equations for a multi component plasma can be found in [7]. Here we wish to specialized theone-dimensional equations to the case of stationary soliton-like structures in an two-component plasma, of arbitrary charge and mass, i.e. we do not consider propagating electromagnetic waves. Let us introduce

$$A_{\perp}(x,t) = a(x)\exp(i\omega t) \tag{12}$$

$$p_{sx}(x) = p_s(x) \tag{13}$$

where ω is the field angular frequency. With these positions, Eqs.(7,8) become

$$\phi_{xx} = N_e - N_i \tag{14}$$

$$a_{xx} + \omega^2 a(x) = \frac{a(x)N_e}{R_e\gamma_e} + \rho Z\frac{a(x)N_i}{R_i\gamma_i}, \tag{15}$$

which have to be accompanied by the equations for the parallel components of the particle momenta:

$$\phi = R_e\gamma_e - R_{e0} \tag{16}$$

$$-\rho Z\phi = R_i\gamma_i - R_{i0}, \tag{17}$$

and the equations for the particle densities

$$N_e = \frac{\phi + R_{e0}}{\left[(\phi + R_{e0})^2 - a^2\right]^{1/2}} \cdot \left\{\frac{1}{\alpha_e\lambda_e}\left[\left[(\phi + R_{e0})^2 - a^2\right]^{1/2} - 1\right]\right\}^{\frac{1}{\Gamma_e - 1}} \tag{18}$$

$$N_i = \frac{-\rho Z\phi + R_{i0}}{\left[(-\rho Z\phi + R_{i0})^2 - \rho^2 Z^2 a^2\right]^{1/2}} \left\{\frac{1}{\rho\alpha_i\lambda_i}\left[\left[(-\rho Z\phi + R_{i0})^2 - \rho^2 Z^2 a^2\right]^{1/2} - 1\right]\right\}^{\frac{1}{\Gamma_i - 1}}. \tag{19}$$

Associated with the system of Eqs.(14,15), we have the following quantities:

$$R_e = \left[(\phi + R_{e0})^2 - a^2\right]^{1/2}, \quad R_i = \left[(-\rho Z\phi + R_{i0})^2 - \rho^2 Z^2 a^2\right]^{1/2} \tag{20}$$

$$\gamma_e = \left(1 + \frac{a^2}{R_e^2}\right)^{1/2}, \quad \gamma_i = \left(1 + \rho^2 Z^2\frac{a^2}{R_i^2}\right)^{1/2}. \tag{21}$$

Then Eqs.(14,15) form a closed system for the real functions $a(x)$ and $\phi(x)$, valid for one-dimensional stationary localized solutions and for arbitrary charges and masses of the two plasma components. In the above equations the following normalization rules have been adopted: $x\omega_{pe} \to x$, $t\omega_{pe} \to t$, $P_{s\perp}/m_e \to P_{s\perp}$, $N_s/N_{s0} \to N_s$, $eA_{\perp}(\phi)/m_e \to A_{\perp}(\phi)$, where $\omega_{pe} = (4\pi N_{e0}e^2/m_e)^{1/2}$ is the electron plasma frequency,

160

N_{s0} is the unperturbed electron density (taken at $|x| \to \infty$, m_s is the particle rest mass, Z the ionic charge, $\rho = m_e/m_i$, $N_{e0} = ZN_{i0}$, and $\lambda_s = T_{s0}/m_e$; finally, $c = 1$.

Let us consider an electron-positron (e^-/e^+) plasma in the frame of the *adiabatic* model developed above [7]. Eq.(15) decouples from Eq.(14) and we get a single equation for the amplitude $a(x)$:

$$a_{xx} + \omega^2 a(x) = \frac{2a}{\left(R_0^2 - a^2\right)^{1/2}} \left\{ \frac{1}{\alpha\lambda} \left[\left(R_0^2 - a^2 \right)^{1/2} \right] - 1 \right\}^{\frac{1}{\Gamma - 1}}, \qquad (22)$$

where: $\rho = Z = 1$, $\alpha = \alpha_e = \alpha_i$, $\lambda = \lambda_e = \lambda_i$, $T = T_e = T_i$.

Studying the Hamiltonian structure of Eq.(22), it is possible to show [7] that localized solutions (which decrease exponentially at infinity) are possible in a well defined range of radiation frequencies and of plasma temperatures. Specifically for a given frequency ω there are two limitations: (*a*) if $(2/\alpha)^{1/2} < \omega \leq \sqrt{2}$, soliton–like solutions exist for arbitrary low plasma temperature, while for $\omega < (2/\alpha)^{1/2}$ a minimum temperature exists, *i.e.* $\lambda > \lambda_1 \equiv 2(2 - \alpha\omega^2)/(\alpha\omega)^2$; (b) for any frequency below the cut-off $\omega < (2/\alpha)^{1/2}$, a maximum plasma temperature exists over which the pressure is too high to give a stable electromagnetic structure, i.e. $\lambda < \lambda_2 \equiv \alpha^{-1}(2/\omega^2 - 1)$.

Eq.(22) has been numerically integrated over a wide range of plasma and radiation parameter values. In Fig.1 we show the spatial distributions of the field amplitude $a(x)$ (*A*), plasma density $N(x)$ (*B*), and plasma temperature $T(x)$ (*C*), for $\lambda = 10^{-4}$, $\omega = 0.9$ (left column), and for $\lambda = 30$, $\omega = 0.1$ (right column). All quantities are normalized according to the rules defined above.

DISCUSSIONS OF THE RESULTS

In this paper the existence of soliton-like spatial distributions of electromagnetic energy in a hot multi-component plasma has been addressed considering the particular case of an electron-positron plasma. First, the general set of relativistic hydrodynamic equations for a two-component plasma has been derived from the conservation laws of particle density and energy-momentum. The adiabatic equation for the temperature has been used to close the system. The obtained equations are consistent with the kinetic formulation of [8], where the relativistic Maxwellian was used to close the system of equations for the moments of the particle distribution function. Then, the equations have been specialized to the one-dimensional geometry and to stationary distributions of circularly polarized electromagnetic energy. Eqs.(14-21) are still completely general and can be applied to a two-component plasma with species of arbitrary charges and masses. These equations have been written for the case of an electron-positron plasma and an extensive analysis of this physical system has been performed. The region of the parameter space where soliton-like solutions are allowed has been characterized and particular solutions have been found. At low temperature, small amplitude solitons are able to produce the full expulsion of the plasma from the region where they are situated (the plasma density being defined non-negative). At ultra-relativistic temperatures, ultra-strong solitons can be trapped in strongly overdense plasmas. Correspondingly, large temperature nonuniformities are produced.

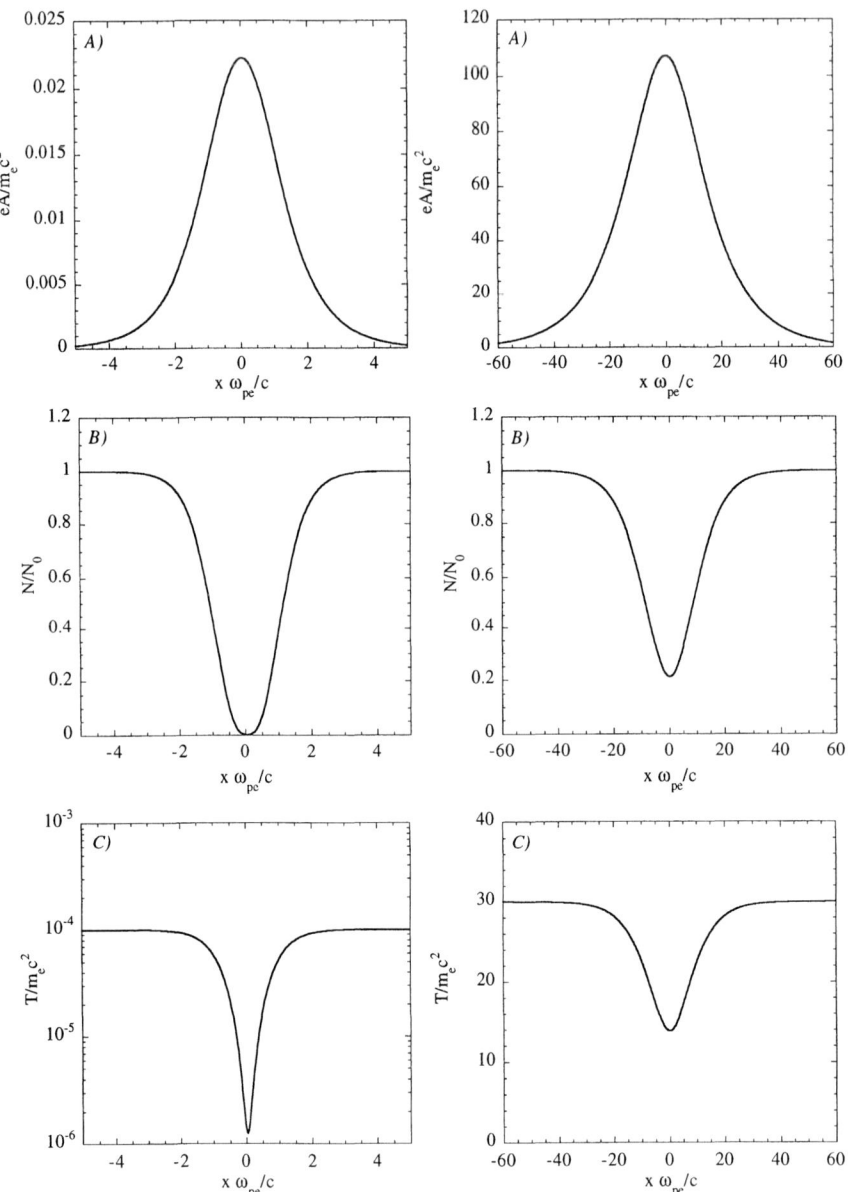

Fig.1 – Spatial distributions of the field amplitude $a(x)$ (A), plasma density $N(x)$ (B), and plasma temperature $T(x)$ (C), for $\lambda = 10^{-4}$, $\omega = 0.9$ (left column), and for $\lambda = 30$, $\omega = 0.1$ (right column). All quantities are normalized as specified in the text.

Inhomogeneous temperature distributions are allowed by the adiabatic closure of the system of equations. It means that over the scale of the soliton Coulomb collisions act to prevent the energy flow. Then the present model applies whenever the collisonal mean free path is shorter than the spatial dimensions of the solitons.

ACKNOWLEDGMENTS

M.L. feels indebted with A. Kim, M. Lisak, A.M. Sergeev, and T.Tajima for several useful discussions, comments and suggestions.

REFERENCES

[1] – S.V. Bulanov, T.Zh Esirkepov, N.M. Naumova, F. Pegoraro, V.A. Vshivkov, *Phys. Rev. Lett.* **82**, 3440 (1999).

[2] – Y. Sentoku, T.Zh. Esirkepov, K. Mima, K. Nishihara, F. Califano, F. Pegoraro, H. Sakagami, Y. Kitagawa, N.M. Naumova, S.V. Bulanov, *Phys. Rev. Lett.* **83**, 3434 (1999).

[3] – T. Tajima and T. Taniuti, *Phys. Rev. A* **42**, 3587 (1990).

[4] – J.H. Marburger, R.F. Tooper, *Phys. Rev. Lett.* **35**, 1001 (1975).

[5] – V.A. Kozlov, A.G. Litvak, E.V. Suvorov, *Sov. Phys. JETP* **49**, 76 (1979).

[6] – S. Weinberg, *Gravitation and Cosmology* (Wiley, New York, 1972).

[7] – M. Lontano, S. Bulanov, J. Koga, *Physics of Plasmas* **8**, 5113 (2001).

[8] – D.I. Dzhavakhishvili and N.L. Tsintsadze, *Sov. Phys. JETP* **37**, 666 (1973).

Fluid Modelling Of Relativistic Laser-Overdense Plasma Interaction

M. Tushentsov[1,2], A. Kim[2], F. Cattani[1], D. Anderson[1], M. Lisak[1]

[1]*Department of Electromagnetics, Chalmers University of Technology, SE-412 96 Göteborg, Sweden*
[2]*Institute of Applied Physics, Russian Academy of Sciences, 603950 Nizhny Novgorod, Russia*

Abstract. The interaction of short, relativistic laser pulses with a cold overdense plasma is investigated using computer simulations based on a fluid code. Depending on the background plasma density, two qualitatively different scenarios were observed for the penetration of the laser pulse, which is normally incident onto an overdense plasma. The first scenario is realized at moderate values of the background density ($N_0 < 1.5 N_{cr}$, where N_{cr} is the critical plasma density) and implies a dynamic regime with moving soliton-like structures, which penetrate deeply into the plasma. The second one takes place at higher densities ($N_0 > 1.5 N_{cr}$) and the laser radiation penetrates over a finite length only. As long as we can neglect the ion motion, in this regime the plasma-field structures consist of alternating electron layers separated by cavities of about half a wavelength with strong charge separation. When the effects of ion motion become significant, the electron slabs are squeezed by ions with formation of a plasma shock wave.

INTRODUCTION

The ongoing development of compact high-intensity lasers, based on the techniques of chirped pulse amplification, is allowing exploration of new regimes of laser-matter interaction, previously not achievable [1]. When such lasers irradiate the plasma, electrons oscillating in the field of the laser wave are strongly relativistic. A host of new physical phenomena has been predicted and observed in this new regime: multi-MeV electron beam generation, excitation of nonlinear plasma waves in the wake of the pulse, generation of megagauss quasistatic magnetic fields, etc. Of particular interest is the ability of anomalous penetration of a strong electromagnetic pulse into the overdense plasma by the relativistic electron inertial mass correction, the so-called self-induced transparency (SIT) [4-7]. Besides its theoretical importance, this effect is extremely attractive in connection with the recently proposed fast ignitor concept for inertial confinement fusion [2, 3].

In this paper, we address the problem of the one-dimensional propagation of an intense short laser pulse in a cold overdense plasma. We simulate laser-plasma interaction with the use of an one-dimensional electromagnetic, relativistic both for electron and ions, fluid code. The numerical results are presented for the two different scenarios of penetration depending on the background plasma density. We elucidate the role of the ion motion in the interaction dynamics and estimate the validity criterion for the immobile ion approximation.

CP611, *Superstrong Fields in Plasmas:* Second Int'l. Conf., edited by M. Lontano et al.

PROBLEM SET UP AND MODEL EQUATIONS

We consider a short laser pulse with relativistic intensity, which is normally impinging on a plasma layer of finite length and with a sharp boundary. We assume that the particle energies of the regular motion are much larger then the electron and ion temperatures i.e. the so-called cold plasma approximation is used. We will exploit a fluid model in our considerations, which represents a significant simplification over a full kinetic treatment of plasma dynamics. It is obvious that such a simplification is achieved at the cost of having discarded most of the kinetic effects (such as fast electron and ion beam production, particle trapping, etc.), but retains enough physics to be qualitatively and quantitatively useful. We also simplify our model by involving an averaging procedure to remove the fast time scale associated with the laser carrier frequency and limit our discussion to the one-dimensional case.

Taking into consideration all above approximations, the normalized governing set of self-consistent equations describing the propagation of an electromagnetic wave through a "cold two-component fluid" can be presented as

$$\frac{\partial^2 a}{\partial x^2} + \left(1 - n_0 \sum_\alpha q_\alpha^2 \frac{n_\alpha}{\gamma_\alpha}\right) a - 2i\frac{\partial a}{\partial t} = 0 \tag{1}$$

$$\frac{\partial^2 \varphi}{\partial x^2} = -n_0 \sum_\alpha q_\alpha n_\alpha \tag{2}$$

$$\frac{\partial n_\alpha}{\partial t} + \frac{\partial}{\partial x}\left(n_\alpha \frac{p_{\|\alpha}}{\gamma_\alpha}\right) = 0 \tag{3}$$

$$\frac{\partial}{\partial t}\left(n_\alpha p_{\|\alpha}\right) + \frac{\partial}{\partial x}\left(n_\alpha p_{\|\alpha} \frac{p_{\|\alpha}}{\gamma_\alpha}\right) = -q_\alpha n_\alpha \frac{\partial \varphi}{\partial x} - \frac{n_\alpha q_\alpha^2}{2\gamma_\alpha}\frac{\partial |a|^2}{\partial x}, \quad \alpha = e,i, \tag{4}$$

where $\gamma_e = \left(1 + |a|^2 + p_{\|e}^2\right)^{1/2}$ and $\gamma_i = \left(\delta^2 + Z^2|a|^2 + p_{\|i}^2\right)^{1/2}$ are the electron and ion relativistic factors correspondingly. Here φ is the electrostatic potential, a is the complex amplitude of the transversal component of the electromagnetic potential and laser radiation is circularly polarized, $\frac{e\mathbf{A}_\perp}{mc^2} = \mathrm{Re}\left[\left(\mathbf{e}_x + i\mathbf{e}_y\right)a\exp(i\omega t)\right]$, where ω is the laser carrier frequency and e is the absolute value of the electron charge. The Coulomb gauge is adopted. The n_α and $p_{\|\alpha}$ are the particle density and longitudinal momentum respectively. The plasma is characterized by three parameters: the overcritical parameter $n_0 = N_{cr}/N_0$, the ion-electron mass ratio $\delta = M/m$ and the ion charge number Z. The normalized electron and ion charges are $q_e = -1$ and $q_i = +Z$ respectively. We have rescaled every quantity to dimensionless form as:

$$\omega t \to t, \; \omega x/c \to x, \; n_{e,i}/N_0 \to n_{e,i}, \; p_{\|e,i}/mc \to p_{\|e,i}, \; e\varphi/mc^2 \to \varphi.$$

NUMERICAL RESULTS AND DISCUSSION

The first set of simulations concentrate on the situation with immobile ions ($\delta = \infty$) in order to emphasize the key role played by the ponderomotive force in the propagation dynamics. We start the simulations with a semi-infinite pulse turning on as a *tanh*-function with the purpose to test the numerical code. In this case, for maximum incident intensities lower than the threshold of penetration, after a transient stage, a stationary regime with the formation of a nonlinear skin-layer is reached, which is in perfect agreement with a previously obtained analytical solutions. Furthermore, good agreement is found with the calculated threshold for laser penetration [6, 7]. For intensities above the threshold, the nonlinear skin-layer regime is broken and the interaction leads to penetration of laser energy into the overdense plasma. The analysis of the dynamical process, above the threshold, has revealed two qualitatively different scenarios of laser penetration, depending on the overcritical parameter n_0. The following simulations were performed for a laser pulse with supergaussian envelope in the form of $a_i(t) = a_0 \exp(-(t-2\tau)^4 / 2\tau^4)$.

If $n_0 < 1.5$ we have only a dynamic regime where laser radiation penetrates into the plasma by means of generation of traveling soliton-like structures.

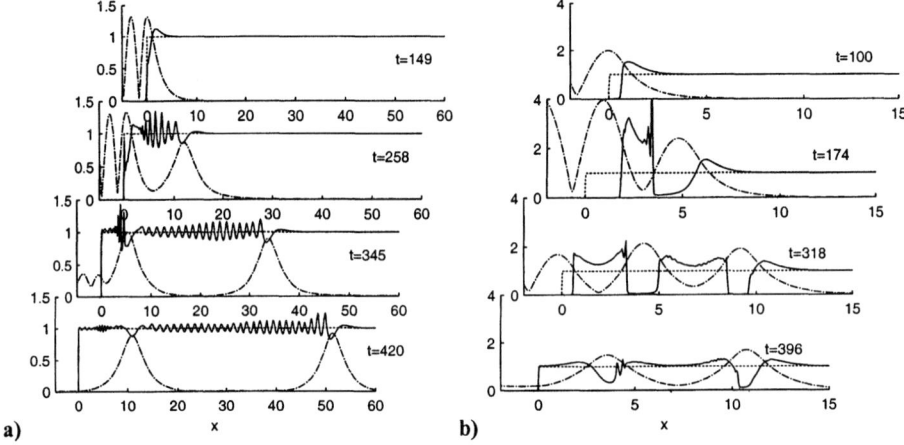

FIGURE 1. Snapshots of the evolution of the electron density (solid line) and the transversal field distribution (dash-dotted line) at a) $a_0=0.74$ and $\tau=100$, $n_0=1.3$; b) $a_0=1.9$ and $\tau=100$, $n_0=1.6$. The dotted line represents the ion density. Ions are assumed immobile.

In Fig. 1a, the temporal evolution of a laser pulse with $a_0=0.74$, interacting with a plasma with $n_0=1.3$, is depicted. At these pulse parameters, two solitary waves with single hump distributions are generated at the plasma boundary and then propagate as quasi-stationary plasma-field structures with a velocity about $0.2c$. It is important to note that such single humped structures, in contrast to distributions with nodes having specific properties corresponding to a discrete spectrum of propagation velocities [8], theoretically should only exist in slightly overdense plasmas, where $n_0-1 \ll 1$ and for low relativistic amplitudes. These structures are pure relativistic solitons where the contribution of the striction nonlinearity to the electron density perturbation is much weaker than the one due to the relativistic nonlinearity. At higher values of the overcritical parameter, $n_0 > 1.5$, the scenario of the interaction is completely different, as shown in Fig. 1b, where the plasma has $n_0=1.6$. At the beginning of the spatial evolution the characteristic distribution of a nonlinear skin-layer is forming. The action of the ponderomotive force at the vacuum-plasma boundary then pushes the electrons into the plasma. When the field amplitude on the real boundary exceeds the threshold, the electrons spill out into the region of the field minimum and start to create an electron slab in that area. As a consequence of this process, a deep electron cavity is created, near the field maximum. The generated plasma-field structure slowly propagates into the plasma and then the same process is later repeated until the creation of the electron slab, closest to the plasma-vacuum boundary, becomes opaque for the given incident intensity. Thus, for comparatively long pulses, after a transient stage, the plasma settles down into a quasi-stationary plasma-field distribution, allowing for penetration of the laser energy over a finite length only, a length which increases with increasing incident intensities (in Fig. 1b to a depth of $x \approx 10$). The electron density distribution becomes structured as a sequence of electron layers superimposed on the ion background and separated by about half a wavelength of depleted regions, which acts as resonators. The peak electron density increases from layer to layer reaching an absolute maximum in the layer closest to the vacuum boundary. At the same time the width of the layers becomes more and more narrow. Such nonlinear plasma structures with strong charge separation act as a distributed Bragg reflector and are quite close to those described analytically in [9]. Finally, when the pulse is turning off, the electrons fill out the cavities and the electron slabs are destroyed one by one. In connection with this, the energy, which is stored in the resonators, is radiated back into the vacuum−a boomerang effect. It is easy to find the validity criterion for the approximation of immobile ions, using the above information. The ions may be considered as immobile for pulse duration smaller than a characteristic time for the ions t^*, which can be estimated as $t^* \approx L/v_i$, where L is the width of the cavity and v_i is the ion velocity. For motionless structures with $L \approx \lambda/2$ and completely uncompensated ion charge (see Fig. 1b) we have $t^* \approx \omega_i^{-1}$, where ω_i is the ion plasma frequency.

The real two-component plasma dynamics also includes the ion motion, which is necessary to take into account for long enough laser pulses. When the ions start to move they tend to decrease the level of charge separation.

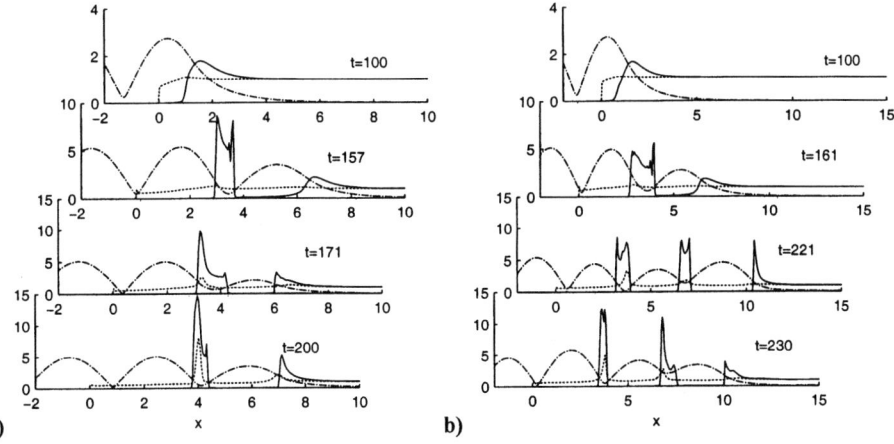

FIGURE 2. Snapshots of the evolution of the electron and ion densities (solid and dotted lines respectively) and the transversal field distribution (dash-dotted line) at $a_0=2.6$, $\tau=100$, $n_0=1.6$ and two different ion-electron mass ratio: a) $\delta=1840$ (hydrogen), b) $\delta=3680$ (deuterium).

A new feature of the plasma dynamics emerges in the case of mobile ions: the possibility of shock wave formation at the plasma-vacuum boundary. It is evident that neither the solitons nor the multi-layer structures can be produced in an electron-positron plasma ($\delta=1$), because the negative and positive charges are equally affected by the ponderomotive force. Only the formation of a plasma shock wave without charge separation can occur in such plasma. Another situation takes place in an electron-ion plasma. Let us consider again the case, where $n_0>1.5$. At the initial stage of interaction the cavity structure can be generated before the ion density becomes inhomogeneous, as is clearly seen in Fig. 2a. Here a laser pulse with $a_0=2.6$ (corresponding intensity is $1.44\times10^{19}W/cm^2$) and 200 time units long (1 fs is about 2 dimensionless time units at wavelength 1 μm) interacts with a hydrogen ($\delta=1840$) plasma, which has the same overcritical index as in Fig. 1b, $n_0=1.6$. For these parameters, the characteristic time for the ions is $t^*\approx40\,fs$. Up to the time $t\approx170$, the penetration proceeds on a smooth ion background, i.e. as in the immobile ion model in Fig. 1b. At subsequent moments of time the ions start to produce an inhomogeneity in the form of an ion layer at $x\approx4$ and an ion peak at $x\approx7$ at the time $t=200$. These aggregations of ions are later transformed into a plasma shock wave, which slowly compresses the plasma. It is important to note that the effective ponderomotive force producing the shock wave corresponds to intensity values less than the maximum incident one and usually is of the order of the threshold of SIT. For the case of heavy ions, which is important for possible applications, deposition of laser energy into the plasma through the excitation of the cavity structures in the SIT regime will be illustrated more clearly. In Fig. 2b we present results for the same parameters as in Fig. 2a, but for $\delta=3680$ (deuterium). This situation roughly describes completely stripped multi-charge heavy ions that can occur at relativistic intensities. As shown in Fig. 2b, multi-electron layers separated by pure ion regions are built up, where strong electrostatic fields due to charge separation are produced. Obviously, these fields are responsible for fast ion beam generation, which is not included in our fluid model.

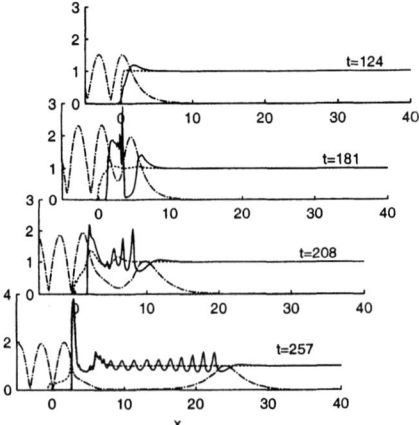

FIGURE 3. Snapshots of the evolution of the electron and ion densities (solid and dotted lines respectively) and the transversal field distribution (dash-dotted line) at $a_0=1.0$, $\tau=100$, $n_0=1.3$ and $\delta=3680$ (deuterium).

However, since the ion beams only absorb a small amount of the laser energy we can expect that the macrodynamics of the interaction will exhibit the same features;

moreover by using these generated plasma-field structures we can estimate possible energetic parameters of the fast ions. On the longer time scale, the electron density distribution will shrink and will be localized around the ion peaks (at $x \approx 4$, 7) and only the last layer at $x \approx 10$ at the moment $t=230$ will form a plasma shock wave. At lower amplitudes $a_0 < 2.6$, the interaction dynamics is completely different and occurs through formation of plasma shock waves with low levels of charge separation.

The regime of interaction, when $n_0 < 1.5$, is not so sensitive to the ion mobility at moderate relativistic intensities, due to the comparatively fast motion of the soliton structures, as is clearly seen in Fig. 3. The generated soliton pass ahead of shock wave formation at the plasma-vacuum boundary and their movement occurs on the unperturbed ion background.

CONCLUSION

We have used a one-dimensional fluid code to explore the interaction of ultraintense laser radiation with an overdense plasma. It has been shown that, depending on the background supercritical density, there are two qualitatively different scenarios of laser energy penetration into overdense plasmas in the regime of relativistic self-induced transparency. For plasma densities less than $N_0 < 1.5 N_{cr}$, the penetration of the laser energy occurs in the form of long-lived soliton-like structures, which are generated at the vacuum-plasma boundary plasma and then propagate into the plasma. At higher densities, the interaction implies the generation of plasma-field structures consisting of alternating electron and subsequent ion layers. The electromagnetic energy then penetrates into the overdense plasma over a finite length only, as determined by the incident intensity. The impact of the ion motion on the penetration dynamics has been elucidated and the validity of the immobile ion approximation for short enough laser pulses has been confirmed.

REFERENCES

1. S.C. Wilks and W.Kruer, IEEE Trans. **QE 33**, 154 (1997); see also in Superstong Fields in Plasmas, AIP Conf. Proc. **426** (1998)
2. M. Tabak *et al.*,Phys. Plasmas **1**, 1626 (1994)
3. S.C. Wilks *et al.*,Phys. Rev. Lett. **69**, 1383 (1992).
4. A.I. Akhiezer and R.V. Polovin, Sov. Phys. JETP **3**, 696 (1956).
5. P. Kaw and J. Dawson, Phys. Fluids **13**, 472 (1970).
6. J. H. Marburger and R. F. Trooper, Phys. Rev. Lett. **35**, 1001 (1975)
7. F. Cattani *et al.*, Phys. Rev. E **62**, 1234 (2000).
8. V.A. Kozlov, A.G. Litvak and E.V. Suvorov, Sov. Phys. JETP 49, 75, (1979)
9. A. Kim et al., JETP Lett. 72, 241 (2000).

Transformation of Laser Radiation into Post-Solitons with Ion Acceleration

N. Naumova[*], S. Bulanov[†], T. Esirkepov[**], D. Farina[‡], K. Nishihara[§],
F. Pegoraro[¶], H. Ruhl[*] and A. Sakharov[†]

[*]Max-Born-Institut, Berlin, Germany
[†]General Physics Institute, Moscow, Russia
[**]Moscow Institute of Physics and Technology, Dolgoprudny, Russia
[‡]Istituto di Fisica del Plasma, CNR, Milano, Italia
[§]Institute of Laser Engineering, Osaka University, Osaka, Japan
[¶]Department of Physics, Pisa University and INFM, Pisa, Italia

Abstract. With 2D particle in cell simulations we show that during the interaction of a high intensity ultra short laser pulse with a collisionless plasma in the wake of the laser pulse electromagnetic relativistically strong solitons are formed, which evolve asymptotically into post-solitons. A post-soliton is a slowly expanding cavity in the ion and electron densities which traps electromagnetic energy. Fast ions are accelerated during the post-soliton formation. The long time evolution of the post-solitons in an inhomogeneous plasma slab is investigated. The spectra of the electromagnetic energy radiated by the post-solitons are analyzed.

INTRODUCTION

As is well known, solitons are of fundamental importance for basic nonlinear science [1]. Relativistic solitons were predicted to occur in laser produced plasmas and were found with particle in cell (PIC) simulations. Relativistic solitons are self-trapped, finite size, electromagnetic (e.m.) waves that propagate without diffraction spreading. Self-trapping appears because the e.m. wave modifies the local refractive index through the relativistic increase of the electron mass and the redistribution of the electron density under the ponderomotive pressure of the radiation. The analytical theory of relativistic e.m. solitons has been developed in Refs. [2, 3, 4, 5]. Relativistic solitons have been seen in multi-dimensional PIC simulations of the laser pulse interaction with a plasma [6, 7, 8, 9]. Often solitons are considered as one of the elementary entities by which the turbulence in plasmas is formed [10].

The theory of these solitons has been developed mainly within the framework of the approximation of unmovable ions. When the ion dynamics is taken into account, solitons have been found to exist only at propagation velocities larger than a critical velocity [5]. On the other hand, PIC simulations [6] demonstrated that the solitons are generated in the wake left behind the laser pulse and that they propagate with a velocity that is well below the speed of light and in a homogeneous plasma is almost equal to zero. The time of the soliton formation is much shorter than the ion response time so that ions can be assumed to be at rest during the soliton formation. Ions can be assumed to be at rest during approximately $(m_i/m_e)^{1/2}$ oscillation periods of the e.m. field inside the

CP611, *Superstrong Fields in Plasmas:* Second Int'l. Conf., edited by M. Lontano et al.
© 2002 American Institute of Physics 0-7354-0057-1/02/$19.00

soliton (the period of the field oscillation inside the soliton is of the order of $2\pi\omega_{pe}^{-1}$). For longer times the ponderomotive pressure of the e.m. field inside the soliton starts to dig a hole in the ion density and the parameters of the soliton change. Rigorously speaking, what was a soliton on the electron timescale ceases to be a soliton on the ion timescale. Nevertheless, a low frequency e.m. wave packet remains in the plasma, being well confined inside the slowly expanding plasma cavity. We call this nonlinear e.m. wave packet a "post-soliton" [11]. It is important to know the properties of post-solitons because in an underdense plasma a substantial part of the laser pulse energy can be converted into solitons [6, 7, 8] which afterwards evolve into post-solitons.

POST-SOLITON FORMATION AND EVOLUTION

Simulation parameters

In the simulations presented here the size of the simulation box is 30λ along x (the direction of the laser pulse propagation) and 30λ along y (transverse direction). The cell size is $1/16 \times 1/16 \, \lambda^2$ with 9 electrons and 9 ions per cell. The total number of particles is approximately $3 \cdot 10^6$. The plasma begins at $x = 5$ and is preceded by a vacuum region 5λ long. The normal incident laser pulse is initialized outside the plasma in the vacuum region $x < 0$. In the run shown here, the dimensionless amplitude of the laser radiation is $a = eE/m\omega c = 1$. The laser pulse has a gaussian form both in the x- and in the y-directions, with full width $l_\perp = 5\lambda$ and length $l_\parallel = 4\lambda$, and is linearly s-polarized with its incident electric field along the z-axis. In the simulations presented here, the ion mass is $1836m_e$. The plasma density is $n_0 = 0.36 \, n_{cr}$, where $n_{cr} = \omega^2 m/4\pi e^2$ is the critical density.

Post-Solitons in a Homogeneous Plasma

The evolution in the (x, y) plane of the z component of the electric field (first row) and of the electron and ion densities (second and third rows) at $t = 30, 70$, and 120 (a, b, and c columns) is shown in Fig 1. Here, and in the following, the time and space-coordinates are measured in laser periods and wavelengths respectively, the electron and ion densities in units of the initial density and the electric field in units of $m\omega c/e$.

In the distribution of the z component of the electric field at time $t = 30$, we see the laser pulse, which has changed its form due to the nonlinear processes, the scattering of the e.m. radiation and the formation of an s-polarized e.m. soliton of the type considered in Refs. [6, 7, 8]. The laser pulse has changed its shape due to nonlinear processes such as self-focusing and energy depletion. The formation of the soliton, as discussed in Ref. [7], is related to the trapping of the e.m. energy due to the pulse frequency downshift caused by the pulse depletion. The soliton is seen at the center of the computation region. In the distribution of the electron density we see a local minimum in the position of the soliton and no substantial modulation in the ion density distribution. In the distribution of the electron density we see also the modulations due to excitation of the wakefield

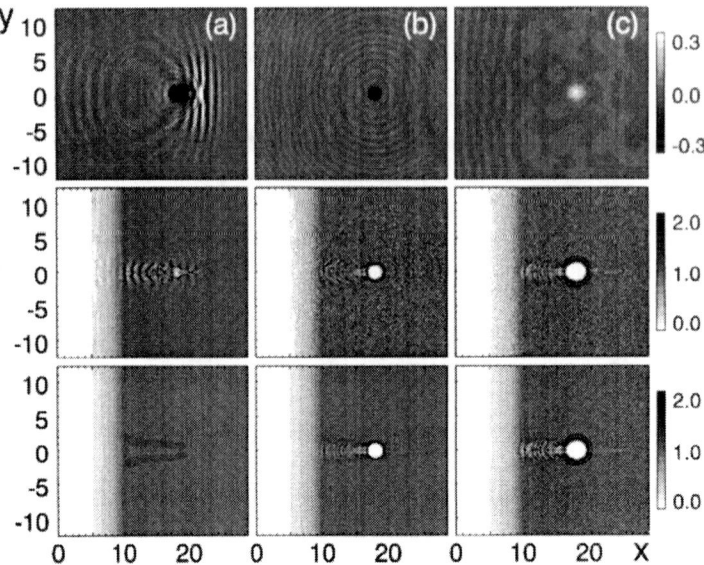

FIGURE 1. Interaction of an *s*-polarized laser pulse with the plasma: the *z* component of the electric field (first row), the electron density (second row) and the ion density (third row) in the x, y plane at $t = 30$ (column *a*), $t = 70$ (column *b*), $t = 120$ (column *c*).

behind the laser pulse near the pulse axis in the region $10 < x < 15$ and a bunch of fast electrons at $x = 22.5$, $y = 0$. However, the wakefield generation and the electron acceleration do not influence the soliton significantly. At time $t = 70$, after the remnants of the pulse have left the simulation box, we see the post-soliton that remains at the place where the soliton was formed. During its evolution the post-soliton radiates away a portion of its energy in the form of an e.m. wave. The electric field inside the soliton oscillates periodically, changing its sign as in the case of electron solitons discussed in Refs. [7, 8]. The hole in the electron density becomes wider and we see a hole in the distribution of the ion density with almost the same radius as that of the electron hole. The largest density corresponds to the rings at the boundary of the post-soliton. The third column in Fig. 1 shows the asymptotic structure of the post-soliton at time $t = 120$. The post-soliton has a perfect circular form, aside from the effect of the interaction with the wake field. The radius of the hole in the electron and ion densities increases. The width of the density ring becomes larger. We see a small scale structure inside the ring, which corresponds to three confocal rings.

We can describe the scenario of the post-soliton formation as follows. Since the soliton formation time is much shorter than the time of the ion response $\approx 2\pi\omega_{pi}^{-1}$, ions can be assumed to be at rest during the soliton formation. The ponderomotive pressure of the e.m. field inside the soliton is balanced by the force due to the electric field which appears due to charge separation. The amplitude of the resulting electrostatic potential is given by $\phi = (1 + a_m^2)^{1/2}$. The ponderomotive pressure displaces the electrons outward and the Coulomb repulsion in the electrically non neutral ion core pushes the ions away.

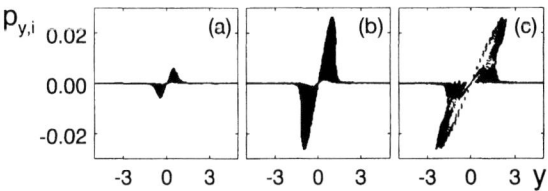

FIGURE 2. The ion phase plane (p_y, y) inside the post-soliton at $x = 18$, $y = 0$ for $t = 30$ (a), 70 (b) and 120 (c).

The typical ion kinetic energy corresponds to the electrostatic potential energy which is of the order of $m_e c^2 a_m$. This process is similar to the so-called "Coulomb explosion" inside of self-focusing channels (see Refs. [12]).

In Fig. 2 we show the ion phase plane (p_y, y) at $t = 30, 70, 120$. We see that the radial component of the momentum of the fastest ions does not depend on time for $t > 30$. The fastest ions are located at the outermost ring seen in the ion density distribution in Fig.1.

The Snow-Plow Model

The ion expansion in the radial direction leads to the digging of a hole in the ion density. The cavity formation in the distribution of the electron and ion densities is shown in Fig. 1. The plasma cavity forms a resonator for the trapped e.m. field. During the cavity expansion, the amplitude and the frequency of the e.m. field decrease. Since the radius of the cavity increases very slowly compared to the period of the e.m. field oscillations we can use the adiabatic approximation to find their dependence on time. From the adiabatic invariant $\int \mathbf{E}^2 dV / \omega_s = const$ we see that the e.m. field amplitude decreases as $E \sim 1/R^2$. We use a snow-plough model to describe the cavity wall motion under the ponderomotive pressure of the trapped mode. The characteristic expansion time of the post-soliton is found to be given by $\tau = \sqrt{6\pi R_0^2 n_0 m_i / \langle \mathbf{E}_0^2 \rangle}$ with n_0 the plasma density and $R_0 \approx d_e$ and $\mathbf{E}_0^2/8\pi$ the initial soliton radius and e.m. energy density. For $t/\tau \gg 1$, the post-soliton radius increases as $R \approx R_0 (2t/\tau)^{1/3}$, and its amplitude and frequency decrease as $E \sim t^{-2/3}$ and $\omega_s \sim t^{-1/3}$. Similarly, in the case of the cylindrical cavity, we find that $R \sim t^{2/5}$, when $t \to \infty$.

In Fig. 3 (a) we present the log-log plot of the dependence of the cavity radius versus time, as obtained from the simulation. We see that this dependence is in fairly good agreement with the $R \sim t^{2/5}$ law. The time dependence of the maximum positive value of the z component of the momentum of different electrons inside the post-soliton is plotted in Fig. 3 (b). The band at the bottom of the figure is due to electron heating. We see that the amplitude and frequency of oscillations decrease in time.

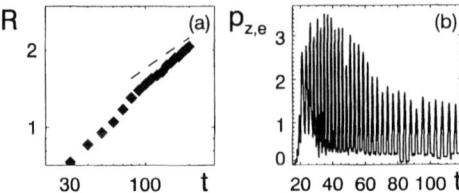

FIGURE 3. Dependence of the logarithm of the cavity radius on the logarithm of time (*a*); time dependence of the maximum positive value of momentum along *z* of different electrons inside the post-soliton (*b*).

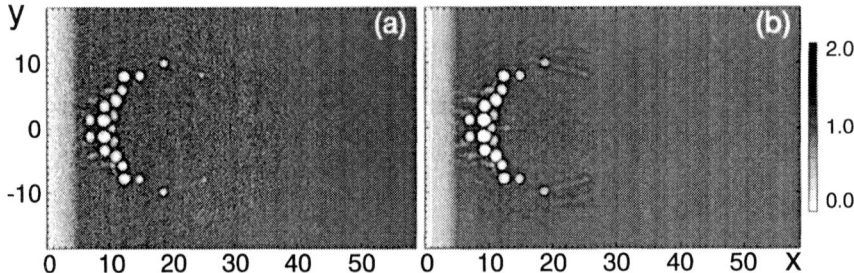

FIGURE 4. Distribution of the electron (*a*) and ion (*b*) density in the (*x*,*y*) plane showing the post-soliton "necklace" formed behind a wide laser pulse at $t = 70$.

Post-Soliton Necklace

Wider laser pulses produce a cloud of solitary waves in their wake [7]. When the ponderomotive pressure of the e.m. field inside the solitons starts to dig holes in the ion density the soliton cloud evolves into post-solitons as shown in Fig. 4 for an s-polarized pulse with amplitude $a = 1$, width 30λ and length 15λ in an underdense plasma ($n/n_{cr} = 0.64$) at $t = 70$. In the distribution of the electron (Fig. 4 (a)) and ion densities (Fig. 4 (b)) we see a "necklace" of the post-solitons.

Effects of the Plasma Inhomogeneity

In a non-uniform plasma, the propagation of a wave is strongly affected by the inhomogeneity of the medium. The propagation of solitons in inhomogeneous media was discussed in Refs. [8],[9]. In order to study the effects of the plasma inhomogeneity on the post-soliton formation and evolution here we present the results of 2D3V PIC simulations of the laser pulse propagation in a plasma slab inhomogeneous in the direction perpendicular to the pulse propagation. In inhomogeneous media a soliton moves with an acceleration that is directed towards the low density side and oscillates in the transverse direction in the density channel formed by the pulse. As a result, the

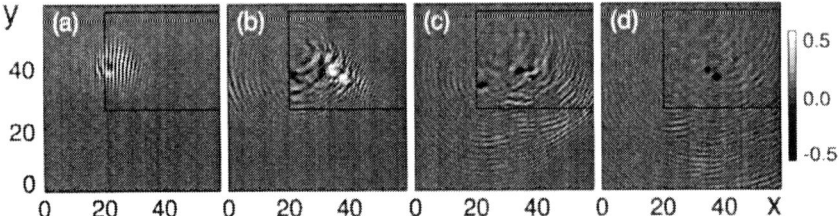

FIGURE 5. Interaction of an *s*-polarized laser pulse with an inhomogeneous plasma: the *z* component of the electric field in the x, y plane at $t = 40$ (a), $t = 60$ (b), $t = 80$ (c), $t = 100$ (d). The region where the plasma slab is located is marked by the frame.

soliton moves towards the plasma vacuum interface, where it radiates its energy away in the form of low frequency e.m. waves, during its non-adiabatic interaction with the plasma boundary. This process has been discussed in Refs.[8],[9]. However in the case of a post-soliton, the conditions for its propagation change since the post-soliton digs a hole in the plasma density and thus modifies the local inhomogeneity.

In the simulation presented in Fig. 5 the plasma slab is located inside the subdomain $20 < x < 60$, $26 < y < 57$, marked by the frame in Fig. 5. The size of the simulation box is 60×60. The plasma density has a linear profile in the y direction and changes from $0.1n_{cr}$ to $0.72n_{cr}$. The laser pulse is *s*-polarized, its amplitude is equal to $a = 1$, the width is $l_\perp = 10$, the length is $l_\parallel = 15$. At time $t = 40$ (a) the pulse has already changed the direction of propagation due to the plasma nonuniformity. Two solitons are formed close to the left hand side plasma boundary. At time $t = 60$ (b) we see a strong burst of the e.m. radiation in the left hand side vacuum region and the beginning of the soliton motion in the y direction. At time $t = 80$ (c) we see that the second soliton is still near the left hand boundary while at time $t = 100$ (d) it abandons the plasma and transforms its energy into low frequency e.m. radiation. In frames (b)-(d) we see that two new solitons are formed in the central region of the plasma. These solitons evolve into post-solitons and remain at the same place where they were formed.

In Fig. 6 we show the spectra of the radiation transmitted through the left hand side boundary (a) and through the lower boundary (b) of the plasma slab. All the radiation has a downshifted frequency except for a small portion, the spot with $\omega = 1$ at $y \approx 45$ in Fig. 6 (a), that is reflected from the plasma. The frequency of the radiation that has left the plasma through its left hand side boundary is $\omega \approx 0.55$ (Fig. 6 (a)). The e.m. bursts radiated through the lower boundary are localized in the region $20 < x < 40$, with frequency $0.45 < \omega < 0.7$, and correspond to the e.m. radiation originated from the solitons.

CONCLUSION

When a relativistically strong e.m. wave propagates in a plasma, a portion of its energy is transformed into solitons which then form slowly expanding cavities in the distribution of the ion and electron densities with e.m. fields trapped inside. During the post-soliton

175

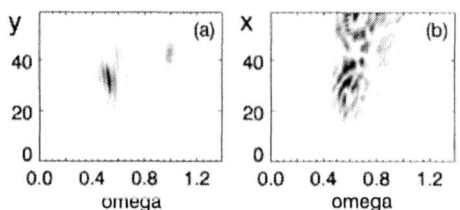

FIGURE 6. Frequency spectra versus y (a) and versus x (b).

formation fast ions are generated. The plasma inhomogeneity can lead to to the soliton propagation towards the plasma vacuum interface and to the emission of low frequency electromagnetic bursts.

ACKNOWLEDGMENTS

We are pleased to acknowledge the use of the Cray T3E supercomputing facility at the Konrad-Zuse-Zentrum für Informationstechnik Berlin (ZIB).

REFERENCES

1. Whitham, G. B., *Linear and Nonlinear Waves*, Wiley, New York, 1974.
2. Gersten, J. I., and Tzoar, N., *Phys. Rev. Letters*, **35**, 934 (1975); Marburger, J. H., and Tooper, R. F., *Phys. Rev. Letters*, **35**, 1001 (1975).
3. Kozlov, V. A., et al., *Sov. Phys. JETP*, **49**, 75 (1979); Kaw, P. K., et al., *Phys. Rev. Letters*, **68**, 3172 (1992).
4. Esirkepov, T. Zh., et al., *JETP Letters*, **68**, 36 (1998); Farina, D., et al., *Phys. Rev. E*, **62**, 4146 (2000).
5. Farina, D., and Bulanov, S. V., *Phys. Rev. Letters*, **86**, 5289 (2001); *Plasma Phys. Reports*, **27**, 680 (2001).
6. Bulanov, S. V., et al., *Phys. Fluids B*, **4**, 1935 (1992); Bulanov, S. V., et al., *Plasma Phys. Reports*, **21**, 600 (1995).
7. Bulanov, S. V., et al., *Phys. Rev. Letters*, **82**, 3440 (1999).
8. Sentoku, Y., et al., *Phys. Rev. Letters*, **83**, 3434 (1999).
9. Bulanov, S. V., et al., "Relativistic interaction of laser pulses with plasmas", in *Reviews of Plasma Physics*, **22**, Kluwer Academic / Plenum Publishers, New York, 2001, p. 227.
10. Mima, K., et al., *Phys. Plasmas*, **8**, 2349 (2001).
11. Naumova, N. M., et al., *Phys. Rev. Letters*, **87**, 185004 (2001).
12. Sarkisov, G. S., et al., *Phys. Rev. E*, **59**, 7042 (1999); Esirkepov, T. Zh., et al., *JETP Letters*, **70**, 82 (1999); Bulanov, S. V., et al., *JETP Letters*, **71**, 407 (2000); Sentoku, Y., et al., *Phys. Rev. E*, **62**, 7271 (2000).

Spatiotemporal Self-Focusing and Splitting of a Femtosecond, Multiterawatt, Relativistic Laser Pulse in an Underdense Plasma

Ivane G. Murusidze, Givi I. Suramlishvili

Institute of Physics, Georgian Academy of Sciences, 380077 Tbilis Georgia

Maurizio Lontano

Istituto di Fisica del Plasma, C.N.R., 20125 Milan, Italy

Abstract. The dynamics of ultrashort, high-power laser pulses in underdense plasmas differs dramatically from that of long laser beams. We present the results of numerical studies of this dynamics within a model, which systematically incorporates finite pulse length (i.e., dispersion) effects along with diffraction and nonlinear refraction in a strongly nonlinear relativistic regime. New spatio-temporal patterns of self-compression and self-focusing typical of ultrashort high-intensity laser pulses are analyzed. The parameters of our numerical simulations correspond to a new class of high-peak-power ($\geq$ 100 TW), ultrashort-pulsed laser systems, producing pulses with a duration in the 10 – 20 femtosecond range.

INTRODUCTION

The advent of short-pulse laser technology, based on the technique of chirped-pulse amplification [1,2], has resulted in laser systems, which are capable of delivering high-peak-power (from multiterawatt to petawatt) pulses on a femtosecond time scale [2-4]. The output intensities reach 10^{19} W/cm^2, whereas focused intensities are in the 10^{20}– 10^{21} W/cm^2 range. At these intensities, the electron motion in the laser fields becomes highly relativistic. These recent achievements are focusing the nonlinear optics of laser pulses in plasmas to a new physical regime of laser-plasma interactions, dominated by strong relativistic nonlinearities and finite-pulse-length effects. In spite of the intensive theoretical studies [5-8], the dynamics of ultrashort relativistic laser pulses in plasmas is not yet completely understood. First, the role of nonlinearities higher than the third order is not well studied, and, second, conventional theories are based on the paraxial wave equation with diffraction and nonlinear refraction terms. Neglecting dispersion effects arising due to the finite pulse duration, this approach is incapable of describing the evolution the pulse's temporal profile and related effects, which become essential for the ultrashort pulse dynamics.

In this paper we study the propagation dynamics of laser pulses within the parameter regime corresponding to a new class of compact, high-power, ultrashort pulsed laser systems, producing pulses in the 10-20 fs range [3,4]. For such pulses, the

CP611, *Superstrong Fields in Plasmas:* Second Int'l. Conf., edited by M. Lontano et al.

conventional paraxial approach is no longer an adequate description. In the subsequent section an adequate physical model of the propagation of an ultarshort, ultraintense laser pulse in underdense plasmas is developed. Within this model, we formulate the so-called pulsed paraxial wave equation (PPWE), which extends the conventional paraxial wave equation, systematically incorporating all relevant finite-pulse-length effects. With the group velocity dispersion (GVD) and the first-order dispersion (FOD) terms, the PPWE is applicable to a wide range of ultrashort laser pulses including single-cycle pulses. The equation describing the dynamic plasma response to such a short pulse completes the set of coupled equations that provides a fully relativistic formulation of the propagation of ultrashort, ultrabright laser pulses in underdense plasmas. Numerical simulations have revealed new self-focusing (SF) and self-compression (SC) patterns, typical of only ultrashort, relativistic laser pulses. Spatiotemporal dynamics of these self-effects and underlying physical mechanisms are discussed.

PHYSICAL MODEL. PULSED PARAXIAL WAVE EQUATION

We describe the ultrashort-pulse laser interaction with a cold, electron plasma within the set of Maxwell's equations in terms of the vector and scalar potentials, $\mathbf{A}$, φ, and the relativistic electron fluid equations

$$\left(\nabla^2 - c^{-2}\partial_{tt}^2\right)\mathbf{A} = k_p^2 n\mathbf{p}/\gamma + c^{-1}\partial_t\nabla\varphi \tag{1}$$

$$\nabla^2\varphi = k_p^2(n-1) \tag{2}$$

$$\partial_t\mathbf{p} = \partial_t\mathbf{A} + c\nabla(\varphi - \gamma) \tag{3}$$

$$\partial_t n + c\nabla\cdot(n\mathbf{p}/\gamma) = 0 \tag{4}$$

Here and throughout the paper the dynamical variables $\mathbf{A}$, φ, the electron momentum $\mathbf{p}$, and the electron density n are normalized as follows: $(\mathbf{A},\varphi) \equiv e(\mathbf{A},\varphi)/mc^2$; $\mathbf{p} \equiv \mathbf{p}/mc$; $n \equiv n/n_0$, where n_0 is the unperturbed electron density. The symbol ∂ stands for partial derivatives, e.g., $\partial_t \equiv \partial/\partial t$. $\gamma = (1+\mathbf{p}^2)^{1/2}$ is the relativistic factor, $k_p = \omega_p/c = (4\pi e^2 n_0/mc^2)^{1/2}$ is the plasma wave number. The ions are assumed to be virtually immobile during the transit time of the ultrashort-pulse laser. The field equation (1) and (2) are in the Coulomb gauge, $\nabla\cdot\mathbf{A} = 0$. The electron equation of motion (3) implies the conservation of the curl of the canonical momentum of electrons, $\nabla\times(\mathbf{p} - \mathbf{A}) = 0$.

Consider a circularly polarized laser pulse propagating in the z direction introducing the transverse component of its vector potential as

$$\mathbf{A} = (1/2)(\mathbf{x} \pm i\hat{\mathbf{y}})A(\mathbf{r},t)\exp(ikz - i\omega t) + c.c., \tag{5}$$

where $A(\mathbf{r},t)$ is the pulse envelope, generally a complex valued function, ω is the central frequency, and $k = (\omega/c)(1 - \omega_p^2/\omega^2)^{1/2}$ is the corresponding wavenumber in a transparent plasma.

Within our model we examine a regime of the ultrashort pulse propagation through an underdense plasma, $\omega_p/\omega \ll 1$, when

$$\ell_L/\ell_\perp \sim \ell_\perp/\ell_\parallel \sim \omega_p/\omega \equiv \varepsilon \ll 1, \tag{6}$$

where ℓ_L is the pulse length, $\ell_\perp$ is the transverse to the propagation direction length scale of the pulse, and $\ell_\parallel$ defines the characteristic length scale over which the laser pulse changes significantly due to the propagation effects. According to Eq. (6), the laser pulse may be identified as a thin, wide disk of light with a sharp intensity gradient along the propagation direction and a smooth transverse profile. The hierarchy of scales implied by (6) naturally leads to the plasma response, which is described self-consistently within the quasistatic approximation [5,9]. Taking these into account, one can expand Eqs. (1)-(4) in powers of the small parameter ε. To the lowest order (zeroth-order in ε) we derive the following set of coupled equations [9,10]

$$2ik\frac{\partial A}{\partial \zeta} - \frac{2}{v_g}\frac{\partial^2 A}{\partial \tau \partial \zeta} + D\frac{\partial^2 A}{\partial \tau^2} + \nabla_\perp^2 A = k_p^2\left(\frac{\beta_g}{\left[(1+\varphi)^2 - \gamma_g^{-2}(1+|A|^2)\right]^{1/2}} - 1\right)A, \tag{7}$$

$$\frac{\partial^2 \varphi}{\partial \tau^2} = \omega_p^2(\beta_g\gamma_g)^2\left(\frac{\beta_g(1+\varphi)}{\left[(1+\varphi)^2 - \gamma_g^{-2}(1+|A|^2)\right]^{1/2}} - 1\right). \tag{8}$$

Here we have employed a moving frame of reference introducing the new variables $\zeta = z$, $\tau = t - z/v_g$, where $v_g \equiv \partial k/\partial\omega = kc^2/\omega$ is the group velocity, $\beta_g = v_g/c$, $\gamma_g = (1-\beta_g^2)^{-1/2}$. The second and the third terms on the left-hand side of the wave equation (7) are due to the finite pulse duration. They represent the first- and the second-order dispersion effects, respectively. The coefficient D is proportional to the group-velocity dispersion parameter, $D \equiv -k\partial^2 k/\partial\omega^2$. In our case $D = v_g^{-2}(\omega_p/\omega)^2$, i.e., the third term in (7) accounts for negative or "anomalous" GVD. The fourth term is the diffraction term. The terms contributing to the nonlinear refraction are on the right side of Eq. (7). The only term neglected on the left of Eq. (7) is the term $\partial_{\zeta\zeta}^2 A$, which is second-order in ε, i.e., the conventional paraxial approximation is naturally assumed within the parameter regime outlined by Eq. (6). In addition, under the same conditions, the GVD and the diffraction terms contribute equally in the pulse evolution. The FOD term extends the applicability of Eq. (7) down to single-cycle pulses. Note that the phase in Eq. (5) $\phi = kz - \omega t$ varies slowly with ζ, Indeed in our case $|\partial_\zeta\phi| = \beta_g^{-2}\varepsilon^2 k \ll k$, which is also one of the prerequisites for the consistent extension of the wave equation down to single-cycle regime [11]. Incorporating the finite-pulse-length effects within the paraxial approximation, the wave equation (7) describes ultrashort pulses propagating in underdense plasmas in the regime when dispersion and diffraction effects play an equally important role, thus it will be referred to as the pulsed paraxial wave equation (PPWE). Equation (8) describes the dynamic plasma response to the laser pulse in terms of the scalar potential φ, which results from the electron density disturbance generated by the ponderomotive force of the laser pulse [9,10]. Within the parameter regime of interest the pulse length is shorter than the plasma wavelength, $\ell_L < \lambda_p$, so that our pulse can not see even a single plasma wave. Therefore its coupling with plasma waves cannot be described in terms of the conventional wave-wave interaction. The shortness of the pulse can suppress parametric Raman processes and significantly limit the number of instabilities. However, the ultrashort pulse experiences a new type of instabilities, which are the subject of our studies in the following section.

SPACE-TIME SELF-FOCUSING AND SPLITTING OF LASER PULSES

The numerical simulations presented in this section model the propagation of an ultrashort, relativistically intense, circularly polarized laser pulse focused at the edge of a preformed underdense plasma. At the input plane $\zeta = 0$ the laser field is given by $A = A_0 \exp(-\tau^2/\tau_0^2 - r^2/w_0^2)$, where $r \equiv (x^2 + y^2)^{1/2}$. This corresponds to a fundamental Gaussian mode with a spot size w_0 and planar wavefront (with infinite radius of curvature) [12]. We have simulated the laser propagation for two sets of laser-plasma parameters. In the first set the normalized laser peak amplitude $A_0 = 1$, the ratio of the laser spot size to plasma skin depth $w_0 k_p = 30$. In the second set of simulations $A_0 = 3$, and $w_0 k_p = 25$, respectively. In both cases the plasma is undercritical with $\omega_p/\omega = 0.1$. These dimensionless parameters, for $\lambda = 800$ nm laser wavelength, assume laser pulses with following characteristics, for $A_0 = 1$ pulse: $w_0 = 38.2$ μm; $\tau_{FWHM} = 15.1$ fs (FWHM of the intensity profile); the vacuum Rayleigh length $Z_R = 5.7$ mm; the peak intensity $I_0 = 4.27 \times 10^{18}$ W/cm^2, though relativistic, it is modest by today's standards; the peak power $P_0 = 98$ TW is within the reach of existing laser systems [4]. The same parameters for $A_0 = 3$ pulse: $w_0 = 31.8$ μm; $\tau_{FWHM} = 12.6$ fs; $Z_R = 3.98$ mm; $I_0 = 3.84 \times 10^{19}$ W/cm^2; $P_0 = 612.7$ TW. In both cases the plasma is underdense with a density $n = 1.7 \times 10^{19}$ cm^{-3}, $k_p^{-1} = 1.27$ μm. Obviously, the peak powers in both cases are well above the critical power for relativistic self focusing, $P_{cr} \approx 1.7$ TW [5].

Figures 1 and 2 show a sequence of contour plots of $|A|$ in (r', z') plane for $A_0 = 1$ and $A_0 = 3$ cases, respectively. Each plot in this sequence represents a part of the simulation box in the moving frame in dimensionless units, $z' = k_p(z - v_g t)$, $r' = k_p r$. Note that in the figures the longitudinal scale is 10 times as large as the radial one. The pulse propagates from the left to the right. Initially the evolution of the pulses in both cases is characterized by a synchronous SC and SF, so that the ratio of the pulse length to its width is almost preserved, and the pulses maintain their initial proportions. Figure 3 clearly illustrates that the SC and SF rates remain close to each other over all the simulation distance. In the following stage, certain asymmetry of the longitudinal profiles becomes evident. In $A_0 = 1$ case, the pulse's center of mass shifts back and its trailing edge becomes steeper, whereas in $A_0 = 3$ case the pulse peak is advanced. This difference results from interplay between the relativistic and the ponderomotive effects, which tend to counterbalance each other at the leading edges of the pulses where the density compression generated by the ponderomotive force decreases the local index of refraction, while the relativistic mass increase acts in the counterdirection increasing the refractive index. For $A_0 = 3$ pulse, the relativistic effect prevails over the ponderomotive effect at the leading half of the pulse, whereas in $A_0 = 1$ case the opposite occurs, as the intensity in the first case is almost an order of magnitude higher than in the latter case, though the intensity gradients (i.e., the ponderomotive effects) are on the same order.

At the next stage the trailing halves of the both pulses develop noticeable modulations (Fig. 1(e), 1(f), Fig. 2(e), 2(f)). Becoming deeper with the intensity dropping almost to zero between successive peaks, these modulations eventually lead to pulse splitting into a sequence of pulses, which follow each other arranged in decreasing order of their amplitudes Fig 1(g), 1(h), Fig 1(g), 1(h)). This splitting can be viewed as a multi-foci regime of the ultrashort pulse propagation. The simulation results presented in Fig. 4 show the role of FOD in splitting of laser pulses.

The spectral characteristics of the pulse presented in Fig. 5 are of particular interest, as they are directly connected to SC, SF and splitting activities. Fig. 5(a) presents a contour plot of the frequency modulation or "chirp" accumulated over a distance of $0.07Z_R$ (278.5 μm) in $A_0 = 3$ case. Corresponding axial and radial profiles are shown in Figs. 5(b) and 5(c), respectively. The large area with a red-shifted spectrum ($\delta\omega < 0$) at the leading half of the pulse results from the nonlinear phase shift induced by self-phase modulation (SPM). Combined with anomalous GVD, this effect accounts for the pulse SC. The well-pronounced small-scale modulations of the frequency chirp at the back of the pulse are induced by FD. Note that these modulations appear before the temporal profile modulations become noticeable. Becoming deeper with time, they initiate changes in the temporal profile of the pulse. The FD is responsible for the radial modulation in the frequency chirp shown also in Fig. 5(c). According to it, the higher frequencies are mainly located on-axis, while the off-axis part of the spectrum is relatively downshifted. Note that, generally, coupling temporal and spatial dynamics of the pulse, FOD accounts for enhanced divergence of lower frequency components [13] that is consistent with the effect observed in Fig. 5(c).

Figure 6 shows axial and radial profiles of the pulse ($A_0 = 3$ case) in the course of its propagation. In Fig 6(d) shows that, during SF, at the trailing half of the pulse its radial profile develops higher order symmetry than the symmetry of the initial Gaussian profile. Note that, within the conventional paraxial wave equation, SF of laser beams or any other modulation connected to the radial transport of energy cannot initiate changes in the radial pattern like one observed in Fig. 6(d).

After some distance (much larger than the pulse lengths) the split pulses show a tendency to coalesce and to form a relatively long, tightly focused structures. Figure 3 shows the onset of these process after the pulse has traveled a distance $\approx 0.12Z_R$ in $A_0 = 1$ simulations and after $\approx 0.10Z_R$ in $A_0 = 3$ simulations, when the effective durations of the pulses start to increase. This is the evidence that we enter a parameter regime beyond the validity of our model for ultrashort pulses ($\ell_L \ll \ell_\perp$).

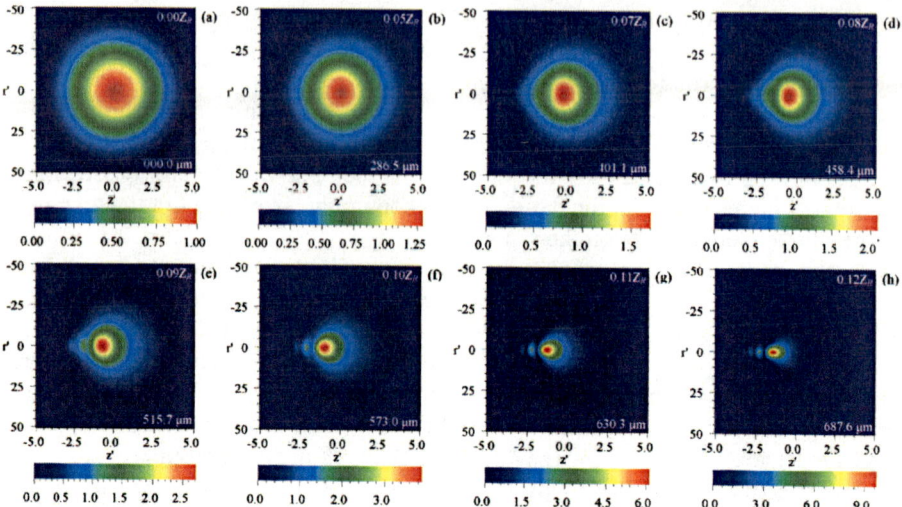

FIGURE 1. Contour plots of the laser amplitude $|A|$ (in units of mc^2/e) from $A_0=1$ simulation show violent self-compression and self-focussing, and subsequent splitting of the laser pulse. The dimensionless radial and longitudinal coordinates in all the figures from 1 to 6 are in units of $1/k_p$.

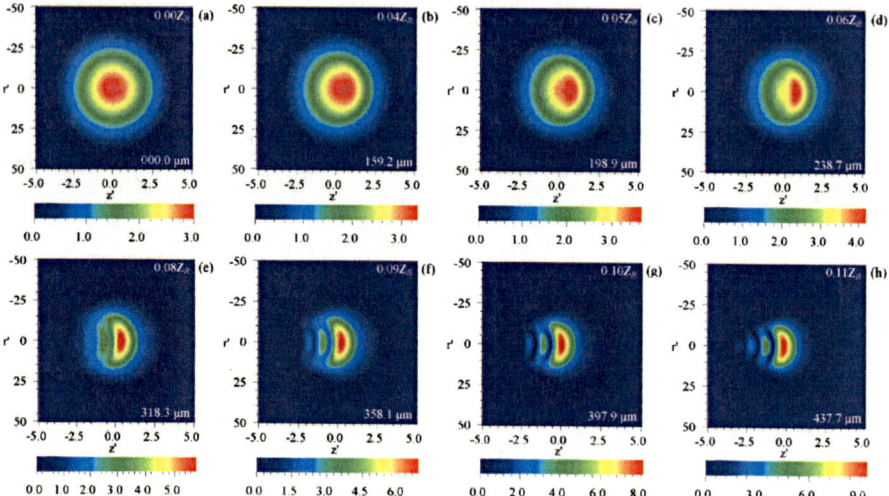

FIGURE 2. Contour plots of the laser amplitude $|A|$ from $A_0=3$ simulation. The spatiotemporal dynamics of the pulse from the input plane up to a distance of $0.11Z_R$ (437.7 μm) shows self-focusing, self-compression and splitting of the pulse in a sequence of pulses with relativistic peak amplitudes.

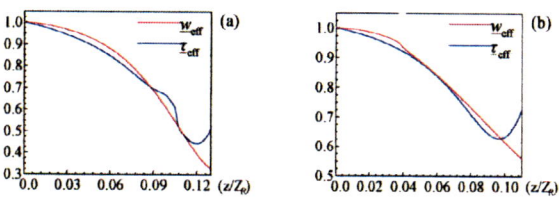

FIGURE 3. Normalized effective duration τ_{eff} and width w_{eff} of the laser pulse vs. the propagation distance, (a) – $A_0=1$ case, (b) – $A_0=3$ case.

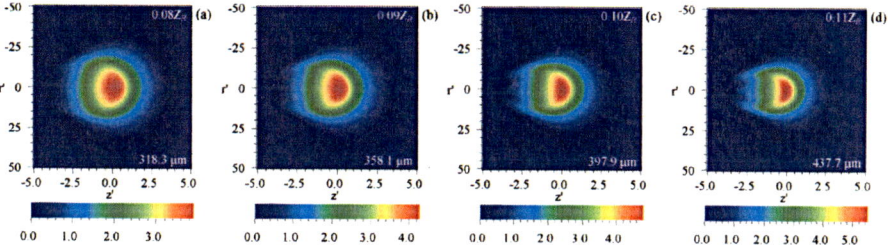

FIGURE 4. Contour plots of the laser amplitude $|A|$ from $A_0=3$ simulation with the same input parameters as in the simulations presented in Fig. 2, but with first-order dispersion "switched-off". No modulation and splitting of the trailing half of the pulse are observed in this case (cf. Figs. 2(e) – 2(h)).

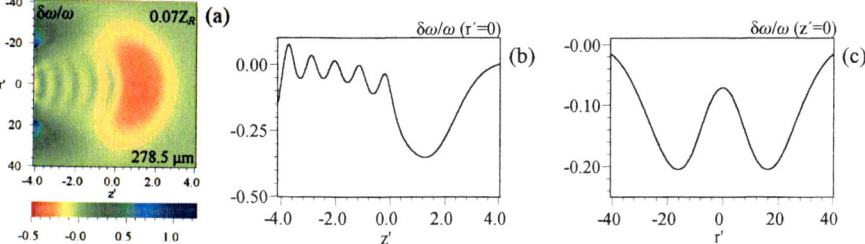

FIGURE 5. Distribution of the nonlinear frequency chirp (a) show a red-shifted spectrum at the leading edge of the pulse. The modulations of its longitudinal (b) and radial profile (b) account for changes in the temporal and radial profile of the pulse.

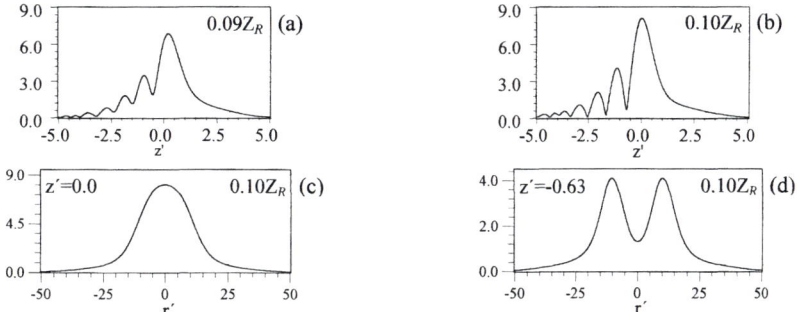

FIGURE 6. Modulations of the pulse's axial profile (or, equivalently, the temporal envelope) – (a), (b), leading to its splitting ($A_0=3$ case). The pulse's head self-focuses with a conventional bell-shaped radial profile (c), while its trailing half develops a radial pattern with minimum intensity at the center surrounded by a ring in which the intensity is concentrated (d).

ACKNOWLEDGMENTS

This work is partly supported by the Grant No. 2-23 of the Georgian Academy of Sciences.

REFERENCES

1. Perry, M. D., and Mourou, G., *Science* **64**, 917-924 (1994).
2. Mourou, G. A., Barty, C. P. J., and Perry, M. D., *Physics Today* **51**, 22-28 (1998).
3. Yamakawa, K., et al., *Optics Letters* **23**, 1468-1470 (1998).
4. Yamakawa, K., Barty, C. P. J., *IEEE J. Sel. Top. Quant. Electronics* **6**, 658-675 (2000).
5. Esarey, E., Sprangle, P., Krall, J., Ting A., *IEEE J. Quantum Electronics* **33**, 1879-1914 (1997).
6. Mori, W. B., *IEEE J. Quantum Electronics* **33**, 1942-1953 (1997).
7. Spangle, P., Hafizi, B., and Peñano, J. R., *Phys. Rev. E* **61**, 4381-4393 (2000).
8. Esarey, E., et al., *Phys. Rev. Letters* **84**, 3081-3084 (2000).
9. Murusidze, I. G., Tsintsadze, L. N., *J. Plasma Physics* **48**, 391-395 (1992).
10. Farina, D., Lontano, M., Murusidze, I. G., Mikeladze, S. V., *Phys. Rev. E* **63**, 056409 (2001).
11. Brabec, T., and Krausz, F., *Phys. Rev. Letters* **78**, 3282-3285 (1997).
12. Siegman, A. E., *Lasers*, Mill Valley, CA: University Science Books, 1986, pp. 626-661.
13. Akhmanov, S. A., Vysloukh, V. A., and Chirkin, A. S., *Optics of Femtosecond Laser Pulses*. New York: AIP, 1992, pp. 53-59, pp. 9-17.

Dependence of Relativistic Self-Guiding and Raman Forward Scattering on Duration and Chirp of an Intense Laser Pulse Propagating in a Plasma

T.-W. Yau[*], C.-J. Hsu[†], H.-H. Chu[**], Y.-H. Chen[†], J. Wang[‡] and S.-Y. Chen[‡]

[*]*Institute of Applied Science and Engineering Research, Academia Sinica, Taipei, Taiwan*
[†]*Graduate Institute of Electro-Optical Engineering, National Taiwan University, Taipei, Taiwan*
[**]*Graduate Institute of Physics, National Taiwan University, Taipei, Taiwan*
[‡]*Institute of Atomic and Molecular Sciences, Academia Sinica, P.O. Box 23-166, Taipei, Taiwan*

Abstract. Relativistic self-guiding and Raman forward scattering of an intense ultrashort laser pulse propagating in an underdense plasma were studied using a terawatt Ti:sapphire laser system. The dependence of these processes on the duration and frequency chirp of the laser pulse was investigated by detuning the laser compressor. We found that the efficiency of Raman forward scattering is enhanced for a positively chirped pulse and diminished for a negatively chirped pulse. In addition, as a result of the dependence on pulse duration and peak power, an optimal duration for Raman forward scattering was found when the pulse energy and spectral bandwidth were fixed. On the other hand, relativistic self-guiding of the laser pulse was affected by the pulse duration while no dependence on the chirp was observed.

INTRODUCTION

For applications such as x-ray sources [1, 2], laser-plasma accelerators [3, 4], and laser fusion [5], relativistic self-guiding [6, 7] and Raman forward scattering [8, 9, 10, 11, 12] are important effects that may hamper or aid the intentional processes. Relativistic self-guiding is advantageous for laser-plasma accelerators because it extends the distance over which the plasma wave is driven and the injected electrons are accelerated. However, the occurrence of Raman forward scattering excites a plasma wave with large shot-to-shot fluctuation and produces an electron beam of broad energy distribution. This is undesirable because it obstructs the more promising schemes that use a stable resonant laser wakefield and laser injection to generate monoenergetic electron bunches. [4] For x-ray sources such as x-ray laser, relativistic self-guiding is favorable for providing a long gain length. In contrast, Raman forward scattering will not only result in laser energy loss but also reduce lasing efficiency by heating up electrons. In the fast ignitor fusion concept, while Coulomb explosion following relativistic self-guiding may be used to bore a channel for guiding the ignition pulse, Raman forward scattering can lead to laser energy loss and preheat the core.

It is predicted theoretically that relativistic self-guiding occurs when the laser power exceeds a critical power determined by plasma density, $P_c = 17(\omega_0^2/\omega_p^2)$GW, and when

CP611, *Superstrong Fields in Plasmas:* Second Int'l. Conf., edited by M. Lontano et al.
© 2002 American Institute of Physics 0-7354-0057-1/02/$19.00

the pulse duration τ_p is much longer than the plasma wave period T_p. In addition, relativistic self-guiding should not occur when $\tau_p \ll T_p$ and the frequency chirp of the laser pulse is not expected to affect the self-guiding. However, no experiment has been reported to examine these theoretical expectations so far. On the other hand, the occurrence of Raman forward scattering was predicted to show a similar dependence on the pulse duration. In addition, it is expected to be influenced by the pulse spectral bandwidth and frequency chirp. An instability like Raman forward scattering has an associated bandwidth that the laser bandwidth may exceed. Those frequencies of light outside of the instability bandwidth will then be unable to drive density perturbations, thereby reducing the effective growth rate when $\Delta\omega > \gamma_0$, for growth rate γ_0 and laser bandwidth $\Delta\omega$. [13] Besides this, when group velocity dispersion is accounted for, it is predicted the same bandwidth may either increase or decrease the growth rate of Raman forward scattering instability, depending on the sign of linear chirp. The amount of chirp needed to eliminate the Raman instability is estimated to be $b = -1/4(\omega_0/\omega_p)(\gamma_0/\tau_0)$, where γ_0 is the unchirped growth rate for Raman forward scattering. [14] Again, no experiments has been presented to study this pulse duration and chirp dependence.

In this experiment we investigated the dependence of relativistic self-guiding and stimulated Raman forward scattering instability on laser pulse duration and frequency chirp. We used a Ti:sapphire laser system that delivers 50-fs laser pulses with a maximal 300-mJ energy, corresponding to 6-TW peak power, at 10-Hz repetition rate. With careful alignment of grating compressor, the system produces nearly transform-limited pulses with 24-nm bandwidth, centered at 800 nm. We focused the laser pulses onto a Helium gas jet target with a 90-deg off-axis parabolic reflector of 152-mm focal length. The gas jet nozzle has an opening of 1.4 mm in diameter and produces a gas jet with a nearly 800-μm-wide flat-top distribution and a sharp gradient at both edges. The gas jet was positioned with its front edge right at the focus of the laser beam in order to reduce ionization defocusing and thus obtain optimal relativistic self-guiding. The focal spot was measured to be nearly symmetric with a diameter of about 4 μm. When fitted to a Gaussian beam profile, 40% of the total laser energy is enclosed within the focal spot.

RELATIVISTIC SELF-GUIDING

For many applications the propagation of laser pulses in a plasma over a long distance while maintaining high intensity is required. However, a high intensity is usually obtained by tight focusing of a laser beam. Since diffraction will spread the laser beam after the focus, the longitudinal size of the effective interaction region is limited by the Rayleigh length, Z_R. One way to extend the propagation distance beyond this diffraction limit is to utilize relativistic-ponderomotive self-channeling (relativistic self-guiding). In this experiment, we observed the relativistic self-guiding of a laser pulse in a plasma. The guided length extended over 8 Z_R. We characterized the propagation of the laser pulse in the plasma simultaneously in forward and side directions. In the forward direction, we split a small portion of the transmitted laser pulse using a fused-silica wedge and sent it into an image-relayed microscope to image the beam profile at and near the laser focus. Fig. 1 shows the beam profiles measured at 600, 900 and 1200 μm after the focus

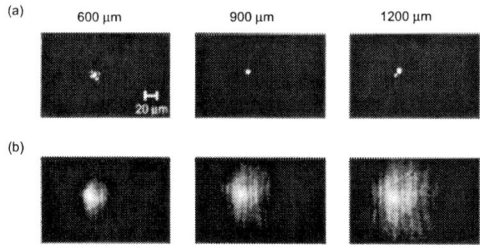

FIGURE 1. Images of the laser beam profile at 600, 900 and 1200 μm after the laser focus for a laser power of 5.1 TW and a plasma density of $3.6 \times 10^{19} cm^{-3}$ when the gas jet is present, (a), and not present, (b).

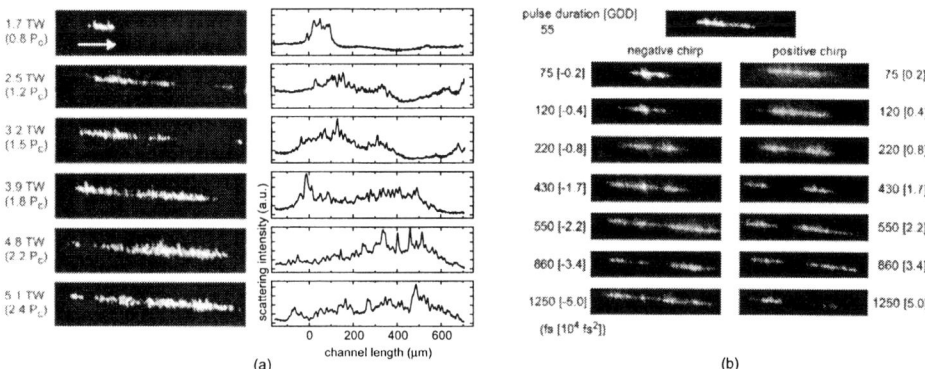

FIGURE 2. The images and lineouts of the laser channel for (a) various laser powers and (b) various pulse durations and chirps at a plasma density of $1.4 \times 10^{19} cm^{-3}$. At this density, P_c =2.13 TW. In (a) the arrow indicates the propagation direction of the laser pulse. In (b) the laser energy is 310 mJ, and GDD denotes the group-delay dispersion of the pulse.

with and without the gas jet by varying the longitudinal position of the first imaging lens. When the gas jet was present, a clearly concentric spot with 10-μm diameter (limited by the imaging resolution) was observed at 900 μm after the focus and diverged to 13 μm in diameter at 1200 μm. In contrast, for the case without gas jet, a widely spread beam profile due to diffraction was observed at the same locations. This comparison gives a clear evidence of beam guiding in the plasma. Using Gaussian-beam approximation, the divergence in free space confirms the guided beam diameter was about 6 μm.

In order to obtain more information on beam guiding, we imaged the side scattering of light, due to Thomson scattering by free electrons, to spatially resolve the laser channel in the plasma. We recorded the light emitted in the direction perpendicular to the axis of propagation and 30-deg above the horizontal plane. The scattered light was collected by a 150-mm lens with 8$\times$ magnification and imaged onto a CCD camera of 16-bit dynamic range. A bandpass filter was located in front of the CCD camera to reject the emission outside a bandwidth of 100 nm centered at 800 nm wavelength. The recorded images for various laser peak powers are presented in Fig. 2(a). Below 1.9 TW, the longitudinal size

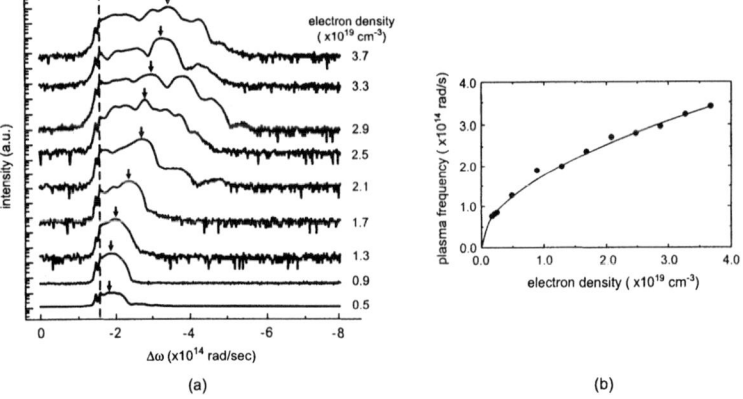

(a)   (b)

FIGURE 3. The Raman spectra for various electron densities and a laser power of 5.6 TW are shown in (a). The dashed line indicates the boundary of an optical attenuator within which the light is attenuated by 10^5 to avoid saturation of the CCD camera by the fundamental beam. The plasma frequency as a function of plasma electron density is plotted in (b), where the theoretical prediction is also drawn as a solid line.

of emission is about 100 μm, comparable to the Rayleigh length in vacuum. For laser power above 2.8 TW, the laser channel extends to 800 μm in length, limited by the gas jet profile. The channel diameter is less than 10 μm, limited by the imaging resolution, and has no significant variation along the propagation axis. The variation of the laser channel with increase of plasma density for a fixed laser power shows similar behavior. These observations are consistent with those reported by other groups. [15] Fig. 2(b) shows the dependence of the laser channel on the laser pulse duration and chirp as the laser energy and plasma density are fixed. As can be seen, better (longer) relativistic self-guiding channel was obtained when the pulse duration was increased even though the peak power was decreased. In addition, frequency chirp seems to have no effect on the results. Both observations are consistent with the theoretical predictions.

STIMULATED RAMAN FORWARD SCATTERING INSTABILITY

It has been reported that the generated Raman satellites should have a larger divergence angle than that of the laser beam, as a result of small cross section of the excited plasma wave. [16] Therefore we placed the concave mirror for collecting the forward scattering satellites at an angle of 30 degrees, larger than the divergence of the laser beam, to reduce the signal of the latter which may obscure the Raman satellite when its spectrum is broadened by self-phase modulation. The emission of Raman scattering was imaged onto the entrance slit of a 1/4-m imaging spectrometer equipped with a 16-bit CCD camera. With this, spatially resolved spectrum in the vertical direction was obtained. The resolution was 4 μm in space and 1 nm in wavelength when a 150-lines/mm grating was used in the spectrometer.

Fig. 3(a) shows the Raman spectra for various electron densities at a fixed laser

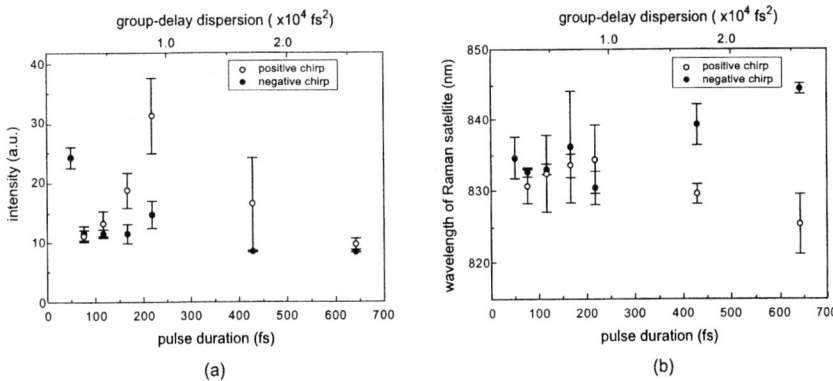

FIGURE 4. The intensity, (a), and wavelength, (b), of the first Stokes Raman satellite for various laser pulse durations and both signs of chirp. The laser energy is 310 mJ and the plasma density is 3.5×10^{18} cm^{-3}.

power of 5.6 TW. Fig. 3(b) reveals that the frequency shift of the first Stokes Raman satellite scales as the square root of plasma electron density as expected. Fig. 4(a) shows the peak intensities of the first Stokes Raman satellite for various pulse durations and both signs of chirp at a fixed laser pulse energy. It is clearly seen that Raman forward scattering for a positively chirped pulse has a higher growth rate than for a negatively chirped pulse with the same duration. During propagation, the group velocity of the laser pulse is determined by local electron density. The presence of an electron plasma wave (plasma density perturbation) results in longitudinal bunching of laser photons, leading to modulation in the laser intensity with a period equal to the plasma wave wavelength. The ponderomotive force arising from such laser intensity modulation then resonantly drives the growth of the electron plasma wave. This forms a feedback loop for the Raman forward scattering instability. Nevertheless, when the laser pulse is chirped, the group-velocity dispersion contracts the positively chirped laser photon bunches and expands the negatively chirped photon bunches. As a result, the local bunching of photons due to positive chirp assists on driving the intensity modulation and thus Raman forward scattering instability, while the stretching of photon bunches due to negative chirp lessens them. [14]

In our experiment, the change of chirp by moving the grating position is accompanied by the increase of pulse duration and reduction of pulse peak power, both of them affecting the growth of Raman instability. As seen in Fig. 4(a), the increase of Raman intensity with pulse duration within 200 fs reveals that the effect of pulse duration prevails on driving a plasma wave over that of laser power for short pulses. However, for longer pulses the effect of laser power is predominant over that of pulse duration. On the other hand, although for the case of negative chirp the Raman satellite intensity shows a similar dependence on the pulse duration, it is reduced with respect to the case of the shortest pulse with no chirp. This proves that a negative chirp indeed prohibits the growth of Raman forward scattering instability. Note that because the scattering by the plasma wave always arises in the pulse tail, the frequency of Raman satellite peak is dominated by the spectral content at the tail of the laser pulse. As seen in Fig. 4(b), since the positively chirped pulse contains a higher frequency toward its tail, the peak of Raman satellite is blue-shifted. In addition, the amount of blue shift increases with larger linear chirp (longer pulse duration) as a result of higher frequency in the tail. In contrast, the Raman satellite of a negatively chirped pulse becomes red shifted and shows the same behavior with increase of linear chirp.

CONCLUSION

We observed that relativistic self-guiding is enhanced for longer pulse duration when the latter is on the order of the plasma wave period. The frequency chirp, regardless of its sign, has no effect on the self-guiding. Furthermore, we verify that Raman forward scattering does increase with longer pulse duration and a negatively chirped pulse can actually reduce the growth rate.

For laser-plasma accelerator, it is beneficial to use a pulse of broader bandwidth and negative chirp when the pulse duration is fixed to drive a plasma wave resonantly. This is because Raman forward scattering can be prohibited under such condition, preventing energy loss and undesirable injection of electrons. In contrast, if the Raman forward scattering itself is used to drive plasma waves and heat up electrons, it is advantageous to stretch the laser pulse duration in the direction of positive chirp for a fixed laser pulse energy. When the lowering of peak laser power is considered, an optimal pulse duration is obtained for efficient Raman forward scattering. Similarly, for the applications of X-ray laser and fast ignitor fusion, stretching the pulse in the direction of negative chirp can lead to better self-guiding while eliminating the undesirable Raman forward scattering. Again, an optimal duration may exist because of the dependence on the laser power. This work was supported by National Science Council under the contract NSC 89-2112-M-001-066.

REFERENCES

1.   R. C. Elton, *X-ray lasers* (Academic Press, San Diego, CA, 1990).
2.   P. Amendt, D. C. Eder, and S. C. Wilks, Phys. Rev. Lett. **66**, 2589 (1991).
3.   T. Tajima and J. M. Dawson, Phys. Rev. Lett. **43**, 267 (1979).
4.   D. Umstadter, J. K. Kim, and E. Dodd, Phys. Rev. Lett. **76**, 2073 (1996).
5.   M. Tabak *et al.*, Phys. Plasmas **1**, 1626 (1994).
6.   G. Z. Sun *et al.*, Phys. Fluids **30**, 526 (1987).
7.   A. B. Borisov *et al.*, Phys. Rev. A **45**, 5830 (1992).
8.   E. Esarey, J. Krall, and P. Sprangle, Phys. Rev. Lett. **72**, 2887 (1994).
9.   W. B. Mori *et al.*, Phys. Rev. Lett. **72**, 1482 (1994).
10.   C. D. Decker *et al.*, Phys. Plasmas **3**, 1360 (1996).
11.   D. Umstadter *et al.*, Science **273**, 472 (1996).
12.   N. E. Andreev *et al.*, Plasma Physics Reports **23**, 277 (1997).
13.   G. Bonnaud and C. Reisse, Nucl. Fusion **26**, 633 (1986).
14.   E. S. Dodd and D. Umstadter, Phys. Plasmas **8**, 3531 (2001).
15.   R. Wagner *et al.*, Phys. Rev. Lett. **78**, 3125 (1997).
16.   S.-Y. Chen *et al.*, Phys. Plasmas **7**, 403 (2000).

Three-Dimensional Electromagnetic Solitary Waves in an Underdense Plasma in PIC Simulations

Sergei Bulanov[a,b], Timur Esirkepov[b,c], Katsunobu Nishihara[c], Francesco Pegoraro[d]

[a]General Physics Institute of RAS, Vavilov str. 38, Moscow, 119991, Russia
[b]Moscow Institute of Physics and Technology
Institutskij per. 9, Dolgoprudny, Moscow region, 141700, Russia
[c]Institute of Laser Engineering, Osaka University,
2-6 Yamada-oka, Suita, Osaka 565-0871, Japan
[d]University of Pisa and INFM, pz.Torricelli 2, Pisa, 56100, Italy

Abstract. A three-dimensional sub-cycle relativistic electromagnetic soliton has been observed for the first time in a 3D Particle-in-Cell simulation of the propagation of an intense short laser pulse in an underdense plasma. The structure of the 3D soliton is identified. It resembles an oscillating electric dipole and has a strong charge separation and toroidal magnetic field component. We call this structure a TM-soliton (transverse magnetic). The 3D TM-soliton resembles a 2D P-soliton in the plane of electric field polarization, and a 2D S-soliton in the perpendicular plane. The core of the soliton is positively charged on average in time, and this results in its Coulomb explosion and in ion heating. Then the soliton evolves into a post-soliton, which is a slowly expanding quasi-neutral cavity in the plasma.

INTRODUCTION

Particle-in-Cell simulations show different types of coherent structures in the wake of an intense short laser pulse propagating in underdense plasmas [1]. These are electrostatic wakefields, quasistatic magnetic fields associated with electron fluid vortices, bubbles in the plasma density associated with electromagnetic solitary waves and soliton traines. In the present paper we discuss the formation and the evolution of electromagnetic solitons in the wake of an intense laser pulse in the 3D case.

The theory of intense electromagnetic soliton was developed in many papers, see references in [1]; their best known application is the use of fast

CP611, *Superstrong Fields in Plasmas:* Second Int'l. Conf., edited by M. Lontano et al.
© 2002 American Institute of Physics 0-7354-0057-1/02/$19.00

electromagnetic solitons with group velocity close to the speed of light for charged particle and photon acceleration [2].

One-dimensional sub-cycle relativistic electromagnetic solitons in an underdense plasma were observed for the first time in a PIC simulation in Ref. [3]. The solitons are cavities in the electron density with trapped electromagnetic field oscillating with frequency below the unperturbed Langmuir frequency. The mechanism of soliton formation and the structure of circularly polarized solitons were investigated in Ref. [4]. An intense laser pulse propagating in an underdense plasma undergoes Raman scattering. The laser pulse loses its energy rapidly while the number of photons in the pulse is conserved. Thus the local frequency of the laser pulse in the rear part of the pulse decreases down to the local Langmuir frequency. As a result, the low-frequency part of the electromagnetic radiation of the laser pulse is trapped inside the cavity in the electron density, and a sub-cycle soliton is formed. Two "pure" types of one-dimensional solitons were found: linearly polarized L-solitons and circularly polarized C-solitons.

The exact analytical solution of the electron hydrodynamic – Maxwell equations representing one-dimensional sub-cycle circularly-polarized relativistic electromagnetic soliton was obtained in Ref. [5] in perfect agreement with the results of 1D PIC simulation.

Two-dimensional relativistic electromagnetic solitons were discovered in the PIC simulations presented in Ref. [6]. Two "pure" types of two-dimensional solitons were found: S-soliton with transverse electric field and azimuthal magnetic field, and P-soliton with the opposite structure — transverse magnetic field and azimuthal electric field. In contrast to the electron vortices, which move across a density gradient, solitons move along the density gradient towards the lowest density level. At some critical density level the soliton emits its energy in the form of a low-frequency short electomagnetic burst [7]. One can imagine "a solitonic gun" when the laser pulse propagates in the direction perpendicular to the plasma density gradient, and solitons are emitted in the direction opposite to the density gradient.

The interaction of two 2D S-solitons leads to their merging. The resulting soliton acquires the total energy of the merged solitons [8]. The interaction of two counter-propagating 1D solitons is completely different: they preserve their shape and velocity but lose their phase, similarly to a well-known "phase shift" in the theory of KdV solitons.

In an electron-ion plasma 1D and 2D solitons evolve into post-solitons on an ion time-scale due to the ion acceleration by the time-averaged electrostatic field inside the soliton. This effect leads to the formation of slowly expanding bubbles in the plasma density [9].

3D SOLITON SEEN IN PIC SIMULATIONS

We present the results of a three-dimensional Particle-in-Cell simulation of laser induced 3D sub-cycle relativistic electromagnetic solitons. We use 2D

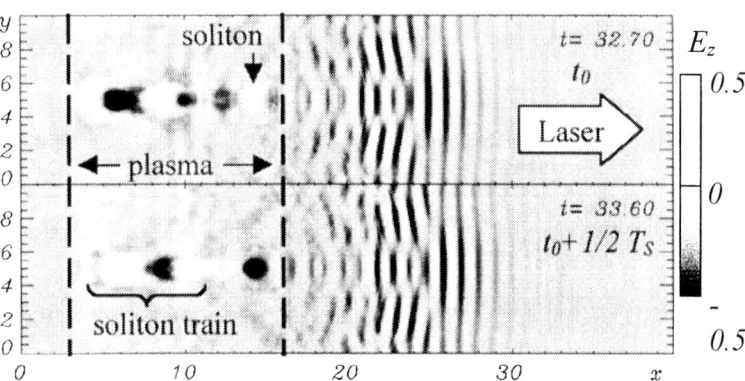

FIGURE 1. Transverse electric field E_z in the S-plane *(z=5)*. T_S is the soliton period.

and 3D PIC relativistic electromagnetic code, written by one of the authors (T.E.) in 1998-1999. This parallel and fully vectorized code exploits a new scheme of current assignment [10], which reduces unphysical numerical effects of the PIC method significantly. In the simulation the laser pulse is gaussian and linearly polarized along the z-axis, its dimensionless amplitude is $a=eE/(m_e\omega c)=1$, corresponding to the peak intensity $I=1.38\cdot10^{18}$ $W/cm^2\cdot\mu m^2/\lambda^2$, its size is $8\lambda\times5\lambda\times5\lambda$. The laser pulse propagates along the x-axis. The focal plane of the laser pulse is placed in front of the plasma slab

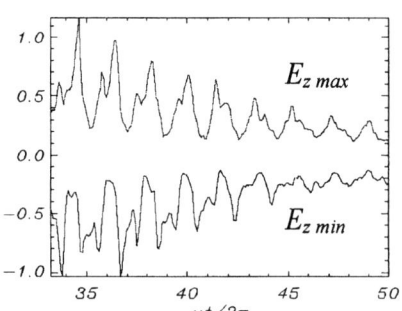

FIGURE 2. Electric field component E_z in the soliton vs time.

at the distance of 3λ. The density of plasma is $n=0.36n_{cr}$. Ions and electrons have the same absolute charge, the mass ratio is $m_i/m_e=1836$. The simulation grid is $800\times200\times200$, grid step is 0.05λ. The plasma slab width and length are 13λ. The total number of quasiparticles is $166.4\cdot10^6$. The boundary conditions along the y- and z-axes are periodic and absorbing for EM radiation and for the particles along the x-axis. The simulations were performed on 8 processors of the NEC SX-5 vector supercomputer in ILE, Osaka University.

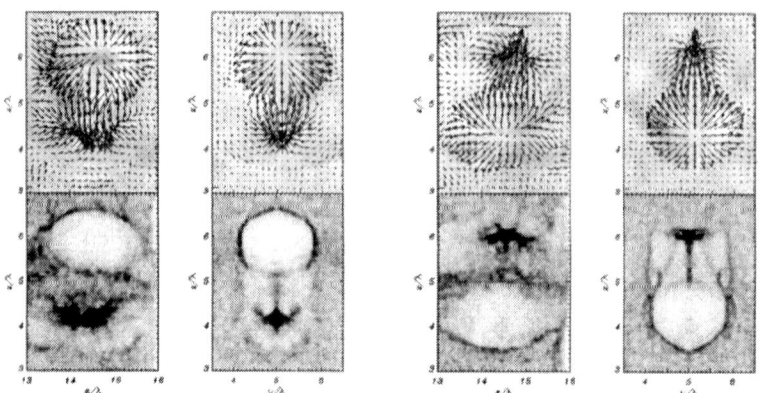

FIGURE 3. The soliton structure. Upper raw: arrows represents planar electric field, background is transverse magnetic field. Bottom: electron density. Columns (a),(c) are cross-sections at $y=5$; (b),(d) – at $x=14.5$. (a),(b) correspond to $t=39.3$;

In the figures the time and space units are the period $2\pi/\omega$ and the wavelength λ of the incident radiation, respectively. The structure of 3D soliton is clearly seen in the animations produced from the data (373 stills with the period 0.075).

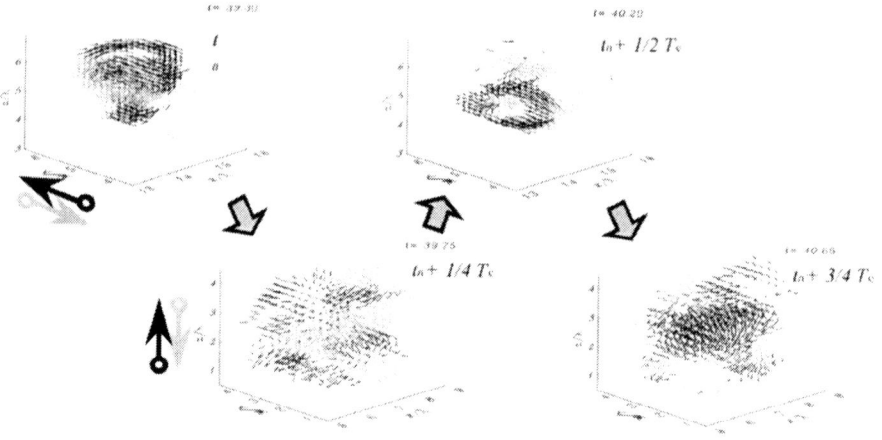

FIGURE 4. Magnetic and electric fields oscillation in the soliton.

The laser pulse undergoes a weak self-focusing in the plasma slab. Behind the laser pulse we see a soliton train and one isolated soliton, Fig.1. These structures have an oscillating electric field polarized in the same direction as in the laser pulse, Fig.2. The frequency of the oscillations is below the surrounding unperturbed Langmuir frequency, $\Omega_S \approx 0.93\,\omega_{pe}$. Therefore the soliton oscillation is not-resonant with the plasma waves and radiation losses due to the wave transformation are almost insignificant.

194

Figures 3-4 show the structure of the isolated soliton. The soliton resembles an oscillating electric dipole. Although we see a strong electrostatic component, the soliton is electromagnetic as one can see directly from the oscillations of the electric and magnetic fields in the soliton, Fig.4. The electric field in the soliton is poloidal and the magnetic field is toroidal. Thus we call this structure a TM-soliton. The soliton dynamics is slightly affected by the quasistatic magnetic field and remnants of density filament induced by the laser pulse.

On the ion time-scale the soliton evolves into a post-soliton. The magnitude of the soliton decreases, Fig.2, and ions acquire momentum, Fig.5. As the last stage we see a slowly expanding ($v \sim 3 \cdot 10^{-3} c$) cavity in the plasma density. In the soliton train almost isolated structures that we can consider as solitons tend to merge and form a foam of bubbles with relatively high-density ($2 \div 5 n_{cr}$) walls.

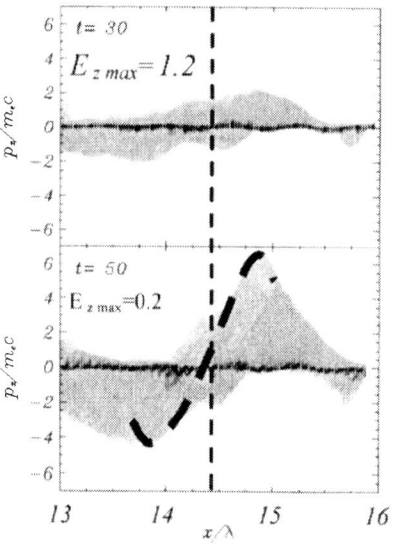

FIGURE 5. Phase space of ions accelerated by the soliton at t=30 and t=50.

CONCLUSIONS

This is the first demonstration of the existence of three-dimensional sub-cycle relativistic electromagnetic solitons in collisionless cold plasma. The soliton is induced by the intense laser pulse. A substantial part of the pulse energy is transformed into solitons. The 3D PIC simulations show the structure of the TM-soliton. In the plane of the electric field polarization it resembles that of the 2D P-soliton in a plane of electric field polarization, and in the perpendicular plane that of the 2D S-soliton. The core of the soliton is positively charged on average in the time and the soliton undergoes the Coulomb explosion on an ion time-scale. The soliton evolves into a post-soliton, which is a slowly expanding quasineutral cavity in plasma. Many merging solitons can form a slowly expanding foam.

ACKNOWLEDGMENTS

We appreciate the help of ILE computer group and Cybermedia Center of Osaka University (Japan). One of the authors (T.E.) thanks Japan Society for the Promotion of Sciense for his grant. This work was supported also by Istituto Nazionale di Fisica della Materia (Italy), Russian Ministry of science and technology, Russian Fund of Basic Research.

REFERENCES

1. Bulanov,S., et al., "Relativistic Interaction of Laser Pulses with Plasmas", in *Reviews of Plasma Physics*, edited by V.D.Shafranov, Kluwer Acad./ Plenum Publ. New York, 2001, Vol. 22., p. 227.
2. Mima, K., et al., *Phys. Rev. Lett.* **57**, 1421 (1986).
3. Bulanov,S., et al., *Phys. Fluids* B **4**, 1935 (1992).
4. Bulanov, S., et al., *Plasma Phys. Rep.* **21**, 600 (1995).
5. Esirkepov,T., et al., *JETP Lett.* **68**, 36 (1998).
6. Bulanov, S., *Phys. Rev. Lett.* **82** , 3440 (1999).
7. Sentoku, Y., et al., *Phys. Rev. Lett.* **83**, 3434 (1999).
8. Bulanov, et al., *Physica D* **152–153,** 682 (2001).
9. Naumova, N., et. al., *Phys. Rev. Lett.,* in press (2001).
10. Esirkepov, T., *Comput. Phys. Comm.* **135**, 144 (2001).

3. SOLID DENSITY PLASMAS, CLUSTER PLASMAS, FAST IONS AND NUCLEAR PHYSICS WITH INTENSE LASERS

Intense Ion Beams Accelerated by Ultra-Intense Laser Pulses

Markus Roth[1], T.E. Cowan[2], J. C. Gauthier[3], J. Meyer-ter Vehn[4], M. Allen[2], P. Audebert[3], A. Blazevic[1], J. Fuchs[3], M. Geissel[1], M. Hegelich[4], S. Karsch[4], A. Pukhov[4], T. Schlegel[1,4]

[1]*Gesellschaft für Schwerionenforschung, 64291 Darmstadt, Germany*
[2]*General Atomics, P.O. Box 85608, San Diego, California 92186-5608*
[3]*Laboratoire pour l'Utilisation des Lasers Intenses, 91128 Palaiseau, France*
[4]*Max-Planck-Institut für Quantenoptik, 85748 Garching, Germany*

Abstract. The discovery of intense ion beams off solid targets irradiated by ultra-intense laser pulses has become the subject of extensive international interest. These highly collimated, energetic beams of protons and heavy ions are strongly depending on the laser parameters as well as on the properties of the irradiated targets. Therefore we have studied the influence of the target conditions on laser-accelerated ion beams generated by multi-terawatt lasers. The experiments were performed using the 100 TW laser facility at Laboratoire pour l'Utilisation des Laser Intense (LULI). The targets were irradiated by pulses up to 5×10^{19} W/cm^2 (~ 300 fs, λ=1.05μm) at normal incidence. A strong dependence on the surface conditions, conductivity, shape and purity was observed. The plasma density on the front and rear surface was determined by laser interferometry. We characterized the ion beam by means of magnetic spectrometers, radiochromic film, nuclear activation and Thompson parabolas. The strong dependence of the ion beam acceleration on the conditions on the target back surface was confirmed in agreement with predictions based on the target normal sheath acceleration (TNSA) mechanism. Finally shaping of the ion beam has been demonstrated by the appropriate tailoring of the target.

INTRODUCTION

The interaction of ultra-intense laser beams[1] with solid targets can produce energetic ions with up to tens of MeV in energy[2]. Relativistic electrons generated from the laser-plasma interaction, having an average temperature of several MeV, envelope the target foil and form an dynamic, electric sheath[3,4] due to charge separation. The electric field in the sheath can exceed 10^{12} V/m, which field ionizes atoms on the surface and accelerates the ions very rapidly normal to the rear surface. Protons from the bulk material or, e.g. in case of metal targets from hydrocarbon and water vapor surface contaminants, are preferentially accelerated in favor of heavier ions due to their higher charge to mass ratio over a distance of a few microns, and within a few picoseconds. The proton energy distribution resembles a Maxwellian with kT = 5-6

CP611, *Superstrong Fields in Plasmas:* Second Int'l. Conf., edited by M. Lontano et al.
© 2002 American Institute of Physics 0-7354-0057-1/02/$19.00

MeV with measured maximum energies exceeding 50 MeV. In contrast to accelerated ions observed in long-pulse laser-matter interaction, ion beams generated by sub-picosecond lasers form a well collimated, directed beam normal to the rear surface of the irradiated targets. We note, that the evolution of the electron sheath is strongly coupled to the accelerated ions, the whole mechanism can not be considered as a static process, and the time for the acceleration is determined by the lifetime of the hot electron component. This mechanism, known as the target normal sheath acceleration mechanism[5], makes these intense ion beams highly interesting for many applications, especially if one can collimate or focus the beam by shaping the target[5,6]. Besides the laser beam parameters, the target material and surface conditions should be crucial for the generation of the ion beams. Therefore we carried out experiments to investigate the influence of the target parameters on the ion beam production.

EXPERIMENTS

The experiments were performed using the 100 TW laser facility at LULI. Pulses of up to 30 J at 300 fs pulse duration at λ=1.05 µm were focused with an f/3 off-axis parabolic mirror onto free standing target foils at normal incidence at intensities up to 5×10^{19} W/cm^2. The $1/e^2$ focal spot radius measured in vacuum was about 8µm. Amplified spontaneous emission (ASE) occurred 2ns before the main pulse at a level of 10^{-7} of the main pulse and preformed a plasma. The target foils consisted of different material compositions and shapes, were of millimeter size and had a thickness varying from a few microns up to hundreds of microns. A scanning electron microscope was used to determine the structure of the target surfaces.

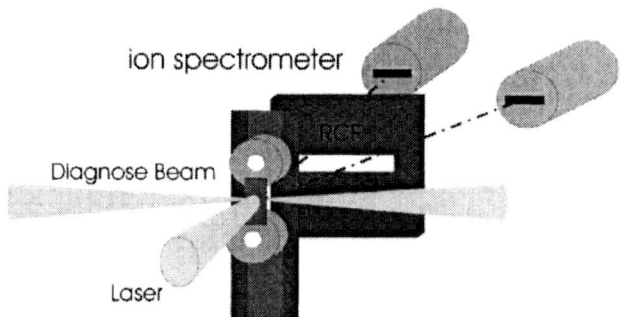

FIGURE 1. experimental setup. The free standing target is irradiated at normal incidence. A slit in the radiochromic film gives a line of sight for the particle spectrometers. The target conditions were probed using an up-converted laser pulse parallel to the surfaces.

The free standing target was probed by a frequency doubled laser interferometry parallel to the surface to determine the plasma conditions on the front and rear surface (Fig.1). A stack of radiochromic film (RCF) was positioned a few cm behind the target to measure the spatial beam profile. The RCF changes through polymerization of a diacetylene active layer, from transparent to dark blue in proportion to the absorbed dose (rads) of ionizing radiation. Due to the pronounced energy loss of ions at the end of their range different layers of the RC film pack allow the imaging of the ion beam at different energies.

A slot in the center of the RCF allowed a free line of sight from the target to the particle spectrometers fielded at 0^0, 6^0 or 13^0 with respect to the target normal (also the axis of the incident high-intensity laser beam) to measure the energy distribution of the emitted particles. Two absolutely calibrated, permanent magnetic particle spectrometers were used to detect the proton energy distribution[7]. The spectrometers were mounted at a distance of about 1m from the target covering a solid angle of 5×10^{-6} sr. Due to protective layers of light-tight paper in front of the detectors, protons below 1.8 MeV were not recordable. To extend the sensitive spectral range to lower proton energies, CR-39 nuclear track detectors were fielded at the low energy range of the dispersion plane. The spectrometers were also able to record the energy distribution of the electrons emitted from the target.

For the detection of heavy ions accelerated from the rear surface of the laser-illuminated targets, we used two high resolution Thompson parabolas in addition to one of the ion spectrometers. The collinear of electric and magnetic fields in the spectrometers discriminate ions with respect to their momentum and charge-to-mass ratio, allowing full determination of the ion species and charge state. To avoid detection threshold effects, the ions were recorded on CR-39 track detectors. By etching the detectors in sodium hydroxide, the material damage caused by the impacts of ions above a threshold energy of several hundred keV become visible. Microscopic scanning provides position as well as the size of the impact, which is proportional to the atomic number (Z) of the ion. So protons lead to much smaller pit-sizes than heavy ions. Careful analysis of the scanned CR-39 detectors also provides absolute numbers of the ions with respect to their energy and charge state.

To precisely estimate the beam emittance we used penumbral imaging of edges at different distances from the target with the magnetic spectrometers to directly measure the emittance core of the proton beam. This technique is closely related to the usual slit-emittance measurements made with simple apertures and screens at conventional accelerators.

To further confirm that the majority of the detected ions for our usual targets were indeed protons we used the nuclear activation of ^{48}Ti. Protons incident on the Ti and having energy above the reaction threshold of ~5 MeV cause ^{48}Ti(p,n)^{48}V reactions in the Ti catcher foil. The yield of ^{48}V was determined by the gamma-ray decay signature (half-life of 15.4 days) using a low-background Ge detector. This allowed a complementary way to determine the total yield of the accelerated protons.

Complementing the laser and ion beam diagnostics we also measured the neutron emission caused by (γ,n), (p,n) and (d,n) reactions from the target. We used several silver activation detectors attached to a photo-multiplier tube (PMT) in order to obtain the angle dependent neutron yield.

For the generation of intense beams of heavy ions we heated the targets in order to clean the surfaces from the proton contaminant layers. We used resistive heating for standard metal targets. For targets where heating was not applicable a second, low intensity laser beam several hundred microseconds before the main pulse was applied to heat a fraction of the rear surface.

RESULTS

The TNSA model predicts an accelerating field strength proportional to T_{hot}/el_0, where T_{hot} is the temperature of the hot electrons and l_0 is the larger of either the hot-electron debye length, or the ion scale length of the plasma on the rear surface[8]. So, for an effective acceleration of ions an undisturbed back surface of the target is crucial to provide a steep ion density gradient in order to maximize the accelerating field. The ASE light arriving prior to the main pulse preforms a plasma that launches a shock wave into the target heating the rear surface and cause it to expand, thereby destroy the steep density gradient of the initially cold, undisturbed target foil. We therefore chose the target thickness to guarantee an undisturbed rear surface based on numerical calculations using the hydrocode MULTI[9]. Figure 2 shows an example simulation for the case of a gold target with the typical prepulse level present at the LULI laser system. The laser irradiates the target from the right. The high temperature corona is visible together with the shock wave running through the target reaching the rear, left surface at about 8 ns. The gold targets were chosen to have a ~50 micron thickness to maintain an undisturbed back surface for a 5 ns ASE level of $5x10^{13}$ W/cm^2. The thickness of other targets (e.g., glass and plastic) were chosen similarly to assure that no shock heating would destroy the rear surface density gradient.

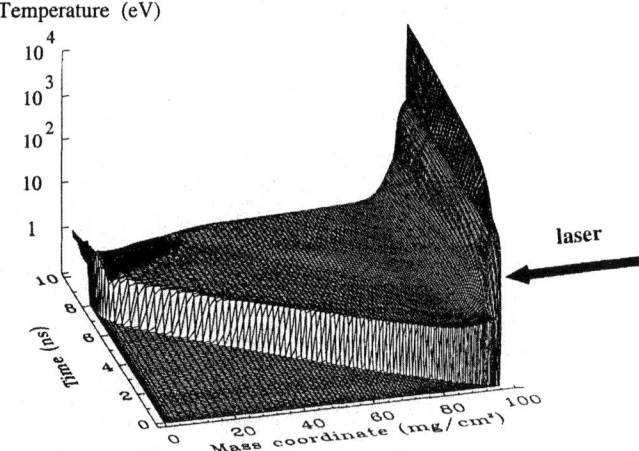

FIGURE 2. MULTI calculation of a shock wave launched by a pre-pulse of $5*10^{13}$ W/cm^2. A target thickness of 50 μm is sufficient to maintain an unperturbed rear surface for a pre-pulse 5 ns ahead of the main pulse.

For standard 48 micron Au targets irradiated with 20 J laser energy we obtained an intense, laminar proton beam. We demonstrated that we could also destroy the proton acceleration by applying an additional pre-pulse to shock heat the rear target surface. With a prepulse 10 ns prior to, and having an intensity of 10^{-7} of the main pulse, the maximum observed energy of the protons dropped from ~20 MeV to < 2 MeV. This is in good agreement with the MULTI calculations which indicate that in 10 ns a shock wave launched by the prepulse penetrates the target and causes a rarefaction wave that diminishes the density gradient on the back and therefore reduces the accelerating field drastically.

The angular dependence of the energy distribution of the proton beam was measured with the two ion spectrometers, positioned at an angle of 0° and 13° respectively. The measured spatial distributions of protons on the dispersion plane were deconvoluted (with respect to the entrance aperture shape) and corrected for the spectrometer dispersion. The energy of protons emitted normal to the target rear surface extended up to about 25 MeV. The maximum energy of the protons dropped to about 13 MeV at an angle of 13°, consistent with a 2-D model of the sheath acceleration process. The spectral shape of each proton energy distribution is generally continuous up to the cut-off energy, in agreement with the electrostatic sheath acceleration mechanism and as well as with previous observations in experiments with the LLNL PETAWATT laser[2]. The best fit to the experimental data was obtained using a two temperature distribution with 2 MeV temperature for the majority of the ions and around 6 MeV for the high energy tail of the spectrum. Details about the angular dependence of the ion beam and the origin of occasionally observed narrow features in the spectral distribution, caused by the segregation of different ion species, are beyond the scope of this paper and will be published elsewhere.

The proton beam ejected from the rear surface is shown in Fig.3. The results showed a clear dependence of the spatial uniformity of the proton beam on the structure of the back surface. In contrast to the homogenous, collimated beam from the flat gold target, protons emitted from the structured gold rear surface showed filaments.

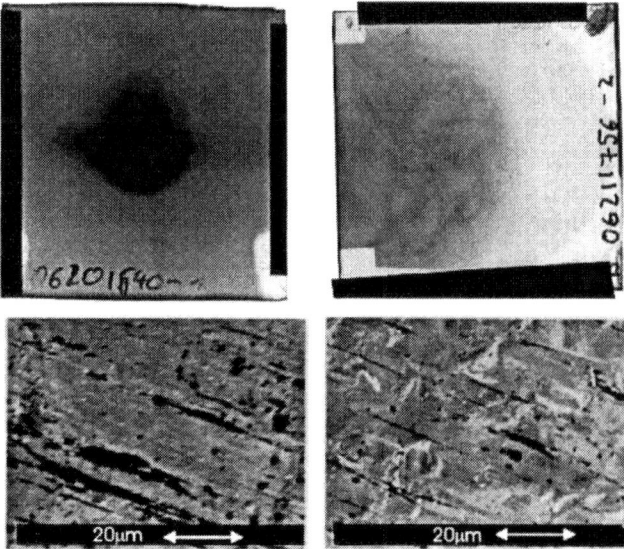

FIGURE 3. Proton emission from smooth and roughened rear surfaces of a Au target. The roughened surface leads to the onset of filamentation. The REM images show the structure of the surface.

The scanning electron microscope images of the respective surfaces are shown in the lower insets in Fig.3 and Fig.4. Structuring the gold surface maintained a smooth surface with hills and valleys, visible as bright shadows on the lower right inset of Fig.3.

To discriminate between conductivity and surface quality effects, we next used ~100 micron plastic and glass targets. The results of the glass and plastic targets were even more pronounced as shown in Fig. 4. While the flat surfaces of glass and plastic yielded a strong, but filamented proton beam, there were no protons detected above 1 MeV from the roughened targets.

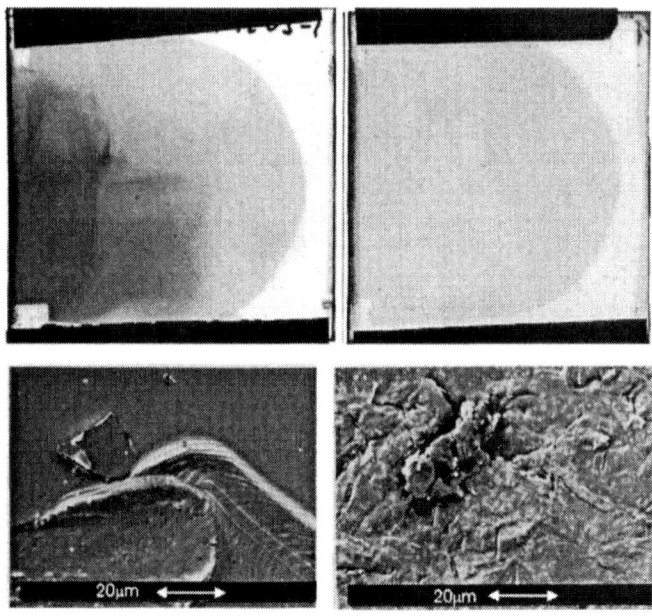

FIGURE 4. Proton emission from smooth (left) and roughened rear surfaces of plastic targets. A strong, filamented proton signal is visible from the smooth target. No protons were detected from the roughened target above 1 MeV. The REM images show the structure of the surface. The structure on the left was an artifact used to focus the REM. The majority of the surface was like the upper right part.

The surface of the plastic and glass targets was largely destroyed by numerous cracks. The different behavior of the structured gold, glass and plastic targets on the accelerated protons can be understood within the context of the Target Normal Sheath Acceleration (TNSA) model. When the material on the rear surface is exposed to the strong electric field generated within the hot electron plasma sheath, it is field–ionized instantaneously. As mentioned above, the accelerating electric field strength is characterized by $E \sim T_{hot}/el_0$, where T_{hot} is the temperature of the hot electron, and l_0 is the scale length of the plasma on the rear surface, and is in the order of megavolts per micrometer. One expects that a shallow, wavelike surface, such as for the roughened gold foils, should lead to a microlensing phenomenon, consistent with the observed filamentation or spatial modulation of the accelerated protons. As long as the hot-electron plasma sheath follows the contours of this surface, acceleration of protons to energies not too much lower than in the case of a very smooth surface, seems plausible. Indeed, such microfocusing effects have been calculated for the case of a single concave depression of the target surface[5]. In the case of a destroyed surface, the cracks and defects on the plastic and glass create many sharp excursions, very different from the rather smooth undulating surface of the gold targets. The ion plasma created during the field ionization is therefore extended over a much larger scale length normal to the (average) surface. We expect this to partially compensating the charge separation sheath created by the hot electrons, and therefore strongly suppressing the ion acceleration. As long as the spatial scale (or spatial frequencies) of the surface damage is comparable to or small with respect to the hot-electron debye length, one expects such a dilution of the average accelerating field.

The second question to be answered was whether the filamentation was due to the origin of the accelerated ions or due to the target conductivity. The origin of the accelerated protons are contaminant layers of hydrocarbons or water vapor for gold and glass targets, whereas in plastic also from the bulk material. The similar results obtained from glass and plastic indicate that the filamentation of the proton beam is insensitive on their origin and mainly supposed to depend on the target conductivity. This has important implications for understanding relativistic electron flow through materials, and will be published elsewhere.

As mentioned above, it was found in recent experiments[2], that the proton yield of plastic targets was higher than the yield of hydrocarbon contaminants on Au targets, whereas the beam quality from metal targets is much better than from glass or plastic targets. We attempted to increase the yield of laser-accelerated protons, while simultaneously maintaining the superior beam quality, by adding a thin hydrogen-containing layer of CH to the rear surface of gold targets. We varied the thickness of a CH layer on the back of gold targets between 5 and 100 nm. The experimental results indicated an increase of the proton yield according to the CH-layer thickness while maintaining the beam quality. However, at a layer thickness of 100 nm we observed the onset of filament-like structures in the spatial distribution of the accelerated protons.

We modeled the response of the RC film package with the SRIM[10] stopping power tables, assuming the beam to be protons, and obtained response functions for the layers of RC film similar to the ones presented in[2]. Taking into account the respective laser pulse energy, we obtained an conversion efficiency of laser energy into protons of 0.1% for the 5 nm coating and 0.25% for the 100nm coated target.

Finally the increase of the proton yield was limited by the laser energy available in our experiments. At a given maximum laser energy of 30 J and a conversion efficiency into hot electrons less than 35%, a layer thickness of several nanometers is sufficient to provide enough protons to be accelerated. This is quite different for petawatt laser systems operating at higher pulse length and accordingly higher total laser energy. In that case an increase in the overall proton yield by at least an order of magnitude can be expected.

The acceleration of protons based on the TNSA mechanism predicts collimated beams always normal to the rear surface. This raises the prospect of tailoring the ion beam by shaping the target surface. Due to the spatial distribution of the electron debye sheath there is an energy dependent angle of divergence, that has to be taken into account.

A cylindrical surface basically constitutes a one dimensional defocusing lens. In the experiments a 60 μm Au wire was used to ballistically defocus the ion beam. Due to the preformed plasma at the front surface, the shape of the front surface is not affecting the orientation of the proton beam while the rear surface geometry should lead to a line shaped imprint in the RCF detector positioned at a few centimeter distance behind the target. The experimental result is shown in Fig. 5.

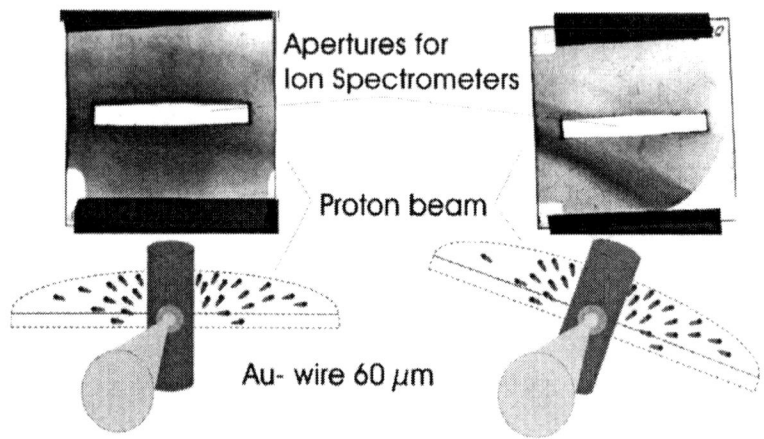

FIGURE 5. Experimental setup and RCF images of experiments with 60μm diameter gold wires. The convex rear surface constitutes a de-collimating cylinder-lens. Accordingly the proton beam was formed into a line.

The line shaped imprint on the detector clearly shows the one dimensional de-collimation of the ion beam due to the target structure. Tilting the target also resulted in a tilted image on the RCF, which results from the radial, fan-shaped expansion of the protons normal to the surface of the wire. For a number of applications like ion-induced material damage research, proton driven fast ignition[11], proton radiography and the use as next generation ion sources, focusing the ion beam is more interesting in order to maximize the energy deposition onto a given sample. Because of the energy dependent angle of divergence the effective focal length is significantly longer and dependent on the proton energy. The targets used in these experiments consisted of 48 μm gold foils of mm size. The curvature of the target surface was changed from a flat target to concave shaped targets with radii of curvature between 10 and 2.5 mm. Fig.6 shows the experimental setup and the corresponding RCF images for a flat target and a target having a 2.5 mm radius of curvature. The RCF detector was mounted 9 mm behind the laser irradiated foil and protected from plasma blowoff by 10 μm of aluminum and 100 μm of titanium. The respective layers of RCF are most sensitive to the energy deposition of stopped protons (Bragg-peak) at 5 MeV for the first and 7.5 MeV for the second layer. The results show a strong reduction in the divergence of the central core of the proton beam representing ballistic collimating of laser produced proton beams.

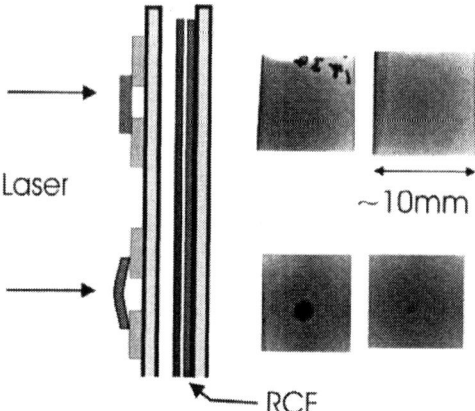

FIGURE 6. Focusing of laser generated proton beams. left: experimental setup. The RCF detector is shielded by 10 μm of aluminum and 100 μm of titanium. right: images of successive layers in RCF for a flat target and a target with 2.5 mm radius of curvature.

The generation of collimated beams of heavy ions is largely suppressed by the accelerated protons. Due to their higher charge-to-mass ratio protons tend to outrun the heavier ion species during the acceleration. For an effective heavy ion beam production the amount of protons has to be reduced significantly. We diminished the hydrocarbon and water vapor contaminants of metal targets heating them resistively. However, this method is not applicable for compound targets that are temperature sensitive. For these targets we used a second, low energy laser beams at intensities below the plasma generation threshold to heat the rear target surface. Cleaning surfaces by laser heating has been explored using electron microscopic surface evaluation[12]. The experimental results using the high resolution Thomson parabolas are depicted in Fig. 7.

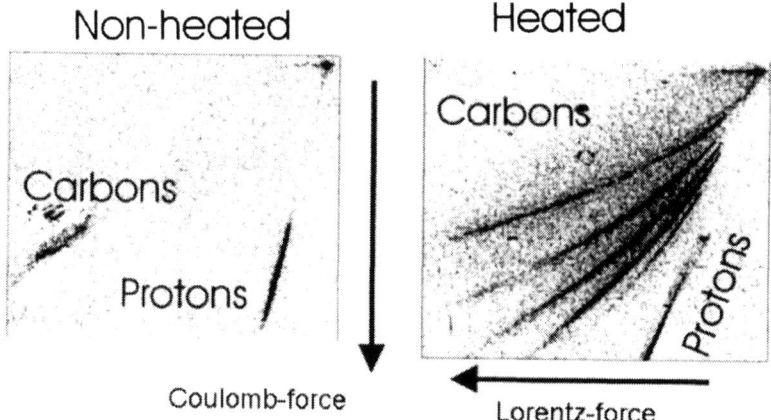

FIGURE 7. Heavy ion beam production. In contrast to the strong proton signal (left) removing the hydrocarbons from the target rear surface results in a strong heavy ion (carbon) signal (right).

The target consisted of a 50 μm aluminum foil coated with 1 μm of carbon at the rear surface. While the non-heated sample leads to a strong proton signal (left spectrum) and only weak structures for heavy ions (carbon) at moderate final energies the heated sample shows a strongly reduced proton signal together with intense traces of carbon ions at much higher energies. A detailed analysis of the ion spectra will be presented elsewhere, but the experiments clearly demonstrate the efficient acceleration not only of protons , but also of heavy ions which is crucial with respect to the use of relativistic laser plasmas as next generation ion sources for ion accelerators.

To precisely estimate the ion beam emittance, penumbral imaging techniques of edges at different distances from the target with the magnetic spectrometers were used to directly measure the emittance core of the proton beam. This technique is closely related to the commonly used slit-emittance measurements made with simple apertures and screens at conventional accelerators. We determined the emittance of protons from flat gold foils to be ~0.2 pi mm-mrad, and a factor of at least two smaller than the resolution-limited measurement we performed on the LLNL PETAWATT laser[11]. Details of the present measurements and systematics of the proton emittance versus energy will be reported elsewhere. The results of this analysis and subsequent modeling, developing a 2-D extension of the model[13], suggest that we observe a rather cold proton beam, which is smoothly diverging and highly laminar. The trace space of the highest energy protons exhibits a tilted ellipse, whose width ultimately is characteristic of the ion temperature. From these data, we deduce that the proton temperature is less than ~1 keV. From simple electron-ion collisional heating during the expansion, one may expect the ion temperature to be even lower, of order ~100 eV. In other words, the protons form a very high quality beam, which may ultimately be suitable for certain applications which now typically use conventional radio-frequency accelerator beams.

Complementing the laser and ion beam diagnostics we also measured the neutron emission caused by (γ,n) and (p,n) reactions from the target. We used a silver activation detector attached to a photo-multiplier tube (PMT). On typical shots, the neutrons are generated by (γ,n) reactions within the target (caused by the bremsstrahlungs photons from the relativistic electrons) and by (p,n) reactions of our proton beam impacting on the RCF screen. We also used an Al-target coated with deuterized plastic (CD_2) on the rear surface, which was heated to produce a beam of deuterons. Fielding a CD_2 catcher foil behind the target we observed the yield of neutrons from (d,d) fusion reactions. Implementing a neutron time-of-flight detector we could also separate neutrons generated by breakup reactions from the ones caused by the (d,d) fusion reaction. We detected a total yield of 2.8×10^7 neutrons in this experiment, which was at least an order of magnitude above the yield on average shots. The fusion neutrons detected via the time-of-flight method showed a strong blue-shift in agreement with the forward peaked reaction kinetics of beam fusion assuming a deuteron beam with a 5 MeV Maxwellian energy distribution.

CONCLUSION

We have presented a detailed investigation of the target conditions on the proton and ion beam production from intense laser solid interactions. The observed strong dependence on the rear surface conditions is in agreement with the target normal sheath acceleration mechanism. The target conductivity appears to have a major influence on the quality of the ion beam, and the quality of the surface finish of the target is very important for maintaining a high gradient sheath and a laminar beam. It has been shown that tailoring the ion beam (yield, shape, composition, homogeneity) by means of target shape and composition is possible, and we present first observations of laser-accelerated proton beam focusing. The successful generation of a heavy ion beam (carbon) further encourages speculation that laser-accelerated ion beams may become a useful tool in a variety of future applications. Finally the detection of an intense, forward peaked burst of neutrons caused by beam fusion reactions was observed.

ACKNOWLEDGMENTS

We would like to acknowledge the LULI laser operations staff and especially thank Ji Ping Zou, Catherine Le Bris, Corinne Felix and Eduard Veuillot for their expert assistance.

This work was supported by EU Programme n^0 HPRI CT 1999-0052.

REFERENCES

1. M. Perry and G. Mourou, *Science* **264**, 917 (1994).
2. R.A. Snavely, et al., *Phys. Rev. Lett.* **14**, Vol 85, 2945 (2000).
3. S.J. Gitomer, et al.,*Phys. Fluids* **29**, 2679 (1986).
4. Y. Kishimoto, et al., *Phys. Fluids* **26**, 2308 (1983).
5. S.C. Wilks, et al., *Phys. Plasmas* **8**, 542 (2001).
6. H. Ruhl, et al., *Plasma Physics Reports* **27**, 363 (2000).
7. Cowan et al., submitted to *Nucl.Instr. and Meth. A*
8. S. Hatchett et al., *Phys. of Plasmas* **7**, 2076 (2000).
9. R. Ramis, R. Schmalz, and J. Meyer-ter Vehn, *Comp. Phys. Comm.* **49**, 475 (1988).
10. J.F. Ziegler, J.P. Biersack, and U. Littmark, *The Stopping and Range of Ions in Solids,* (Pergamon, New York, 1996).
11. M. Roth et al., *Phys. Rev. Lett.* **3**, Vol. 86, 436 (2001)
12. M. Weingärtner et al.,Appl. Surf. Science, 138-139, p. 499 (1999)
13. L.M. Wickens and J.E. Allen, *Phys. Fluids* 24, 1984 (1981).

Surface Effects in Laser Interaction with Overdense Plasmas

A. Macchi[*], F. Cornolti[*], T. V. Liscikina[*], F. Pegoraro[*], F. Califano[*],
H. Ruhl[†] and V. A. Vshivkov[**]

[*]*Dipartimento di Fisica and INFM (sezione A), Università di Pisa, Italy*
[†]*Max-Born Institut für Quantenoptik, Berlin, Germany*
[**]*Institute of Computational Technologies of SD-RAS, Novosibirsk, Russia*

Abstract. The dynamics and generation of surface structures in intense laser interaction with overdense plasmas is studied numerically and analytically. The dynamics of two-dimensional (2D) electron surface oscillations (ESOs) has been resolved in particle-in-cell simulations, showing a "period doubling" effect with respect to the driving magnetic force of the laser. A new parametric process, in which a 1D electrostatic oscillation decays into two electromagnetic surface waves, has been investigated analytically and provides a mechanism for the generation of ESOs. For oblique laser incidence and p-polarization generation of subharmonic surface waves is also predicted. The possible impact of surface deformations on fast electron generation and disruption of plasma "moving mirrors" for high harmonic generation is discussed.

INTRODUCTION

The interaction of sub-picosecond, high-intensity laser pulses with solid targets is of great relevance to the generation of bright sources of energetic radiation as well as a test bed for Fast Ignitor physics. Since for a solid target the electron density $n_e \gg n_c$, where $n_c = 1.1 \times 10^{21}$ cm$^{-3}/[\lambda/\mu\text{m}]^2$ is the cut-off density for laser propagation at the wavelength λ, the laser-plasma coupling occurs at the target surface over a narrow region with a depth of the order of the skin length $d_p \ll \lambda$. The laser force on the plasma has both secular components (leading to plasma acceleration, profile steepening and hole boring) and oscillating components; at high laser intensity, these latter drive an oscillatory motion of the "critical" surface where $n_e = n_c$, that acts as a "moving mirror" leading to the appearance of high harmonics in the reflected light [1].

It is known that pre-imposed deformations of the target surface have a strong impact on the laser-plasma coupling, at least for pulse durations short enough that ion motion is negligible. Therefore, properly microstructured targets may be used to optimize laser absorption and X-ray emission at moderate laser intensity [2, and references therein]. At intensities exceeding 10^{18} W cm^{-2}, experiments [3] and simulations [4, 5, 6] show that self-generated deformations of the target surface grow and strongly affect the laser-plasma coupling. For instance, a "smooth" density deformation wide as the laser spot, as that produced during the hole boring process, leads to high absorption and helps the collimation of fast electrons into a narrow jet due to a geometrical "funnel" effect [5]. Due to the appearance of smaller scale deformations at higher intensities, several electron jets may appear [6].

Surface rippling plays a detrimental role in high harmonics generation since the collimation of the radiation reflected from the plasma moving mirror is destroyed at high intensities. Experimental evidence comes from the wide spreading of the reflected radiation observed in experiments [7, 8, 9] at intensities $\approx 10^{18}$ W cm^{-2} and even for pulse durations as short as 35 fs [8]. This suggests that surface rippling involves some "fast" mechanism related to electron motion rather than Rayleigh-Taylor-like hydrodynamic instabilities driven by the strong target acceleration and occurring on time scales of ion motion. Seeding of density "ripples" by electron instabilities in the underdense plasma region in front of the target was evidenced in simulations [4] and investigated theoretically in [10].

In this paper, we investigate analytically and with particle-in-cell (PIC) simulations the generation of surface structures in steplike density profiles (no underdense region present). Two-dimensional (2D) numerical experiments show that electron surface oscillations (ESOs) may grow much faster than the typical time scale of ion motion, and suggest that a parametric mechanism, involving the excitation of transverse surface modes, is at play [11]. A theoretical model shows that the 1D electrostatic surface oscillation driven by the magnetic force of a normally incident laser wave may excite two electromagnetic, electron surface waves (ESWs) by a parametric process [12]. The analytical model is also extended to the case of oblique incidence of a p-polarized laser pulse, showing that subharmonic ESWs may be generated. This represents a new nonlinear mechanism of surface wave excitation by laser different from previously investigated models (see discussion below). Different mechanisms of generation of surface density inhomogeneities and oscillations by intense laser pulses, not involving ESW generation, have been studied by other authors [10, 13, 14],

TWO-DIMENSIONAL ELECTRON SURFACE OSCILLATIONS

We use 2D PIC simulations to study the dynamics of the ESOs with proper spatial and temporal resolution. In particular, in order to evaluate the frequency of the ESOs, the complete output of 2D fields was produced eight times for each laser cycle. In the simulations reported, the laser pulse is normally incident, s-polarized with a wavelength $\lambda = 0.25\mu$m, a uniform spatial profile in the transverse (y) direction and a temporal profile that rises for three cycles and then remains constant. The plasma has immobile ions and a step-like density profile. The cut-off density is $n_c = 1.6 \times 10^{22}$ cm^{-3}. A numerical box $12\lambda \times 4\lambda$ is taken with a spatial resolution equal to the Debye length corresponding to the initial density n_o and the temperature $T_e = 5$ keV, and 25 particles per cell are used. The plasma fills the right part of the box ($x > 0$) and the laser pulse impinges from the left on the x-axis. For s-polarization, the laser electric and magnetic field are in the z- and y-direction, respectively. We will focus on two typical simulations (run 1 and run 2). Introducing the dimensionless irradiance $a_o = 0.85(I\lambda^2/10^{18}$ Wμm^2/cm$^2)^{1/2}$ in run 1 we take $a_o = 1.7$ and $n_o = 5n_c$, while in run 2 we take $a_o = 0.85$ and $n_o = 3n_c$.

The contours of the electron density $n_e(x, y)$ in Fig. 1 for run 1 at times $t = 8, 10, 12,$ and 14 laser cycles from the run start, show the evolution of the surface dynamics over several laser cycles. Correspondingly, the space-time contours of $n_e(x, y = y_i, t)$

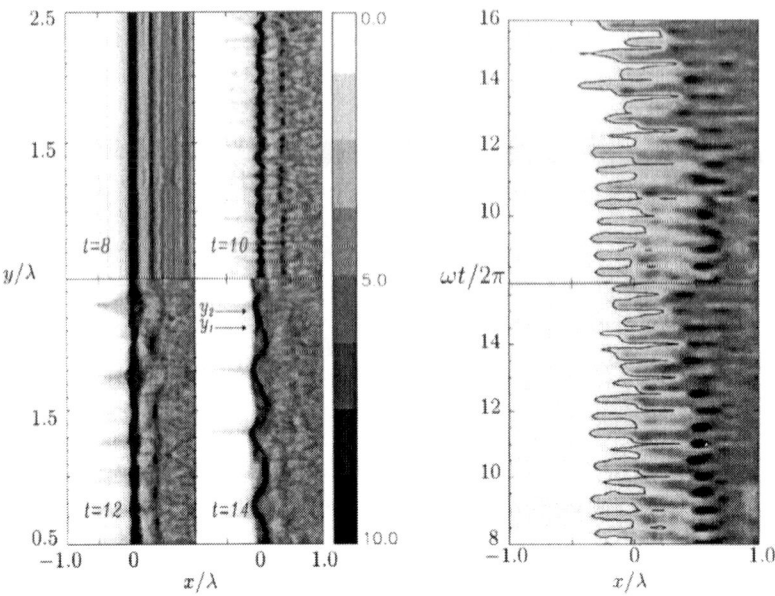

FIGURE 1.   Left: Contours of normalized electron density n_e/n_c for run 1 ($a_o = 1.7, n_o = 5n_c$) at various times (see plot labels) in laser cycle units. Only a small portion of the simulation box around the target surface is shown. Right: Space-time evolution of $n_e(x, y = y_l, t)$ at $y_1 = 1.875\lambda$ (left) and $y_2 = 2.0\lambda$ (right) for run 1. The position of the $n_e = n_c$ surface is evidenced by a black contour line.

at $y_1 = 2.0\lambda$ and $y_2 = 1.875\lambda$ in Fig. 1 show the temporal behavior of the surface oscillations. Initially, the surface oscillation is planar, i.e. uniform along y, and has a frequency 2ω, being $\omega = 2\pi c/\lambda$ the laser frequency. It is natural to identify this 1D motion as the "moving mirror" driven by the longitudinal $\mathbf{v} \times \mathbf{B}$ force at 2ω. In the compression phase, electrons pile up in a narrow layer where the peak density is $n_e \simeq 2n_o$; the electrostatic field $E_x^{(2\omega)}$ (not shown) is positive and counteracts the $\mathbf{v} \times \mathbf{B}$ force. In the expansion phase electrons are dragged out into vacuum, forming a "cloud" of negative charge with a negative electrostatic field.

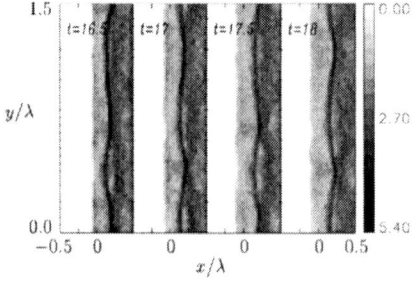

FIGURE 2.   Same as Fig. 1 for run 2 ($a_o = 0.85, n_o = 3n_c$).

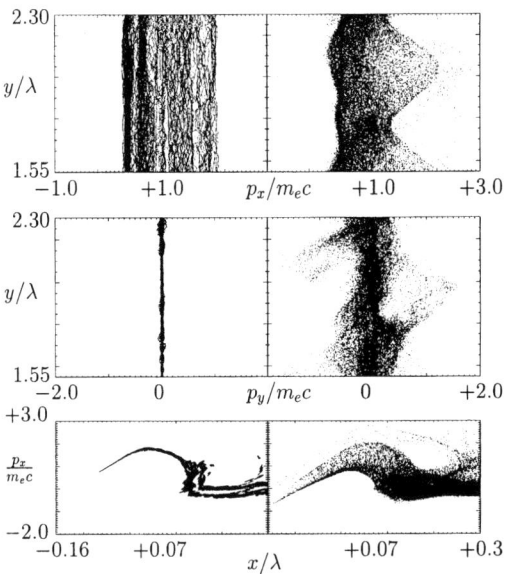

FIGURE 3. Phase space projections for run 1 in the (y, p_x) (top), (y, p_y) (middle), and (x, p_x) (bottom) planes for $t = 8.5$ (left) and $t = 13.5$ (right) laser cycles.

The growth of surface "ripples" can be observed in Fig.1. At $t = 10$, they have small wavelengths ($\simeq 0.1\lambda$), while at $t \geq 12$ they evolve into a steady oscillation with wavelength $\lambda_s \simeq 0.5\lambda$ (Fig.1) and frequency $\simeq \omega$, which is superimposed to the oscillation at 2ω (Fig. 1). This "period doubling" effect is also observed in run 2, for which the density contours are shown between $t = 17$ and $t = 18$. in Fig. 2. The transverse "snaking" of the plasma layer in time at the laser frequency is evident. In run 2 the oscillation amplitude and the density compression is lower than in run 1, and the deformation wavelength is larger ($\lambda_s \simeq 0.75\lambda$). In the following we will discuss only the long-wavelength structures oscillating at ω and refer to them as (2D)ESOs. The 2DESOs are "standing", i.e. not propagating along y. From Fig.1 we see that while at y_1 the amplitude is close to its maximum, at $y_2 = y_1 - \lambda_s/4$ there is no evident growth of the oscillation at ω, while a weakening of the 2ω oscillation is observed. Results for p-polarization (not shown) at normal incidence are qualitatively similar, the main difference being that the plasma "plumes" extending into vacuum are bent by the laser field in the plane.

The ESOs have a substantial impact on fast electron generation. Fig. 3 shows phase space distributions at times $t = 8.5$ and 13.5 laser cycles for run 1. At early times, the momentum distribution is uniform in y, with no accelerated particles in p_y and most energetic electrons having $p_x \simeq 2m_ec$. At later times, when the 2DESOs have grown, stronger forward acceleration occurs near the maxima of the oscillation, showing that most oscillatory energy has been transferred to the unstable 2D modes. Correspondingly, strong acceleration in p_y also occurs. The momentum distribution in p_x is very regular, with only a minority of electrons "outrunning" the oscillation. This suggests that the generation of fast electrons is correlated with the nonlinear evolution and "breaking" of the density oscillations. This may give a spatial "imprint" on the transverse structure of the fast electron currents. In previous simulations one may clearly observe a spatial correlation between electron jets and "corrugations" at the surface [15, 16]. When penetrating into the bulk the jets may either merge or drive current filamentation instabilities [16, 17], thus producing different spatial scales.

PARAMETRIC EXCITATION OF SURFACE WAVES

The simulation results suggest that the driving mechanism of the 2DESOs is a parametric process involving the decay of of the forced 1D oscillation with frequency 2ω and transverse wavevector $k_y = 0$ into a couple of transverse modes (ω_1, k_1) and (ω_2, k_2). In fact, the matching conditions for this process $2\omega = \omega_1 + \omega_2$ and $0 = k_1 + k_2$ immediately give $k_1 = -k_2$ and $\omega_1 = \omega_2 = \omega$. The two overlapping transverse waves thus form a standing oscillation with frequency ω and wavevector $k = k_1 = -k_2$.

The parametric process is resonant if the transverse waves are normal modes of the plasma. In a step-boundary profile, such modes are "H" electron surface waves (ESWs) [18]. The "pump" mode, however, is not a normal mode but a forced oscillation driven by the strong $\mathbf{v} \times \mathbf{B}$ force. We have developed a cold fluid, non-relativistic 2D model of this novel mechanism which we may call either parametric surface instability (PSI) or two surface wave decay (TSWD) [11]. Preliminary analytical calculations show that the mechanism is also at play for oblique incidence, although with significant differences. An essential description of the calculations is reported in the following, for the two cases of normal and oblique incidence.

Normal incidence

We consider normal incidence of a plane wave on a step-boundary plasma with ion density $n_i(x) = n_i\theta(x)$. Using the conservation of the transverse canonical momentum $p_\perp = eA_\perp/c$, a system of 2D Maxwell-Euler equations is obtained where the laser action enters via the ponderomotive force $f_p = -(e/mc)^2 A_\perp \partial_x A_\perp = f_o(x)(1 + \cos 2\omega t)$. Secular terms are not important in the following and we drop them for simplicity. The electron fluid density and velocity are thus written as $n_e = n_o(x) + \varepsilon \delta n_e^{(2\omega)}(x,t) + \varepsilon^2 \delta n_e^{(\omega)}(x,y,t)$ and $\mathbf{v}_e = \varepsilon V_x^{(2\omega)}(x,t)\hat{\mathbf{x}} + \varepsilon^2 \mathbf{v}^{(\omega)}(x,y,t)$, where $\varepsilon \sim a_o^2(n_c/n_e)$ is a small expansion parameter. The terms at 2ω describe the electrostatic, 1D "moving mirror" oscillation, that acts as a pump for the instability, and is assumed to be unperturbed by the ESWs. The terms at ω are the superposition of two ESWs:

$$\mathbf{v}^{(\omega)} = e^{-i\omega t}\left(\tilde{\mathbf{v}}_{+k}e^{iky} + \tilde{\mathbf{v}}_{-k}e^{-iky}\right)/2 + \text{c.c.}, \tag{1}$$

where $\tilde{\mathbf{v}}_{\pm k} = \tilde{\mathbf{v}}_{\pm k}(x,t)$ varies slowly in time. To order ε^2, the coupling between 1D and 2D modes may be neglected, so that one obtains the usual dispersion relation for "H" surface waves propagating along a density discontinuity:

$$k^2 c^2 = \omega^2 (\omega_p^2 - \omega^2)/(\omega_p^2 - 2\omega^2), \tag{2}$$

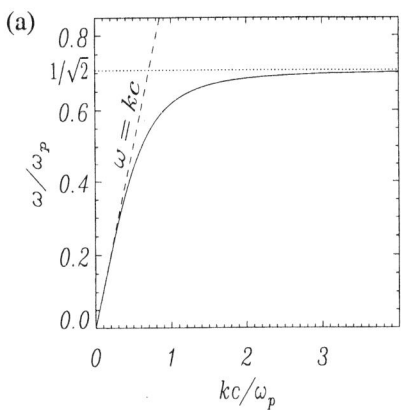 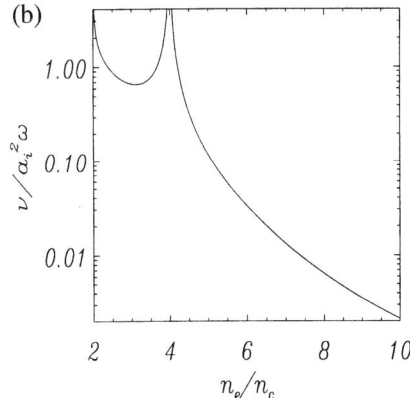

FIGURE 4. (a) Dispersion relation of electron surface waves, Eq.(2). (b) The growth rate for the parametric growth of electron surface waves at normal laser incidence as a function of n_e/n_c, Eq.(5).

where $\omega_p^2 = 4\pi n_o e^2/m_e$ is the plasma frequency. The dispersion relation is reported in Fig.4. The evanescence length of the ESWs in the plasma is $L_{SW} = (c/\omega)(1 - 2\omega^2/\omega_p^2)^{1/4}(1 - \omega^2/\omega_p^2)^{-1/2}$. Notice that for electron surface waves $\nabla \cdot \mathbf{E}^{(\omega)} = -4\pi e \delta n_e^{(\omega)}(x,y,t) = 0$, and that their maximum frequency is $\omega_{max} = \omega_p/\sqrt{2}$, so that the matching conditions may be satisfied only if $\omega < 2\omega_{max}$, i.e. $n_e > 2n_c$.

By keeping only terms up order ε^3 in the Euler equation and neglecting feedback effects on the 1D motion, the Euler equation for the ESW velocity is $m_e \partial_t \mathbf{v}^{(\omega)} = -e\mathbf{E}^{(\omega)} + \varepsilon \mathbf{f}_{NL}^{(\omega)}$ where $\mathbf{f}_{NL}^{(\omega)}$ includes the nonlinear coupling with the 1D motion:

$$\mathbf{f}_{NL}^{(\omega)} = -m_e V_x^{(2\omega)} \partial_x \mathbf{v}^{(\omega)} - m_e v_x^{(\omega)} \partial_x V_x^{(2\omega)} \hat{\mathbf{x}} + e/c V_x^{(2\omega)} B_z^{(\omega)} \hat{\mathbf{y}}. \tag{3}$$

Using this equation and Poynting's theorem the rate of growth of the surface energy for the 2DESOs was evaluated as $\Gamma \equiv U^{-1} \partial_t U$, where U is the energy density per wavelength of the two ESWs and the cycle-averaged variation is

$$\partial_t U = \int dx \left\langle \mathbf{v}^{(\omega)} \cdot \left(e \delta \tilde{n}_e^{(2\omega)} \tilde{\mathbf{E}}^{(\omega)} + m_e n_o \partial_t \tilde{\mathbf{v}}^{(\omega)} \right) \right\rangle$$

$$= \frac{1}{4} \int_0^{+\infty} dx \tilde{\mathbf{v}}_{+k}^* \cdot \left(e \delta \tilde{n}_e^{(2\omega)} \tilde{\mathbf{E}}_{-k}^* - m_e n_o \tilde{v}_{x,-k}^* \partial_x V_x^{(2\omega)} \right.$$

$$\left. - m_e n_o \tilde{V}_x^{(2\omega)} \partial_x \tilde{\mathbf{v}}_{-k}^* + n_o \frac{e}{c} \tilde{V}_x^{(2\omega)} \tilde{B}_{z,-k}^* \hat{\mathbf{y}} \right) + \text{c.c.} \tag{4}$$

Substituting in the integrand for the expressions of the (unperturbed) ESW fields, finally one obtains the growth rate as

$$\nu \simeq 4\omega a_o^2 \frac{(\alpha - 1)^{3/2}}{\alpha |\alpha - 4| [(\alpha - 1)^2 + 1](\alpha - 2)^{1/2}} \tag{5}$$

where $\alpha = n_e/n_c = \omega_p^2/\omega^2$. The denominator $(\alpha - 4)$ is actually due to the resonant excitation of longitudinal plasmons at $\omega_p = 2\omega$, which makes $V_x^{(2\omega)}$ very large and invalidates our expansion procedure near resonance. We note that v diverges also for $\alpha \to 2$; however, in this limit the ESW wavelength is very small and thus one expects a strong damping by thermal effects neglected in the cold fluid model.

Oblique incidence and subharmonic generation

We now discuss the case of oblique incidence and p-polarization of the laser pulse. The main physical difference with the previous case is that now the laser field has a component normal to the plasma surface, and the "pump" oscillation is at the laser frequency ω. Thus, the two ESWs have frequencies $\omega_{\pm} = \omega/2 \pm \delta\omega$. Due to conservation of the transverse component of the laser wavevector $k_0 = (\omega/c)\sin\theta$, being θ the incidence angle, the matching condition $k_+ + k_- = k_0$ also holds.

For a cold fluid plasma and in linear, non-relativistic approximation, all EM fields may be found from Fresnel formulas for refraction and transmission at a boundary between vacuum and a medium with (imaginary) refractive index n, given by $n^2 = 1 - \omega_p^2/\omega^2 < 0$. For instance, the magnetic field inside the plasma is given by

$$B_z^{(\omega)}(x,t) = B_z(0^+)e^{iky\sin\theta - x/l_p - i\omega t} + \text{c.c.}, \tag{6}$$

$$B_z(0^+) = 2B_{z,i}n^2\cos\theta \left(\sqrt{n^2 - \sin^2\theta} + n^2\cos\theta\right)^{-1} \equiv B_{z,i}F(\alpha,\theta), \tag{7}$$

where $B_{z,i} = E_i$ is the incident field amplitude in vacuum and the screening length is given by $l_p = (c/\omega_p)\left(1 - \omega^2\cos^2\theta/\omega_p^2\right)^{-1/2}$. The other EM fields are obtained from Maxwell's equations. One finds that to lowest order there is no volume charge oscillation, i.e. $\delta n_e^{(\omega)} = 0$, but a spatially periodic, oscillating surface charge density.

In our approach, similarly to the case of normal incidence, we now adopt the following expansion

$$u = U^{(\omega)}(x,y,t) + \varepsilon[u_+(x,y,t) + u_-(x,y,t)], \tag{8}$$

where, again, ε is a small expansion parameter. Here, U and u stand for a generic EM or velocity field, and $u_{\pm}$ are the ESW fields at the frequencies (corresponding to "blue" and "red" waves, respectively) $\omega_{\pm} = \omega/2 \pm \delta\omega$, with wavevectors $k_{\pm}$ and dispersion relation $\omega_{\pm} = \omega(k_{\pm})$ given by Eq.(2). The shift $\delta\omega$, obtained from the matching conditions and Eq.(2), is shown in Fig.5 (a). One finds the following equation for the velocity of the ESW:

$$m_e\partial_t \mathbf{v}_{\pm} = -e\mathbf{E}_{\pm} + \varepsilon\mathbf{f}_{\pm}^{(NL)}, \tag{9}$$

where $\mathbf{f}_{\pm}^{(NL)}$ are the resonant parts (at the frequencies ω_+ and ω_-, respectively) of the force term

$$\mathbf{f}^{(NL)} = -m_e(v_x\partial_x\mathbf{V}^{(\omega)} + V_x^{(\omega)}\partial_x\mathbf{v}) - m_e(v_y\partial_y\mathbf{V}^{(\omega)} + V_y^{(\omega)}\partial_y\mathbf{v})$$
$$-(e/c)(\mathbf{V}^{(\omega)} \times \mathbf{b} + \mathbf{v} \times \mathbf{B}^{(\omega)}), \tag{10}$$

where small caps are used for the ESW fields.

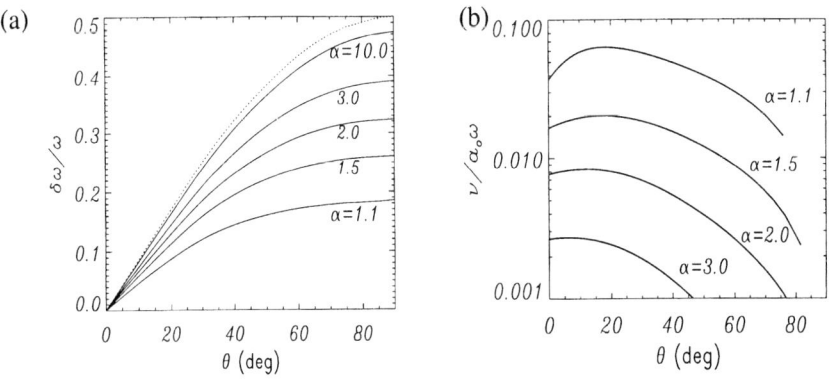

FIGURE 5. (a) The frequency shift $\delta\omega/\omega$ as a function of θ and $\alpha = n_e/n_c$. The shift increases for high densities and approaches the limiting value $\sin\theta/2$. (b) The growth rate for oblique incidence as a function of θ and α, Eq.(11).

Evaluating the growth rate of the surface energies as in the case of normal incidence, one finds again an energy growth of the two subharmonic ESWs. The growth rate may be written as

$$\nu \simeq a_o\omega\alpha^{-5/2}\,|F(\alpha,\theta)|\,G(\alpha,\theta), \tag{11}$$

where $G(\alpha,\theta)$ is a dimensionless factor which scales weakly with α. The growth rate ν is shown in Fig.5 (b) as a function of θ and for four values of n_e/n_c. We note that ν scales with density as $\approx (n_e/n_c)^{-5/2}$ and is proportional to the laser field amplitude a_o (and not to the irradiance a_o^2). In addition, ν has a maximum near $\theta \simeq 20$ degrees for n_e close to n_c, but ν is non-vanishing even at $\theta = 0$. The rate is quite low, so one may expect that subharmonic ESWs do not grow at all during the laser pulse or when, e.g., damping effects are included. However we notice that our expansion is limited to rather low intensities, for which there is almost no "moving mirror" motion. On the basis of the present calculations nothing can be said about the growth in the high-intensity regime. Nevertheless the present calculations show in principle that the mechanism works for oblique incidence also and suggest that subharmonic ESWs may be generated.

DISCUSSION

Our numerical and analytical results both indicate that the "parametric" generation of surface oscillations occurs strongly only for moderate plasma densities, i.e. a few times the cut-off density n_c. This observation, however, does not necessarily lowers the relevance of these effects to experiments on laser interaction with solid targets. In fact, both theory and experiment indicates that processes such as generation of high harmonics and fast electrons require a moderate density "shelf" with a non-vanishing gradient at the critical surface. Such a shelf is often incidentally or even intentionally produced by a low intensity prepulse. The strong radiation pressure of the short interaction pulse steepens the density profile near the critical surface, so that near the laser intensity peak the plasma profile may well look rather similar to that of our simulations. Note also that recent measurements [19] showed that even prepulses with intensity below the ionization threshold may effectively lead to formation of a moderately dense preplasma by preheating the target and stimulating evaporation at the surface, so that the number of experiments affected by preplasma formation may be effectively higher than previously estimated.

The case of normal laser incidence is directly relevant to the numerical experiments reported above. In this case, the parametric excitation of surface waves leads to a superposition of field oscillations at frequencies ω and 2ω, as observed in the simulations. Thus, one expects to see a "snaking" of the moving mirror at the frequency ω and with a spatial period equal to the ESW wavelength. This is what is observed in run 2, Fig.2. Inserting the value of the laser frequency in (2) we find that the expected wavelength of deformations in run 2 is $\lambda_s = 2\pi/k \simeq 0.71\lambda$, in good agreement with the simulation, where the laser and plasma parameters are such that the expansion procedure of our analytical model are marginally valid. This suggests that the 2D ESOs are due to the parametric excitation of ESWs.

For run 1 one finds $\lambda_s \simeq 0.87\lambda$, quite larger than the numerical result. This is not surprising since our expansion procedure is not applicable for the parameters of run 1, where the interaction is in the relativistic regime, and the magnetic force drives a surface density increase of the order of the background density. The appearence of sharp 2D structures in the plasma density is also an indication that, altough we believe that a parametric excitation of surface modes is still at play, the process occurs in a strongly nonlinear and relativistic regime so that spatial scales may be significantly different from those predicted by the cold fluid, non relativistic model. One important relativistic effect may be the lowering of the effective plasma frequency by the (time-averaged) relativistic factor $\gamma_o \simeq \sqrt{1+a_o^2/2}$, due to the relativistic quiver motion with $v_\perp \simeq c$, while $v_x \ll c$ usually holds for $a_o \approx 1$. By replacing ω_p^2 by ω_p^2/γ_0 in (2), for run 1 we obtain $\lambda_s \simeq 0.55\lambda$, much closer to the numerical result. One may also expect that, due to the strong dependence of the growth rate (5) on the ratio $n_e/n_c = (\omega_p/\omega)^2$, there is a sharp increase of the growth rate when $a_o \geq 1$.

Note that our model for normal incidence is valid for both s- and p-polarizations, since even if the laser electric field is parallel to the k-vectors of the ESWs there is no resonant coupling of the quiver motion with the ESWs. The transverse oscillations are however observed more clearly for s- rather than for p-polarization, since in the latter case the motions in the laser and the ESW fields overlap in the polarization plane.

The case of oblique incidence and p-polarization has a more direct relevance to experiments on high harmonic generation. In this case, in addition to the 2ω motion the laser drives the moving mirror at the frequency ω. We have shown analytically that, to lowest order, this may lead to the pumping of two ESW sidebands around the frequency $\omega/2$. Numerical simulations of the oblique incidence case are presently in progress in order to confirm this prediction as well as to study the process in the strongly nonlinear regime and to evaluate the possible role of ESW excitation in driving a "fast" rippling of the moving mirror. However, the set-up and the analysis of these simulations is more involved (for instance, different mechanisms of generation of surface oscillations are at play [13, 14]) than the case of normal incidence and s-polarization and results will be presented in a forthcoming paper. We note, however, that the s-polarization, normal incidence results already give some interesting indication for the study of fast moving mirror disruptions. First, the growth rate is maximum in conditions "optimal" for IIII generation, i.e. when the moving mirror oscillation is driven at high velocities, which requires a moderate density "shelf" rather than solid densities. In such conditions the ESO instability can be much faster than RT instabilities. In fact, even for accelerations of order $g \simeq 10^{20}$ cm/s^2 as measured in this regime [20, 21], the typical RT growth rate $\Gamma_{RT} \simeq \sqrt{k_{RT}g} \simeq (140 \text{ fs})^{-1}$ for $2\pi/k_{RT} \simeq \lambda/2 = 0.125\mu m$, is much slower than the rise of the 2DESOs which occurs over a few fs in the simulations. We also note that our numerical results suggest that the surface rippling becomes stronger in the relativistic regime $a_o \geq 1$, a condition close to the experimentally observed threshold for surface rippling.

Parametric excitation of ESWs also provides an additional mechanism for energy absorption. So far, structured targets have been used to increase short pulse absorption by ESW excitation [22, and references therein]. More in general, in previous studies [23, 24, 25, and references therein], one finds that some "special" conditions such as microstructuring of the target surface, special preformed plasma profiles, presence of an external magnetic field, or temperature discontinuities are needed to allow the coupling of the laser pulse with an ESW. Our study shows indeed that it may be possible to excite a couple of ESWs parametrically in the "simplest" case of a laser pulse impinging on an overdense plasma, even at normal incidence and for s-polarization. Our numerical simulations also show that, for relativistically strong intensities, strong surface density oscillations are driven and these in turn affect fast electron generation. This connection between surface instabilities and the production of fast electron jets, of particular relevance for Fast Ignitor studies, will be further studied in future work.

ACKNOWLEDGMENTS

The PIC simulations were performed at the CINECA supercomputing facility (Bologna, Italy), sponsored by the INFM supercomputing initiative.

REFERENCES

1. Bulanov, S. V., Naumova, N. M., and Pegoraro, F., *Phys. Plasmas*, **8**, 745 (1994).
2. Kulcsar, G., et al., *Phys. Rev. Lett.*, **84**, 5149 (2000).
3. Feurer, T., et al., *Phys. Rev. E*, **56**, 4608 (1997).
4. Wilks, S. C., Kruer, W. L., Tabak, M., and Langdon, A. B., *Phys. Rev. Lett.*, **69**, 1383 (1999).
5. Ruhl, H., Macchi, A., Mulser, P., Cornolti, F., and Hain, S., *Phys. Rev. Lett.*, **82**, 2095 (1999).
6. Macchi, A., Cornolti, F., and Ruhl, H., *Las. Part. Beams*, **18**, 375 (2000).
7. Norreys, P. A., et al., *Phys. Rev. Lett.*, **76**, 1832 (1996).
8. Tarasevitch, A., et al., *Phys. Rev. E*, **76**, 023816 (2000).
9. Dietrich, C., Tarasevitch, A., Blome, C., and von der Linde, D., (2000), 2nd ULIA conference, Pisa (Italy).
10. Cadjan, M. G., Ivanov, M. F., and Ivlev, A. V., *Phys. Lett. A*, **222**, 325 (1996).
11. Macchi, A., Cornolti, F., and Pegoraro, F., (2001), preprint physics/0105017.

12. Macchi, A., Cornolti, F., Pegoraro, F., Liseikina, T. V., Ruhl, H., and Vshivkov, V. A., *Phys. Rev. Lett.* (2001), in press [preprint physics/0105019].
13. von der Linde, D., and Rzazewski, K., *Appl. Phys. B*, **63**, 499 (1996).
14. Plaja, L., Roso, L., and Conejero-Jarque, E., *Astrophys. J. Supp. Ser.*, **127**, 445 (2000).
15. Lasinski, B., et al., *Phys. Plasmas*, **6**, 2041 (1999).
16. Sentoku, Y., Mima, K., Kojima, S., and Ruhl, H., *Phys. Plasmas*, **7**, 689 (2000).
17. Califano, F., Prandi, R., Pegoraro, F., and Bulanov, S. V., *Phys. Rev. E*, **58**, 737 (1998).
18. Landau, L. D., Lifshitz, E. M., and Pitaevskij, L. P., *Electrodynamics of Continuous Media*, Pergamon Press, New York, 1984.
19. Wharton, K. B., et al., *Phys. Rev. E*, **64**, 025401(R) (2001).
20. Sauerbrey, R., *Phys. Plasmas*, **3**, 4712 (1996).
21. Häßner, R., et al., *AIP Conf. Proc.*, **426**, 213 (1998).
22. Kupersztych, J., and Raynaud, M., *Phys. Rev. E*, **59**, 4559 (1999).
23. Dragila, R., and S, V., *Phys. Rev. Lett.*, **61**, 2759 (1988).
24. Gamaliy, E. G., *Phys. Rev. E*, **48**, 516 (1993).
25. Magnitskii, S. A., Platonenko, V. T., and Tarasishin, A. V., *AIP Conf. Proc.*, **426**, 73 (1998).

Hot Solid-State Aluminum Plasmas, Positrons, and Neutrons Generated with the Garching Laser Facility ATLAS

Klaus J. Witte*, Ulrich Andiel*, Klaus Eidmann*, Christoph Gahn*, Peter Hakel¶*, Stefan Karsch, Roberto Mancini¶, and George Tsakiris*

*Max-Planck-Institut für Quantenoptik, Hans-Kopfermann-Straße 1, D-85748 Garching

¶Department of Physics, University of Nevada, Reno Nevada 89557-0058, USA

Abstract. We report on time-integrated and time-resolved measurements of the K-shell emission from aluminum plasmas at solid-state density isochorically heated with 2-ω ATLAS pulse of high contrast. We compare the measured spectra with simulated ones. We investigate both plane aluminum and layered targets. The latter consist of a top carbon layer upon an aluminum layer of variable thickness deposited on a sigradur (glass-like carbon) substrate. The layered targets are well suited to study electron beam transport through an overdense plasma. In a different type of experiment, we have produced 10^6 positrons per laser shot by the interaction of an MeV-electron jet emerging from a relativistically self-focused laser channel in an underdense helium plasma whose density is close to the critical one using a 2-mm thick lead disk. We report about details of the measurement and discuss the propsects of this new table-top positron source for a variety of applications when near-future laser systems are envisaged as a driver. For the neutron generation, we used 790-nm/130-fs/1-J ATLAS pulses focused onto fully deuterated polyethylene targets at intensities of up to 10^{19} W/cm^2. We observe neutron yields of up to 10^5 per shot. We discuss how the measured neutron spectra can be related to the ion energy distribution.

INTRODUCTION

Present-day high-repetition table-top laser systems using titanium sapphire as the amplifying material deliver energetic femtosecond pulses focusable to intensities approaching 10^{20} W/cm^2. Due to these favorable characteristics, titanium sapphire lasers have become an important workhorse in the new field of high-intensity physics. This includes the propagation of femtosecond pulses through underdense or overdense plasmas, the generation of MeV electron jets, MeV γ-rays, positrons, neutrons employing either deuterated planar targets or deuterium clusters, and XUV radiation in form of many harmonics of the fundamental emerging from a solid surface when irradiated with a short laser pulse. Furthermore, these pulses can create strongly

CP611, *Superstrong Fields in Plasmas:* Second Int'l. Conf., edited by M. Lontano et al.
© 2002 American Institute of Physics 0-7354-0057-1/02/$19.00

coupled plasmas at solid-state density ($\sim 10^{24}$electrons/cm^3) and electron temperatures of a few hundred eV. Such plasmas are encountered in stars and in laser fusion shortly before ignition. The high-repetition rate of titanium:sapphire lasers has made possible their systematic investigation.

In the present paper, not all subjects just mentioned will be addressed. The focus will be on three of them: Strongly coupled plasmas, positrons and neutrons. The ATLAS (Advanced Titanium:Sapphire Laser) facility at the Max-Planck-Institut für Quantenoptik at Garching has been used for these investigations. ATLAS runs at 10 Hz and can be operated at two different power levels. The low-power version ATLAS-2 consisting of an oscillator and two amplifiers releases 790-nm/130-fs/250-mJ pulses corresponding to a power of ~2 TW. Employing a 2-mm thick KDP crystal, these pulses can be frequency doubled with 30% conversion efficiency whereby the pulse duration increases to 150 fs. The high-power version ATLAS-10 has three amplifiers hence producing stronger pulses of 1–J energy and 8-TW power at the same wavelength and pulse duration. ATLAS-2 is the workhorse for the experiments related to strongly coupled plasmas and the production of positrons; ATLAS-10 serves for the neutron generation.

STRONGLY COUPLED ALUMINUM PLASMAS

Principle of Isochoric Heating of Matter

Fig. 1 illustrates the scenario when a short laser pulse irradiates a metal target. In order to avoid preplasma formation by prepulses prior to the arrival of the main pulse, the contrast ratio of the pulse must be rather high. We meet this requirement by frequency doubling and simultaneous suppression of the fundamental by a factor of 10^8. We thereby achieve a 2ω-intensity contrast ratio of at least 10^{12}:1 up to a few ps before the pulse maximum. In the final picosecond before the peak is reached, the intensity increases by 6 orders of magnitude. The laser pulse is focused by an f/3 off-axis parabolic mirror at an incidence angle of 45° resulting in a peak intensity of $\sim 2 \times 10^{18}$ W/cm^2. There is hence no plasma formation at times $\geq$3ps ahead of the peak.

When the pulse impinges on the target surface, a thin front layer is first heated the

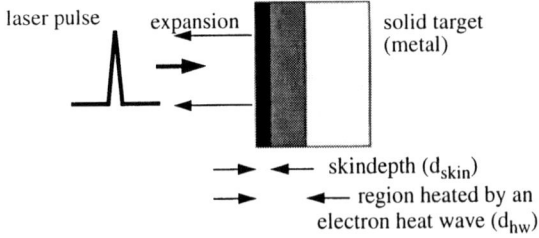

Figure 1. Interaction scenario occurring when a femtosecond laser pulse irradiates a solid target.

thickness of which is given by the skin depth

$$d_{skin}=\frac{\lambda_L}{4\pi\,\mathrm{Im}(\hat{n})} \qquad \text{where} \qquad \hat{n}^2 =1-\frac{\omega_p^2}{\omega_L(\omega_L-i\nu_c)} \quad .$$

Here, $\hat{n}$ is the complex refractive index of the plasma, $\omega_p^2= e^2n_e/(\varepsilon_0 m_e)$ is its frequency, $\omega_L=2\pi c/\lambda_L$ is the laser frequency, ν_c is the electron collision frequency, e is the electron charge, n_e is the electron density, $\varepsilon_0=8.86\times10^{-12}$As/Vm is the permittivity of free space, λ_L is the laser wavelength, and $m_e=9.1\times10^{-31}$kg is the electron mass. For cold aluminum, we have $\nu_c=8.5\times10^{14}$/s [1] and $n_e=1.5\times10^{23}$/cm^3 leading to $d_{skin}=7$nm at $\lambda_L=395$nm. The typical expansion time is d_{skin}/ν_i where $\nu_i =[(ZT_e+T_i)k_B/m_i]^{1/2}$ is the ion sound velocity in a plasma with electron temperature, T_e, and ion temperature, T_i, at the average ionization degree, Z; $k_B=1.38\times10^{-23}$ J/K is the Boltzmann constant. At an intensity of $\sim10^{15}$ W/cm^2 present in the leading pulse edge, T_e and T_i reach 100 eV at a value of $Z\sim7$. For these conditions, the expansion time amounts to 140 fs which is no longer negligible versus the ATLAS pulse duration of 150 fs.

This estimate shows that a surface layer with the thickness of the order of the skin depth cannot be isochorically heated to high temperatures. However during the interaction of the laser pulse with the expanding plasma, fast electrons are generated which penetrate into the target beyond the skin depth whereby they heat the solid by partially depositing their energy due to collisions with ions. This process can be described in terms of a diffusive heat wave whenever its penetration depth, d_{hw}, exceeds the mean free path of the electrons. When this condition is met, the electrons largely thermalize their energy in the region traversed by the heat wave. Fig. 2 presents an example of such a situation. It is a numerical simulation performed with the hydrodynamic code MULTI-fs [1, 2]. The p-polarized laser pulse (this state of polarization is chosen to obtain efficient absorption; see also below) irradiates a solid aluminum target at an incidence angle of 45°. The pulse duration (FWHM) is $t_p=150$fs and the pulse shape is taken as $I=I_0 sin^2(0.5\pi\,t/t_p)$ with $I_0=1\times10^{17}$W/cm^2. Fig. 2 is a snapshot of the situation occurring 250fs after the front edge of the pulse has hit the target surface. During this 250-fs time interval, the expansion has created a plasma with a steep density gradient. At the position where the electron density equals the critical density, $n_{cr}=\omega_L^2\varepsilon_0 m_e/e^2$, the laser energy is effectively coupled into the plasma due to resonance absorption. This process is accompanied by the generation of keV-electrons which penetrate into the target and drive the heat wave. In Fig. 2, it has traversed a distance of ~200nm and thereby converted a ~100-nm thick piece of unexpanded matter into a hot plasma at original solid-state density. Since this plasma is almost fully ionized, $Z=11$-12, its electron density is approximately given by $n_e\sim Z\cdot n_{s,cold}\sim7\times10^{23}$/cm^3, where $n_{s,cold}=6\times10^{22}$/cm^3 is the density of cold aluminum. The characteristic signature of this hot dense plasma is its K-shell emission from which the transition processes contributing to the emission as well as the plasma density and temperature can be inferred.

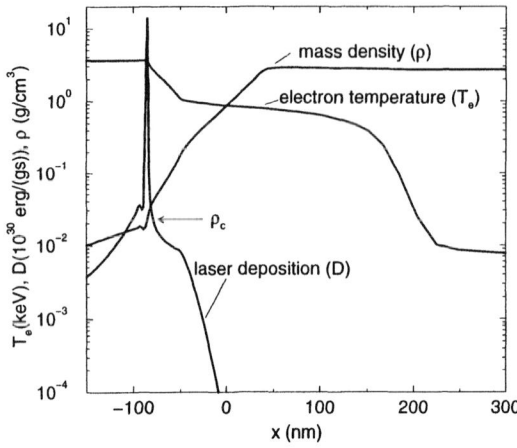

Figure 2. MULTI-fs simulations for the spatial dependence of the mass density, the electron temperature, and the laser energy deposition in an aluminum plasma 250fs after the front edge of the pulse has touched the target surface. The laser pulse comes from left. Its intensity shape is $I=I_0 sin^2(0.5\pi t/t_p)$ with $t_p=150$fs and $I_0=1\times10^{17}$W/cm^2. Prior to interaction, the target surface is at $x=0$. Note the short plasma scale length of ~8nm at the critical density

When the laser pulse is over, the heat wave is no longer fed by electrons coming from the resonance absorption zone but it still propagates into the target resulting in a lifetime of the dense plasma of a few ps before it disappears by expansion and cooling. The time scale of the plasma dynamics is hence *decoupled* from the laser pulse duration.

When I_0 exceeds a few times 10^{17}W/cm^2, the energy deposition changes. It turns out that at these higher intensities a few hundred nm thick layer of unexpanded matter can be homogeneously heated, i. e. in contrast to the *low-intensity case* there is no temperature gradient in the direction normal to the target surface. This was found from experiments with multi-layer targets consisting of a top carbon layer (the choice for carbon is due to the fact that the carbon plasma has no lines in the aluminum K-shell spectrum and is optically very thin in this wavelength range) followed by a thin aluminum layer (on the one hand as thin as possible, on the other hand thick enough to provide sufficient photons for detection; the compromise is at ~25nm) deposited on a substrate material either SIGRADUR (glass-like carbon) or titanium whose radiation is also outside the aluminum K-shell window. The purpose of the top carbon layer is to restrict the expansion to within the carbon plasma so that the aluminum can stay at solid-state density. When this happens is found by systematically varying the thickness of the carbon layer. This technique is hence well suited in establishing very clean conditions for the radiating aluminum layer characterized by the absence of expansion and a temperature gradient. This is an attractive feature for comparing the profiles of the resonance lines of the experimental results with those of theory. Very good agreement for the Ly-α and He-β lines was observed in [3]. On the other hand when the thickness of the top carbon layer is fixed to a value of 45 nm ensuring that most of the expansion takes place in the carbon plasma and the thickness of the aluminum layer is varied, it was found that for a massive target layers located up to slightly more than 100nm below the aluminum surface contribute to the emission [3].

At high irradiation intensities and normal incidence, efficient absorption of the laser pulse energy by the plasma cannot be achieved since the electron-ion collision frequency strongly decreases with temperature. Less than 10% have been measured at intensities $>10^{17}$W/cm^2 [4]. However in case of resonance absorption occurring at oblique incidence with p-polarized light, the situation changes. The absorption is larger due to the excitation of an intense longitudinal electric field and electron-density oscillations at the critical density. As measured in [5], the absorption increases for p-polarized light with increasing incidence angle, ϑ, up to a maximum of 65% at $\vartheta=70°$. In contrast, the absorption remains low at a level of ~10% for s-polarized light. The large difference in absorption between p- and s-polarized light correlates with findings for the aluminum K-shell emission which is stronger by a factor of 10 for p-polarized pulses. The measurements compare well with simulations from MULTI-fs and the PIC code EUTERPE [5, 6].

Time-Integrated Spectra

The aluminum K-shell emission from the Ly-β to the K-α lines covers the wavelength range from 6.0 to 8.4Å or from 2070 to 1450eV. A low-density spectrum $(n_e\sim10^{22}$/cm$^3)$ and a high-density spectrum $(n_e\sim7\times10^{23}$/cm$^3)$ are shown in Fig. 3a, b. These were measured with a von Hàmos crystal spectrometer equipped with a cylindrically bent pentaerythritol crystal (PET) and recorded on photographic x-ray film. The resolving power of the spectrometer, $\lambda/\Delta\lambda$, is >1500 corresponding to ~1eV in energy resolution. The low-density spectrum is from a massive aluminum target and was generated by adding a prepulse to the main pulse. The prepulse energy is 20% of that of the main pulse. The prepulse precedes the main pulse by 30 ps. The weak prepulse creates a plasma which expands into vacuum. The main pulse hence interacts with a low-density soft-gradient plasma whose characteristic feature is the *small width of all lines*. The high-density spectrum was generated with a single pulse only. The target consisted of massive aluminum covered with a 45-nm thick carbon layer in order to inhibit the expansion of the aluminum plasma. The *first* characteristic feature of this spectrum is that *all lines have become very broad* which is mainly due to Stark broadening. In addition, *strong satellite lines* show up.

The tabulated spectral positions of the resonance lines are indicated by the dotted vertical lines in Fig. 3a. The Lyman lines are emitted by hydrogen-like aluminum ions with one bound electron only, the helium lines are emitted from ions with two bound electrons (one in the ground state, the other in an excited state). The K-α line is from singly ionized atoms located in the cold bulk material. This line is hence neither affected by Stark- and/or Doppler-broadening nor shifted. Its width is therefore given by the natural width of 0.4eV (not resolved by our spectrometer). These features qualify the K-α line for wavelength calibration. In the low-density plasma, the K-α line is stronger than that in the high-density plasma. This is explained with the behavior of the electrons arising from resonance absorption: the softer the density gradient

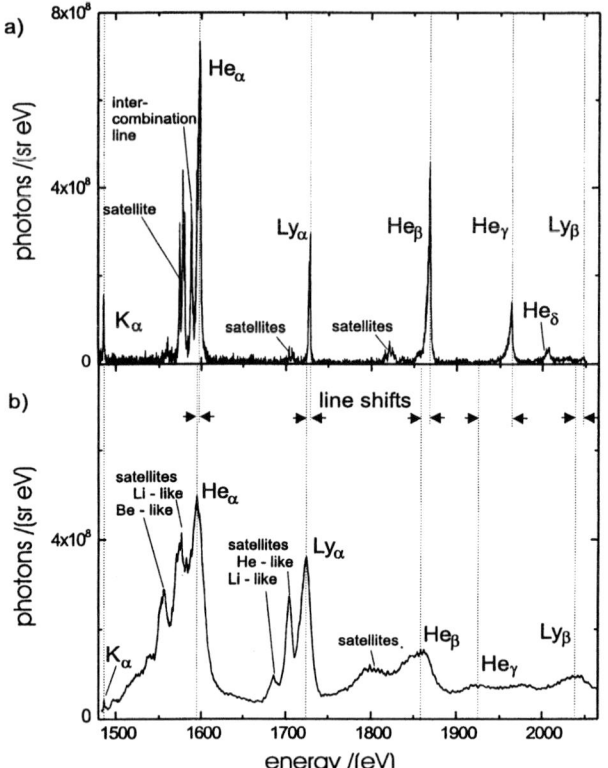

Figure 3. Aluminum K-shell spectra at (a) low (~10^{22}/cm^3) and (b) high (~7x10^{23}/cm^3) density. The low-density spectrum is from a massive aluminum target and was generated by a weak prepulse preceding the main pulse by 30 ps. For the high-density case, the aluminum was covered with a 45-nm thick carbon layer and a single pulse was used only.

the faster the electrons become. Hence in the low-density plasma, the fast electrons penetrate deeper into the *cold* material thereby producing a stronger signal.

The *second* characteristic feature of the high-density spectrum is that *all lines exhibit a red-shift* (compare Figs. 3a, b). Since these so-called plasma-polarization shifts are all at ≤10eV, their precise measurement needs a wavelength fiducial neither shifted nor broadened. As mentioned above, these conditions are met by the aluminum K-α line at 1486.7eV. In addition, we used the silicon K-α line at 1740.1 eV which is much closer to the Ly-α doublet (1728.9 and 1727.6eV). Two x-ray films each of which was subject to a double exposure were employed. The first film is exposed to aluminum K-shell emission from a *low-density* plasma. The aluminum target is then exchanged by a silicon target without touching the spectrometer. In a second shot, the silicon emission, again from a *low-density* plasma, is recorded on the same film which is afterwards developed. Silicon has no lines in the region of the aluminum Ly-α and He-α lines. The second film is first exposed to aluminum K-shell emission from a *high-density* plasma, then after target exchange in a second shot to silicon emission from a *low-density* plasma and finally developed. By superimposing the spectra from both films, the two aluminum K-α lines are found to fall upon each other when the two silicon K-α lines are brought to coincide. This method hence provides an excellent wavelength calibration. Fig. 4 shows an example of the two-film double-exposure technique. For the low-density aluminum plasma, the double structure of the Ly-α line is clearly resolved which disappears in the high-density emission. The red-shift amounts to ~3.4eV.

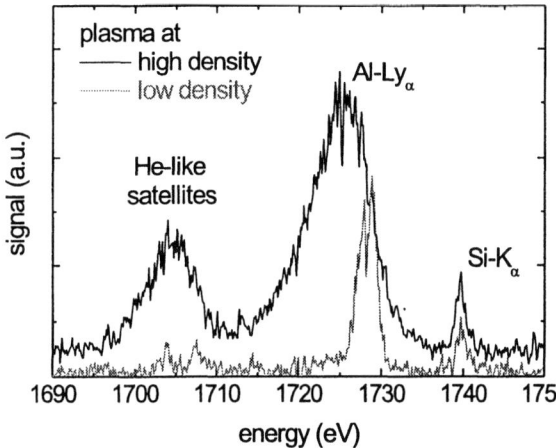

Figure 4. Comparison of time-integrated emission spectra from low- and high-density aluminum plasmas. The low-density plasma is generated with a double pulse: a prepulse followed by the main pulse delayed by 30ps. The silicon K-α line serves as the wavelength normal. The red-shift between the two centers of gravity of the low- and high-density aluminum Ly-α lines amounts to 3.4eV.

Rigorous quantum mechanical impact theory [8] predicts that in a plasma at solid-state density with a few bound electrons the free electrons are non-homogeneously distributed around the nucleus thereby causing level shifts. There is also a contribution from the ions. For the aluminum Ly-α line where the electron contribution dominates, theory predicts a line shift of 3.2eV assuming $n_e=7 \times 10^{23}/cm^3$ and $T_e=350eV$.

A complete theoretical model for the calculation of the total spectrum as shown in Fig. 3 or at least portions of it is rather complicated because a huge number of levels are involved whose populations have to be determined by solving the pertinent collisional-radiative rate equations taking into account all relevant processes. These are electron collisional excitation and de-excitation, ionization and recombination, autoionization and electron capture, photoexcitation and photoionization, spontaneous and stimulated emission, and radiative recombination. Furthermore for the final composite spectrum, Stark broadening due to electrons and ions as well as line shifts have to be considered. Finally to account for opacity effects, the radiative transfer equation has to be solved. A numerical simulation taking all these effects and processes into account was performed for the Ly-α line with its He-like and Li-like satellites as depicted in Fig. 3b. The experimental spectrum was almost perfectly reproduced assuming $n_e=8 \times 10^{23}/cm^3$ and $T_e=380eV$.

Time-Resolved Spectra

For recording the time-resolved spectra, the PET crystal was replaced by a conically curved MICA crystal. It concentrates the x-rays emitted from the target spectrally resolved to a line focus coincident with the cone symmetry axis. The spectrometer is coupled to the AXIS-PX1 ultrafast x-ray streak camera [9]. The spectral resolution, $\Delta\lambda/\lambda$, in the window covering the K-α, He-α, and Ly-α lines was 500. Due to the low energy of a single x-ray pulse, 1000 shots were usually accumulated to achieve a good signal-to-noise ratio. For that purpose, the sweep voltages of the deflection plates of the streak camera were triggered at high temporal precision by irradiating a photoconductive GaAs Auston switch with a small fraction of the fundamental laser pulse. Thereby a time resolution of 0.9ps was achieved.

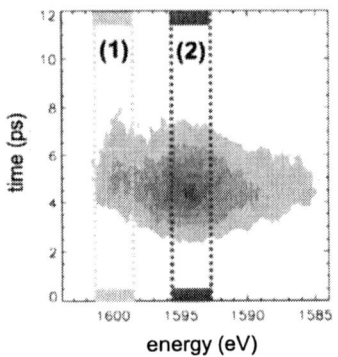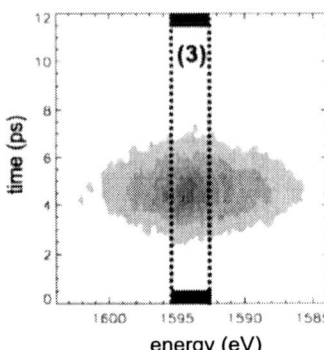

Figure 5. Time-resolved He-α emission from aluminum. (Left) massive target. (Right) target overcoated with a 45-nm thick carbon layer.

Figs. 5 and 6 show the effect of the carbon top layer on the spectrum and duration of the aluminum He-α line. The emission from the massive aluminum target has an asymmetric line profile. The spectral portion denoted by (1) is due to the rapidly expanding front layers of the target. Since this plasma is thin, no shift occurs. Portion (2) is from a denser plasma and hence red-shifted. The emission from the aluminum target with a 45-nm thick carbon layer (3) is spectrally symmetric and red-shifted indicating that the expansion took place in the carbon layer but not in the aluminum and that the aluminum plasma hence remained at solid-state density. With 1.7ps, this emission is shortest (Fig. 6). For the same target, we also measured the duration of the Ly-α line, He-like satellites to Ly-α, Li-like satellites to He-α, and K-α line finding 1.3, 1.2, 1.8, and 1.2ps, respectively.

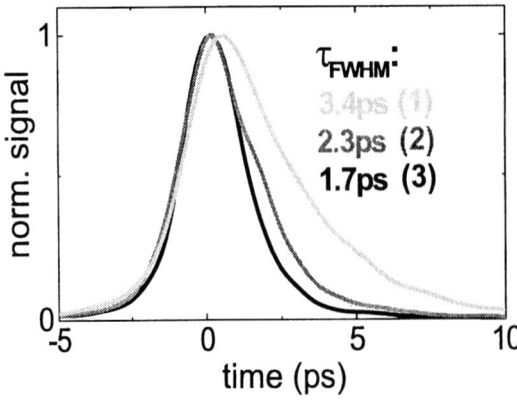

Figure 6. Comparison of the duration of He-α lines emitted from massive and coated aluminum targets. The numbers in parentheses refer to the spectral portions designated in Fig.5.

228

POSITRONS FROM A TABLE-TOP LASER

There is a distinct peculiarity associated with this new laser-based positron source. Unlike common radioactive sources which undergo disintegration with a given rate and emit continuously over a long period of time, the laser-based source emits a burst of positrons within a ps. Although this is in principle advantageous since the source can be turned *on* and *off* at will, the coincidence techniques developed for the analysis of the cw sources are not applicable to this new source.

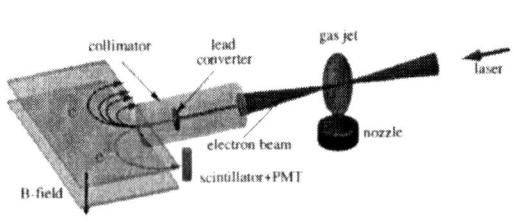

Figure 7. Schematics of the miniaturized arrangement used for the production of positrons. A fraction of the MeV-electrons created in the helium gas jet enter the 1-cm bore in the plastic collimator and produce γ-photons in the 2-mm thick lead converter. These then generate e$^+$-e$^-$ -pairs. The positrons are separated from the primary and secondary electrons via a static magnetic field and detected by the scintillator-photomultiplier tube-combination.

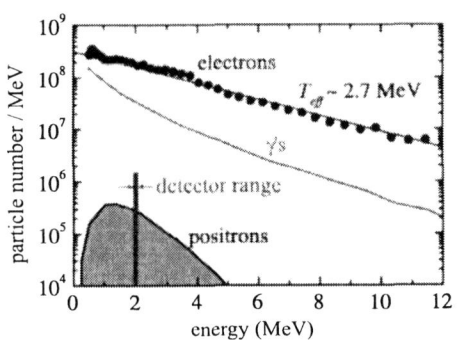

Figure 8. Measured energy spectrum of the primary electrons used to generate the positrons (full dots; exponential fit as dashed line). With this distribution, the spectra of the γ-photons and positrons are calculated assuming a 2-mm thick lead converter.

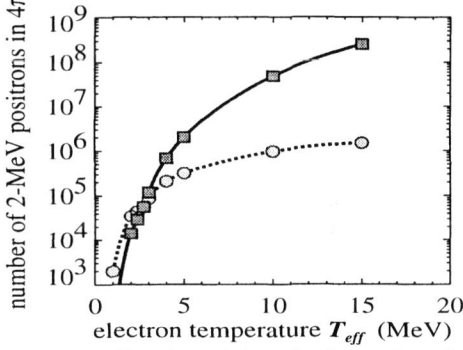

Figure 9. GEANT simulations for the number of 2-MeV positrons per shot produced in 4π sterad as a function of T_{eff} for a 2-mm thick lead converter. The dotted line is for n_e=*const.* and the solid line for $N_e \propto I^{3/2} \propto T_{eff}^3$.

The scheme employed by us for the generation of positrons from high-Z converters is analogous to that in linear accelerators. The MeV-electrons emerging from the helium gas jet (see Figs. 7, 8) along the laser beam direction within a cone of apex angle of 16° [10] hit a 2-mm thick lead slab after having freely propagated a distance of 16cm. In the interaction with the lead nuclei, they generate γ-photons via bremsstrahlung which in turn produce electron-positron pairs. The trident process in which the electrons directly produce electron-positron pairs via electron-nucleus collisions is by far less important under our conditions [11]. Positron emission has also been reported from the direct interaction of PW laser pulses with gold targets [12]. However, such pulses can presently be generated only by huge single-shot facilities whereas titanium:sapphire lasers like ATLAS run at 10 Hz..

The lead converter is embedded in a plastic block with a 1-cm bore. Plastic is a low-Z material that can stop electrons without producing undue bremsstrahlung. The bore diameter of 1cm reduces the number of electrons, N_e, from a total of 2×10^{10} to 8×10^8. This reduction was made in order to be able to perform a clean demonstration experiment. The positrons emanating from the converter have a quasi-isotropic distribution. Those travelling in e-beam direction enter a magnetic field of 150mT provided by permanent magnets. Due to the magnetic field, the positrons describe an orbit of 180° and are then detected by a light-tight, 1.5-cm thick plastic scintillator coupled to a photomultiplier tube. To suppress the strong background signal due to stray γ-photons in favor of the weak positron signal, the detector had to be carefully shielded by an appropriate arrangement of lead bricks (not shown in Fig. 9). The absolutely calibrated detector covers the energy range of (2±0.08) MeV (see Fig. 10) and subtends a solid angle of 7msterad to the converter. With this setup, we measured a yield of 30±15 positrons per shot.

In order to further substantiate our experimental results, we have performed detailed Monte-Carlo-type simulations using the code GEANT [13]. This code allows to exactly simulate the experimental setup, i. e. plastic, converter, lead shielding, magnet, vacuum chamber wall, and detector. The simulations reproduce the measured value of 30±15 positrons very satisfactorily. Scaling the number of positrons detected within the energy interval of 0.16MeV to the full energy range and the solid angle of 7msterad to the full solid angle of 4π, a total number of 10^6 positrons per laser shot is obtained. Using the full electron beam gives a positron number of 2×10^7 per laser shot which corresponds to an activity of 2×10^8 Bq for 10-Hz operation. These values pertain to the ATLAS-2 facility running in these experiments at a power level of 1.2-TW/pulse. However, laser systems delivering pulses up to 100 times more powerful are being available now. Assuming the experimentally confirmed scaling law, $T_{eff} \propto I^{1/2}$, much larger positron yields can then be expected. This is demonstrated in Fig. 9, where predictions of the GEANT code for higher values of T_{eff} and the same type of converter are given. Two cases pertaining how the number of electrons in the e-beam, N_e, may scale with the laser intensity, $N_e = const$ (conservative) and $N_e \propto I^{3/2} \propto T_{eff}^3$ (optimistic), are considered. At higher values of T_{eff}, saturation occurs in both cases. This is attributed to the higher energy of the electrons and γ-photons and the correspondingly smaller interaction cross sections with the lead nuclei. For 100-TW lasers running at 10 Hz, activities of 10^{12}Bq appear feasible when the converter thickness is optimized.

In recent years, positrons have played an important role in applications in diverse fields of physics, e. g., surface and crystal physics [14], positronium spectroscopy [15], and electron-positron plasmas [16]. Although for these applications conventional radioactive sources like ^{22}Na or ^{58}Co are presently used, the distinct possibility exists to replace them by laser-driven positron sources. Extrapolating our results based on the performance of the next laser generation, positron yields comparable to those from linear accelerators can be expected but from a facility orders of magnitude smaller in size. Other obvious advantages of laser-driven positron sources are: They can be turned off when they are not needed thereby considerably reducing the radiation hazard associated with large sized radioactive sources. It is for this reason that replacing the ^{22}Na-source by a laser-driven positron source in the planned production of cold anti-hydrogen [17] is worth to be considered. The pulsed character of the laser-driven positron source can also be exploited in positron annihilation spectroscopy [14] to eliminate the cumbersome timing electronics since the start signal is well defined by the laser pulse. It is beyond doubt that the laser-based positron source will be advantageous in other applications as well.

NEUTRON GENERATION

The neutron production from solid deuterium targets using the fusion reaction $D+D\rightarrow{}^3He(0.82\text{MeV})+n(2.45\text{MeV})$ driven by high-repetition rate femtosecond laser pulses is attractive for two reasons. There is first the potential for a point-like source delivering neutron pulses of ultrashort duration suited for investigating dynamic material damage processes occurring on the time scale of a few ten picoseconds. Secondly, measuring the neutron emission spatially and spectrally resolved provides access to clarifying the question of how the ions are accelerated [18, 19].

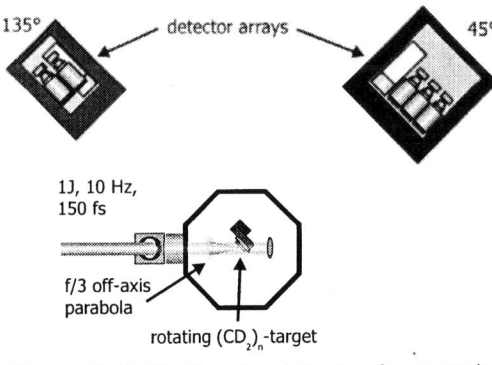

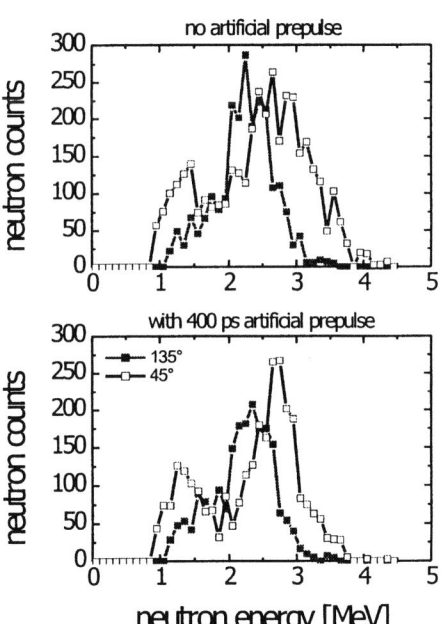

Figure 10 (Left). Experimental setup for measuring the neutron spectra along the target surface (135°) and normal to it (45°). The target is made from fully deuterated polyethylene

Figure 11 (Right). Neutron spectra for the two directions as indicated in Fig. 10 with and without a prepulse.

The contrast ratio of the laser pulse is a parameter of paramount importance in this respect It strongly affects the neutron spectra as was recently shown in numerical simulations [20]. We have started a series of measurements (for the experimental setup see Fig. 10) to study this effect. A few results are shown in Fig. 11. Only the spectra normal to the target surface are significantly different from each other in that the high-energy edge of the spectrum pertaining to the case with no prepulse is blue-shifted. Obviously, there are considerably more ions accelerated normally into the target than in the case with a prepulse. From the quantitative analysis of the spectra which is in progress we expect a more detailed picture of the ion acceleration mechanism being dominant in either case.

REFERENCES

1. Eidmann, K.,. Meyer-ter-Vehn, J., Schlegel, T., and Hüller, S., Phys. Rev. E **62,** 1202-1214 (2000).
2. Ramiz R., Schmalz R., and Meyer-ter-Vehn, J., Comput. Phys. Commun. **49**, 475-505 (1988).
3. Andiel, U., *Isochore Heizung von festem Aluminium mit Femtosekunden-Laserpulsen: eine rönt-gensgenspektroskopische Untersuchung der K-Schalenemission,* Dissertation LMU München 2001
4. Price,D. F., et al., Phys. Rev. Lett. **75**, 252-254 (1995).
5. Eidmann, K., Rix, R., Schlegel, T., and Witte, K. J., Europhys. Lett. **55**, 334-340 (2001).
6. Schlegel,T., Bastiani, S., Gremillet, L., Geindre, J. P., Audebert, P., Gauthier, J. C., Lefevre, E., Bonnaud, G., and Delettrez, J., Phys. Rev. E, **60**, 2209-2217 (1999).
7. Kruer, W. L., *The Physics of Laser-Plasma Interactions*, Addison-Wesley Publishing Company,7 Inc., 1988, pp. 37-43.
8. Nguyen, N., Koenig, M., Benderjen,D., Caby, M., and Coulaud, C., Phys. Rev. A **33**, 1279-1290 (1986).
9. Coté, C. Y., et al., SPIE Conference Proceedings Vol. **2869**, 956-961 (1997).
10. Gahn, C., Tsakiris, G., Pukhov, A., Meyer-ter-Veh, J., Pretzler, G., Thirolf, P., Habs, D., and Witte, K. J., Phys. Rev. Lett. **83**, 4772-4775 (1999).
11. Gahn, C., Tsakiris, G., Pretzler, G., Witte, K. J., Delfin, C., Wahlström, C., and Habs, D., Appl. Phys. Lett. **77**, 2662-2664 (2000).
12. Cowan, T., et al., Laser&Particle Beams**17**, 773-783 (1999).
13. Brun, R., et al., *GEANT User's Guide,* CERN Report DD/EE/82 (1982).
14. Szeles, C., and Lynn,K., in *Encyclopedia of Applied Physics,* G. Trigg Ed., VCH Publishers, Inc. New York, Vol.14, pp.607-632 (1996).
15. Rich, A., Rev. Modern Phys., **53**, 127-165 (1981).
16. Greaves, R., and Surko, C., Phys. Plasmas **4**, 1528-   (1997).
17. Gabrielse, G., et al., ATRAP Proposal SPSC 97-8/P306 *The Production and Study of Cold Antihydrogen,* presented to the CERN SPSLC on 25 March 1997.
18. Pretzler, G. et al., Phys. Rev. E **58**, 1165-1168 (1998)
19. Hilscher, D., Berndt, O., Enke, M., Jahnke, U., Nickles, P., Ruhl, H., and Sandner, W., Phys. Rev. E **64**, 016414-1–016414-9 (2001).
20. Toupin, C., Lefebre, E., and Bonnaud, G., Phyics of Plasmas **8**, 1011-1021 (2001).

Nuclear Processes in Dense Plasma Produced By Femtosecond Laser Pulses At Sub-Relativistic Intensities

Gordienko V.M., Chutko O.V., Golishnikov D.M., Mikheev P.M., Savel'ev A.B., and Volkov R.V.

International Laser Centre & Physics Faculty M.V.Lomonosov Moscow State University, Vorobyevy gory, Moscow, 119899, Russia, Tel:+7(095)9394719; Fax:+7(095)9393113; e-mail:gord@femto.phys.msu.su

Abstract. We overview our recent results on experimental observation of nuclear process in hot dense plasma created by focusing of femtosecond laser pulses with intensity below 10^{17} W/cm^2 on the target surface. In the first part of the paper we introduce special technique of target properties modification allowing us to increase both hot electron temperature and ion temperature. We successfully applied this technique to rise ion temperature in D-enriched Ti target up to 10-20 keV, thus observing 2.5 MeV neutron yield at intensity of as small as 10^{16} W/cm^2. We further discuss some possibilities arising from our recent experiments on low energy nuclear level excitation in hot dense plasma created by femtosecond laser pulses of moderate intensity.

Introduction

A number of publications on the experimental observation of nuclear processes accompanying the interaction of subpicosecond superintense laser pulses with solid state targets have appeared recently [1–3]. These processes become possible primarily because of the efficient generation of suprathermal electrons in the plasma formed when a target is irradiated with an ultrashort laser pulse with intensity $I > 10$ PW/cm^2. These investigations essentially open up a new field associated with laser stimulation of nuclear reactions [4]. Thus, for intensities greater than the so-called relativistic limit $I\lambda^2 > 5 \cdot 10^3$ (PW/cm^2 µm^2), when the electron "temperature" reaches hundreds of kiloelectronvolts, direct stimulation of a nucleus by an electron impact or photostimulation of the nuclear reaction become possible. Fundamentally different situation arises for sub-relativistic or "moderate" intensities ranging from 10 to 100 PW/cm^2 [5, 6]. On the one hand, the "temperature" of suprathermal electrons in this case ranges from 3 to 10 keV, which is sufficient for direct excitation of low-lying nuclear levels of stable as well as metastable isotopes (the standard methods of nuclear spectroscopy of such levels are based on direct excitation via "normal" states with energies above 100 keV) [6,7]. On the other hand, microctructuring of the target allows one creating plasma differs greatly from that of flat solid ones. Such a microstructuring can be done either using cluster targets [8], or by direct structuring of

CP611, *Superstrong Fields in Plasmas:* Second Int'l. Conf., edited by M. Lontano et al.
© 2002 American Institute of Physics 0-7354-0057-1/02/$19.00

thin surface layer of the solid target [9]. Among other features, such a plasma exhibits high neutron yield being made of D atoms [10, 11]. Finally, to obtain intensities 10– 100 PW/cm² it is possible to use relatively cheap and commercially accessible tabletop femtosecond laser systems.

In this paper we report on our recent progress in investigation of interaction of moderate intensity femtosecond laser pulses with micro-structured solids, and in designing new methods for the nuclear spectroscopy of low energy nuclear levels.

Experimental Setup & Methods

We used dye femtosecond laser system [12] provided us with 200 fs, up to 1 mJ, 610 nm laser pulses at 1 Hz repetition rate. Tight focusing provided us with intensity up to 50 PW/cm² at intensity contrast ratio better than 10^5. In the most experiments (if other not specified) p-polarised light was used with 45 glancing angle with respect to target surface normal. The target was placed inside the vacuum chamber with residual pressure of less than 10^{-5} Torr. Experimental diagnostics (see Fig. 1) used in experiments included: (i) double channel hard x-ray detection scheme, allowed us to estimate plasma hot electron temperature at each laser shot [13]; (ii) ion time-of- flight (TOF) spectrometer; and (iii) neutron detector. TOF spectrometer was placed in one of three position all 22 cm apart from the target (see Fig. 1). This enables to view roughly spatial distribution of the plasma plume. While being at the position B or C the TOF spectrometer covers angle range from 0 up to 20 from the target surface. Shevron-type MCP was used as the detector. Negative potential of input MCP plate prevents it from detecting electrons with energy less than 2.4 keV. Faster electrons reach MCP within less than 1 ns delay from laser pulse that corresponds to ions velocity of $2 \ 10^9$ cm/s. This value is far beyond expected ones for our laser intensity. Special grounded grid 1 cm before MCP served to prevent electric field to penetrate inside vacuum tube connecting MCP and vacuum chamber. MCP exit plate was loaded by 50 Ohm and connected to the oscilloscope with 100 MHz sampling rate. Hence we obtain temporal resolution of 10 ns corresponding to 10% resolution by velocity at $u \sim 10^8$ cm/s. To set "zero" point on the time axes fast optical detector was used measuring laser light reflected from the plasma surface.

To detect thermonuclear fusion in plasma we chose ^{2}H(d, n)^{3}He channel yielding neutrons of 2.5 MeV energy (at zero energy of deuterium ions). Spectrometric neutron detector filled with Ar and ^{3}He gas mixture operated in the regime of ionising chamber. It was placed 12 cm apart from the target outside the vacuum chamber in the polyethylene moderator. Electric cur- rent from single thermal neutron detection was amplified by charge sensitive amplifier and measured by the fast ADC. In our experiments we measured 10–15 background neutrons per second. Taking into account 5 ms temporal window with respect to the laser pulse, the probability to detect 1 neutron in 1 laser shot was deduced as $5 \ 10^{-4}$. To make absolute calibration of the detector reference neutron source was used emitting 11200 neutrons per second. It was placed at the same distance from the detector as the target was. In so doing we estimated the detection efficiency of our scheme as 0.04%. Target preparation consists of the two steps: deuterium enrichment and structuring. D-enriched Titanium target was standard one used in ion accelerator neutron generators. Mean deuterium density was 1.65 in accordance with its certificate. As it was obtained from RBS measurements deuterium concentration at the depth of 0.3 μm was at least 0.6.

Target Structuring With Femtosecond Laser Pulses

By the contrast to our previous experiments making use of pre-formed target structuring with electrochemical processing (highly porous Si) [14], in this research we tried to structure the target surface by the laser pulse treatment. Initial targets were flat wafers of crystalline silicon, and Ti film 5 μm in thick sputtered onto Mo substrate 300 μm thick. By the contrast to the commonly used experimental scheme, there target moves providing fresh flat surface for each laser shot, we fired a few shots at the same target position. It was found that well- developed surface is formed under action of the femtosecond laser pulse itself. Namely, first laser shot was used to prepare a crater with structured walls and bottom. The next shot creates plasma from these structures. It was checked (measuring hard X-ray yield from plasma) that up to 6–7 shots can be applied without target displacement. As expected, the crater diameter was slightly larger than plasma spot size, and rough surface forms on the periphery of the crater. Crater depth for single laser shot was estimated from SEM images as 1–2 μm. Subsequent laser shots lead to formation of the "hole" deep into the target tilted with respect to the target normal by the same angle as the laser beam.

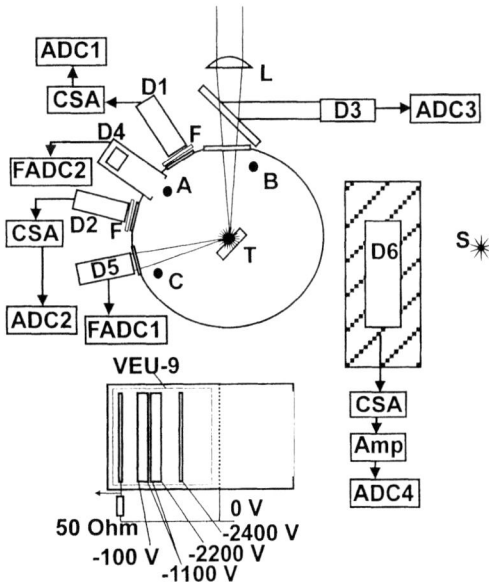

Fig. 1. Scheme of the experimental diagnostics: L , focusing objective; D1, D2, hard X-ray detectors; D3, laser pulse energy calorimeter; D4, ion time-of-fight spectrometer; D5, fast photo-electric optical detector; D6, neutron detector; T, target; F, X-ray filters; CSA, charge sensitive amplifiers; ADC1–4, four channel peak ADC; FADC1,2, fast ADC with internal data buffer; Amp, amplifier; A, B, C, three possible positions of the detector D4; S, neutron source. On the inset. simplified scheme of TOF detector D4 with applied voltages.

We investigated hard X-ray yield X dependence on intensity I for different targets: flat silicon (FS), laser modified silicon (LMS), flat (FT) and laser modified (LMT) titanium. To choose the specific spectral range different X-ray filters were applied made of Be, Al, Ta and Cu foils (see [13] for more details on a filter choice). It was obtained for all the spectral ranges that hard X-ray yield for all the modified targets is higher than for the flat ones. The harder X-ray cut-off of the filter E the higher slope . of the X(I) curve were deduced assuming power law in the form $X \sim CI^{\alpha}$: for $E \sim 4$ keV $\alpha \sim 2$ while for $E > 15$ keV $\alpha \sim 4$. Figure 2 presents distribution of hot electron temperature deduced from our data for FS and LMS species. Each point for histogram was obtained in a single laser shot. Average intensity was kept at the level of $I \sim 20$ PW/cm^2. Much higher width of the distribution in the case of LMS tar- get is to be attributed to bad reproducibility of structure formed under laser pulse action. Mean hot electron temperature for the FS target $T_h \sim 5$ keV coincides well with known theoretical and experimental data. The temperature value for LMS target is $T_h \sim 8$ keV. For LMT target fluctuation of the hard X-ray yield from shot to shot were even larger than in the case of LMS target. This obstacle prevents us from getting any systematic data for consecutive analysis and extracting the mean values of T_h. At the same time both hard X-ray yield and hot electron temperature undergone significant increase for the LMT as compared to the FT one.

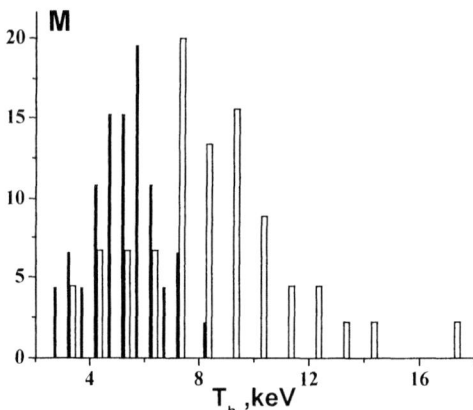

Fig. 2. Distribution of single-shot-measured hot electron temperature T_h for flat Si (black bars) and laser modified Si (open bars).

The results of TOF ion measurements also support hard X-ray yield measurements. Lets start by considering results for the position A of the TOF spectrometer. Shown in Fig. 3a three records of the ion current from FS target all demonstrating two component constitution of plasma ions: fast ions with velocities in excess of 10^8 cm/s and slow ions with velocities of $10^7 - 10^8$ cm/s. The same double-peak structure was observed previously at lower intensities in [15, 16]. The kinetic energy of fast ions K_f increases with intensity as $I^{0.4}$ that corresponds well to the ion TOF data at higher intensities discussed in [17]. It is worth noting here that we did not observed any fast ions for s-polarised laser radiation. The picture changes drastically for LMS target (see

236

Fig. 3b). Here we observed a few (up to 4–5) maxims in ion current with stochastic positions in the range 10^7 $3 \cdot 10^8$ cm/s. Nearly for all the realisations recorded we detected peaks corresponding to the fast ions with even higher velocity than in the case of the FS target. This fact also coincides well with hard X-ray data showing further increase in hot electron temperature. A few peak structure of "thermal" component could be due to well developed structuring of the initial surface of the LMS target. By the contrast to FS target polarisation state of the laser light had no influence of the ion current from LMS target. This fact could be easily understand keeping in mind high curvature and structuring of the crater surface. Interesting results were obtained while placing the TOF spectrometer to the positions B and C. In the case of FS target we observed only slow ions with velocity $u \sim 10^7$ cm/s independently on the polarisation state of the laser light. By the contrast in the case of the LMS target we observed the same multi-peak structure with lower amplitudes as in the case of the TOF detector at the position A. Besides upper ions velocity is slightly lower as well. Nearly the same spatial and temporal behaviour we observed for the FT and LMT targets, but again, as in the case of hard X-ray yield measurements the reproducibility of the results were worse. Hence laser modified targets exhibit nearly isotropic but inhomogeneous plasma ablation. For D-enriched laser modified targets this should lead to collisions of fast and slow ions with each other and cold atoms of the crater thus initiating thermonuclear acts. In accordance with our TOF measurements deuterium ion energy ranges from 1 to 10 keV—the value enough to produce a few thermonuclear neutrons in each laser shot.

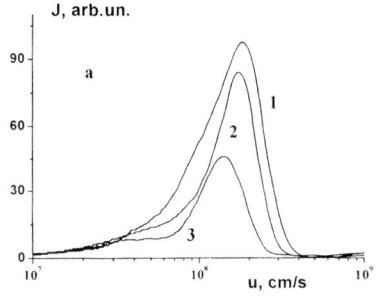 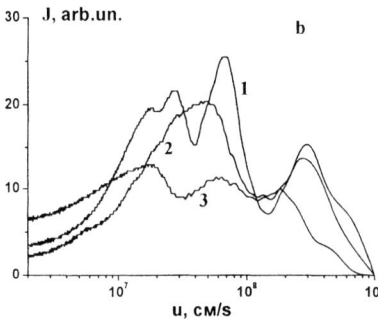

Fig. 3. Time-of-flight ion spectra for flat Si (a) and laser modified Si (c) targets for different laser intensities of 5 *(1)*, 10 *(2)*, and 25 PW/cm 2 *(3)*. The TOF detector is at the position A.

Thermonuclear Reaction In Laser Structured Targets

During experiments we used LMT target to observe thermonuclear reaction in plasma at moderate intensity. A few set of experimental runs were made for LMT target with 500–2000 laser shots in each. The mean value of neutron yield N was, $N \sim$ 1 neutron for 400 laser shots. The probability $P_b(5)$ to detect 5 background neutrons in 2000 laser shots is small at our experimental conditions: $P_b(5) \sim 0.04$. Hence taking into account neutron detector efficiency we can deduce the neutron yield of 5 neutron per shot in plasma volume. Let us compare the obtained value of neutron yield with simple theoretical predictions. Neutron yield from plasma can be assessed as:

$$N_n \approx 0.5 n_D^2 \langle \sigma v \rangle_{DD} \tau V \qquad (1)$$

where n_d is the deuterium concentration, τ is the duration of neutron pulse and V is the plasma volume. Taking for the deuterium ions the average value obtained from TOF measurements $v_d \sim 1\text{--}2 \ 10^8$ cm/s we come to the reaction velocity of $\langle \sigma v \rangle_{DD} \sim 3.5 \ 10^{-18} - 1.5 \cdot 10^{-19}$ cm^3 s^{-1}. Plasma volume V and deuterium concentration n_d can be estimated supposing that the crater serves as the thermonuclear "boiler". Thus we have for the crater volume $V \sim 10^{-10}$ cm^{-3}, and for deuterium concentration $n_d = 3 \cdot 10^{20}$ cm^{-3}.

We calculated the neutron pulse duration from the following. We divided plasma expansion and ablation process by the two parts. At first plasma fills the crater volume. This takes place within 1 ps and leads to formation of plasma with deuteron density $n_d = 3 \cdot 10^{20}$ cm^{-3}, electron temperature of 300-500 eV and ion temperature of 2-3 keV. On the second stage this plasma undergoes planar ablation into vacuum. It cools down due to electron-ion relaxation, and its density falls with time. To account for these complex plasma behaviour we made computer simulations with the help of 1D single-fluid two-temperature hydrodynamic model [18], incorporating formula (1) inside to calculate temporal dependency of neutron yield. Calculations were made within broad range of plasma parameters (we varied plasma layer thickness, density, electron and ion temperatures, see [19] for more details). For the parameters of our experiment we conclude that neutron pulse duration does not exceed 5 ps.

Finally neutron yield calculated on the basis of the equation (1) is 0.1-10 neutron per pulse, value agrees well with experimental data.

Isotope Separation Through Excitation Of Low Energy Nuclear Levels In Femtosecond Plasma

Recent experiments [1, 7] have demonstrated that, in principle, low-lying isomer nuclear levels (with an energy level less than 20 keV) can be efficiently photo-excited in a plasma produced by a femtosecond laser pulse of moderate intensity. Various schemes of isotope separation have been discussed for several decades (see the review [19]). Morita [20] proposed an original method of isotope separation through the excitation of a nuclear level with an energy of 75 eV in a uranium-235 isotope with subsequent chemical separation. The main drawback of this method is that a nucleus is excited through the NEET process, whose efficiency is overestimated by the authors by many orders of magnitude. We deal with a radically different situation when a plasma created by a sub-picosecond laser pulse with an intensity higher than 10 PW/cm^2 is employed as a medium of excitation. A high luminosity of the emission characteristic of such a plasma within the range up to 100 keV in combination with a solid-state plasma density, a high ionisation degree of atoms, and, as a consequence, an electron density at the level of 10^{24} cm^{-3} ensure an efficient excitation of a large number of nuclei due to the photoexcitation by plasma emission, as well as inverse inner electron conversion and inelastic electron scattering [7]. Low-lying nuclear levels possess rather high coefficients of inner conversion (about 70 for the 6.238-keV level of tantalum-181). The key point of the method for isotope separation described here is the production of singly ionised ions of a single isotope due to the selective photoexcitation of its low-lying level with subsequent decay of this level through the inner electron conversion. This scheme implies that other isotopes with the same nuclear charge do not possess a low-lying isomer nuclear level around the photoexcitation range.

Let us consider in greater detail the applicability range of the proposed approach [22]. Bremsstrahlung and photorecombination X-ray emission of hot plasma electrons play the main role in the photoexcitation of low-lying nuclear transitions in a hot dense plasma. The inner electron conversion in an excited nucleus of a multiply ionised ion does not result in noticeable changes in the balance of the charge state of an isotope mixture. Only if the time of complete recombination of an ion down to the atomic state is less than the total lifetime of the excited isomer nuclear state, the process of inner electron conversion occurs in an atom. It is only in this case that the chosen isotope has an ionisation degree other than the ionisation degree of other isotopes. Recombination of multiply ionised ions in a plasma occurs in the bulk of the plasma, in the expanding plasma jet, and in plasma areas colliding with the inner surfaces of the interaction chamber. The rate of three- body recombination in a dense plasma is rather high. For characteristic parameters of an expanding plasma, $Z \sim 30$, $T \sim 1$ eV, $N_e \sim 10^{15}$ cm^{-3} the recombination time is on the order of 0.1–10 ns. In reality, this rate is even higher due to the influence (in the case when the electron density is low) of two-electron recombination and photorecombination, as well as recombination through charge transfer in ion–ion and ion–atom collisions. Ions also efficiently recombine when they collide with a surface. Each collision of this type considerably lowers the ionisation degree.

The characteristic velocity of the leading front of an expanding plasma clot is 3–5 $\times 10^7$ cm/s. Thus, it takes ions from 10 to 30 ns to cover a characteristic distance of 1 cm, which implies that the excited isomer state will mainly decay in the atomic state provided that the total lifetime of this state exceeds 10–30 ns. Analysis of characteristics of low-lying isomer nuclear levels with transition energies less than 100 keV for stable isotopes allows us to speak of the whole class of isotopes with atomic numbers from 26 (Fe-57) to 92 (U-235) whose parameters meet the requirements considered above. A high coefficient of inner conversion, which determines the relative number of decays through the channel of inner electron conversion as compared with gamma decay and which varies from 8.56 for Fe to 1120 for Ge, is important from the viewpoint of achieving a sufficient yield of the final product. Another, much larger group of isotopes includes nuclei with energies of the first excited levels ranging from 30 to 100 keV. For such isomer nuclei, the coefficients of inner conversion typically vary from 2 to 1000, and the lifetime usually falls within the nanosecond range. This group of isotopes includes Ag-107, Te-125, Os-187, etc.

Let us estimate roughly the efficiency of the photoexcitation process, which is the main factor determining the overall efficiency of the isotope-separation process. For a solid-state hot plasma with a concentration of ions $N_i \sim 10^{23}$ cm^{-3} displaying a Planck spectrum of X-ray emission, we can easily find an estimate for the number of excited nuclei:

$$\zeta = \frac{\alpha \tau}{\upsilon} \frac{1}{e^{E/T} - 1} \xi N_i V \qquad (2)$$

where E, α, and υ are the energy, the coefficient of inner conversion, and the total lifetime of the nuclear level; ξ is the atomic concentration of an isotope in the sample; $\tau \sim 1$ ps is the duration of the pumping pulse; and T is the temperature of the pumping source. The plasma volume $V \sim 10^{-8}$ cm^{-3} is determined by the size of the focal spot of laser radiation and the depth of heating of the material in the target. Setting $E/T \sim 1$, we find that $\zeta \sim 10^9$ in a single laser pulse for an Fe-57 isotope ($\alpha \sim 9$, $\upsilon \sim 100$ ns, and $\xi \sim 0.02$). Thus, with a femtosecond laser with a pulse energy on the order of 1 J and a pulse repetition rate of 30 Hz, one can obtain microscopic amounts of isotope up to 10 ng per hour. The enrichment degree of an isotope will be determined by the efficiencies of the recombination of plasma ions to the atomic state and the extraction of singly ionised ions from the volume of the gas after the recombination. The approach described above can be also extended to the separation of unstable long-lived isotopes. Specifically, Pb-205 isotopes (with a ground-state lifetime on the order of a minute) possess a low-lying level with an energy of 2.329 keV, which meets the requirements of the scheme discussed above.

INTERNAL CONVERSION BLOCKING IN FEMTOSECOND LASER PLASMA

Compared to the method outlined in [23], where an ion accelerator was applied to obtain highly ionised atoms with an excited state of the nucleus, the use of an femtosecond laser plasma simplifies the production of these atoms and, in addition, permits controlling the decay rate of the nuclear states [24]. The most probable channels of decay of excited nuclei are the gamma decay and the internal electronic conversion (IEC). The probability of the latter exceeds the radiative transition probability by an order of magnitude and even more, depending on the energy of the nuclear transition and its multipolarity. A numerical analysis of the dependencies of coefficients of conversion to different electron shells on the ionisation degree was conducted by the example of the 201-Hg isotope. Up to a certain point, the partial coefficients vary only slightly with increasing ion degree Z. Upon reaching a certain ionisation degree, the partial conversion coefficients decrease due to ionisation and an increase in the electron binding energy. In particular, the conversion to the N1 shell terminates because the binding energy rises from 834 eV for $Z \sim 26$. Therefore, mercury atoms should be ionised to $Z \sim 26$-35 in order for the conversion decay channel to be significantly suppressed.

Let us estimate the parameters of the plasma and the laser pulse at which mercury atoms of such ionisation degree can be produced. The ionisation kinetics of plasma was calculated in the mean charge approximation for a 200-fs laser pulse with an intensity of $2 \cdot 10^{16}$ W cm^{-2}. In this case, the temperature of thermal electrons is estimated as 800 eV. It follows from the Fig. 4 that the ion ionisation degree ($Z > 35$) during a period of 1 ps is high enough for a complete conversion suppression. On the other hand, a rapid three-body recombination of the plasma ions causes Z to decrease below 20 in a time significantly shorter than the expected half-life of the nuclear state

240

(1- 10 ns). Nevertheless, this effect can be observed for the fraction of multiply charged ions which rapidly escape from the dense plasma during its expansion in vacuum. Indeed, these ions exhibit freezing of the ion composition. Estimates show that the adiabatic expansion of a 201- Hg plasma with the initial parameter Z~ 41 and the electron temperature T~ 400 eV results in freezing of Z at a level Z ~35 for t < 100 ns. Note also that higher-charged ions escape from the plasma more rapidly, thereby increasing the probability of a nuclear isomer retaining in the excited state. The total number of excited mercury nuclei in the portion of the plasma with a volume of 10^{-12} cm^{-3} expanding into vacuum is ~100 nuclei per laser pulse, which is sufficient for observation of the IEC suppression.

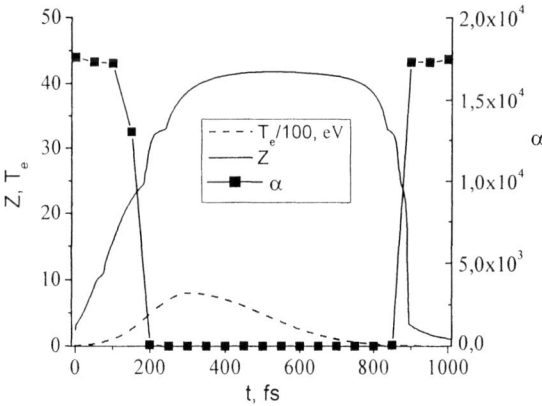

Fig.4. Time dependencies of the electron temperature T $_e$, the degree of ionisation Z of laser-produced mercury plasma, and the total coefficient of conversion α.

Conclusions

Hence, modification of the thin surface layer of the solid target lead to significant changes in the properties of electronic and ionic femtosecond plasma components as well as to the manifestation of new phenomenon unobservable for flat targets at non-relativistic intensities. Using such targets one can effectively induce nuclear processes in extremely small plasma volume at low laser pulse energy. Our investigation adds the new type of modified targets - laser modified targets, besides previously used porous, "smoke" and other targets. In modified targets hot electron temperature increases with intensity faster than for the flat one reaches 8–10 keV at 20 PW/cm². This increase in hot electron temperature leads to the shift of hard X-ray spectrum toward shorter wavelengths. Detection of extremely hard (for our experimental conditions) X-ray quanta with energy above 40 keV points at strongly non-Maxwellian tail of electronic distribution in plasma. Ionic plasma component in solid modified targets undergoes principal changes. From one hand, increase in hot electron temperature leads to the higher ion kinetic velocity up to $2-3 \cdot 10^8$ cm/s. Energy of fast ions grows up with intensity as $I^{0.4}$. From the other hand, ion jet collisions inside plasma determines abrupt jump of plasma ion temperature from 50 eV for flat tar- get to tenths of keV for solid modified ones. This phenomenon is the characteristic feature of solid modified targets, while for lower density targets, such a cluster jet, ion-ion collision time is high as compared to the ablative time. For laser modified targets nearly isotropic but turbulent ion flow forms, and this behaviour is characteristic both for slow and fast ion component. High ion velocity in laser modified targets allows us to observe thermonuclear neutrons in deuterium enriched titanium target undergone laser modification. We counted 5 neutrons in each laser shot irradiating as low as 0.5 mJ laser pulse energy in plasma volume of 10^{-12} cm^{-3} . Due to the higher plasma density neutron production efficiency by the energy input in plasma was the same as in cluster jet experiments. To get more neutrons in each laser shot nanometer-size metal brushes can be used as it was proposed in our papers [9, 14] and experimentally checked by X-ray measurements in [25]. In this case one can really achieve solid plasma density simultaneously heating up ion component to multi-keV regime.

Excitation of low energy nuclear level with femtosecond plasma X-rays and electrons may provide new approaches in different fields. Some stable and unstable isotopes may be effectively separated provided their nuclear life-time is longer, than plasma recombination time. Special measures could help fasten the latter process. From the other hand, high ionisation degree of the free expanding plasma may sustain the nucleus in the excited state much longer than its full lifetime due to charge state blocking [24]. A 100- 500-fs laser pulse with moderate intensity can be used to control such a basic property as the decay rate of nuclear states. For 201- Hg, in particular, a total suppression of the conversion decay channel should be observed in the plasma fraction expanding into vacuum. In this situation even new nuclear decay channels may manifest their selves. Finally, search for yet unknown nuclear levels and their characteristics is feasible due to the broad energy spectrum of plasma X-rays and electrons.

Acknowledgments

This work was supported by the Russian Foundation for Basic Research (project nos. 99-02-18343, 00-02-17302, 01-02-06463, 01-02-06458, 01-02-06459) and State Scientific Program "Fundamental metrology".

References

1. Andreev, A.V., Volkov, R.V., Gordienko, V.M., et al., *JETP Lett.* **69**, 371 (1999).
2. Ledingham, K.W.D., Spencer, I., McCanny, T., et al., *Phys. Rev. Lett.* **84**, 899 (2000).
3. Cowan, T.E., Hunt A.W., Phillips, T.W., et al., *Phys. Rev. Lett.* **84**, 903 (2000).
4. Andreev A.V., Gordienko V.M., Savel'ev A.B., to be published in *Quantum electronics.*
5. Andreev A. V., Gordienko V. M., Dykhne A. M., et al., *JETP Lett.* **66**, 331 (1997).
6. Andreev A. V., Volkov R. V., Gordienko V. M., et al., *Quantum electronics* (Moscow) **26**, 55 (1999).
7. Andreev A. V., Volkov R. V., Gordienko V. M., et.al., *JETP*, **91**, 1163 (2000).
8. Ditmire T., Springate E., Tisch J.W.G., et al., *Phys. Rev. A*, **57**, 369 (1998).
9. Gordienko V.M., Savel'ev A.B., *Phys. Uspekhi*, **42**, 72 (1999).
10. Zweiback J., Smith R. A., Cowan T. E., et al., *Phys. Rev. Lett.* **84**, 2634 (2000).
11. Volkov R. V., Golishnikov D. M., Gordienko V. M., et.al., *JETP Letters*, **72**, 401 (2000).
12. Volkov R.V., Gordienko V.M., Dzhidzhoev M.S., et al., *Quantum Electron.*, **27**, 1081 (1997).

13. Volkov R.V., Gordienko V.M., Mikheev P.M., and Savele'v A.B., *Quantum Electron.*, **30** (2000).
14. Volkov R.V., Gordienko V.M., Dzhidzhoev M.S., et al., *Quantum Electron.*, **28**, 3 (1998).
15. Meyerhofer D.D., Chen H., Delettrez J.A., et al., *Phys. Fluids*, **B**5, 2585 (1993).
16. Andreev A.A., Bayanov V.I., Vankov A.B., et al., *Quantum Electron.*, **26**, 884 (1996).
17. Clark E.L., Krushelnik K., Zepf M., et al., *Phys.Rev. Lett.*, **85**, 1654 (2000).
18. V.M.Gordienko, M.S.Dzhidzhoev, M.A.Joukov, et.al., Superstrong Fields in Plasmas, Eds.: M.Lontano, et.al., AIP Conf.Proc. **426**, AIP, New-York, p.241 (1998).
19. Gordienko V.M., Rakov E.V., Savel'ev A.B., to be pubplished in *Quantum electronics.*
20. Karlov N.V., Prokhorov A.M., *Usp. Fiz. Nauk*, **118**, 583 (1976).
21. Morita M., *Prog. Theor. Phys.*, **49**, 574 (1973).
22. Andreev A. V., Gordienko V. M., and Savel'ev A. B., *Laser Physics*, **10**, 557 (2000).
23. Attallah F., Aiche V., et al. *Phys. Rev. C: Nucl. Phys.* **55** 1665 (1997).
24. Andreev A.V., Chutko O.V., V.M.Gordienko, et.al., *Quantum electronics*, (2001).
25. Kulcsar G., AlMawlawi D., Budnik F.W., et al., *Phys. Rev. Lett.*, **84**, 5149 (2000).

Neutron Generation by Laser Irradiation of CD$_4$ Clusters

Ph. Balcou, G. Grillon, S. Moustaizis*, J.P. Chambaret, M. Pittman, I. Sanchez-Molinero, A. Rousse, J.-Ph. Rousseau, S. Sebban, D. Hulin

*Laboratoire d'Optique Appliquée, ENSTA-Ecole Polytechnique
CNRS UMR 7639, Palaiseau, France*

O. Sublemontier, M. Schmidt

*CEA,DSM/DRECAM/SPAM, Centre d'Etudes Nucléaires de Saclay,
F-91191 Gif-sur-Yvette cedex, France*

Th. Pussieux and J. Martino

*CEA,DSM/DAPNIA/SPhN, Centre d'Etudes Nucléaires de Saclay,
F-91191 Gif-sur-Yvette cedex, France*

Abstract. It was shown in 1999 that D$_2$ cluster explosion under ultrashort intense laser irradiation can lead to ion energies sufficient to drive nuclear fusion reactions [Ditmire et al. *Nature* **398**, 491]. We show how the use of molecular clusters allows to further enhance the ionic acceleration, allowing one to reach optimal fusion cross sections. A new low density high energy regime is described, in which fusion occurs via a spallation-like process. The process increases with laser intensity up to the relativistic threshold, at which the neutron production is overwhelmed by gamma production resulting from electron acceleration.

INTRODUCTION

Observations of neutron production during the interaction of powerful lasers with matter have long been common in the field of Confined Inertial Fusion, using very energetic nanosecond lasers, like Nova, Phoebus or Gekko XII. Neutron production in the short pulse regime (picosecond or femtosecond) was only observed in 1998 [1].

Two kinds of processes can be considered to produce neutrons with lasers: γ-ray induced fission, and deuterium fusion.

In the first process, the intense short pulse laser first creates a plasma and accelerates some electrons to very high velocities; these electrons then collide with ions, yields large amounts of γ-rays by Bremstrahlung. Those may eventually induce fission of nuclei in the core of the target, with (γ,n) or (γ,f) reactions, such the well-known reaction : ^{63}Cu (γ,n) ^{62}Cu.

In the second process, the laser interaction should accelerates deuterium ions from the target to velocities high enough to induce fusion. This was first observed by Pretzler et al. and Disdier et al. on solid targets. More recent results can also be found

CP611, *Superstrong Fields in Plasmas:* Second Int'l. Conf., edited by M. Lontano et al.
© 2002 American Institute of Physics 0-7354-0057-1/02/$19.00

in the works of Nickles et al. [2] and Gordienko *et al.* (same issue). On cluster targets, a breakthrough was obtained by Ditmire et al. who observed fusion from deuterium (D_2) clusters irradiated by a femtosecond laser [3]. The basic idea for cluster targets is to take advantage of the large kinetic energies of ions expelled in the Coulomb explosion of clusters irradiated by intense laser fields.

Fusion is obtained in a step-wise process : the target is initially composed of a gas jet of large clusters, with typically 10^4 to 10^5 atoms per cluster. The laser will ionize the clusters very strongly, stripping off a large number of electrons, and leaving the remaining ions in dense non neutral plasma balls. These subsequently undergo rapid Coulomb explosion, transferring large kinetic energies to the ions. During binary collisions between deuterium ions, fusion processes may occur by two main channels : $^2H(d,p)^3H$, and $^2H(d,n)^3He$; the neutron produced in the latter case has a characteristic energy of 2.45 MeV.

The Livermore experiment has demonstrated the production of about 10^4 neutrons per shot, even though the ion kinetic energies remained rather low, in the range of a few keV [3]. Let us emphasize the important parameters of that experiment, which will differ from those used in the results presented below. It made use of a cryogenated D_2 gas jet, operating in a sonic high density regime (average gas density about $2. 10^{19}$ cm^{-3}). Strong propagation effects were observed, as the laser had to be focused exactly at the jet entrance; the neutron emission was observed to be isotropic.

Two main parameters can be considered in order to optimize the neutron yield. The first is of course the gas density; however, the density used in [3] is already so high that the laser could not propagate beyond a few hundred micrometers, so that further increase in the density would only induce premature absorption of the laser energy. The second is the increase of the ionic average temperature, in order to boost the fusion cross section or thermal plasma reactivities, which increase sharply between 2 and 30 keV, and start saturating between 50 and 100 keV [4].

In order to optimize this ionic energy, we have proposed to use molecular clusters, and in particular CD_4, instead of D_2. CD_4 has indeed a higher polarizability, allowing in turn strong Van der Waals forces to bind the cluster. This allows a very good clustering in the CD_4 gas jet, even at room temperature. The cluster fusion experiment can thus be attempted even without cryogenation. Another possibility could be simply of increasing the laser intensity to boost the electron ejection from the cluster.

As a result from these considerations, we have performed two series of experiments : one at moderate intensities (10^{17} W/cm^2), and with a low density supersonic gas jet ($2. 10^{17}$ cm^{-3}); and one with a higher density jet (10^{18} cm^{-3}), submitted to much higher intensities (around 10^{19} W/cm^2). These experiments will now be presented in turn, starting from the experimental setups used in each configuration.

LOW DENSITY / MEDIUM INTENSITY REGIME

Experimental setup

The experiment was performed within the European laser facility of LOA, on the multiterawatt "Salle jaune" system [5]. This Ti-Sapphire laser delivers pulses of 35 fs at a wavelength of 820 nm. For those experiments, the laser was operated at a power level of 15 TW, with an output beam diameter of 5 cm. The laser was focused with a two meter focal length, yielding a maximum intensity of nearly 10^{18} W/cm^2. The contrast ratio between ASE and the main pulse can be varied between 10^{-5} and 10^{-6}.

The target is a cluster gas jet of CD_4, operated in a supersonic configuration with a throat of 500 μm, followed by an expansion cone up to 5 mm diameter. The backing pressure could be increased up to 55 bars. From the geometry of the jet, the backing pressure, we can estimate the average density to be $2. 10^{17}$ cm^{-3}, the cluster size to about 10^5 molecules per cluster, with a corresponding cluster density of $2. 10^{12}$ cm^{-3}.

The emitted particles were detected by a lead- shielded ne213 scintillator, coupled to a photomultiplier (PM) and to a pulse counting electronic system. A "racket" detector detected gamma rays following each shot, and acted as a trigger to validate the subsequent signal on the PM. The electronics included a Pulse Sensitive Discriminator to distinguish between neutrons and gamma rays.

The energy of the ions resulting from the interaction could be measured for each shot by a Thomson Parabola (TP) mass spectrometer, positioned at 90° from the laser axis. Ions enter the TP by a 200 μm pinhole, after which they are submitted to parallel electric and magnetic fields. A drift zone follows, until the ions fall onto a detector consisting of a MultiChannel Plate, a phosphor screen and an 8-bit CCD camera. Due to the deflection by the electric and magnetic fields, each ion species impinges on the detector on a parabola-shaped curve, determined by the q/m ratio; the position within the parabola gives a direct access to the energy or velocity distribution : the closer the impact is to the top of the parabola, the higher the energy.

Figure 1 represents a typical example of Thomson Parabola image thus obtained. One can see clearly the parabolas corresponding to the D$^+$ and C$^+$ ions, as well as a faint one for C^{2+} ions. The image also shows a bright central spot due to direct X-rays, which forms a pinhole imaging of the active plasma region. One can therefore benefit from two diagnostics from the same image.

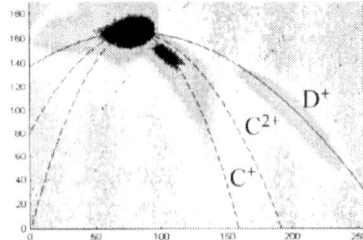

FIGURE 1. Typical image of the Thomson Parabola mass spectrometer.

Experimental results

Figure 2 shows Time of flight spectra of the signal measured, which has already be recognized as probable neutrons by the PSD discriminator. The two spectra were taken at distances from the chamber separated by 30 +/- 1 cm. From Fig. 2, one measures a time difference of 13 +/- 1 ns between the maxima of the distribution, corresponding to a spatial separation of 28.3 +/- 2.1 cm for 2.45 MeV neutrons. This is completely consistent with the value measured, so that we can conclude that the TOF gives a strong indication that we observed 2.45 MeV fusion neutrons.

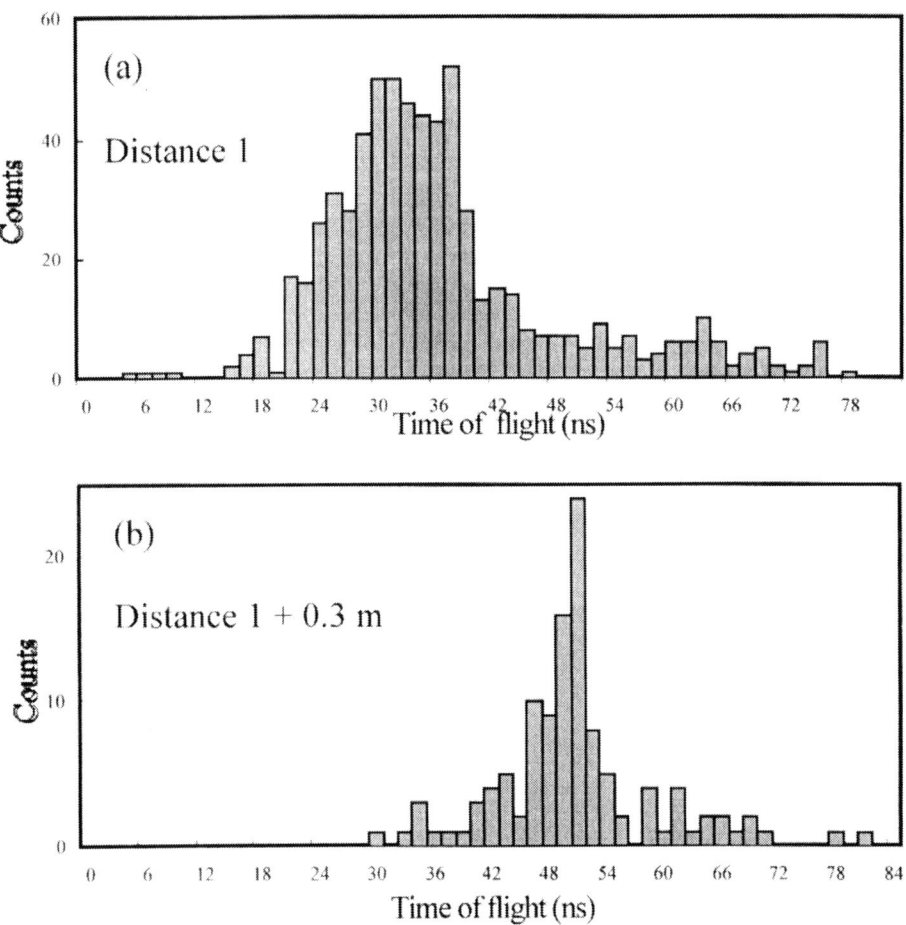

FIGURE 2. Neutron Time of flight spectra at two distances from the jet, separated by 30 cm.

Additional tests included putting a paraffin block before the scintillator, which brought the signal down by a factor of 40; and replacing CD_4 by CH_4, which completely suppressed any signal.

The neutron signal was detected starting from laser energies around 150 mJ, and increased steadily up to a maximum value of 7000 detected at peak power. The intensity dependence of the neutron signal is shown in Figure 3. The last 3 points are best fitted with an $I^{2.2}$ power law. No sign of saturation can be seen.

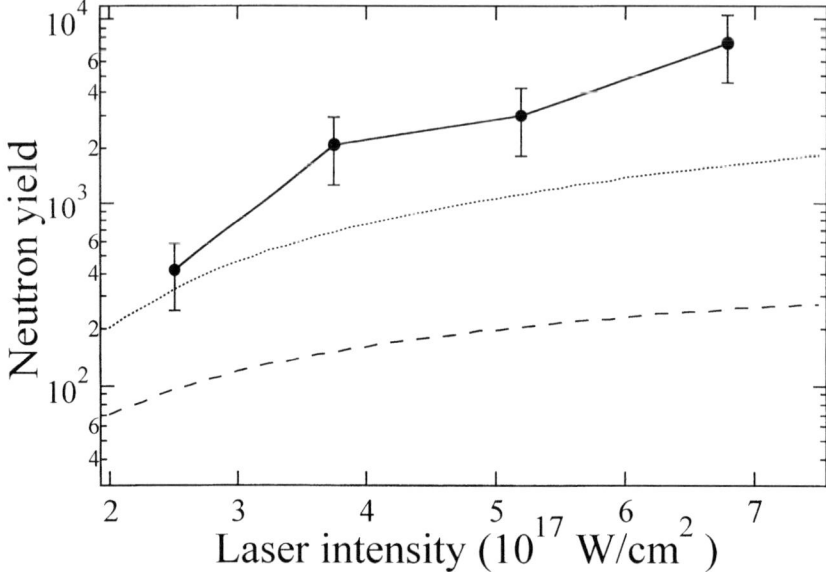

FIGURE 3. Experimental neutron yield as a function of laser intensity (points). Predictions of the central plasma model (Dashed line, Equ. 1), and of the lateral collision model (dotted line, Equ. 2).

In order to understand and model these results, one needs to know the energy distribution of the deuterons involved in the reaction. This is achieved thanks to the TP. Figure 4 shows the energy distributions of C^+ and D^+ obtained at 5. 10^{17} W/cm^2.

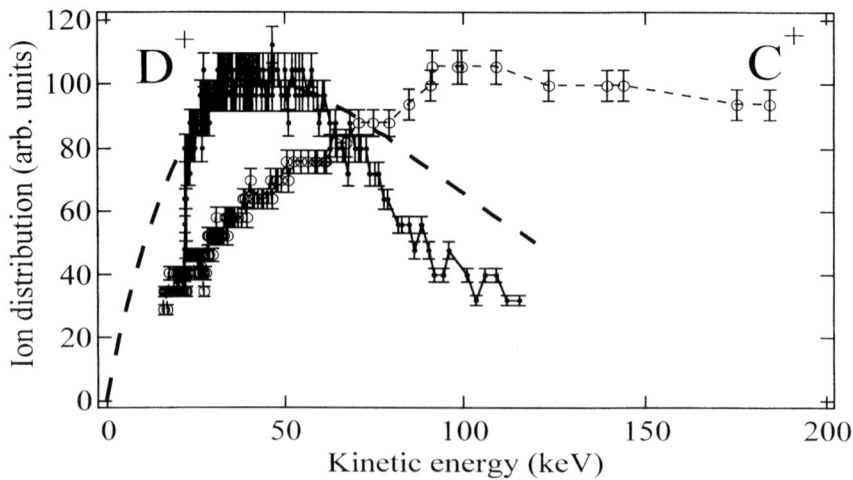

FIGURE 4. Energy distribution measured by the Thomson Parabola at 5. 10^{17} W/cm^2 for D^+ and C^+.

The sharp fall of the energy distribution of D+ on the lower energy side is simply related to the physical edge of the detector; the overall distribution is seen to be close to a Maxwellian, with a slightly faster decay on the high energy side. The C+ distribution can be similarly fitted with a Maxwellian at 100 keV. From similar measurements at different laser energies, one can infer a quasi-linear dependence of the most probable energy as a function of laser intensity.

From the knowledge of the intensity-dependent average kinetic energies of the ions, one can deduce estimates for the neutron yields. The number N of fusion events expected in the core of the hot plasma is :

$$N = n^2 <\sigma v> V t \tag{1}$$

where n is the deuterium ion density, $<\sigma v>$ is the Maxwellian averaged product of fusion cross sections times ion velocity, V is the hot plasma volume, and t the plasma disassembly time. The dashed curve in fig. 2 shows that this estimate comes short by a factor of 100 to account for the observed neutron yields; this is simply due to the quite low gas densities used. We have therefore to consider another process, in which rapid ions originated from the plasma core collide with colder ions from the outer plasma regions, where the intensity is still enough to cause dissociation but not to heat up as efficiently as in the central region. The number of fusion events due to this "lateral collision model" can very easily be estimated as :

$$N = n^2 \langle\sigma\rangle VR \tag{2}$$

where $<\sigma>$ is the Maxwellian-averaged fusion cross section, V is still the hot plasma volume, and R is the cold plasma radius. The dotted line of figure 3 shows that result of this estimate is much closer to the experimental data. The lateral collisions play therefore a very important role in the present case, whereas they were negligible in the Livermore experiment. The difference arises from the much higher energy range of the deuterons : at 5 keV average energy (Livermore case), only head on collisions between fast deuterons can produce significant fusion probability; at , say, 50 keV, the deuterons are energetic enough to produce fusion in the course of a collision with a motionless ion. As a result, the time scale of the process will also be much longer : whereas collisions in the hot plasma core can only happen during roughly 50 ps, lateral collisions can be delayed by a few ns. Finally, the two kinds of processes are not antagonistic : both kinds of processes will of course happen, plus all the intermediate cases. Careful modeling would then require extensive computer simulations.

Another consequence of the role of the lateral collisions can be seen on the angular distribution of the emitted neutrons. The DD fusion process is indeed very anisotropic, especially at high deuteron energies : the neutron are mostly emitted in the forward and backward directions of the impinging deuteron. Experimentally, we have indeed measured a small anisotropy, with a minimum of neutron yields in the laser forward direction, as would be expected from lateral deuteron collisions.

In conclusion for those experiments, the use of CD_4 to induce clustering allows one to obtain easily fusion processes, without having recourse to cryogenation. The kinetic energies of the ions is very high, so that collisions between energetic deuterons, and still deuterons from the plasma corona, start playing an important role. This in turn lengthens quite a lot the intrinsic duration of the process.

The intensity dependence (Fig. 3) does not show any sign of saturation ; in following experiments, we have therefore attempted to use much higher intensities, at the expense of a reduced interaction volume.

HIGHER DENSITY -- HIGH INTENSITY REGIME

Modified setup

In a second series of experiments, we tried to optimize the neutron yield by increasing both the average gas density, and the peak laser intensity at focus. This was achieved, first, by shortening the expansion cone of the cluster gas jet, so that the gas density could now be estimated to about 10^{18} cm^{-3}; and second, by focusing the laser beam by a 50 cm focal length, thereby reaching an intensity of 5 10^{18} W/cm^2.

When the laser beam is focused at the entrance of the gas medium, the interaction takes clearly place in a relativistic regime.

The detection principle was also modified with respect to the previous non-relativistic experiment. The scintillator/photomultiplier was directly connected to a fast digitizing oscilloscope, of 1 GHz bandwidth, triggered by a photodiode. The traces were then recorded by a PC. Figure 5(a) shows one illustrative example of the photomultiplier signal as a function of flight time (in ns) :

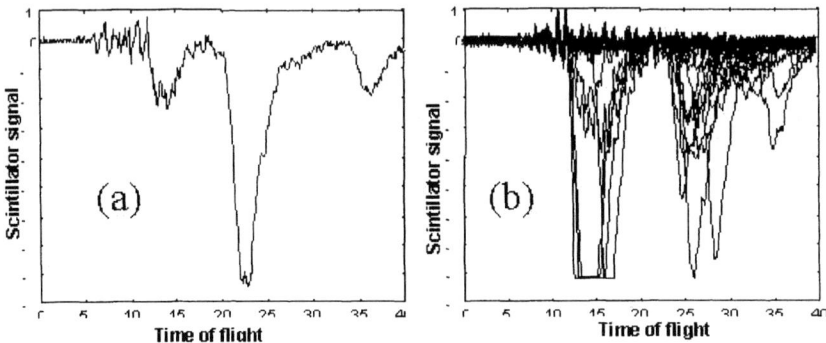

FIGURE 5. Oscilloscope traces (a) : from a single shot, and (b), with many shots superposed.

The first peak on the left corresponds to the impact of a gamma ray; the second and third ones correspond to neutron impacts. Fig. 5(b) displays many traces superposed, to illustrate the bunching of the arrival times of gammas, followed 10ns later by neutron peaks. From this direct time of flight measurement, the neutron energy can easily be derived, allowing one to check again that the peak corresponds to the fusion 2.45 MeV neutrons. However, it is also obvious from fig. 5 that many neutron impacts occur at longer delays. These are attributed to scattered neutrons, that have undergone a collision process on their way to the detector. Such delayed events can be found even microseconds after the interaction; they indicate a gradual thermalization of the neutron yield in the experimental hall.

In terms of neutron numbers, about 10^4 neutrons / shot were still observed, when the gas jet was not cooled. In spite of the similarity of the yield with the previous experiment, we observed qualitatively different behaviours of the data in this very high intensity regime, indicating that the mechanisms at stake were different. For instance, the intensity dependence of the neutron yield was observed to be linear. Most striking was the observation of copious amount of gamma ray radiation, whereas the previous results were almost free of gammas. This occurs mainly when the laser beam is focused right in the gas jet; on the other hand, focusing the laser before or after the jet results in high neutron yields and few gammas.

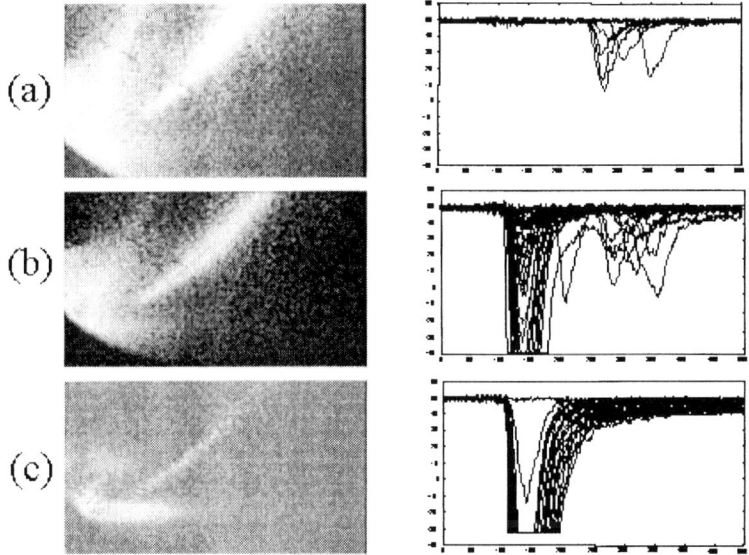

FIGURE 6. Thomson Parabola and oscilloscope traces, at three different focus positions : (a): focus before jet, (b), focus close to jet entrance, (c), focus within gas jet.

Figure 6 shows comparisons of Thomson parabola images, and superposed oscilloscope traces, in three cases : (a), with the focus before the gas jet, (b), with the focus close to the entrance of the gas jet, and (c), with the focus within the gas jet. The neutron yield increases slightly from (a) to (b), which is consistent with a small increase in the D^+ ion energies; however, the gamma ray signal is already dominant in (b). In (c), the neutrons have completely disappeared; the parabola corresponding to the fast deuterons impinging on the Thomson parabola also seem to vanish, even though the maximum kinetic energy has increased even further. The Thomson image now shows a relativistic guiding of the laser pulse. This interaction regime is known to induce significant electron acceleration up to multi-MeV energies; the fast electrons created then induce bremsstrahlung in the interaction chamber walls, resulting in the very strong gamma ray signal observed.

The most surprising feature here is the parallel disappearance of the neutron and fast ion signals. The strong magnetic field induced by the fast electron beam, together with the cold electron return current, could actually deflect the deuterons off the direction of the Thomson Parabola. However, this scenario would not explain why the neutrons simply are no longer produced. One other possibility would link the decreased neutron yield to a contrast problem : if the laser has a contrast of 10^{-6}, then clusters are subjected to 10^{13} W/cm^2 for few nanoseconds, which may be quite enough to induce a slow cluster explosion, and spoil the Coulomb explosion after the interaction with the main pulse. However, worsening slightly the laser contrast changed little to the results. Significant improvements on laser contrast are now required to proceed on this subject.

Finally, we attempted to cool down the gas down to about −30 C, in order to enhance the neutron yield [6]. The neutron signal was indeed boosted to about 10^5 neutrons/shot.

PERSPECTIVES

Further optimization is of course strongly needed before this pulsed neutron source can be used for experiments. A minimum of 10^9 to 10^{10} rapid neutrons per shot would seem necessary for most applications, especially if one considers the comparisons with the common reactor sources, or the new spallation sources. Such an improvement would now require to confine the laser-produced plasma for nanoseconds. Eventually, it may be interesting to note that confinement for 10 ms would be necessary to get efficiencies of 10^{13} neutrons per Joule of incoming energy, in order to balance the input and output energies. Such a long confinement time is of course far beyond reach of any present day technique.

ACKNOWLEDGMENTS

We would like to acknowledge the technical assistance of J. Ball.

REFERENCES

1. Pretzler, G., *et al.*, *Phys. Rev. E* **58**, 1165 (1998).
2. Hilscher, D., *et al.*, *Phys. Rev. E* **64**, 016414 (2001).
3. Ditmire, T., *et al.*, *Nature* **398**, 491 (1999).
4. Brown, R.E., and Jarmie, N., *Phys. Rev. C* **41**, 1391 (1990).
5. Antonetti, A., *et al.*, *Appl. Phys. B* **65**, 197 (1997).
6. Zweiback, J., *et al.*, *Phys. Rev. Lett.* **84**, 2134 (2000).

* : Permanent address : Institute of Material Science and Laser Physics, Technical University of Crete, Kounoupidiana-Campus, Chania, Crete, Greece

Nuclear Diagnostics of High Intensity Laser Plasma Interactions

K. Krushelnick[1], M. I. K. Santala[1], K. W. D. Ledingham[2], F. N. Beg[1], E. L.Clark[1], R. J. Clarke[3], A. E. Dangor[1], T. McCanny[2], P. A. Norreys[3], I. Spencer[2], M. Tatarakis[1], I. Watts[1], M. S. Wei[1] and M. Zepf[1],

1 Imperial College of Science, Technology and Medicine, London SW7 2BZ, UK
2. Department of Physics and Astronomy, University of Glasgow, Glasgow G12 8QQ UK
3 Rutherford Appleton Laboratory, Chilton, Oxon, OX11 0QX , UK

Abstract. Nuclear activation has been observed in materials exposed to energetic protons and heavy ions generated from high intensity laser-solid interactions (at focused intensities up to 5×10^{19} W/cm^2). The energy spectrum of the protons is determined through the use of these nuclear activation techniques and is found to be consistent with other ion diagnostics. Heavy ion fusion reactions and large neutron fluxes from the (p,n) reactions were also observed. The reduction of proton emission and increase in heavy ion energy using heated targets was also observed.

INTRODUCTION

Recent experimental work studying high intensity laser plasma interactions at intensities greater than 10^{19} W/cm^2 has shown that such plasmas can potentially be an efficient source of very energetic electrons [1] , gamma rays [2] and ions [3-6]. At such energies one of the most important methods for diagnosis of these high energy particles is through the use of nuclear activation techniques [2]. Previous work has shown that much of the physics of the production of electron beams and proton beams can be uncovered using (γ,n) [2] and (p,n) reactions. There are clearly many potential applications for these energetic particles such as compact accelerators and as diagnostic probe beams of high density plasmas. The results of these experiments are particularly interesting for technological applications since such sources should be easily scalable to "table-top" sizes [1, 4].

Here we discuss experiments to examine the properties of energetic ions produced during the high power laser-plasma interaction by measuring the nuclear activation induced in different materials exposed fast ions emitted from the plasma. Mechanisms for generating fast ions are presently not well understood and are the subject of some controversy. A well understood mechanism is the expansion of hot plasma, where ion-accelerating space-charge fields are created by the hot electron population which escape the target. The nuclides observed include ^{63}Zn, ^{13}N and ^{11}C. These are positron

CP611, *Superstrong Fields in Plasmas:* Second Int'l. Conf., edited by M. Lontano et al.

emitting nuclides with half-lives of the order of tens of minutes, which makes the measurement of induced activity practical. The ion-induced production reactions have reaction thresholds of approximately 4 MeV.

Previously, nuclear activation has been observed in laser-solid interactions from γ ray -induced reactions [2]. In comparison to (γ,n) reactions, it should be noted that (p,n) reaction thresholds are typically only a fraction of the (γ,n) thresholds, especially for light elements, and the cross-sections may be nearly an order of magnitude larger. However, the range of energetic ions in matter is very short compared to that of γ-rays. Consequently, activation samples for (p,n) reactions are usually thin foils while large pieces of solid matter are used for (γ,n) studies. This is beneficial since the significance of target self-absorption in the measurement of the induced activity can be reduced.

ION ACTIVATION MEASUREMENTS

These experiments were carried out using the CPA (chirped pulse amplification) beam of the VULCAN Nd:glass laser system at Rutherford Appleton Laboratory (RAL). The laser wavelength was 1.054 μm, pulselength 0.9 – 1.2 ps and energy incident on target was 20-50 J. The laser beam (111 mm x 200 mm) was focused on to the target surface using a $f.l.$=225 mm on-axis parabolic mirror (Figure 1). The laser beam was p-polarised and was incident at an angle of 45°. The peak intensity was measured to be about 2 x 10^{19} W/cm^2. The targets were 0.5 - 1.5 mm thick slabs of BK-7 glass, some of which were coated with a 2-3 μm layer of polyethylene as well as thin foils of aluminum and titanium. The target and all of the optics for the beam after compression were placed in a vacuum chamber evacuated to <10^{-4} mbar pressure.

For the nuclear plasma blow-off diagnostic the sample materials were placed in the direction of the target normal 15 cm away from the target. The sample materials used were Cu-foil, Pyrex glass (which contains boron) and PVC-foil. The activated samples were removed from the target chamber as soon as possible after each shot and taken for radio-activity measurements. This typically causes a delay of 10 minutes before the activity can be measured. The principal reaction products studied are nearly 100% positron emitters. The activity can thus be measured by counting 511

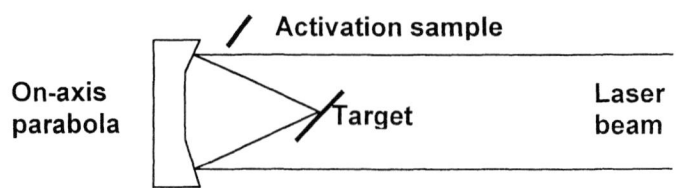

FIGURE 1. Basic experimental set-up

keV positron annihilation gammas in coincidence. This results in very high sensitivity and low background even with minimal detector shielding. A set-up with two 3" x 3" NaI(Tl) scintillators shielded with 2 cm of lead was used for detection. The set-up has about 10 % positron detection efficiency, and background count rate less than 0.01 cps.

Samples were counted initially for 200 s which gives approximate detection threshold of 0.4 Bq. For nuclide identification, the counting was repeated several times over a period of several hours to measure the half-life. Late in time, samples were often counted for much longer times to improve the statistical significance and the signal-to-background ratio.

The average initial activity of the copper foils was 600 Bq, corresponding to 2 x 10^{6}

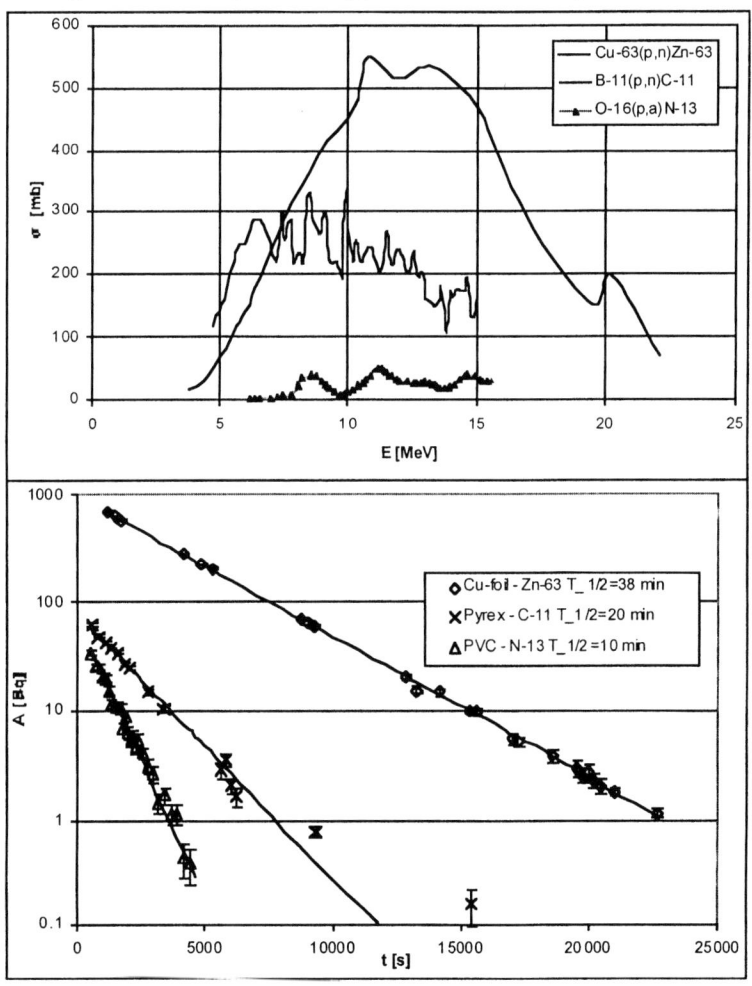

FIGURE 2. Cross-section and measured half-life for various (p,n) reactions.

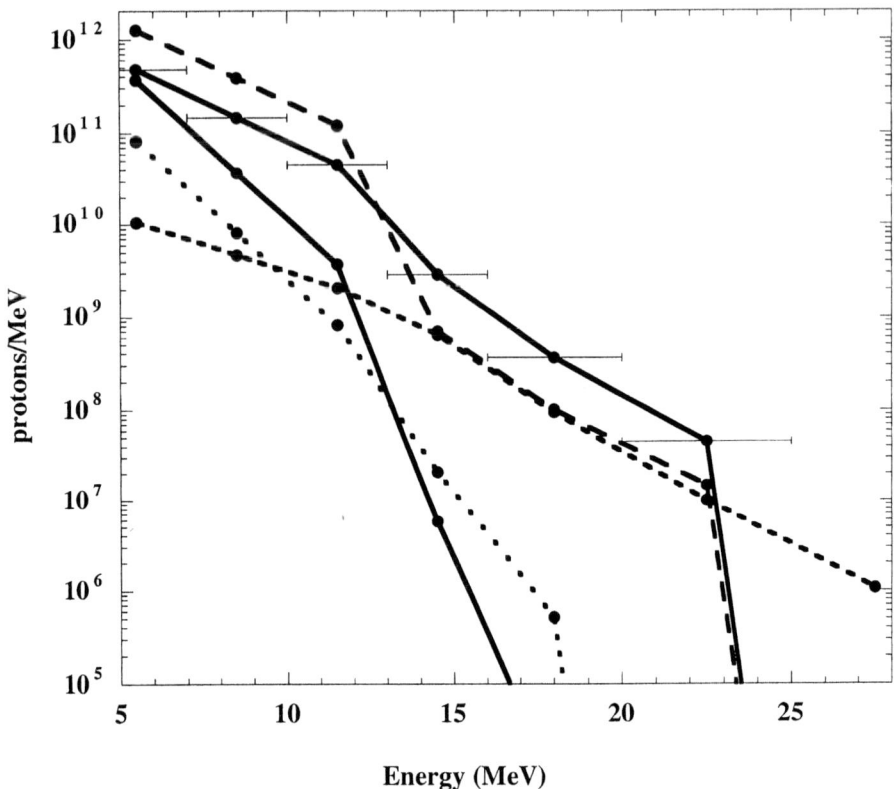

FIGURE 3. Proton spectrum from use of activation stacks. There are significant shot-to-shot fluctuations. All of this data was taken at ~ 10^{19} W/cm^2.

^{63}Zn nuclei produced in a single shot. The size of Cu foils was typically 25 by 25 mm, 100-250 μm thick, and they were 15 cm from the target. Results from the half-life measurements are shown in Figure 2. Very pure exponential decays are observed, indicating that only one nuclide is present. The solid lines display least-squares fit to the measured activities. The best-fit half-lives are 38.7, 19.9, and 10.0 min for Cu-foil, Pyrex-glass and PVC foil, respectively. This would indicate that ^{63}Zn, ^{11}C, and ^{13}N are produced in these target materials. The literature values for their half-lives are 38.47, 20.38, and 9.96 min. The likely production reactions for the first two are ^{63}Cu$(p,n)^{63}$Zn, ^{11}B$(p,n)^{11}$C. There are two reaction candidates for the production of ^{13}N: ^{16}O$(p,\alpha)^{13}$N and ^{13}C$(p,n)^{13}$N, both of which are likely to contribute.

By stacking several thin (12 μm) Cu foils, information about the incident spectrum can be deduced by utilizing the classical stopping ranges of ions in solid matter. The preliminary results indicate that, on a large scale, the spectrum can be described reasonably well by an exponential distribution with a temperature parameter varying

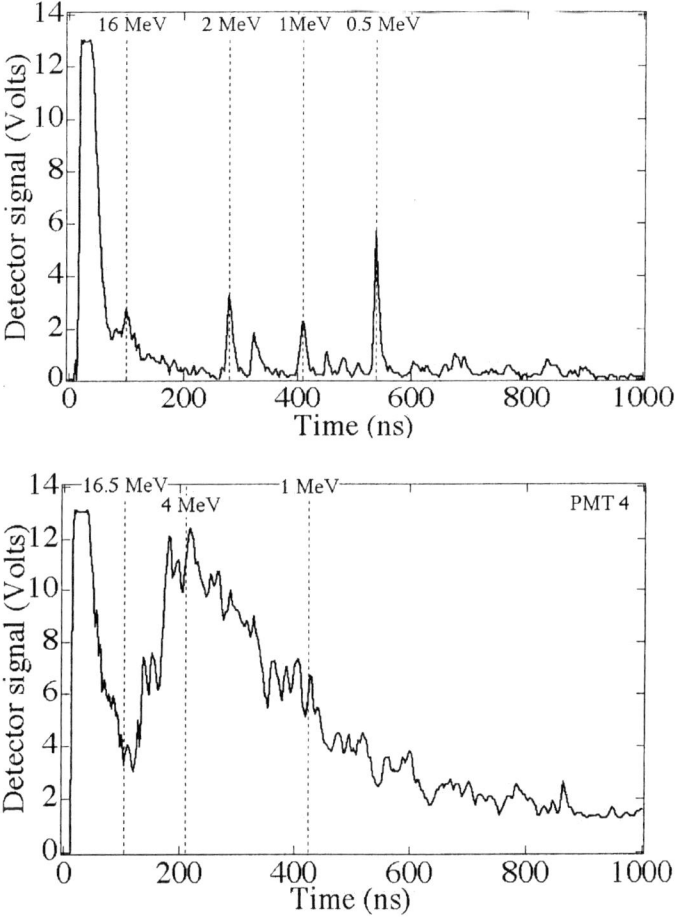

FIGURE 4. PMT signal from neutrons resulting from (p,n) reactions in copper detector stack. Upper spectrum shows PMT signal without activation stack, lower spect7rum shows PMT signal with Cu stack. There are ~ 10^8 neutrons /shot.

on shot-to-shot basis in the range of 1-2 MeV (see Figure 3). The results suggest that the total number of protons is $>10^{12}$/sr, and $>10^{11}$/sr above 5 MeV. Fine-scale modulations in the ion energy spectrum that are observed in Thomson parabola measurements do not appear in the activation measurements [6]. This is likely to be due to the large degree of smoothing caused by the wide cross-section distribution in copper. However activation measurements are the only low background method for determining the entire high energy spectrum of energetic ions.

The plastic coating on glass surface has only a small effect on the properties of blow-off protons. This suggests that a large source of protons must exist at the surface,

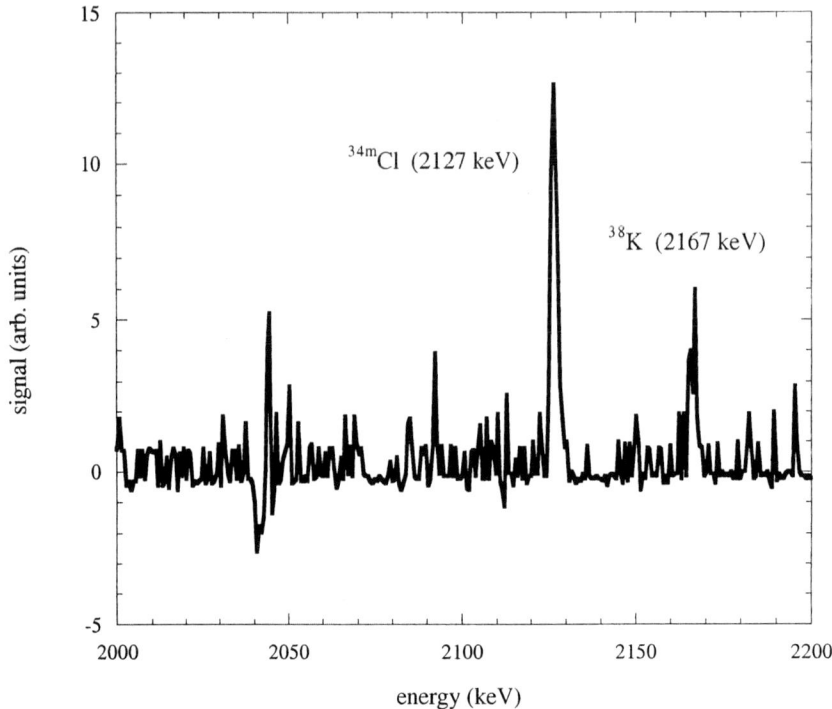

FIGURE 5. Gamma ray spectrum resulting from carbon/aluminum fusion reactions in the blow-off plasma.

probably due to surface contamination. The targets were exposed to normal atmosphere for long periods after fabrication, and no special precautions were taken to remove possible surface contamination. The number of protons observed corresponds to all the protons in 10-100 μm^3 of CH plastic.

Compared to ion spectrometer diagnostics it has a much larger sensitive solid angle. The results are thus more likely to reflect better the large-scale properties of the emitted ions. Angularly resolved measurements are possible by exposing several spatially independent samples to the blow-off plasma. One of the main advantages of the ion diagnostic is the complete insensitivity to other particles or lower-energy processes. It is also immune to electrical noise generated by the laser-plasma interaction as counting is carried out off-line. This diagnostic can have a very large dynamic range, however, at low proton numbers the counting statistics can be poor.

To measure the heavy ions produced in these experiments aluminum foils were exposed to the ion blow-off. After the interaction these foils were examined with Germanium detector gamma ray detector to measure the characteristic x-rays produced by any radioactive isotopes produced during experiment. In several samples the characteristic emission lines of ^{34m}Cl (146, 1177, 2127, and 3304 keV) and of ^{38}K (2167 keV) were found (Fig. 5). In addition, positron emission was also detected. Decay of these lines was found to be consistent with the half-lives of these nuclides (32 min and 7.6 min). These observations prove conclusively that ^{34m}Cl and ^{38}K are produced in the Al foil. This can be explained by fusion of fast ^{12}C ions in the blow-off with ^{27}Al in the foil. This forms highly excited ^{39}K, and subsequently several lighter nuclides through nucleon evaporation. ^{34m}Cl and ^{38}K are the only ones with long enough half-lives (~10 min) to be seen in the off-line counting.

The initial activities can be found to be 15-20 Bq for ^{34m}Cl and 20-30 Bq for ^{38}K using the measured line intensities. This corresponds to production of about 5×10^4 ^{34m}Cl nuclei and 2×10^4 ^{38}K nuclei. The reaction thresholds are approximately 40 MeV and the typical reaction cross-sections are ~300 mb. The stopping power of ~50 MeV ^{12}C ions is about 0.6 MeV/μm, so they exceed the reaction threshold for about 15 μm. 2×10^9 ^{12}C ions are then needed to produce the observed activation. This is consistent with measurements of the heavy ion spectrum using Thomson parabola ion spectrometers [6].

In these experiments thin aluminum foils were placed on both the front and the rear of thin solid targets. These measurements indicated that, although there was significant proton emission in both directions, energetic carbon ions were only measurable at the front of the target. This is because either: (a) the source of both the energetic carbon ions and protons is the front surface and carbon ions are stopped before they can pass through the target or (b) the mechanism for accelerating ions on the rear surface is not as efficient for accelerating carbon ions as it is for protons.

EXPERIMENTS WITH HEATED TARGETS

Generally, the majority of the energy in the ions is carried by protons which are accelerated preferentially with respect to heavier ions. The lighter protons move at much higher velocities and consequently tend to "short-circuit" the accelerating electric fields – which therefore limits the peak energy of other ion species. In solid target interactions these protons originate from a surface contamination layer on the target – which is largely composed of water and/or vacuum pump oil. Both of these surface contaminants contain hydrogen – which provides the source of the observed protons in the experiments [5].

We have also performed experiments which attempt to remove the contamination layer through heating of the target. Since the boiling/melting points of the thin targets used (aluminum or gold) are typically much higher than that of water or pump oil, we show that this is an effective way to remove the proton component from the accelerated ion spectrum. Using a Thomson parabola ion spectrometer we also show that the total number of protons is significantly reduced and that consequently, the energy of the heavier ions which remain is significantly increased.

These experiments were carried out using the CPA beam of the VULCAN Nd:Glass laser system. The laser wavelength was 1.054 μm, the pulselength was 0.9 – 1.2 ps and the energy incident on target was 20 – 50 J. The laser beam (111 mm x

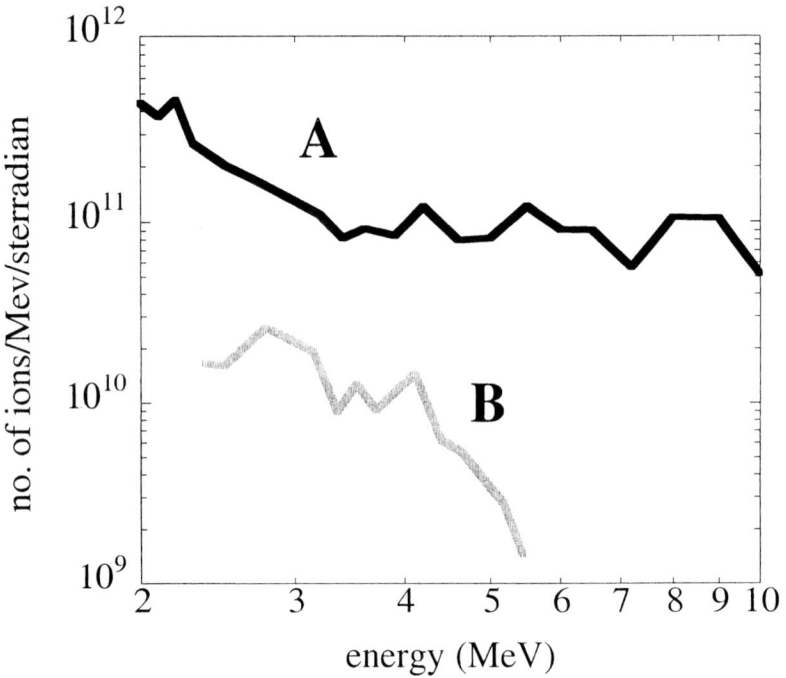

FIGURE 6. A) Proton spectrum from unheated target B) from heated target

200 mm) was focused onto the target surface using a $f.l.$=225 mm on-axis parabolic mirror. The laser beam was p-polarised and was incident on the target at an angle of 45°. The peak intensity was measured to be about 2 x 10^{19} W/cm^2. The targets were typically thin foils of gold and the target and all of the optics for the beam after compression were placed in a vacuum chamber evacuated to less than 10^{-4} mbar pressure.

A specially designed target mount was used which was capable of heating a target to a temperature of greater than 400 C. The target was held in position over a small heating element (filament) which was able to heat the target radiatively. The target temperature was monitored using a thermocouple and was maintained at a constant level prior to each shot.

Energetic ions were measured using a Thomson parabola ion spectrometer which was placed approximately 1 meter from the target and was aligned to the front of the target at an angle of approximately 20 degrees from target normal. CR39 – a nuclear track detector – was used as the detector.

Figure 6 shows the measured proton spectrum from the Thomson parabola diagnostic from an interaction at an laser intensity of 2×10^{19} W/cm^2 with 100 μm thick gold targets. The number of protons from heated targets (Fig. 6B) was reduced by more than an order of magnitude when compared to unheated gold targets at a similar laser intensity (Fig 6A). The peak energy of the measured protons was also significantly reduced and does not approach the diagnostic limit of 11 MeV as in the unheated target .

Although these targets were heated to temperatures greater than 400 C there were still significant amounts of carbon ions remaining and although the number of carbon ions was reduced – the energy of these carbons was increased significantly. Figure 7 shows the spectrum of the carbon ions from two consecutive shots – one which used a heated target (Fig. 7B and 7D) and one which did not (Fig. 7A and 7C). Clearly, from the spectra there are still carbon ions present however in the heated targets the peak energy has increased by more than 10 MeV. While the number of low energy carbon ions has been reduced the entire distribution has been shifted to higher energies. The spectra of the C^{6+} ions are shown in Fig 7A and 7B and the C^{5+} ion spectra are shown in Fig 7C and 7D. This is in contrast to the fact that the proton number and maximum energy have been radically reduced.

From the Thomson parabola measurements it was also apparent that the number, energy, and ionisation stage of the gold ions also significantly increased – although it was not possible to make spectra of this data due to the low signal level and the uncertainty in the ionisation stage (i.e., the parabolas from the many ionisation stages were observed to overlap).

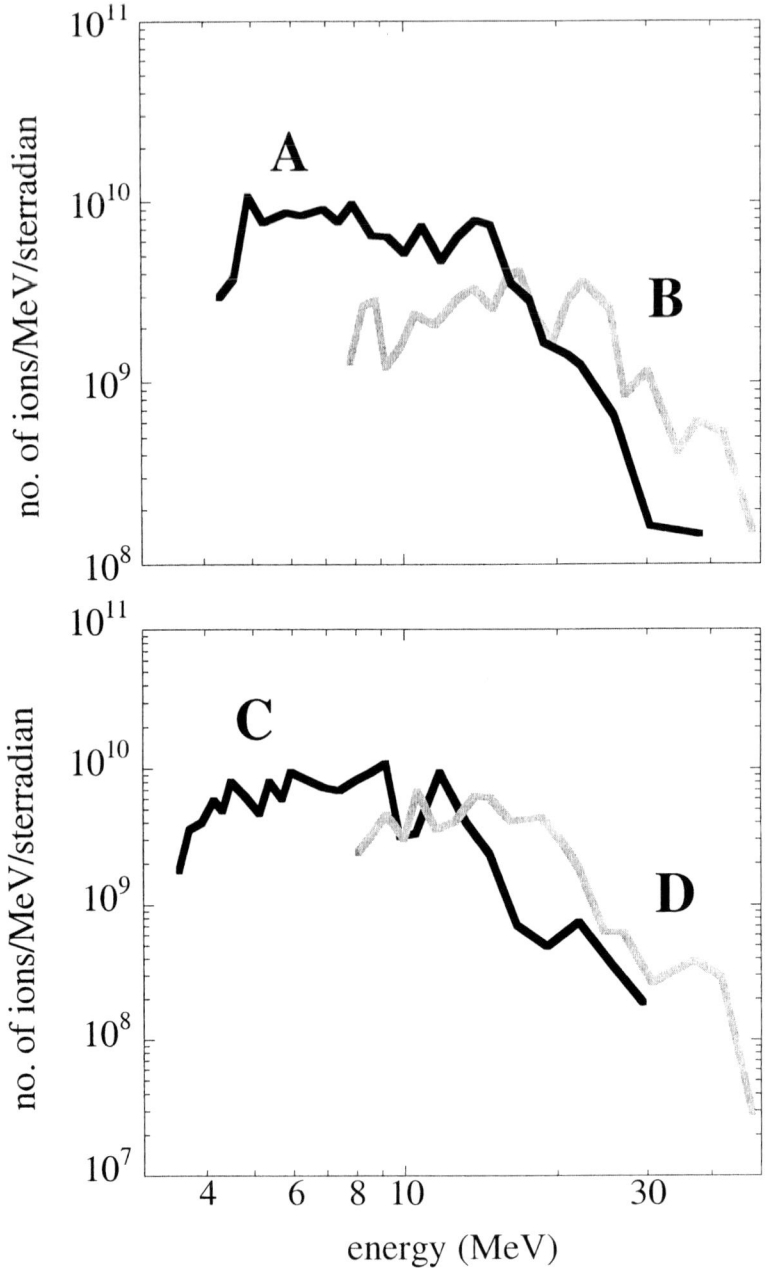

FIGURE 7. A) Carbon 6+ spectrum from unheated target B) Carbon 6+ from heated target. C) Carbon 5+ spect7rum from unheated target D) Carbon 5+ spectrum from heated target

CONCLUSIONS

Ion-induced activation by high-intensity lasers may prove to be an attractive source of short-lived radionuclides. The total activation could be increased by an order of magnitude by using samples that span a larger solid angle. If the energy spectrum of ions depends only on the incident intensity and number of ions scales with laser pulse energy small high-repetition rate lasers with similar focused intensity to Vulcan could produce similar samples in a few seconds. Integrating activity over minutes to hours samples with activities in the 10^6 Bq range could conceivably be produced. Further improvement could almost certainly be achieved by optimising the laser prepulse, the target material and surface treatment. It should be noted that as ion beams can activate many materials this could be a useful techniques for producing a wide range of isotopes.

In conclusion, it appears that he use of heated targets provides a relatively easy way to increase the energy of the heavy ions produced from intense laser produced plasma. This may allow a more efficient ion source for future heavy ion accelerators – and may perhaps avoid the use of pre-acceleration stages.

ACKNOWLEDGMENTS

We would like to acknowledge the technical assistance of the VULCAN operations team.

REFERENCES

1. A. Modena Z. Najmudin, A. E. Dangor *et al. Nature* **377**, 606 (1995).
2. M.I.K. Santala, M. Zepf, I. Watts *et al., Physical Review Letters,* **84**, 1459 (2000); P.A. Norreys, M. Santala, E. Clark *et al.* Physics of Plasmas, **6**, 2150 (1999).
3. E. L. Clark K. Krushelnick, J.R. Davies *et al., Physical Review Letters,* **84**, 670 (2000); R. Snavely, M.H. Key, S.P. Hatchett *et al., Physical Review Letters*, **85**, 2945 (2000); A. Maksimchuk, S. Gu, K. Flippo, D. Umstadter, V.Y. Bychenkov, *Physical Review Letters,* **84**, 4108 (2000).
4. K. Krushelnick, E. Clark, R. Allott, et al., *IEEE Transactions in Plasma Science* **28**, 1184 (2000).
5. S. J. Gitomer, R. D. Jones, F. Begay, A. W. Ehler, J. F. Kephart, R. Kristal,, *Physics of Fluids,* **29**, 2679 (1986).
6. E.L. Clark , K. Krushelnick, M. Zepf *et al., Physical Review Letters,* **85**, 1654 (2000).

Laser-Cluster Interaction for Nuclear Fusion

Y. Kishimoto[†1] , T. Masaki[†] and T. Tajima[*]

† Naka Fusion Research Establishment
Japan Atomic Energy Research Institute, Naka, Ibaraki 311-01, Japan
* Department of Physics and Institute for Fusion Stidies
The University of Texas at Austin, Austin, Texas 78712, USA

Abstract

The key physical processes of laser-cluster interaction essential to understand and opimize the cluster fusion are investigated by using numerical simulation. By properly choosing the cluster size, spatial packing fraction, and laser field amplitude, cluster ions are efficiently accelerated in a controlled manner to high energy in such a way for the fusion cross-section to be maximized. The producton of fusion neutrons is expected to be enhanced by taking into account the spatial propagation of explosion front toward the surrounding fuel cluster region. It is also found that the average ion energy can exceed the Coulomb energy stored originally in the cluster by obtaining the laser energy through ambi-polar electrostatic field around the vacuum-medium interface. Such high energy ion generation may enhance the neutron yield by introducing the solid fuel collar that surrounds the cluster medium. Although the area of neutron irradiation is tiny, the resultant neutron intensity with this method may rival that of the conventional much larger system of neutron sources.

I INTRODUCTION

A cluster medium is known as a *small particle system* from the statiatical view point where the level of the fluctuation is high comparted with other preparation of the same materials such as gas and/or plasma. A number of salient phenomena have been observed in the laser irradiation of clusters. These include : the Coulomb explosion of clusters [1,2], enhanced emission of X-rays [3,4,5], generation of energetic electrons [6], and energetic ions [7]. There are indications that some strong nonlinear interaction of laser and clusters do occur upon the intense short pulse irradiation [8,9] . It has been pointed out [10,11] that the polarization induced on the cluster surface by the laser plays an essential role in the interaction. Prior to the contemporary interest, there has been a long standing research on the properties of cluster materials since Doremus[12] and Kubo et. al.[13]. Perhaps the most spectacular demonstration of the enhanced laser-cluster interaction among the series of latest experiments is the report of copious fusion neutron generation upon the laser irradiation of deuterium clusters[14,15]. The theoretical and computational studies for the cluster based fusion reaction and a possible reactor design were performed by the present authors [16].

In this paper, we discuss the feasibility to employ such laser-cluster interaction for high flux fusion neutron generation in the range of $N \geq 10^{13}/cm^2$ per shot by using a short pulse laser with total energy around 10J. This would be comparable to an intense neutron source required for studying the first wall neutron damage for fusion reactor. In order to realize such a high flux neutron source, we need to optimize the cluster medium and laser system, so that cluster ions are efficiently accelerated and fusion reaction is optimally enhanced. For this purpose, we investigate key physical processes of laser-cluster interaction and try to control the interaction and optimize the neutron yield.

---

[1] kishimoy@fusion.naka.jaeri.go.jp : also Advanced Photon Research Center, JAERI

CP611, *Superstrong Fields in Plasmas:* Second Int'l. Conf., edited by M. Lontano et al.
© 2002 American Institute of Physics 0-7354-0057-1/02/$19.00

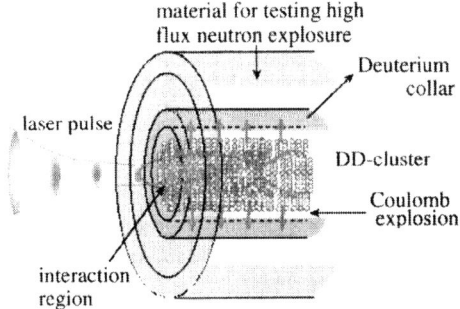

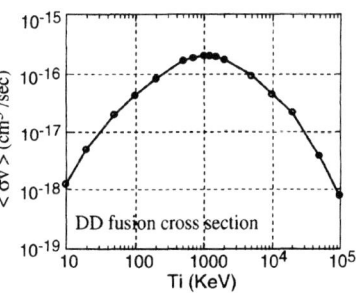

Figure 1: Schematic picture of the set-up of laser-cluster interaction for producing fusion neutrons. Some material for testing high flux neutron exposure is placed in the outside region.

Figure 2: DD fusion cross-section $\langle \sigma v \rangle$ averaged over Maxwellian velocity distribusion as a function of deuterium ion temperature.

II   Estimate of fusion neutron by laser-cluster interaction

A difference of the cluster medium from the conventional plasma is the variety of controllable parameters which characterize the interaction. A cluster medium is characterized by : (1) its material (the charge state Z, and density $n_e = Zn_{cl}$, n_{cl} is the cluster ion density), (2) spatial size (the radius a), (3) surface and internal structures (for example, multi-layer coatings using different materials), (4) packing fraction (ratio that clusters occupy to the system volume $f \equiv 4\pi a^3 N_{cl}/3V$, N_{cl} is the number of cluster in the volume V) (5) spatial configuration or distribution of clusters (ordered or disordered/random distribution). In relation to the interaction with laser field, following parameters such as : (6) cluster size to laser wavelength (a/λ_l), (7) electron skin depth to cluster size (δ_e/a, $\delta_e \equiv \omega_p/c$), (8) electron excusion length to cluster size (ξ_e/a), (9) inter-cluster distance to electron excusion length [ξ_e/R, $R = (3V/4\pi N_{cl})^{1/3}$], (10) laser cut-off density to local cluster density (Zn_{cl}/n_c), etc, regulate the interaction. Such a large number of parameters widens applications utilizing the laser-cluster interaction. We have tried to utilize such laser-cluster interaction for efficient neutron production by employing the deuterium-deuterium(DD) nuclear fusion.

In the laser-cluster interaction, there exist two possible regimes depending on the above relations (specifically relaion (7) and (8)), i.e. *Coulomb explosion* that is triggered for $\xi_e >> \delta_e \geq a$ and alternatively *ambi-polar expansion*[17] that is triggered for $a >> \delta_e \sim \xi_e$. In the Coulomb explosion, since the electron skin depth is greater than the cluster size ($a < \delta_e$) and the laser intensity is high enough to fulfill the relation $a < \xi_e$, the laser field penetrates to the interior of the cluster and a large fraction of cluster electrons leave the host cluster within its optical cycle and set up large charge separation. Such charge separation causes a strong electrostatic field and stochastic and many electrons are heated to high energies so that they no longer returns to their host cluster. When this happens, the cluster becomes highly non-neutral and ions explode due to its Coulomb repulsive force. On the other hand, in the latter case, when the skin depth is smaller than that of the cluster size ($a > \delta_e$), the laser field does not fully penetrate to the interior of the cluster, but interacts only with a limited peripheral region of the cluster. Then, the coronal plasma with a hot electron temperature is produced at the ion front. As a result, the ion front is accelerated by the strong ambi-polar field and expanded into vacuum.

An experimental set-up of the laser-cluster interaction we are proposing is schematically illustrated in Fig.1. In this picture, high intensity short pulse laser is irradiated to the surface of the cluster medium, and the small spot with high temperature and density is produced through direct laser-cluster interaction. The hot spot expands triggering further cluster explosion to the surrounding region so that cluster ions are accelerated to high energies to induce nuclear fusion. Here, we restrict our discussion to deuterium-deuterium (DD) fusion by employing deuterium clusters. However, the idea developed here is straightforwardly applied to deuterium-tritium(DT) fusion where the cross-section is higher than that of the DD fusion. The deuterium fusion cross-section σ is a smooth function of the deuteron energy E_i and the average of the cross section over the Maxwellian velocity distribution, i.e. $\langle \sigma v \rangle^{(M)}$, is illustrated in Fig.2, as a function of the deuterium temperature. The peak of $\langle \sigma v \rangle^{(M)}$ apears at around $T_i = 1 \mathrm{MeV}$, and the cross-section around there is approximately $\langle \sigma v \rangle \simeq 3 \times 10^{-16} \mathrm{cm}^3/\mathrm{sec}$. As seen in Fig.2, a wide range of deuterium temperature around $0.1\mathrm{MeV} \leq T_i \leq 10\mathrm{MeV}$ is found to contribute for producing neutron yield. The total neutron yield is roughry estimated by

$$
\begin{aligned}
N^{(cluster)} &\sim \int dt \int \langle \sigma v \rangle n_{av}^2 d\mathbf{X} \\
&\simeq \frac{\tau_d}{2} \langle \sigma v \rangle f^2 n_{cl}^2 V \ ,
\end{aligned}
\tag{1}
$$

where $\langle \sigma v \rangle$ is the average over the arbitrary deuterium energy distribution function, V and τ_d are the effective interaction volume and confinement and/or disassembly time. Here, we assume that $\langle \sigma v \rangle$ and n_{av} are spatially uniform and the average deuterium density n_{av} is given by $n_{av} = f n_{cl}$. The packing fraction f is described as $f = a^3/R^3$. The disassembly time τ_d is approximately estimated by $\tau_d \sim V^{1/3}/C_s$ where $C_s = \sqrt{T_e/M}$ is the deuterium ion sound speed with the eletron temperature T_e. Note that since the deuterium energy distribution resulting from the laser-cluster interaction is not generally described by the Maxwellian distribution, the direct velocity space integral has to be performed to evaluate the cross section. From Eq.(1), in order to increase the neutron yield, besides $\langle \sigma v \rangle$, we need to increase the packing fraction f, the effective intetaction volume V, and also the disassembly time τ_d. Note that the interaction volume is not restricted to the region where the laser directly interacts with clusters, but is extended to the surrounding region that is secondly heated by accelerated high energy particles.

The number of deuterium ions which contribute to nuclear fusion is quite small compared with the total number of heated deuterons (i.e. $n_{av}V$). Therefore, in order to efficiently utilize heated ions, it is desirable to introduce a solid deuterium (and/or deuterium doped) collar that surrounds the cluster medium. This is also schematically illustrated in Fig.1. In this case, the neutron yield produced by the collar is estimated as

$$
N^{(collar)} \sim \langle \sigma l_m \rangle n_D n_{av} V \ ,
\tag{2}
$$

where $n_D (= n_{cl})$ is the solid deuterium collar density and l_m represents the velocity dependent mean free path of an accelerated deuterium ion in the solid collar. The mean free path of the energetic deuterium ion is roughly evaluated from the stopping power in the plasma given by $dE_i/dt = -Z^2 e^4 n_e \ln \Lambda / 4\pi\epsilon_0^2 m v_i$ as

$$
l_m \simeq \int_0^{E_i} \frac{dE_i}{dE_i/dx} \sim -\frac{4\pi\epsilon_0^2}{Z^2 e^2 n_e \ln \Lambda} \left(\frac{m}{M} \right) E_i^2 \ ,
\tag{3}
$$

assuming that ions are primarily stopped by electrons. Here, the MKS unit is used, and $E_i \equiv M v_i^2/2$ and $\ln \Lambda$ represent the ion kinetic energy and Coulomb logarithm, respectively. When the surrounding collar is introduced, the total neutron yield is a sum of Eq.(1) and Eq.(2), i.e. $N^{(cluster)} + N^{(collar)}$. The ratio of the neutron yield resulting from the cluster region and collar one is

$$
\frac{N^{(cluster)}}{N^{(collar)}} \sim \frac{\langle \sigma v \rangle \tau_d}{\langle \sigma l_m \rangle} \frac{n_{av}}{n_D} \sim \left(\frac{v_i \tau_d}{l_m} \right) f \sim \left(\frac{V^{1/3}}{l_m} \right) f \ ,
\tag{4}
$$

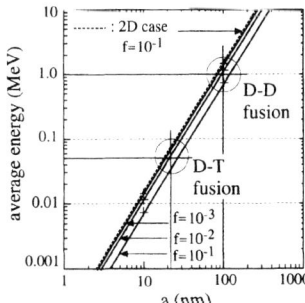

Figure 3: Maximum average deuterium kinetic energy obtained by the Coulomb explosion as a function of cluster radius for different packing fraction. Circles in the figure denote the region where DD and DT fusion cross-sections are maximized.

assuming that the cross-section is almost the same both regions. Since the mean free path is propotional to E_i^2, the contribution from the collar fusion is expected to become large for higher laser power regime.

Here, we estimate the fusion neutron yield for Coulomb explosion regime realized by using a high power laser. From Fig.2, the neutron yield is maximized when cluster ions are accelerated up to the energy of MeV. Since the obtainable ion energy is related to the Coulomb energy stored in the cluster, here we evaluate the average potential energy assuming that all electrons are expelled from the host cluster and the cluster consists of pure ions. However, it is natural to consider such that the expelled electrons from the cluster are redistributed in the system and plays a role to partially cancel the electrostatic field. As a simple case, here we assume uniformly distributed electrons in the system and this corresponds to the case such that the electron temperature is extremely high as $n_e = n_0 \exp(e\phi/T_e) \simeq n_0(n_0 = Zn_{av})$. Then, the charge density is given by $\rho = Ze(n_{cl} - n_{av})$ for $0 \le r < a$ and $\rho = -Zen_{av}$ for $a \le r < R$.

In the 3-dimensional case, the electrostatic field is given by

$$E_r(r) = \frac{4\pi Ze}{3}(n_{cl} - n_{av})r \qquad \text{for} \qquad 0 \le r < a, \tag{5}$$

$$E_r(r) = \frac{4\pi Ze}{3}\left(n_{cl}\frac{a^3}{r^3} - n_{av}\right)r \qquad \text{for} \qquad a \le r < R, \tag{6}$$

and the potential energy per ion that corresponds to the average ion energy is given by

$$w_{ion}^{(3D)} = \frac{3}{4\pi a^3 n_{cl}}\frac{1}{8\pi}\int_0^R E_r^2(r)4\pi r^2 dr$$

$$= \left(\frac{2\pi Z^2 e^2}{5}\right)n_{cl}a^2\left(2 - 3f_{3D}^{1/3} + 3f_{3D}\right) \tag{7}$$

Here, the inter-cluster distance R is regarded as a boundary beyond which the electrostatic potential is set to be zero. The result is shown in Fig.3 where $w_{ion}^{(3D)}$ is plotted as a function of the cluster radius for different value of packing fraction. For example, the cluster size of $a \simeq 100$nm is required to obtain the kinetic energy around $E_i \sim 1$MeV. Note that such a cluster size is larger by an order of magnitude than that achieved so far in experiments using gas jet technique[14,15]. This suggest that the neutron yield can be drastically increased by enlarging the cluster size and utilizing strong laser field to expel electrons from the cluster.

In order to perform a rough estimate of the neutron yield, we assume a plasma filament of the ion and electron temperature given by $T_i \sim T_e \sim 0.5$MeV and the average density of $n_{av} = 4.56 \times 10^{21}/$cm^3($f = n_{av}/n_D = 0.1$), which diameter and axial length are given by $d \sim 20\mu$m and $l \sim 40\mu$m (the interaction volume : $V = 1.26 \times 10^{-8}$cm^{-3}). The stored energy in such plasma filament is around $W_p \simeq 10$J and then the disassembly time is estimated as $\tau_d \sim V^{1/3}/C_s \sim 5\mathrm{psec}$. Assuming the cross section $\langle \sigma v \rangle \sim 10^{-16}$ at $T_i \sim 0.5\mathrm{MeV}$, the neutron yield is estimated as $N^{(cluster)} \sim 1.3 \times 10^8$ per laser shot, which corresponds to 1.3×10^7 neutrons per joule. When we introduce the solid deuterium collar as illustrated in Fig.1, the stopping range is about $l_m \sim 20\mu\mathrm{m}$ from Eq.(2) and the neutron yield is estimated as $N^{(cluster)} \sim 5.2 \times 10^8$, which is the same order as that obtained from the inside cluster fusion.

If we surround the interaction region including the collar by some material at the radius $r_m = 50\mu\mathrm{m}$ to which we like to expose neutrons as illustrated in Fig.1, the average neutron flux on the material surface is $N/\pi r_m^2 l \sim 5.2 \times 10^{12}/\mathrm{cm}^2$ per laser shot. To obtain high energy ions around 1MeV, we need to employ a large size cluster around $a \sim 100\mathrm{nm}$ and choose the packing fraction $f \simeq 10^{-2} - 10^{-1}$. The laser field will be determined to expell electrons from the cluster to trigger the Coulomb explosion.

As found in our numerical simulation, ions are efficiently accelerated in a low packing fraction case where the effect of inter-cluster interaction is rather small. In this case, the laser pulse can deeply propagate along the axial direction since the average density is low. On the other hand, when the packing fraction is high, although the interaction region as well as the ion accelelation are suppressed, electrons are efficiently heated to high temperature, so that such high energy electrons penetrate to surrounding clusters and secondarily heat and explode the clusters. This is an electron-triggered explosion of clusters, as opposed to the direct laser-triggered explosion. As a result, we can expect wider interaction region. The propagation of such a heated spot to surrounding clusters is a complicated problem where we need to take into account the collision effect and this remains a feature works. In the above example, the stored energy of the plasma filament is $W_p \sim 10\mathrm{J}$ and this corresponds to 500TW laser with $\tau_l \simeq 20\mathrm{fsec}$. In order to explode the large cluster as $a \simeq 100\mathrm{nm}$, the excursion length of heated electrons has to satisfy the relation $\xi_e = (c/\omega)\arcsin(a_0/\sqrt{1+a_0^2}) \geq a(\simeq 100\mathrm{nm})$. Once the cluster which size is around $a \simeq 100\mathrm{nm}$ is exploded in the Coulomb explosion regime, the average temperature becomes $T_e \sim T_i \sim 1\mathrm{MeV}$. Assuming that high power laser around $I_0 \simeq 10^{21}\mathrm{W/cm}^2(a_0 \simeq 20)$ is required to explode the cluster with $a = 100\mathrm{nm}$ by considering the relation $a > \delta_e$, the interaction area and the radius are estimated as $S_l(= \pi r_l^2) \simeq W_P/I_0\tau_l \simeq 5 \times 10^{-7}\mathrm{cm}^2$, and then $r_l \simeq 4\mu\mathrm{m}$. Then, as we discussed, we expect that this small spot expands to the above plasma filament of the volume V. Further numerical simulation is nessessary to precisely predict the plasma filament and the ion energy distribution.

III   Local simulation of laser-cluster interaction

In order to analyze the fundamental process of laser-cluster interaction as well as to explore an operating regime optimal for cluster fusion, we carry out 2 dimentional particle in cell (PIC) simulation where laser pulse propagates in the x-direction, while the polarization of electric filed is in the y-direction[18]. The packing fraction is then given by $f_{2D}(\equiv n_{av}/n_{cl}) = a^2/R^2$ with $R = (S/\pi N_{cl})^{1/2}$ where N_{cl} represents the number of cluster in the area $S(= L_x L_y)$.

In this section, we consider two types of simulation. One is the local simulation that is presented in this section. Here, we employ a periodic boundary condition in both x- and y-directions and we choose the system size and laser wavelength as $L_x = n\lambda_l(= 820\mathrm{nm})$, where n is an integer. The other one is the nonlocal simulation that is presented in the next section (Sec.IV). Here, we employs the open boundary condition in the x-direction where the laser pulse is irradiated from the vacuum region toward the cluster medium.

Note here that the laser-cluster interacion and resultant expansion process are qualitatively different between 2D and full 3D cases. The average potential energy per ion is given

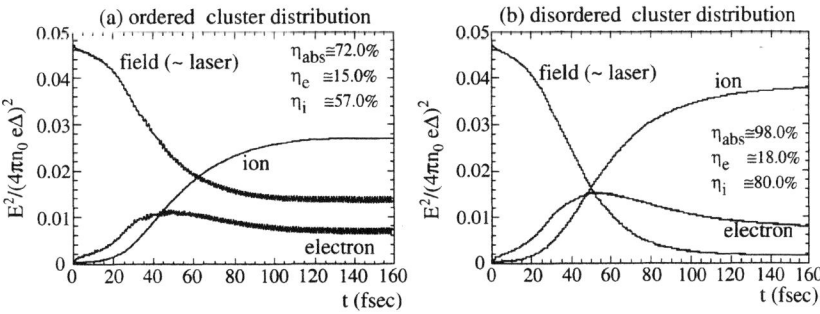

Figure 4: Time history of normalized laser field energy, electron and ion kinetic energies in the case of $a = 32$nm and $f \simeq 4.78 \times 10^{-3}$ for ordered cluster distribution (a) and disordered one (b). Absorption efficiency and conversion to electron and ion energies are estimated at the end ($t = 160$fsec).

in the 2D case as

$$w_i^{(2D)} = \left(\frac{\pi Z^2 e^2}{4} \right) n_{cl} a^2 \left(f_{2D} - \ln f_{2D} - 1 \right) . \tag{8}$$

Namely, in the 2D case, due to the rod structure of clusters that infinitely extends in the z-direction, the potential energy diverges with logarithmic dependence as $R \to \infty$ (or $f_{2D} \to 0$, i.e. infinite free space). However, as discussed in Ref.[9], although the potential energy and corrsponding ion kinetic energy in the 2D case overestimates compared with that in the 3D case (about 3-6 times), the ratio of potential energy between 2D and 3D cases, i.e. $w_i^{(3D)}/w_i^{(2D)}$, weakly depends on the packing fraction f over a wide range of interest. This suggests that the 2D simulation may be used to grasp fundamental features of the laser-cluster interaction.

At first, we present the typical result of numerical simulation for the periodic boundary condition in Sec.III A. In Sec.III B, we investigate the parametric dependence of the interaction, especially with respect to the laser field amplitude and try to optimize the neutron yield.

A   Fundamental feature of interaction

Here, we perform two types of simulation, i.e. one is an ordered cluster distribution with lattice structure and the other is a disordered one where the cluster locations are probabilistically determined. Note that since we assume the periodic boundary condition, even when we introduce the disordered distribution, a periodicity is assumed beyond the simulation box. In the present simulation, we assume the irradiation of continuous laser whose pulse length is sufficiently long compared with the typical time scale of inter-cluster interaction across the simulation boundary.

We employ a solid deuterium cluster with the radius of $a = 32$nm and the density $n_{cl} \simeq 4.56 \times 10^{22}$cm^{-3}, where the ratio with respect to the cut-off density for 820nm laser wavelength is $n_{cl}/n_c \simeq 27.5$ and the skin depth is $\delta_e = 25$nm. The laser amplitude is fixed for the moment at $a_0 = 0.25$ ($\xi_e \simeq 32$nm), and the packing fraction and average density are given as $f \simeq 4.78 \times 10^{-3}$ and $n_{av}/n_c = 0.13$ ($n_{av} = 2.19 \times 10^{20}$cm^{-3}) for both ordered and disordered distributions. The relative relation among the cluster radius, electron excursion length, and skin depth is given as $\delta_e < a \sim \xi_e$.

Figures 4 illustrates the time history of field energy E_l, electron and ion kinetic energies, E_e and E_i, in the case of (a) ordered cluster distribution and (b) disordered one. Here, we define the conversion ratio from the initial laser energy $E_l^{(0)} = E_l(t = 0)$ to electron and ion

269

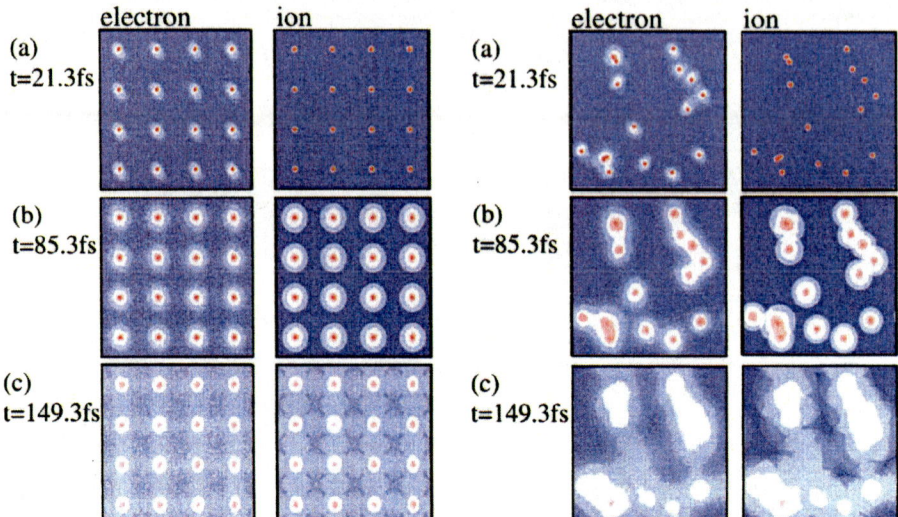

Figure 5: Electron and ion number density distribution in configuration sp ace for ordered distribution case that corresponds to Fig.4(a) at three different snap times.

Figure 6: Electron and ion number density distribution in configuration sp ace for disor- der ed distribution case that corresponds to Fig.4(b) at three different snap times.

kinetic energies, by $\eta_e \equiv E_i/E_l^{(0)}$ and $\eta_i \equiv E_e/E_l^{(0)}$, and the total absorption rate by $\eta_{abs} \equiv (E_i + E_e)/E_l^{(0)}$. In both cases, the laser energy is absorbed first by cluster electrons ($t \leq$ 20 fsec) in the early phase of interaction, and then the ion energy gradually increases, and at around $t = 50$fsec the electron energy gain saturates, while the ion kinetic energy keeps increasing and surpasses the electon energy gain. It is found that the disordered distribution reveals a strong coupling with the laser field, so that the total absorption and the conversion to the ion energy provide higher values at the end. Note that the maximum electron energy gain that appears at around $t = 50$fsec is larger in the disordered distribution case ($\eta_e^{max} \sim 32\%$) than that of the ordered one ($\eta_e^{max} \sim 23\%$). However, electrons are cooled in higher rate during the ion expansion and the electron energy gain remains almost the same for both cases at the end, suggesting that the extra energy absorbed from the laser field is mainly transferred to ions. The expansion process in this case belongs to a weak hydrodynamic and/or ambipolar expansion where the removal of electrons from the cluster core is not complete. Figures 5 and 6 shows the electron and ion number density distribusions in configuration space at three different times for the ordered and disordered distributions, which correspond to Fig.4 (a) and (b), respectively. As seen in Fig.5(a), some part of cluster electrons leaves their host cluster due to the laser heating. Resultantly, a large electrostatic field is formed near the vacuum-cluster interface and then the ion expansion is triggered [see Fig.5(b)]. In later time, although the electron core still exists, cluster electrons are distributed over a wide region and the expanded ion distribution starts to overlap with that of neighboring clusters [see Fig.5(c)].

The feature of the interaction in the disordered distribution case is somewhat different as seen in Fig.6. In this case, heated electrons from one particular cluster immediately interacts with adjacent clusters that are probabilistically located within a short distance. As a result, localized colony-like structure in which several and/or more clusters are contained are formed as seen in Fig.6(a). Such structure becomes more prominent with time as the ion expansion and resulting overlapping are developed as seen at Figs.6(b) and (c), and some heated electrons becomes meandering among clusters within each colony. The electron

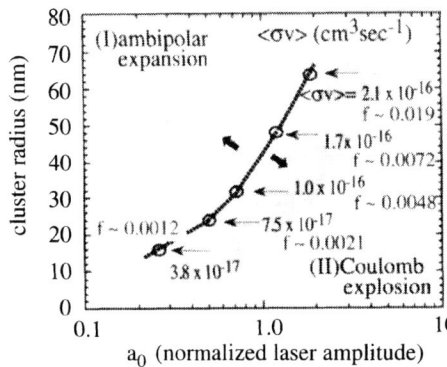

Figure 7: Critical line on the plane of cluster radius a and normalized laser field amplitude a_0 that distinguishes between the ambipolar expansion regime and Coulomb explosion one for a given simulation system size ($L_x = L_y = 820$nm). The packing fraction increases with the cluster radius according to $f \propto a^2$. The fusion cross-section at different five cluster sizes are described in the figure at the saturated stage in the Coulomb explosion regime.

orbits are considered to be more complex and stochastic, and resultantly the coupling between laser and clusters becomes strong so that the absorption efficiency increases as discussed in Fig.4(b).

B Parametric dependence of laser-cluster interation

Here we investigate the dependence of expansion characteristics on the cluster radius a and laser amplitude a_0. In Fig.7, we illustrate the critical line on the plane of (a, a_0), which distinguishes between the ambipolar expansion (I) and Coulomb explosion (II).

In the simulation, we fix the system size to $L_x = L_y = 820$ nm and only change the cluster size. Therefore, the packing fraction and average density increase in proportion to a^2. It is found from Fig.7 that the laser field amplitude above which the Coulomb explosion is triggered increases with the cluster radius. In order to optimize the fusion cross-section by employing the laser field amplitide as low as possible, it is desirable to choose the cluster size and the laser amplitide to the right-hand side (RHS) of the critical line as schematically indicated by a hatched region in Fig.7. The saturated value of the fusion cross section is described in Fig.7 for each cluster radius. It is found that the fusion cross section increases approximately in proportion to the cluster radius This weak dependence of the cross-section results from the fact that the ion kinetic energy increases with an increase of the cluster size and/or laser amplitude, but decreases with the packing fraction. Since the neutron yield is propotional to n_{av}^2 or f^2, even if the dependence of the cross section is weak ($\langle \sigma v \rangle_{DD} \propto a$), we expect the significant yield which is propotional to $\langle \sigma v \rangle_{DD} n_{av}^2 \propto a^5$.

IV Nonlocal simulation with open boundary condition

A Fundamental feature of interaction

Here, we show the results of nonlocal simulation. The configuration of the simulation is shown in Fig.8, where clusters with $a = 64$nm are uniformly distributed in the localized region with a packing fraction of $f = 0.11$. We assume that the p-polarized laser field is spatialy localized in the x-direction modeling an ultrashort pulse laser. Note that the laser field is uniform in the y-direction along which the boundary condition is periodic. Here, the laser amplitude is fixed at $a_0 = 5$ where the electron excursion length is approximately $\xi \simeq 180$nm. The average density in the cluster region is $n_{av} \simeq 5.02 \times 10^{21}$ /cm^3. Figure 8 shows electron and ion density distributions in configuration space at each snap time. It is found that the electrons are disassembled by the laser field irradiated from the left-hand side (LHS). Such disassembling propagates from the LHS to the right-hand side (RHS) together with the expansion into the vacuum. Accordingly, since a large amount of electrons is

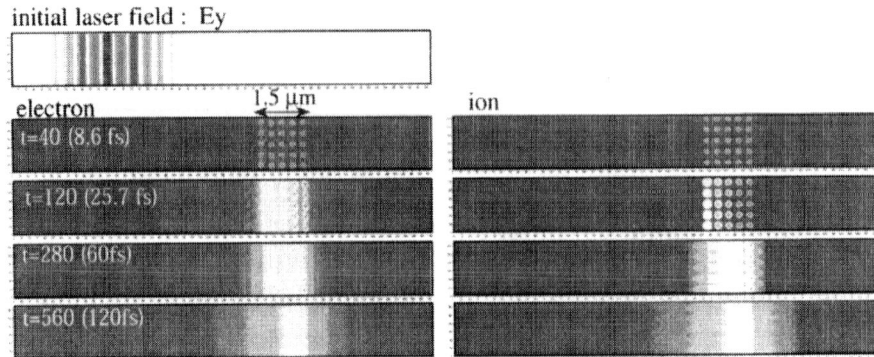

initial laser field : Ey

electron ion

t=40 (8.6 fs)

t=120 (25.7 fs)

t=280 (60fs)

t=560 (120fs)

Figure 8: Spatial configuration of the initial laser pulse that propagates in the x-direction, and electron and ion cluster density distribution after the interaction at four different snap times. Laser field amplitude $a_0 = 5$, cluster radius $a = 64$nm, the packing fraction $f = 0.11$ are chosen.

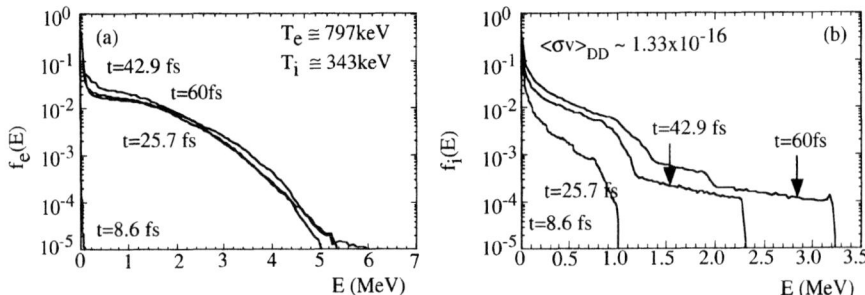

Figure 9: (a)Electron and (b) ion energy distribution that corresponds to Fig.8 for different snap times. The average electron and ion temperatures, and fusion cross-section at the end ($t = 60$fsec) are indicated in the figure.

expelled from the cluster due to its large excursion length, ions start to explode as seen in Fig.8 (t=25.7fsec). It is also found that the ion explosion also propagates from the LHS to the RHS, suggesting that high energy electrons produced by the laser-clusters interaction at the LHS boundary heat and explode the surrounding clusters where the laser field does not penetrate. Finally, ions also expand into the vacuum from the RHS boundary as seen in Fig.8.

Figure 9 illustrates the electron and ion energy distribution at each snap time. It is found that a significant amount of electrons are heated up to a few MeV and accordingly ions are accelerated beyond 1 MeV as expected in the discussion in Sec.II. In order to understand the effect of laser-cluster interaction more clearly, we perform the simulation that employs the plane target with corrugated surface as shown in Fig.10. The time history of the field energy and electron/ion kinetic energy are shown in Fig.11(a) together with the result that corresponds to Fig.8, i.e. Fig.11(b). The absorption efficiency is found to be high, giving rise to the same level as that of the cluster medium. However, it is clearly seen that the partition of the kinetic energy between electrons and ions are drastically different and the conversion to ion energy is small. It is further worthwhile to investigate the energy distribution functions as shown in Fig.12. Althogh the laser energy is absorbed in the same level for both cases, the amount of high energy electron is suppressed, and the lower energy component is dominantly heated as seen in Fig.12(a). Thus, the laser-cluster interaction

electron   ion

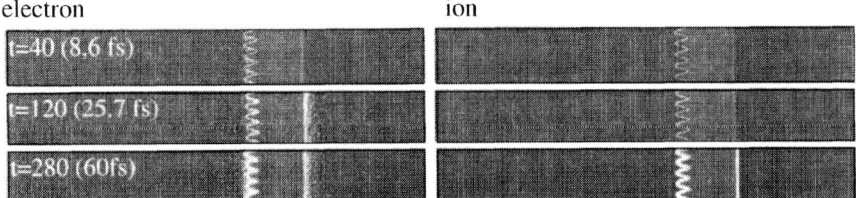

Figure 10: Electron and ion ion density distribution of plane target where the surface of the left-hand side is corrugated with the wavelength of 64nm and the depth of 64nm. The laser field is irradiated in the same manner as Fig.8.

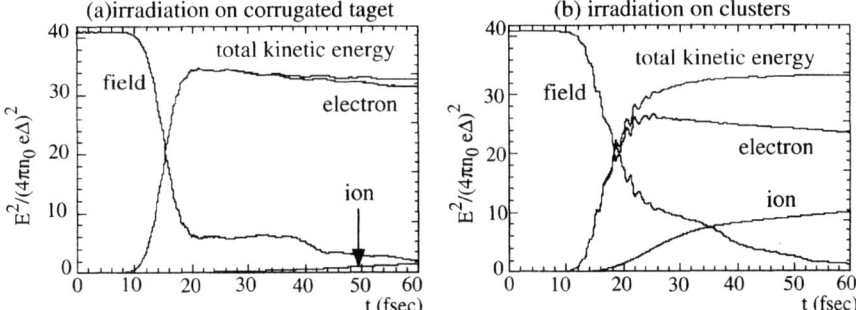

Figure 11: Time history of laser field energy, and electron and ion energies for laser irradiation (a)on the corrugated plane target in Fig.10 and (b)on the cluster medium in Fig.8. Note that the reflected laser field on the left-hand side of the medium is going out from the system.

is found to be superior for producing high energy electrons and ions. The high energy ion component is also found to be suppressed as seen Fig.12(b).

B   Parametric dependence of the interaction

Figure 13 illustrates the obtained electron and ion kinetic energies as a function of the laser field amplitude for the cluster size of $a = 64$nm and the packing fraction of $f = 0.11$. At first, as seen in Fig.13, the average electron kinetic energy increases in propotion to $E_e \propto a_0^2$ up to $a_0 \simeq 10$, but gradually tends to saturate. The average and maximum ion kinetic energies also increase with a slightly weaker dependence as $E_i \propto a_0^{1.6}$. As found from Fig.2, the average maximum ion energy obtained from the Coulomb explosion for the cluster of $a \simeq 64$nm is smaller than $E_i = 1$MeV (i.e. around 0.6MeV). However, it is found that the average ion energy exceeds this level without observing a clear saturation This suggests that the laser energy is converted into ions exceeding the Coulomb energy stored in the cluster that is evaluated from Fig.2. This is considered to arise from the acceleration of the ion expansion front formed at the interface between vacuum and cluster medium at both the LHS and RHS. This is clearly seen in the ion energy distribution of Fig.9, where the high energy tail component of $E_i \geq 1$MeV is established besides the medium energy of 0.2MeV$< E_i \leq 1$MeV resulting from the Coulomb explosion. This high energy component may have some additional contribution for neutron generation when the external collar is introduced as illustrated in Fig.1, because the stopping length becomes longer for high energy ions according to Eq.(3). The fusion cross-section $\langle \sigma v \rangle$ is also indicated in the figure in the unit of 10^{-16}. The maximum value, $\langle \sigma v \rangle \simeq 1.8 \times 10^{-16}$, is obtained around $a_0 \simeq 10$ that is consistent with the discussion in Sec.II. Further, the cross-section is found to be

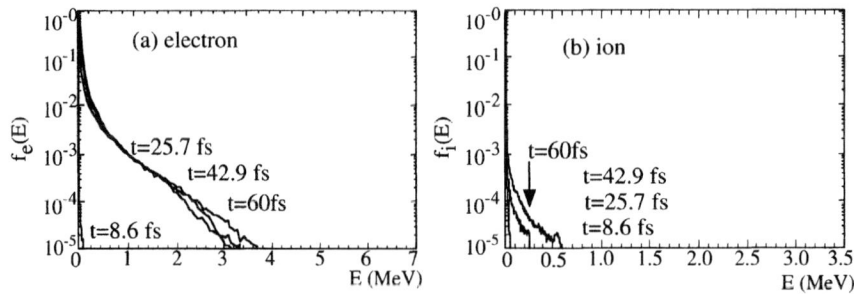

Figure 12: (a)Electron and (b) ion energy distribution that corresponds to the int eraction with corrugated plane target of Fig.10.

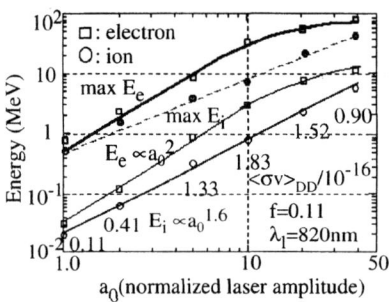

Figure 13: Electron and ion kinetic energies both average energy E_i and maximum one $E_i^{(max)}$ as a function of normalized laser field amplitude a_0. The configuration is same as Fig.8 where the open boundary condition is used. The cluster size of 64nm and the packing fraction of $f = 0.11$ are employed. The fusion cross-section $\langle \sigma v \rangle$ is indicated corresponding to each laser field amplitude in the unite of 10^{-16}.

decreased at around $a_0 \simeq 20$, since the average deuterium energy exceeds the one that provides a maximum cross-section, i.e. $E_i \simeq 1\text{MeV}$.

V CONCLUDING REMARKS

We have investigated the laser-cluster interaction to clarify key physical issues essential to understand and optimize the cluster fusion. The laser-cluster interaction is found to be essentially different from the conventional preparation of laser-plasma and/or laser-solid interaction. One of the crucial advantage to use the cluster material is that there exists the freedum to control the interaction, or more specifically to control electron and ion energy distributions. The ion energy distribution is controlled by the cluster size if we employ the high power laser field so that the expansion dynamics is dominated by the Coulomb explosion. By properly choosing the cluster size, packing fraction and laser field amplitude, we can efficiently accelerate cluster ions to high energy where the fusion cross-section is maximized.

In order to increase the neutron yield, we need to increase the volume where the fusion reaction takes place, and also confinement and/or disassemly time. The volume is not restricted to the region where the laser directry interacts with clusters, but is extended to the surrounding region where the high energy electrons and also ions propagate. This is schematically illustrated in Fig.1. Namely, once the high temperature tiny spot is formed due to the laser-cluster interaction, such region expands to the surrounding one so that the secondary Coulomb explosion is triggered as a chain reaction. During such expansion to the surrounding region, the average ion energy is decreased and eventually passes through the energy that provides the maximum fusion cross-section. Thus, the effective volume is expected to be increased so that the total neutron yield is also increased. In this sence, the initial small hot spot where the laser directly interacts with cluster is considered to be an "ignitor" that triggers the propagation of the expansion front of the Coulomb explosion.

From the simple (and also rough) estimate, we evaluate about 10^7 neutrons per joule. It is expected that this value may be further enhanced by properly adjusting the paramerers and by incorporating with the surrounding clusters and collar. For example, a laser with 10J energy with repetition rate 100Hz irradeating DD clusters surrounded by a collar produces the intensity of neutron $10^{14}/cm^2/sec$ at a distance of $50\mu m$. If we adopt DT clusters, the intensity of nertrons is expected to be $\sim 10^{15}/cm^2/sec$. It is worthwhile to try to increase the confinement and/or disassembly time, for example, by introducing strong confinement magnetic field. These, however, remain as future works.

ACKNOWLEDGEMENTS

One of authors, Y. Kishimoto, acknowledges Dr. A. Kitsunezaki and Dr. H. Ninomiya of Naka Fusion Research Establishment, and Dr. Y. Kato and Dr. T. Arisawa of Advanced Photon Research Center, of JAERI. One of the authors, T. Tajima, was supported in part by the Grant USDoE DE-FG03-97ER54439.

References

1. N. G. Gotts, D. G. Lethbridge, and A. J. Stase. J. Chem. Phys. **96**, 408 (1962).

2. J. Purnell, E. M. Snyder, S. Wei, and A. W. Caslemand, Chem. Phys. Lett. **229**, 33e (1994).

3. T. Ditmire, T. Donnelley, R. W. Falcone, and M. D. Perry, Phys. Rev. Lett. **75**, 3122 (1995).

4. S. Dobosz, M. Lezius, M. Schmidt, P. Meynadier, M. Perdrix, and D. Normand, Phys. Rev. A **56**, R2526 (1997).

5. C. Deutsch, Plasma Phys. Control. Fusion **41**, A195 (1999).

6. Y. L. Shao et. al. Phys. Rev. Lett. **77**, 3343 (1996).

7. T. Ditmire, J.W.G. Tisch, E. Springate, M.B. Mason, N. Hay, R.A. Smith, J. Marangos, and M.H.R. Hutchinson, Nature **386**, 54 (1996).

8. Y. Kishimoto and T. Tajima: in *High Field Science*, eds. T. Tajima, K. Mima, and H. Baldis (Kluwer, NY, 2000) p.83.

9. Y. Kishimoto, T. Masaki and T. Tajima, Phys. Plasmas, Vol.9, No.1(2002).

10. T. Ditmire, T. Donnelley, A. M. Rubenchuk, R. W. Falcone, and M. D. Perry, Phys. Rev. A**53**, 3379 (1996).

11. T. Tajima, Y. Kishimoto, and M.C. Downer, Phys. Plasma **6**, 3759 (1999).

12. R. H. Doremus, J. Chem. Phys. **40**, 2389 (1964).

13. A. Kawabata and R. Kubo, J. Phys. **40**, 1765 (1966).

14. T. Ditmire, J. Zweibeck, V.D. Yanovsky, T. E. Cowan, G. Hays, and K. B. Wharton, Nature **398**, 489 (1999).

15. J. Zweibeck, R.A. Smith, T. E. Cowan, G. Hays, K. B. Wharton, V.P. Yanovsky, and T. Ditmire, Phys. Rev. Lett. **84**, 2634 (2000).

16. T. Tajima, Y. Kishimoto, and T. Masaki, Physica Scripta. **T89**, 45 (2001).

17. Y. Kishimoto, K. Mima, T. Watanabe, and K. Nishikawa, Phys. Fluids **26**, 2308 (1983).

18. Y. Kishimoto (private communication): 2-dimensional fully relativistic electromagnetic particle-in-cell code, EM2D-EB.

Fullerenes and Molecules in Strong Laser Pulses: Collective Dynamics, Field Ionization and Harmonic Generation

D. Bauer[*], F. Ceccherini[*], A. Macchi[†] and F. Cornolti[†]

[*]Theoretical Quantum Electronics, Darmstadt University of Technology, Germany
[†]Dipartimento di Fisica and INFM (sezione A), Universitá di Pisa, Italy

Abstract. Computational studies of many-electron systems in strong laser fields are presented. The ionization and nonlinear dipole response of the ball-shaped C_{60} fullerene molecule are investigated with a time-dependent density functional approach and a jellium approximation for the ionic background. We find that C_{60} ionization at 800 nm wavelength occurs multiphoton-like rather than via excitation of a "giant" resonance. Harmonic generation from the interaction of a circularly polarized laser field and a molecule with a discrete rotational symmetry is also studied, and spectra and selection rules are interpreted via a general group theory approach.

Many-electron systems such as big molecules in strong laser fields have great physical interest as mesoscopic objects where the transition from a quantum-like, single particle dynamics to a collective, semiclassical behavior can be studied. They are a challenging subject both from the theoretical point of view, due to the need of non-perturbative approaches, and from the numerical side, because the solution of the complete time-dependent Schroedinger equation for more than two electrons is far above present-day computing power. In the following we present selected results on the strong-field dynamics of the soccer-ball-shaped C_{60} fullerene, using a time-dependent density functional theory (TDDFT) approach, and on *ab initio* studies of harmonic generation in systems having a dynamical symmetry (e.g., cyclic, ring-shaped molecules).

C_{60} IN STRONG LASER PULSES

The time-dependent Kohn-Sham (TDKS) equation for the orbital $\Psi_{i\sigma}(\mathbf{r},t)$ reads (atomic units are used unless noted otherwise)

$$i\frac{\partial}{\partial t}\Psi_{i\sigma}(\mathbf{r},t) = \left(-\frac{1}{2}\nabla^2 + V(r) + V_I(t) + V_{ee\sigma}[n_\uparrow(\mathbf{r},t),n_\downarrow(\mathbf{r},t)]\right)\Psi_{i\sigma}(\mathbf{r},t). \quad (1)$$

Here, $\sigma = \uparrow,\downarrow$ indicates the spin polarization, $V(r)$ is the potential of the ions, $V_I(t)$ is the laser in dipole approximation, $V_{ee\sigma}[n_\uparrow,n_\downarrow]$ is the effective electron-electron interaction potential which is a functional of the electron spin densities $n_\sigma(\mathbf{r},t) = \sum_{i=1}^{N_\sigma}|\Psi_{i\sigma}(\mathbf{r},t)|^2$, N_σ is the number of orbitals occupied by KS particles with spin σ. The total electron density is $n(\mathbf{r},t) = \sum_\sigma n_\sigma(\mathbf{r},t)$. The electron-electron potential is splitted, $V_{ee\sigma}[n_\uparrow,n_\downarrow] =$

CP611, *Superstrong Fields in Plasmas:* Second Int'l. Conf., edited by M. Lontano et al.

$U[n] + V_{xc\sigma}[n_\uparrow, n_\downarrow]$, where $U[n]$ is the Hartree potential $U[n] = \int d^3r\, n(\mathbf{r}',t)/|\mathbf{r} - \mathbf{r}'|$ and $V_{xc\sigma}[n_\uparrow, n_\downarrow]$ is the exchange correlation (xc)-part. Although the Runge-Gross theorem [1] ensures that, in principle, the time-dependent KS scheme could yield the exact density $n(\mathbf{r}, t)$ on which all observables depend, in practice an approximation to the exchange-correlation potential $V_{xc\sigma}[n_\uparrow, n_\downarrow]$ has to be made. We chose the Slater expression $V_{xc\sigma}^{Slater}(\mathbf{r}, t) = \sum_{i=1}^{N_\sigma} (n_{i\sigma}(\mathbf{r}, t)/n_\sigma(\mathbf{r}, t)) u_{xci\sigma}(\mathbf{r}, t)$, where $u_{xci\sigma} = V_{xc\ \sigma}^{XLDA}[n_\uparrow, n_\downarrow] - U[n_{i\sigma}] - V_{xc\ \sigma}^{XLDA}[n_{i\sigma}, 0]$, i.e., the self-interaction is removed, and the exchange-only local density approximation (XLDA) was employed. In TDDFT simulations of non-sequential ionization in laser atom interaction [2] we found that, compared to plain XLDA, the Slater potential improved the ionization potentials and the fulfillment of Koopman's theorem significantly.

The C_{60} is treated in a jellium approximation with the ionic background as an uniform charge shell. [3]. Parameters of the ion potential are "tuned" in order to match the results of first-principle calculations with the "real" soccer-ball structure of C_{60} [4]. Ground state features are shown in Fig.1 and details are discussed in [3].

We discuss the nonlinear response of our C_{60} model to a kick at time $t = 0$ by a delta-like electric field $E(t) = \hat{A}\delta(t)$. Such a kick is equivalent to the method proposed by Yabana and Bertsch in Ref. [5]. From the Fourier-transform of the dipole $d(t) = \int zn(\mathbf{r})\, d^3r$ the frequency dependent response is obtained. In Fig. 2 the Kohn-Sham orbital resolved dipole (KSORD) response for a delta-kick of magnitude $\hat{A} = 0.01$ a.u. is presented. On the left a contour plot of the dipole vs. orbital number and time is shown. On the right-hand-side the corresponding spectrum is plotted, obtained by Fourier-transforming the dipoles of the individual KS orbitals. The total dipole and its Fourier transform are also included in Fig. 2. Looking at the dipoles vs. time one easily identifies

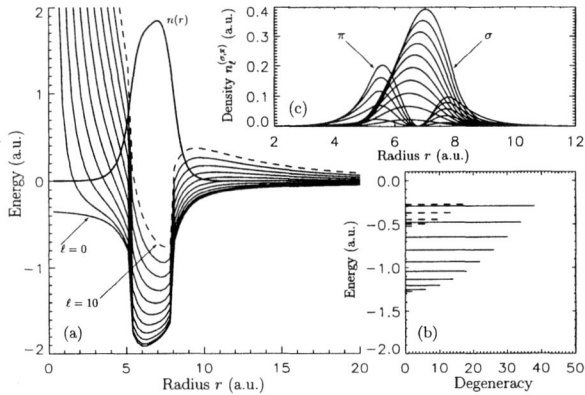

FIGURE 1. Ground state properties of the jellium C_{60}-model. (a) The total potential $V(r) + U[n(r)] + V_{xc}[n(r)]$ depends on the angular quantum number ℓ (through the centrifugal barrier). For $n = 1$ (σ-electrons) orbitals from $\ell = 0$ up to $\ell = 9$ are occupied in the ground state situation, for $n = 2$ (π-electrons) orbitals from $\ell = 0$ to $\ell = 4$ are occupied. The potential for the empty $\ell = 10$ orbitals is drawn dashed. The radial shape of the total ground state density $n(r)$ is also plotted. (b) The single Kohn-Sham particle energy levels corresponding to the potentials in (a). σ-states are drawn solid while π-states are plotted dashed. The degeneracy is $2(2\ell + 1)$. (c) The orbital densities $n_\ell^{(\sigma,\pi)}$. The sum of those is $n(r)/4\pi$.

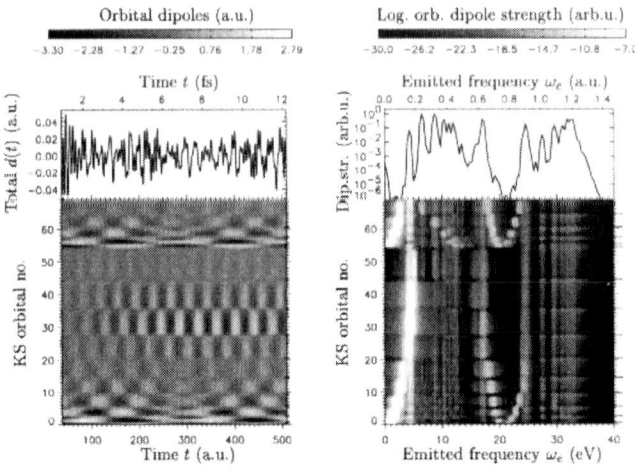

FIGURE 2. Kohn-Sham (KS) orbital resolved dipoles (KSORD, left) and the corresponding spectrum (right) after a delta-kick with an electric field $E(t) = \hat{A}\delta(t)$, $\hat{A} = 0.01$. The total dipole $d(t)$ and the total dipole strength (Dip.str) are given also. See [3] for a detailed discussion.

the different ℓ-shells. The σ-electrons are KS orbital numbers 0–54, the π-electrons range from 55–69. Examining the KSORD spectra on the right one clearly identifies rather narrow vertical lines. Each of those vertical lines can be understood as a single particle transition between ground state KS levels. For a certain KS orbital the dipole strength is particularly high for those transitions where one of the two levels involved is the one which is occupied in the ground state configuration of this KS orbital. This explains the parabola-like structures of strong dipole emission visible in the KSORD spectra. A detailed discussion of the KSORD spectra is reported in [3].

We calculated the interaction of our jellium model with ten cycle $\sin^2$-shaped (with respect to the electric field) laser pulses in the frequency range from 6.8 up to 47.6 eV at The intensity times pulse duration (energy per unit area) was held constant, i.e., $\hat{A}^2\omega_i = 1.8375 \times 10^{-5}$. After the laser pulse was over we continued the propagation of all KS orbitals to allow for delayed ionization and free oscillations. The result is presented in Fig. 3. The contour plot shows the logarithmically scaled dipole strength vs. incident frequency ω_i and emitted frequency ω_e. Left to the contour plot the number of removed electrons $N_{rem} = 250 - \int_{grid} d^3r\,n(\mathbf{r}, t_{end})$ is plotted as a function of the incident frequency ω_i. In all cases delayed ionization was negligible compared to the electron density which was freed owing to the laser pulse.

The maximum dipole strength is along the diagonal $\omega_i = \omega_e$ where the excitation was resonant. However, at $\omega_i = \omega_e = 20.5$ eV there is a relatively weak dipole response but the ionization yield has a local maximum (dashed line). Instead, the local minima in ionization at 17 and 24 eV coincide with emission in two strong lines near the same values for ω_e. For $\omega_i = 30$ eV one observes both, strong ionization *and* a strong dipole response. For even higher ω_i ionization drops and the jellium system does not provide modes > 40 eV. The (somewhat surprising) mutually exclusive behavior of ionization and dipole emission at selected frequencies is discussed in [3].

We simulated the interaction of our C_{60} jellium model with 800 nm, 26 fs pulse duration, $\sin^2$-shaped laser pulses. Fig. 4 shows the removed electron density N_{rem} (e.g., the single ionization probability if $N_{rem} \ll 1$) after the pulse vs. the peak intensity. In the case of perturbative off-resonant multiphoton ionization of atoms the ionization probability is $\sim I^n$ where n is the number of photons needed to reach the continuum from the initial state. In our jellium model the first ionization potential is $I_p^+ = 0.279$, and thus we expect $n = 5$ photons necessary for ionization if C_{60} behaves multiphoton-like. Fig. 4 shows that this is the case, in agreement with recent experiments [6] but in contrast with earlier experimental results in [7] where excitation of the 20 eV Mie-resonance, corresponding to $n = 13$ photons, was concluded to be the dominant ionization mechanism. The $n = 13$ slope is depicted in Fig. 4 in the upper left corner and is much too steep to fit with our numerical result. This appears reasonable because the incident laser frequency 1.6 eV lies energetically far below 20 eV so that in short pulses an efficient excitation of the latter is unlikely. The experimental fragmentation onset and the C_{60} saturation intensity as observed in [6] is also indicated in Fig. 4.

DYNAMICAL SYMMETRIES AND HARMONIC GENERATION

High harmonic generation has interesting properties in the case of systems which Hamiltonian is invariant under so-called dynamical symmetry operations, such as cyclic molecules in a circularly polarized field, or single atoms in a two-color field, with the two pulses having opposite circular polarizations. A general Hamiltonian including both

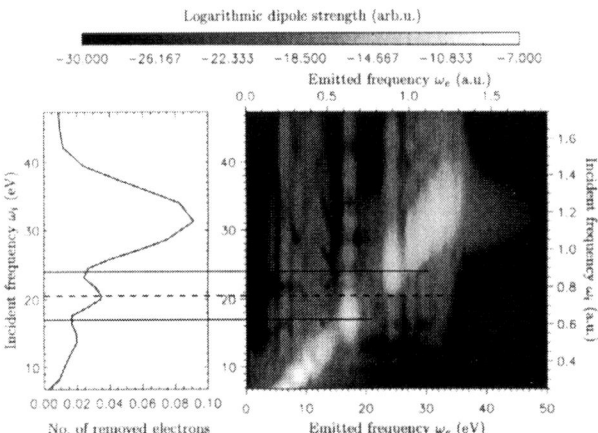

FIGURE 3. Right-hand-side: contour plot of the Fourier-transformed dipole $d(t)$ after ten cycle pulses of frequency ω_i. The dipole strength vs. incident frequency ω_i and emitted frequency ω_e is logarithmically scaled (cf. color bar ranging from 10^{-30} to 10^{-7}, in arbitrary units). Left-hand-side: number of removed electrons. The horizontal solid (dashed) lines (line) indicate frequencies ω_i where ionization was relatively low (high) and excitation was efficient (inefficient).

279

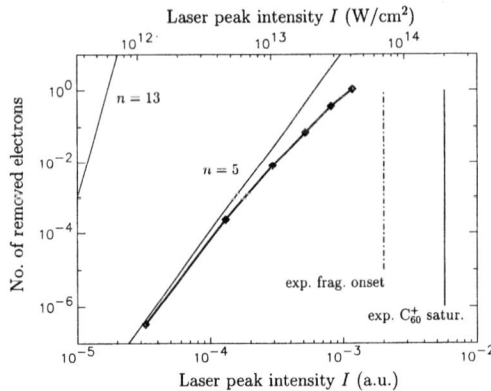

FIGURE 4. Removal of the first electron vs. the peak intensity of a ten cycle 800 nm $\sin^2$-pulse (solid curve). A multiphoton-like behavior $\sim I^n$ with $n = 5$ is evident. Instead, $n = 13$ photons would be necessary to excite a 20 eV Mie-resonance. Fragmentation threshold and C_{60}^+ saturation intensity from [6] are also indicated (vertical lines).

cases may be written in cylindrical coordinates as

$$H = T + V(\rho, \varphi, z) + \frac{\varepsilon_1}{\sqrt{2}}\rho\cos(\varphi - \omega t) + \frac{\varepsilon_2}{\sqrt{2}}\rho\cos(\varphi - \eta\omega t). \tag{2}$$

where $T = -(1/2\rho)\partial_\rho(\rho\partial_\rho) - (1/2\rho^2)\partial_\varphi^2$ is the kinetic energy operator. In the case of a cyclic molecule with N identical atoms, $V = V(\rho, \varphi, z)$ is the molecular potential having a symmetry described by the group $\mathcal{D}_{Nh}$ and the second laser field is absent ($\varepsilon_2 = 0$). In the case of an atom in a two-color laser field, V is the atomic potential. In any case, H is invariant under the transformation

$$\hat{P}_L = \left(\varphi \to \varphi + \frac{2\pi}{L}, t \to t + \frac{2\pi}{L\omega}\right), \tag{3}$$

where $L = N$ for the cyclic molecule and $L = \eta + 1$ for the two-color case. This invariance leads to the selection rule for the harmonic order $n = kN \pm 1$, with $k \in \mathcal{N}^+$. This leads to a "filter" effect which increases both the frequency and intensity of the first allowed harmonic [8].

We used a single active electron model in either one [9] and two spatial dimensions [10] to perform an *ab initio* study of harmonic generation in ring-shaped molecules (representing a model for benzene and other aromatic compounds or even nanotubes) for finite duration laser pulses, beyond the single-state, infinite pulse duration approximation of [8]. We obtain that, together with the harmonics expected from the selection rules, other lines are present [10]. Depending on the laser frequency the intensity of the additional lines becomes comparable to those of the harmonics, due to resonances or recombination processes with states of a symmetry different from the initial one, making the single-state approach inadequate. Nevertheless, also the "enriched" spectrum can be fully understood within a general group theory approach, providing information on the

280

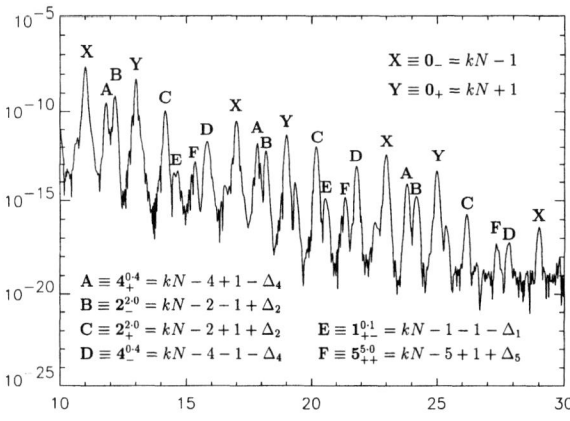

FIGURE 5. Dipole spectrum for a benzene model molecule. The laser frequency was 0.0942. Together with the lines expected by the single state approach (**X** and **Y**) other lines due to recombinations with states of different symmetries are present. The second excited state plays an important role.

field-dressed level scheme [11]. As an example the dipole spectrum of a benzene model molecule is shown in Fig.5. More details are reported in [10, 11]

ACKNOWLEDGMENTS

This work was partly supported by INFM through the supercomputing initiative (FU-MOFIL project) and by the Deutsche Forschungsgemeinschaft within the SPP "Wechselwirkung intensiver Laserfelder mit Materie."

REFERENCES

1. Runge, E., and Gross, E. K. U., *Phys. Rev. Lett.*, **52**, 997 (1984).
2. Bauer, D., and Ceccherini, F., *Optics Express*, **8**, 377 (2001).
3. Bauer, D., Ceccherini, F., Macchi, A., and Cornolti, F., *Phys. Rev. A*, **64**, 054112 (2001).
4. Puska, M. J., and Nieminen, R. M., *Phys. Rev. A*, **47**, 1181 (1993).
5. Yabana, K., and Bertsch, G. F., *Phys. Rev. B*, **54**, 4484 (1996).
6. Tchaplyguine, M., et al., *J. Chem. Phys.*, **112**, 2781 (2000).
7. Hunsche, S., et al., *Phys. Rev. Lett.*, **77**, 1966 (1996).
8. Alon, O., Averbukh, V., and Moiseyev, N., *Phys. Rev. Lett.*, **80**, 3743 (1998).
9. Bauer, D., and Ceccherini, F., *Las. Part. Beams*, **19**, 85 (2001).
10. Ceccherini, F., and Bauer, D., *Phys. Rev. A*, **64**, 033423 (2001).
11. Ceccherini, F., Bauer, D., and Cornolti, F., *J. Phys. B: At. Mol. Opt. Phys.* (2001), in press [preprint physics/0107024].

Cluster Dynamics at Different Cluster Size and Incident Laser Wavelengths

Tara Desai and Andrea Bernardinello

Dipartimento di Fisica, Università degli Studi di Milano-BicoccaVia Emanueli, 15-20216, Milano, ITALY, e-mail:tara.desai.@mi.infn.it

Abstract: X-ray emission spectra from aluminum clusters of diameter ~0.4 μm and gold clusters of dia. ~1.25 μm are experimentally studied by irradiating the cluster foil targets with 1.06 μm laser, 10 ns (FWHM) at an intensity ~10^{12} W/cm². Aluminum clusters show a different spectra compared to bulk material whereas gold cluster evolve towards bulk gold. Experimental data are analyzed on the basis of cluster dimension, laser wavelength and pulse duration. PIC simulations are performed to study the behavior of clusters at higher intensity $I \geq 10^{17}$ W/cm² for different size of the clusters irradiated at different laser wavelengths. Results indicate the dependence of cluster dynamics on cluster size and incident laser wavelength.

INTRODUCTION

Clusters have opened a new branch for the study of matter. Current advanced technologies require small and high speed micro electronic devices for improved processing and clusters may be advantageous when a cluster starts showing the properties of the bulk material. Clusters are an aggregate of large number of identical atoms/molecules and occupy an intermediate state between atoms/molecules and condense phase [1]. They can be produced by various techniques like, as a tiny fragment from solid material, gas jets, laser ablation etc. In the first case, cluster size can be large up to few microns while in the later cases cluster size is small of the order of few tens of Angstroms. Although clusters are classified as an intermediate state of matter, many of their properties differ from quantum objects (atoms/mol) and the bulk material. Some of these properties are discussed in ref [2]. One of the property we have studied is the X-ray emission spectra from clusters and compare with the respective bulk material. The comparison of the spectra provides the information about the process of X-ray emission in clusters. Response of an individual atom to the laser radiation depends on the laser field besides the electron configuration of the target atom; while that of the bulk material is the collective response of all the atoms as a whole. Therefore, the study of clusters provides a

CP611, *Superstrong Fields in Plasmas:* Second Int'l. Conf., edited by M. Lontano et al.

valuable information about how the properties of matter changes as one progresses from single atom to that of solid material [3].

In the recent years there has been a significant interest to understand the physics of ultra short (~fs) and intense laser (I >10^{15} W/cm^2) interaction with clusters generated by a gas jet. Here cluster size is few tens of A°. However, the purpose of all these research have been directed in the generation of higher order harmonic conversion [4,5], energetic ion generation [6,7], anomalous X-ray generation [5,8,9], as a diagnostic tool for high density, high temperature plasma etc. However, the study of large size clusters having diameter (~µm) which are not presently produced in the gas jets, offer an unexplored field of cluster physics.

EXPERIMENTS

Experiments were performed using a laser radiation of wavelength λ_L = 1.06 c with optical energy E_L<1 J in 10 ns (FWHM) duration. Intensity on the target surface was ~10^{12} W/cm^2. Two types of targets were used viz. 1.Slab targets of aluminum and gold. 2. Cluster targets; clusters of aluminum (~0.4µm) and gold cluster (~1.25µm) were embedded in polymer. Fabrication of cluster targets is discussed in [10] and metallic clusters are commercially available. Here we consider two cases where cluster diameter is less than the laser wavelength ($C_d < \lambda_L$ like aluminum cluster) and secondly, of the order or larger than laser wavelength ($C_d \geq \lambda_L$ as in the case of gold cluster). X-ray emission spectra in 5-22 nm range were recorded at 45° to target normal using grazing incidence spectrometer and compared with respective slabs under identical conditions.

Since, we have used 50 mg of aluminum clusters by weight per cc of polymer, the no. of clusters ~5.5x10^{11} /cc and number of atoms per aluminum clusters ~ 8x10^9. Inter cluster spacing (center to center) A= $(3/4\pi n)^{1/3} \approx 0.755$ µm. Similarly density of gold cluster of dia~1.25µm is ~2.6x10^9/cc, and number of atoms in each of the gold clusters were ~ 10^{11} with interspacing of gold clusters at ~ 4.0 µm. Since A > dia of the clusters in both the cases, we assume clusters behaves independently dring the process of X-ray emission. With the onset of laser beam, each clusters within the focal spot area will experience uniform radiation field.

RESULTS

Aluminum clusters of average diameter ~0.4 µm embedded in polymer solution show negligible X-ray intensity in 5-22nm spectral range (fig.1 a) compared to pure aluminum slab target (fig.1b). It is clear that X-ray emission spectra for aluminum cluster target do not match in shape and intensity with that of the aluminum slab target. Shape of the spectra of aluminum cluster is not identical to bulk material indicates, the plasma ionization level in cluster is different than that of the bulk material under identical experimental conditions. Contrary to this gold cluster target (fig.2a) with average cluster diameter ~1.25 µm, emit continuum which is nearly similar in shape to that of bulk gold target (fig.2b) but has lower intensity. This means the process of X-ray emission from gold cluster of diameter~1.25 μm >λ_L, is evolving towards the bulk material.

DISCUSSION

We assume the laser interaction with a cluster is similar to that of the laser interaction with a small ball of solid material and the mechanism of plasma production, heating etc are similar bulk material [11]. The hydrodynamic expansion of the plasma from cluster surface results in the decrease of plasma density. For planar targets [12], plasma density falls to 1/e times the critical density at a distance L~2R where R is laser focal spot radius but in the case of clusters, this distance depends on 2 factors viz1).Plasma density decreases exponentially from the cluster surface due to hydrodynamic expansion of the surface plasma, similar to slab target with a velocity V~$(KT_e)^{1/2}$ where KT_e is temperature of the plasma on cluster surface. 2). Simultaneously, cluster as a whole expands in all directions due to its internal pressure $P_i + P_e = (nkT)_i + (nkT)_e$ and increases in size. This leads to cluster disassembly. Time required for a cluster to disassemble is known as disassembly time $t_d = C_d / 2C_S$ where C_d is cluster dia. and $C_S = (ZKT_e / m_i)^{1/2}$. Maximum temperature T_e depends on the cluster dimension, it's mass density, laser intensity, pulse duration etc., and effective laser energy deposition on the cluster surface on a time scale comparable to the cluster disassembly time.

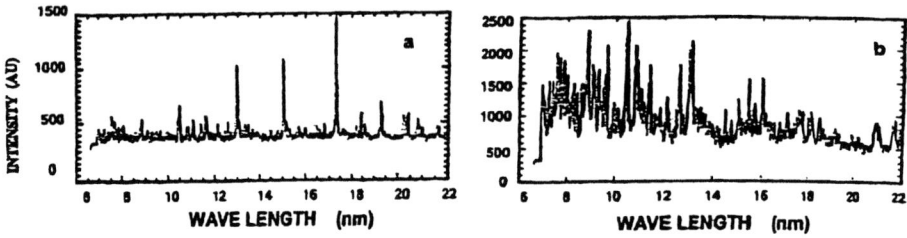

FIGURE 1. X-ray emission spectra from a) Aluminum cluster targe and b) Aluminum slab target.

These experimental results raise a question; why a cluster, a fragment from a bulk material, having a dimension smaller than the incident laser wavelength behave differently than the bulk in the present investigation reported here? While with increased cluster diameter, more than the laser wavelength, cluster behavior evolves towards the bulk material. Therefore, we have analyzed our results on the basis of cluster dimension, laser wavelength and pulse duration. We believe there can be several reasons to explain these results.

$C_{da} < \lambda_L$, Aluminum Cluster

In our present case aluminum cluster dia, $C_{da} < \lambda_L/2$. At the beginning of laser pulse, electrons of the entire cluster as a whole will respond to the electric field. Since clusters are independent entities, free electrons will oscillate in the direction of electric field in half the cycle and in the opposite direction in the remaining half of the cycle of the wave. This will create a giant dipole (3). Evanescence of the laser radiation is at a distance larger than the cluster dimension, $\delta > C_{da}$. Ablation depth [13] for aluminum slab target at $I \sim 10^{12}$ W/cm^2 is ~ 0.4 μm which is comparable to cluster diameter. Since the laser pulse is still on during the cluster disassembly, the remaining laser radiation penetrates the cluster plasma without being absorbed. Therefore no significant laser energy is deposited on the cluster surface and a large fraction of laser radiation transmits through the diluting cluster plasma.

Thus in the case of aluminum cluster of dia~0.4 μm,1) plasma density and temperature are not very high hence a low X-ray yield. 2) Radiation mean free path [14] for a slab target at $I \sim 10^{12}$ W/cm^2 is $\lambda_R \sim 3$ μm which is larger than the dimension of the cluster; implies initial X-rays will transmit through the plasma. Thus, further X-rays can not be generated due to down conversion mechanism similar to solid targets. For the above mentioned reasons, we believe that X-ray yielding process in cluster plasma is not same as that in the solid aluminum target. Therefore, the shape of the X-ray emission spectra is different and X-ray intensity is less than that of solid target.

$C_{dg} \geq \lambda_L$, Gold Cluster

Here the gold cluster diameter is $C_{dg} > \lambda_L$. i.e. Laser wave experiences evanescence inside the cluster within the skin depth $\delta \approx \lambda_L$. We assume, the behavior of such cluster is like that of bulk target for the following reasons. Ablation depth for planar gold target at $I \sim 10^{12}$ W/cm^2 is ~1.2 μm, which was estimated from ablation pressure reported in our earlier work [13,15] and is comparable to cluster dimension. Therefore, the laser interaction process and the process of X-ray emission from gold cluster could be similar to that of bulk material. Here each cluster behaves as a source of X-rays. The number of gold clusters which are responsible for X-ray emission within the laser irradiated area being ~2.6x10^9/cc, X-ray intensity from gold clustered target is lower than the bulk gold target.

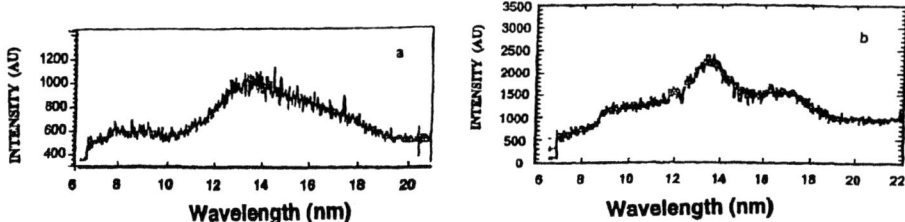

FIGURE 2. X ray emission spectra from a) Gold cluster target and b) Solid gold target.

PIC SIMULATIONS

To study the behavior of clusters at higher laser intensity, PIC simulations are performed and the results indicate the correlation between cluster size and laser wavelength. We have performed preliminary particle-in-cell (PIC) simulations of clusters irradiated by intense laser pulses. For simplicity we used a two-dimensional (cartesian) geometry and assume a fully ionized, collisionless deuterium plasma at an initial temperature of 4 keV. In principle this model is appropriate to study the collective dynamics of a cluster plasma irradiated by a high intensity laser pulse, after that the leading edge of the pulse has already ionized and heated the cluster. The most serious limitation comes from the different asymptotic behavior of the electrostatic force in 2D and 3D clusters [16] so that for instance only qualitative features of Coulomb explosion may be descrived in 2D. Particle simulations of clusters in different regimes and geometries are reported in Refs.[16,17].

The important parameter which discriminates cluster behavior with respect to laser irradiation is $a=k(n_e/n_c)(r/\lambda)$ where k is the dimensionality of the cluster (2 in the present case of simulations), n_e is the electron density, n_c is the critical density, r is the cluster radius and λ the laser wavelength. For $a \gg 1$ the cluster is opaque to the laser radiation and behaves as a small piece of bulk plasma, so that cluster heating may occur only via surface interaction, i.e. via collisionless mechanisms which are well known in the irradiation of solid targets (anomalous skin effect or vacuum heating). This is the case for the simulation of Fig.3a, ($n_e/n_c=10$, $I=10^{17}$ W/cm^2, $r=\lambda=0.5$ microns) where we clearly see that the clusters screens the laser light. No significant energy transfer to the ions is observed. In contrast, for a case with $r=0.1$ microns, $\lambda=0.4$ microns, and $I=4 \times 10^{17}$ W/cm^2, the laser field penetrates the clusters and drags out electrons. We find that within a few laser cycles an electron cloud is formed outside the ion cluster as in fig 3b. Due to the unscreened electrostatic field of ions the Coulomb explosion begins leading to a radially symmetric acceleration and expansion of the ions.

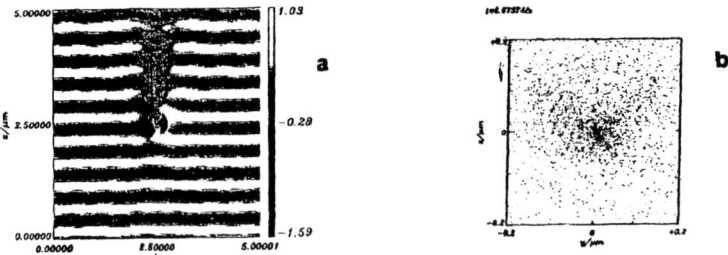

FIGURE 3. Laser-Cluster interaction with a) large cluster and b) small cluster

CONCLUSION

Clusters offer an interesting and important behavior. At moderate laser intensities $I \sim 10^{12}$ W/cm^2, small clusters with dimension less than the laser wavelength show a significantly different behavior in X-ray emission spectra in 5-22 nm range than the bulk material when irradiated with long pulse length laser(10ns). Contrary to this cluster spectra depicts the evolution towards the bulk property for the clusters larger than laser wavelength. Pic simulations at higher laser intensity $I \sim 10^{17}$ W/cm^2 predict surface interaction for large size clusters and Coulomb explosion for small clusters. We believe there is a direct correlation between the size of the cluster and the wavelength and pulse duration of the incident laser radiation. Further study using various size of the cluster and laser parameters is necessary to under stand the cluster properties and their evolution towards the bulk.

Although substantial work has been done in laser cluster interaction, we believe a lot of their characteristic behavior are yet to be revealed. The cluster size (no.of atoms per cluster),electron configuration combined with various laser parameters (like laser wavelength, pulse duration and intensity) may offer an important study on cluster properties.

ACKNOWLEDGMENTS

We sincerely acknowledge the cooperation received from Drs. Andrea Machhi (Univ. of Pisa, Italy) and H. Ruhl (MBI for Nonlinear Optics, Berlin, Germany) in PIC simulation.

REFERENCES

1. Hagena,O.F. *Rev. Sc.Instru.* **63**, 2374- (1992)
2. .Mark, T.D, *Int.J of mass spectrometry and ion processes.* **79**, 1 (1987)
3. Hutchinson H., *Science,* **280**, 693 (1998)
4. Jeffrey L.,et al, , *Phy. Rev. Letts.* **68**, 3535 (1992)
5. Tisch J., et al, *J.Phys.B, At.Mol.Opt.Phys.* **30**, L709 (1997)
6. Ditmire T., et al, , *Phys. Rev. A,* **57**, 369 (1998)
7. Eloy M.,et al, , *Proc.of IFSA meeting, Bordeaux,* 1003 (1999)
8. Thomson B.D., et al, , *J. Phys. B, At. mol. opt. phys.* **27**,4391 (1994)
9. McPherson A., et al, , *Nature,* **370**, 631 (1994)
10. Desai T. and Pant H.C., *Laser and Part.beams,* **18**, 119 (2000)
11. Ditmire T., et al, , *Phy. Rev.* A. **53**, 3379 (1996)
12. Max C.E., , *Phys. of Laser Fusion*, Vol.**1**, UCRL-53107 (1982)
13. Desai T., *Ph. D thesis, Univ. of Bombay* (1992)
14. Zeldovich Ya. and Raiser Yu., , *Phys. of Shock waves and high temperature hydrodynamics* (1966)
15. Godwal B.K., et al, *J. Appl.Phys.,* **65**, 4608 (1989)
16. Esirkepov T., et al., Laser and Part. Beams 18, 503 (2000).
17. Eloy M., et al., Phys. Plasmas 8, 1084 (2001).

Absolute X-ray yield studies from Xe-clusters with ultrashort Ti:Sa laser pulses at 2×10^{18} W/cm^2

P.V. Nickles[1], S. Ter-Avetisyan[1,2], H. Stiel[1], W.Sandner[1], M. Schnürer[1]

[1]Max-Born-Institut, Max-Born-Str. 2a, D12489 Berlin, Germany
[2]Institute for Physical Research, Ashtarak-2, 378410, Armenia

Abstract. High intensity ($\sim 2\times10^{18}$ W/cm^2) laser excited large Xe–clusters ($10^5 \ldots 10^6$ atoms per cluster) have been studied concerning scaling and absolute EUV–emission in a wavelength range between 7 nm and 15 nm. Ultrashort (50 fs) pulses from a Ti:Sa multi–TW laser at 800 nm wavelength were applied in the experiments. Characterized cluster target variations in combination with different laser irradiation have been used for EUV-yield optimization. Maximum emission as a function of the backing pressure and a spatial emission anisotropy covering a factor of 2 is discussed with a simple model of the source geometry and EUV–radiation absorption. High charge states and strong x-ray yield from laser heated clusters are considered due to collisional and optical field ionization processes. Circularly polarized laser light instead of linear polarization results in a factor of 2.5 higher emission in the 11 nm to 15 nm wavelength range. This indicates the initial influence of optical field ionization for the interaction parameter range used and contrasts to collisional heating which seems to influence preferentially higher ionization states. An absolute emission efficiency at 13.4 nm of up to 0.5% in 2π sr and 2.2% bandwidth has been measured.

INTRODUCTION

Several laser driven target systems are studied in view of sufficient flux in the EUV–region which is a prerequisite for key technologies and intriguing applications, such as Extreme Ultra Violet (EUV) lithography, spectroscopy and microscopy [1 – 3]. Among them the interaction of high-intensive laser radiation with "van der Waals" clusters has received increased attention both regarding fundamental excitation processes and a possible application potential. Due to the near solid density of a single cluster object which consists of several hundred up to million atoms in a nanometer-sized volume and a gaseous average atomic density of a cluster ensemble within the extension of a micrometer–sized laser focus, an attractive combination of different properties results. Because the disassemble time of laser heated "van der Waals" clusters is on a picosecond time scale an effective high intensity laser interaction is anticipated during pulse widths similar or shorter as this characteristic time scale. Meeting this condition high laser absorption efficiency could be demonstrated [4] which leads to large energy deposition in the target. Therefore a lot of work was devoted to study laser–cluster interaction phenomena, which give significant emission of short wavelengths light in the spectral range of the EUV [5] up to hard x–rays [6].

CP611, *Superstrong Fields in Plasmas:* Second Int'l. Conf., edited by M. Lontano et al.
© 2002 American Institute of Physics 0-7354-0057-1/02/$19.00

Moreover energetic ions from exploding clusters [7, 8] have been analyzed and attributed to different excitation mechanisms [5, 9]. Furthermore laser driven rare gas cluster targets provide a low debris, which is a significant problem for continuously operating sources.

Here we direct our view to the spectral region between 7 nm and 15 nm emitted by Xe–clusters which is currently under intensive investigation as a highly attractive wavelength region for future lithographic systems [10]. Different laser and target parameters have been varied for studying the relevant excitation mechanisms and maximum EUV-yield.

EXPERIMENTAL SETUP

The experimental setup in the target chamber is shown in Fig. 1. The experiments have been carried out with 50 fs laser pulses at 800 nm center wavelength from a 10 Hz repetition rate multi–TW Ti:Sa laser [11] at Max–Born–Institute. We used a 4 TW

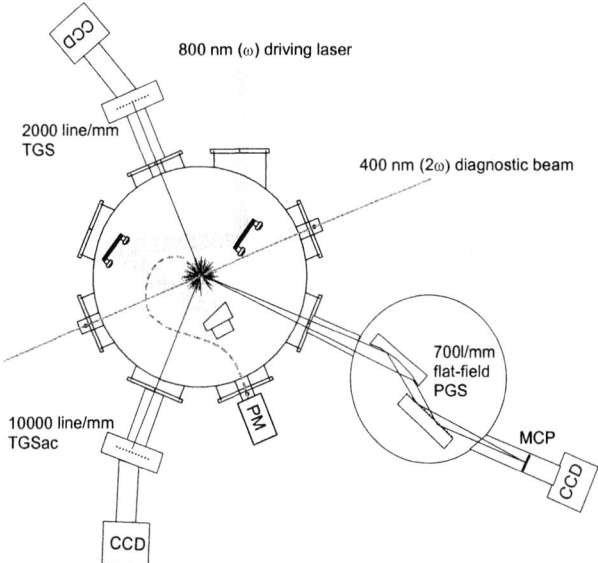

FIGURE 1. Experimental setup at the vacuum interaction chamber including the Xe-cluster jet target: PM - photomultiplier coupled to an optical fiber for scattered light detection using the 400 nm diagnostic beam, TGS - transmission grating spectrograph (on - axis, which refers to the incident laser beam axis), TGSac - absolute calibrated transmission grating spectrograph (off - axis), PGS - plane (flat field) grating spectrograph with MCP - micro channel plate detector.

(200 mJ) beam of 60 mm in diameter. The beam is focused with an f/2 off axis parabolic mirror having a focal length of 155 mm. From focus measurements an interaction intensity of $\sim 2\times10^{18}$ W/cm^2 was inferred. This intensity level was verified under comparable conditions in single atom ionization experiments. Varying the laser

pulse grating compressor also 2 ps pulses focussed to 5×10^{16} W/cm^2 have been applied.

The Xe–cluster target was formed by expanding the gas from pressures up to 55 bar and temperature of $T_0 = 300$ K through the hypersonic conical nozzle with a cone angle $2\theta = 7°$, throat diameter of 400 μm and an 8 mm long conical section. The laser focus is placed approximately 1 mm below the orifice of the nozzle.

Soft x–ray spectra were measured with two transmission spectrometers and one flat field grazing incidence spectrometer allowing us the simultaneous registration of x–ray emission at different angles relative to the incident laser beam axis. One of the transmission grating spectrometer (grating 10000 l/mm, resolution $\lambda/\Delta\lambda = 300$ at 13 nm, comparable to those described in [12]) was calibrated in absolute terms in the wavelength range between 10 nm and 15 nm at the synchrotron facility BESSY II (PTB – beamline). Optical light is blocked with 200 nm Zr filters.

CLUSTER SIZE

With our nozzle at T = 300 K for Xe already at 5 bar a parameter $\Gamma^* \geq 10^5$ results (see e.g. [13]). Employing Hagenas formula [13 and ref.] a value of $\approx 2 \times 10^6$ atoms/cluster at 5 bar is derived. Because the actual values may be influenced by further specifics, which are not included in Hagenas formulas, the presence of clusters in the gas jet was analyzed with a Rayleigh scattering technique. Our measurements between 5 and 55 bar indicate that the scattered signal scales with backing pressure as $\sim p_0^{1.9}$ that gives an exponent somewhat lower than in [14].

A quantitative estimation of the cluster size based on scattered photon numbers and assuming 100% condensation into clusters gives about $(2 - 6) \times 10^6$ atoms/cluster at 20 bar backing pressure. With a linear relation between the number of cluster particles and backing pressure [5] we calculate on the basis of our measured dependence of the scattered signal $S \propto p_0^{1.9}$ cluster sizes of $(3 - 10) \times 10^5$ atoms at 5 bar and of $(2 - 8) \times 10^6$ atoms at 50 bar.

An optical thick target is one prerequisite for good laser light absorption and efficient ionization for short wavelengths emission of desired ionic species. The absorption behavior was determined by a measurement of the laser energy transmitted through the target as a function of the backing pressure. However, this method implies a low level of scattering and reflection of the high intensity laser beam. Several work [e.g. 4] have relied on this. The obtained laser light transmission curve [15] showed, that the cluster target forms an dense object for powerful optical femtosecond laser pulses [4] and already at low pressures of about 5 bar a sharp decrease in the transmission is evident.

EUV-EMISSION CHARACTERISTICS

In Fig.2 we depict spectra from the on-axis spectrometer obtained with different backing pressures. Three broadband emission peaks were observed in the 1 nm to 16 nm wavelengths range. The emission between 10.5 nm and 16 nm is well known from literaturc [e.g.16]. Four emission peaks were observed in this region which come from Xe^{9+} to Xe^{11+} ions. Most of the intensive EUV–emission occurs at certain Xenon ion stages and relates generally to transitions from resonance levels to the ground state of the respective ion. The emission region from (6 ÷ 9) nm ((138 ÷ 206) eV) has been

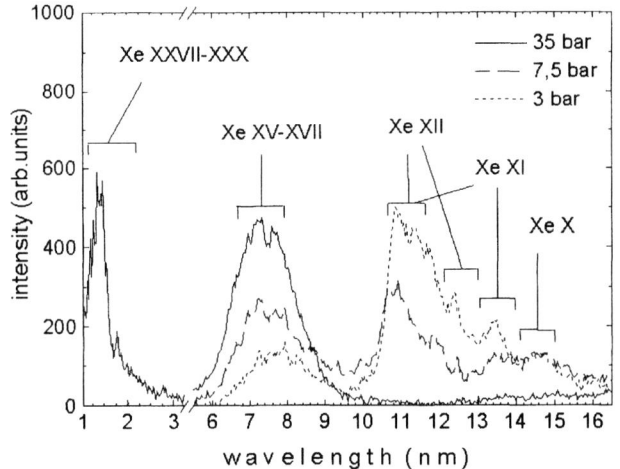

FIGURE 2. Spectra at 3, 7.5 and 35 bar Xe backing pressure registered with the on-axis TGS and marks of emission line bands from different Xe ion stages. Emission between 1 nm and 2 nm is not shown for low pressure values.

observed in some spectral measurements [e.g.17], but it was not addressed in further discussions. Between 6 nm and 7 nm the recorded signal decreases due to the transmission characteristic of the Zr–filter. Strong emission in the 1 nm to 2 nm wavelengths region (Xe XXVII to Xe XXX) [18] was observed too (see Fig.2) but not investigated further in this work.

In Fig.3 the pressure dependent emission signals from the on-axis and off-axis spectrometer are given. Our gas–cluster target has an extension of approximately 1.4 mm (orifice diameter). Inside we produce the heated plasma emission source. Registered signals may be influenced if absorption comes significantly into play. Different registered signals can result if the radiation passes different length of the absorbing medium. In our target such a situation can occur if the radiation emitting plasma object is not placed in the center of the ambient cluster gas similar to the study of Miura et al. [19]. We set up a simple source model which is based qualitatively on the experimental findings of Miura. We regard only a one dimensional plasma column in a 1.4 mm diameter cluster gas plume. Our detection geometry is schematically depicted in the insert of Fig.3. The signal of the plasma source is scaled linearly with the average atom density and the absorption of the ambient cluster gas is calculated in dependence of the viewing angle of each source point. Because of inadequate knowledge of the EUV–absorption in the ionized plasma it is set to zero. In Fig.3 we depict the registered emission at 13.4 nm versus backing pressure together with a function plot of our simple target model. Both curves are fitted together in dependence of one parameter for the signal height and one for the source extension. From Fig.3 it is visible that the general behavior of signal increase and decrease can be reflected with the simple model, which of course can not account for all the details of the experiment.

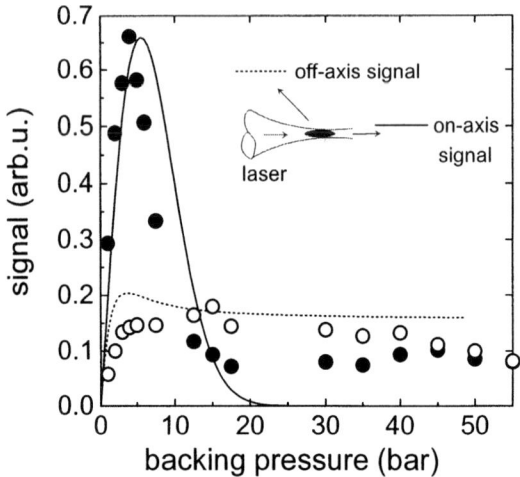

FIGURE 3. Registered on-axis (closed square) and off-axis (open square) emission data at 13.4 nm in comparison to a computed signal (solid line – on-axis, dotted line – off-axis) on the basis of a simple geometric target model which is supported by the experimental findings of [19](details see text), insert shows detection geometry.

INFLUENCE OF LASER POLARIZATION

Several experiments [e.g. 4,8] have revealed that collisional processes can play an important role for cluster heating resulting in the release of highly energetic ions up to MeV energies and hard x-ray emission. Therefore it is of interest to what extend collisional heating of the electrons by inverse bremsstrahlung and stimulated Raman scattering [5] influence the ion species being responsible for the EUV–emission looked at this work. From theoretical considerations, however, EUV spectra generated with linearly and elliptically polarized laser light should show a significant difference due to resulting higher energies of electrons field ionized with circularly polarized light. Up to now experiments have verified this mechanism in low density gas targets, where other heating mechanisms are ineffective. Contrary to this in case of high density gas jets or in cluster targets the initial field ionization is thought to be followed by further collisional processes which in turn modify the ionic state distribution. We observed [15,20] that in the whole range of relevant Xe backing pressures the emission around 13.4 nm is about 2 times higher if circularly instead of linearly polarized light is used. At 10 bar the ratio is about three times.

CONVERSION EFFICIENCY

Averaging the off-axis and on-axis data we can infer a maximum laser light into EUV-emission conversion efficiency (CE) of 0.5 % in 2π sr solid angle. This value could be increased up to 0.8 % when 2 ps pulses have been used [20]. At around 11.4 nm in respect to Mo:Be multilayer mirrors with highest reflectivity the CE amounts to 2 % in in 2π sr solid angle. Moreover in the wavelength region from 10 nm to 15 nm a CE up to 7.5 % in 2π sr has been achieved.

ACKNOWLEDGMENTS

A part of this work was supported by BMBF - project 13N7784 . We thank A. Egbert (LZH Hannover) and C.Reinhardt (U. Hannover) for setting up partly the Rayleigh scattering experiments and the laser team headed by M. Kalashnikov.

REFERENCES

1. Emerging Lithography Technologies IV, SPIE **3997** (2000).
2. H. Kondo, T. Tomie, H. Shimizu, Appl. Phys. Lett. **72**, 2668 (1998).
3. J. Thieme, G. Schmahl, D. Rudolph, E. Umbach (Ed.s) X-ray Microscopy and Spectromicroscopy, (Springer, Heidelberg, 1998).
4. T. Ditmire, R. A. Smith, J. W. G. Tisch, M. H. R. Hutchinson, Phys. Rev. Lett., **78**, 3121 (1997).
5. T. Ditmire, T. Donnelly, A. M. Rubenchik, R. W. Falcone, M. D. Perry, Phys. Rev. **A 53**, 3379 (1996).
6. A. McPherson, B. D. Thompson, A. B. Borisov, K. Boyer, C. K. Rhodes, Nature, (London) **370**, 631 (1994).
7. T. Ditmire, J. W. G. Tisch, E. Springate, M. B. Mason, H. Hay, R. A. Smith, J. Marangos, M. H. R. Hutchinson, Nature (London) **386**, 54 (1997).
8. M. Lezius, S. Dobosz, D. Normand, M. Schmidt, Phys. Rev. Lett. **80**, 261 (1998).
9. K. Boyer, B. D. Thompson, A. McPherson, C. K. Rhodes, J. Phys. B. At. Mol. Opt. Phys., **27**, 4373 (1994).
10. J. Underwood, in Extrem Ultraviolet Lithography, G. Kubiak, D. Kania, eds., Vol.4 of *Trends in Optics and Photonics Series* (Optical Society of America, Washington, D. C., 1996), pp.162-166; G. D. Kubiak, L. J. Bernardez, and Kevin Krenz, SPIE **3331**, 81(2000).
11. P.V. Nickles, M. Kalachnikov, P. J.Warwick, K. A. Janulewicz, W. Sandner, U. Jahnke, D. Hilscher, M. Schnürer, R. Nolte, A. Rousse, Quantum Electronics (Russia), **29**, 444 (1999).
12. T. Wilhein, S. Rehbein, D. Hambach, M. Berglund, L. Rymell, H. M. Hertz, Rev. Sci. Instrum., **70**, 1694 (1999).
13. J. Wormer, W. Guzielski, J. Stapelfeldt, T. Moller, Chem. Phys. Lett. **159**, 321 (1989).
14. E. Springate, N. Hay, J. W. G. Tisch, M. B. Mason, T. Ditmire, M. H. R. Hutchinson, J. P. Marangos, Phys. Rev. **A 61**, 063201 (2000).
15. S. Ter-Avetisyan et al. acc. for publ. in Phys. Rev. E.
16. J. Blackburn, P. K. Carroll, J. Costello, G. O'Sullivan, JOSA **73**, 1325 (1983).
17. M. McGeoch, Appl. Opt., **37**, 1651 (1998).
18. H. Honda, E. Miura, E. Katsura, E. Takahashi, K. Kondo, Phys. Rev. A, **61,** (2000), B. A. M. Hansson, L. Rymell, M. Berglund, H. M. Hertz, Microel. Engin. **53**, 667 (2000).
19. E. Miura, H. Honda, K. Katsura, E. Takahashi, K. Kondo, Appl. Phys., **70**, 783 (2000).
20. M. Schnürer, S. Ter-Avetisyan, H. Stiel, U. Vogt, W. Radloff, M. Kalashnikov, W. Sandner, and P.V. Nickles, Eur. Phys. J. D **14**, 331 (2001).

Plasma Mirror Distortions and Parametric Instabilities Induced by High Intensity Femtosecond Pulses on Solid Targets

A. Tarasevitch, C. Dietrich, and D. von der Linde

Institut für Laser-und Plasmaphysik, Universität-GHS-Essen, D-45117 Essen, Germany

Abstract. The interaction of femtosecond laser pulses with a plasma "mirror" in a weak relativistic regime with a controlled plasma gradient is investigated. The characteristic scale lengths for the decay of high order harmonic generation, increase of the divergence of the reflected radiation, and development of plasma instabilities are determined.

INTRODUCTION

The development of high-power femtosecond lasers has made possible experiments on laser-plasma interactions in the relativistic regime. Experiments at this intensity level are of great interest because of a number of applications such as fast ignition of laser fusion targets [1], particle acceleration [2], hard x-ray [2] high order harmonic generation (HOHG) [3], etc. During the laser-solid interaction at such an intensity the target is highly ionized by the leading edge of pump pulse, and a reflecting layer of supercritical plasma is formed. Neglecting the details of the electron density distribution, the motion of the electrons in the field of the incident electromagnetic wave can be considered as an "oscillating mirror" [4]. At high intensities the "oscillating mirror" can be distorted by plasma instabilities and surface deformations induced by ponderomotive forces. Unfortunately, different processes contributing to the mirror distortions are generally difficult to identify. These instabilities and deformations strongly depend on the plasma scale length, the parameter which is difficult to control in experiments. The main reason for this is a poor pulse contrast in most modern high-power femtosecond lasers.

In this paper we report on the experiments on laser-solid interaction in the weak relativistic regime using high-contrast–ratio femtosecond pulses. The plasma scale length L was changed in a controlled way in order to explicitly investigate its influence on the mirror distortions. The dependences of the efficiency of HOHG from the "oscillating mirror", the divergence of the reflected fundamental beam, and the onset of the plasma instabilities on L were studied. In order to produce a controlled plasma scale length a two-pulse technique was used. The first relatively low intensity femtosecond pulse (controlled "prepulse") prepared the plasma with the desired L for the interaction with the second high intensity "main" pulse. The plasma density

CP611, *Superstrong Fields in Plasmas:* Second Int'l. Conf., edited by M. Lontano et al.
© 2002 American Institute of Physics 0-7354-0057-1/02/$19.00

gradient was controlled by changing the delay of the "main" pulse with respect to the "prepulse".

EXPERIMENTAL

High Order Harmonic Generation

In our experiments we used a titanium sapphire laser producing 200 mJ pulses of 120 fs duration at a wavelength 800 nm. The ratio of the pulse peak intensity to the intensity at 1 ps (intensity contrast) was determined to be about 10^5. The ASE intensity contrast was approximately $10^9 - 10^{10}$. The incident laser beam was p- polarized and focused onto the target with the help of an off-axis parabolic mirror (Fig. 1). The angle of incidence was 55^0. The diameter of the "main" beam in the focal plane of the parabolic mirror was 6 μm (FWHM). It was verified that 40% of the pulse energy was concentrated within this circle. The peak intensity on the target surface obtained from these data was approaching 10^{18} W/cm². The light reflected from the target was imaged onto a phosphor screen with the help of a toroidal grating, and the phosphorescent light was recorded with a CCD camera. The targets were optically polished glass substrates, which were raster-scanned to provide a fresh surface for each laser pulse. The experiments were performed in a vacuum chamber at an ambient pressure of 10^{-3} Torr.

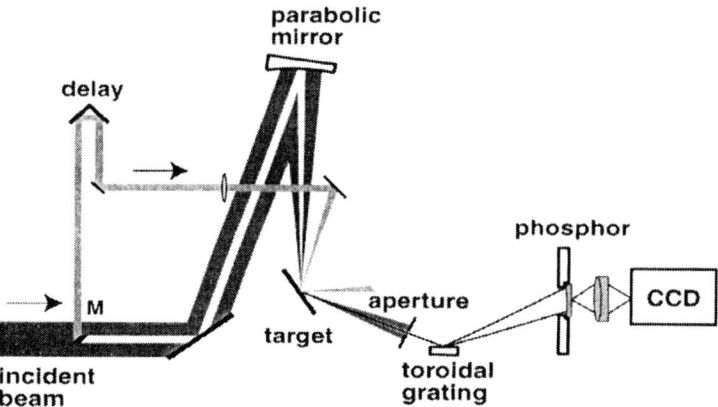

FIGURE 1. Schematic of the experimental setup.

The beam providing the controlled "prepulse" was formed from the "main" beam with the help of the small mirror M and an optical delay as shown in Fig. 1. For the estimation of the plasma scale length a simple isothermal model of plasma expansion [5] was used. The intensity of the "prepulse" beam on the target surface was 5×10^{14} W/cm². For this intensity the electron temperature T_e and the expansion velocity v_T were estimated to be 100 eV and 6×10^6 cm/s respectively which gives $L =$

$0.07\lambda_0 \times \tau[ps]$, where λ_0 is the laser wavelength and τ is the time delay of the "main" pulse with respect to the "prepulse".

An example of the recorded harmonic spectrum at $\tau = 0$ is depicted in Fig. 2. In a single laser pulse harmonics up to 17th can be easily seen. The actual image of the output plane of the spectrometer is also shown. The measured dependence of the harmonic energy on τ of several harmonic orders is presented in Fig. 3. For all harmonic orders the data can be fitted by an exponential function, $\exp(-\tau/\tau_s)$, with $\tau_s = 200$ fs which corresponds according to our estimations of the plasma scale length to $\exp(-L/L_s)$ with $L_s = 0.015\ \lambda$.

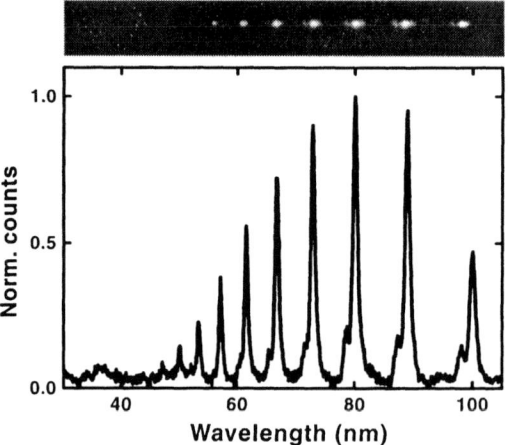

FIGURE 2. Harmonic spectrum recorded in a single laser pulse. The actual image recorded with the CCD camera is shown above.

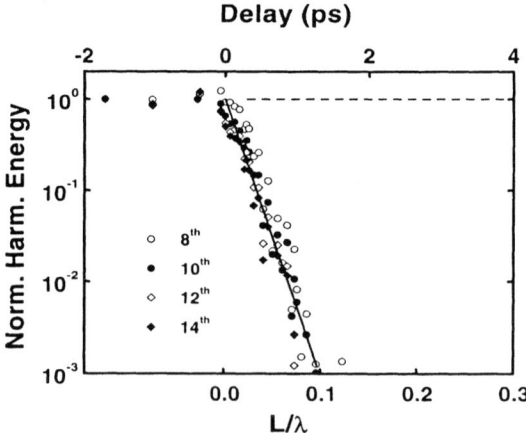

FIGURE 3. Energy of different harmonics as a function of the time delay (upper horizontal scale). Lower horizontal scale: estimated plasma scale length. Solid line: $\exp(-L/0.015\lambda)$. Dashed line corresponds to the calculated dependence of the harmonic energy on the scale length.

The mechanism of HOHG is normally explained as a phase modulation of the incident pump wave by the "oscillating mirror" [4,6]. So harmonic generation must be very sensitive to the mirror distortions. PIC simulations show that the dependence of the harmonic efficiency on L itself is relatively weak. Indeed, the dashed line in Fig. 3 represents the dependence of the harmonic efficiency on L calculated for the exponential plasma profile and maximum plasma density of n_e=27.5625 n_{cr}. The decay of the efficiency can be hardly seen on the given scale. A full relativistic PIC Codes LPIC++ has been used in our calculations [7]. The fact that the harmonic efficiency measured in the experiments drops down very fast with L suggests that the "oscillating mirror" is destroyed due to some surface distortion which cannot be taken into account by 1D PIC calculations.

Divergence of the Reflected Beam and Parametric Instabilities

In the following experiments the angle of incidence was about 40^0 and the intensity of the "prepulse" on the target surface was 5×10^{15} W/cm^2. The plasma scale length as a function of the delay time, was estimated using the "Medusa" hydrocode [8] The electron temperature after the "prepulse" was about 200 eV.

First, the shape of the "oscillating mirror" was examined at zero delay time. The spatial distribution of the phase of the radiation reflected from the "plasma mirror" was investigated with the help of a femtosecond interference microscope. Noticeable "deformations" of the mirror was recorded only at the pump intensities above 10^{18} W/cm^2 (see Fig. 4). All experiments were carried out below this level.

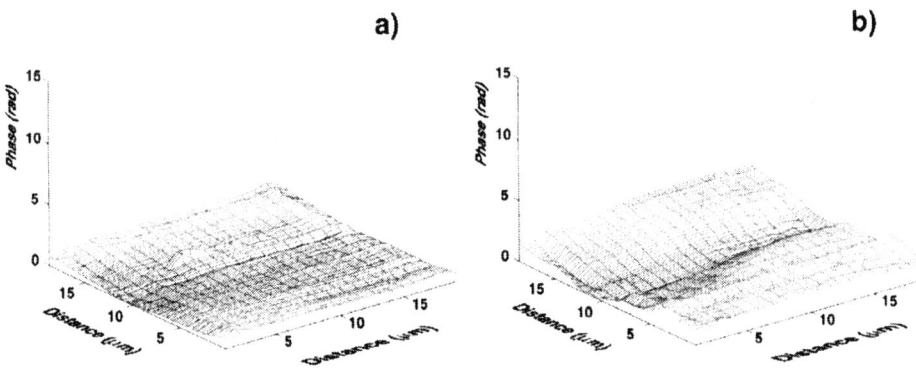

FIGURE 4. Spatial distribution of the probe beam phase acquired on the reflection from the plasma. The distribution corresponds to the target plain for different pump intensities: a) 5×10^{17} W/cm^2, b) 2×10^{18} W/cm^2. The center of the pump beam lies in the middle of the X-Y plane with the dimensions ~5μm × 7μm (intensity FWHM) respectively.

In order to investigate the spatial distribution of the radiation reflected from the plasma mirror a screen was placed in the reflected beam instead of the spectrometer (Fig. 1). The angular distribution of the reflected light was observed on the screen with the help of a CCD-camera. It was found that the noticeable broadening of this distribution starts only when the plasma scale length L exceeds the laser wavelength λ_0 (Fig. 5). We have found that this scale length also corresponded to the threshold of plasma parametric instabilities. At short delays only the collimated beam of the reflected fundamental and second harmonic radiation in the direction of the specular reflection is present. However, starting from a certain delay time green 3/2 harmonic emission could be seen. It is interesting that in a particular range of delay times the 3/2 harmonic emission is well collimated and directed at $\sim 25^0$ and $\sim 75^0$ to the target normal [9].

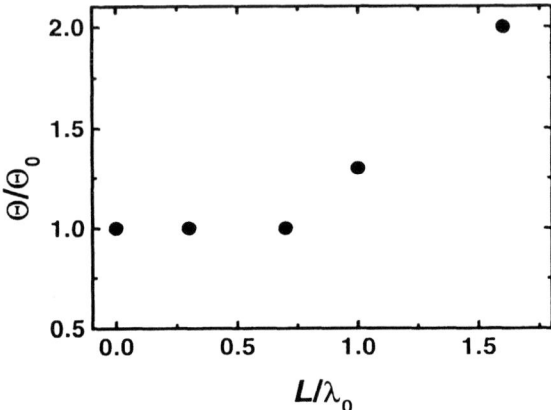

FIGURE 5. Divergence of the reflected fundamental beam normalized to the divergence at $\tau = 0$ as a function of the plasma scale length.

The emission of the 3/2 harmonic is an indication of the development of the two-plasmon decay (TPD) instability in plasma. The TPD instability has the lowest threshold with respect to plasma inhomogeneity [5]. In Fig. 6 the 3/2 harmonic energy is plotted as a function of delay time (lower horizontal scale) and scale length (upper

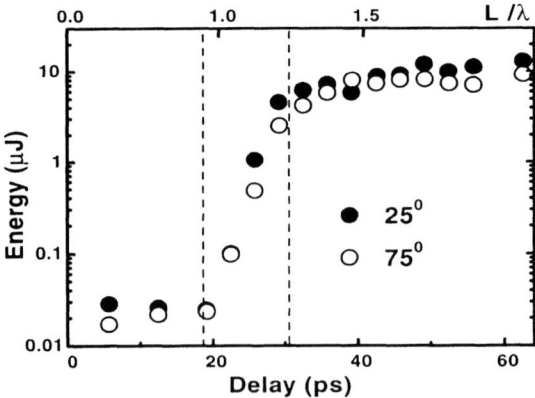

FIGURE 6. Dependence of the energy of the 3/2 harmonic for the angles of 25^0 and 75^0 on the delay time and L. The energy of about 20-30 nJ at small delays corresponds to noise level. The dashed vertical lines show the range in which the 3/2 harmonic is emitted as two collimated beams.

horizontal scale). The energy was measured with the help of two calibrated photodiodes. It can be seen that the TPD instability already develops at a very steep plasma-vacuum interface. The energies of the 3/2 harmonic at the angles of $\sim25^0$ and $\sim75^0$ were approximately equal. The total energy of the 3/2 ω emission is about 10 μJ at the delay time of 30 ps (L $\approx$ 1.2\λ_0), resulting in the conversion efficiency of 5×10^{-4}.

CONCLUSIONS

In conclusion, our experiments with *controlled* plasma scale length have made it possible to "plasma-scale-length resolve" the "oscillating mirror" distortions and instabilities. In contradiction with the predictions of the PIC simulations the efficiency of HOHG drops down on a very short scale length: $L \sim 0.015\lambda$. This suggests the onset of small scale mirror distortions which cannot be taken into account by 1D PIC simulations. Judging by the reflection of the pump beam the noticeable deformations of the plasma mirror take place at much longer scale length $L \sim \lambda_0$. We have obtained *first* evidence that $L \sim \lambda_0$ gives also the threshold for the onset of the plasma parametric instabilities.

ACKNOWLEDGMENTS

We would like to thank P. Gibbon and R. Lichters for giving us access to the "Medusa" and PIC codes, and the Deutsche Forschungsgemeinschaft for its financial support.

REFERENCES

1. Tabak, M. et al. , Phys. Plasmas **1**, 1626-1634 (1994).
2. Umstadter, D., Phys. Plasmas **8**, 1774-1785 (2001); Schnürer, M. et al., Phys. Rev. E **61**, 4394-4401 (2000); Kruschelnick, K.et al., Phys. Plasmas **7**, 2055-2061 (2000).
3. Tarasevitch, A. et al., Phys. Rev. A **62**, 023816-1-6 (2000).
4. Bulanov, S.V., Naumova, N.M., Pegoraro, F, Phys. Plasmas **1**, 745 (1994).
5. Kruer, W. L., *The Physics of Laser Plasma Interactions*, Addison-Wessly, Redwood City, CA 1988.
6. R. Lichters, J. Meyer-ter-Vehn, A. Pukhov, Phys. Plasmas **3**, 3425-3437 (1996); D. von der Linde and K. Rzàzewski, Appl. Phys. B **63**, 499-506 (1996).
7. R. Lichters et al., Phys. Plasma 3, 3425 (1996)
8. J. Christiansen, D. Ashby, and K. Roberts, Comput. Phys. Commun. **7**, 271 (1974); P. A. Rodgers, A. M. Rogoyski, and S. J. Rose, RAL-89-127, Dec 1989.
9. Tarasevitch, A. et al., to be published.
10. B. Quesnel et al., Phys. Plasmas **4**, 3358 (1997); H. C. Barr, P. Mason, and D. M. Parr, ibid. **7**, 2604 (2000).

4. APPLICATIONS OF ULTRA-INTENSE FIELDS: PARTICLE ACCELERATION, INTENSE X-RAY SOURCES, INERTIAL CONFINEMENT FUSION, AND LABORATORY ASTROPHYSICS

Overview of laser particle acceleration in an underdense plasmas

V. Malka

Laboratoire pour l'Utilisation des Lasers Intenses (LULI) Unité mixte n 7605 CNRS - CEA - Ecole Polytechnique - Université Pierre et Marie Curie, victor.malka@polytechnique.fr

Abstract. The use of underdense plasmas to accelerate efficiently particles has been recently demonstrated. Such scheme as Laser Wake Field, Laser Beat Wave and Self Modulated Laser Wake Field, which were first theoretically investigated have now been experimentally improved. An overview of these results as well as some perspectives will be presented here.

INTRODUCTION

A Plasma is an ideal medium for particle acceleration since (i) plasmas can support very high longitudinal electric fields (100 GV/m), which are four orders of magnitude higher than presently used in RF cavities. (ii) Plasmas converte the transverse electromagnetic field of the laser into longitudinal space-charged oscillations or "plasmas waves" which can trap and accelerate charged particles. (iii) Phase velocities of these electron plasma waves can be relativistic and in the order of the speed of light.

How are such electric field generated? The first step is to generate large scale homogenous underdense plasmas. This is done by focusing a laser beam into a gas filled chamber, a gas cell or a gas jet. The front pulse of the laser ionized directly the gas. Depending on the laser intensity the ionization mechanism is multiphoton (in the low intensity case) or tunnel ionization (in the high intensity case). After ionization processes have occurred, part of the rest of the laser energy will generate plasma waves with relativistic phase velocities.

The first idea was proposed by Tajima and Dawson in 1979 [1]. In their letter two schemes were proposed to generate these large amplitude plasma waves : the LWF and LBW. A third scheme, the SMLWF was proposed more recently by three groups in the 1992 [2–4]. These three schemes have been demonstrated in the last ten years. The acceleration of relativistic particles in a plasma using high-power laser beams has been the subject of several experiments in the past few years. In all cases a plasma wave with high amplitude and relativistic phase velocity is generated by the interaction between a plasma and one or two laser beams. The

CP611, *Superstrong Fields in Plasmas:* Second Int'l. Conf., edited by M. Lontano et al.
© 2002 American Institute of Physics 0-7354-0057-1/02/$19.00

electric field associated with these plasma waves has been used to accelerate injected electrons in the first two cases LWF [5] and LBW [6,7]. In the SMLWF relativistic wave breaking regime has been obtained and electrons from the plasma itself are directly accelerated, generating a bright and energetic electron beam [8,9]. We will present here a summary of these results. It should be noted that preformed plasmas can also be achieved focusing a laser beam on thin foil targets. This process can achieve higher plasma densities but with relatively short (1mm) and inhomogeneous plasma.

LASER WAKE FIELD

In the low density wakefield regime [1], the plasma wave is excited by a single short laser pulse with a duration in the order of the plasma period. The pondero-motive force associated with the laser pulse pushes the electron out from the high to the lower intensity region generating in the wake of the laser pulse a relativistic plasma wave. This scheme has been addressed in several experiments after the appearance of short pulses laser. These plasma waves have been optically [10,11] or in the THz range [12] measured. We present here the first results we obtained on the acceleration of relativistic electrons in this regime [5]. This experiment has been set-up at Ecole Polytechnique in 1998. The experimental set-up was very similar to the one we used for beat-wave experiments [7]. The pulse duration was $\tau_{1/2} = 400\,fs$ Full Width Half Maximum, with an laser intensity in the order of $10^{17}\,W/cm^2$. This short pulse laser beam was focused by a 1.4 m focal length off-axis parabola into a vessel filled with helium gas. This laser pulse ionized the gas and excited a relativistic plasma wave. Electrons with a 3 MeV total energy were injected into this wave. A magnetic spectrometer together with scintillators and photomultiplier tubes are used to detect the accelerated electrons. Special care has been taken to minimize the noise coming from deflected electrons either in the plasma wave or in the gas itself. As a first step we measured the number of acceler-ated electrons in the first channel as a function of the gas pressure, or equivalently as a function of the electron density for a moderate laser energy up to 1.5 J. This is plotted in Fig.1. This first result show an effective acceleration of the order of 1 MeV, with a maximum when the electron density is close to the optimum value for which the laser pulse length is about half the plasma wavelength.

The signal starts from 0 at very low pressure, then increases up to a maxi-mum near 0.4-0.5 mbar before decreasing slowly at higher pressures. These results are consistent with the acceleration in the plasma wave generated by laser wake-field. This is indicated in Fig.1 where the maximum theoretical longitudinal electric field is plotted vs gas pressure for a pulse duration of 400 fs. This field scales as $\omega_p \tau exp(-(\omega_p \tau)^2/4)$, and has a maximum for $\omega_p \tau = 2$. Here $\tau = \tau_{1/2}/2\sqrt{(Ln2)}$ and the intensity is Gaussian in time : $I = I_{max}exp(-(t/\tau)^2)$ and ω_p is the plasma frequency. On Fig.2 we present a typical spectrum obtained in the resonant condi-tion. To check the energy of the electrons impinging on a given channel, we have

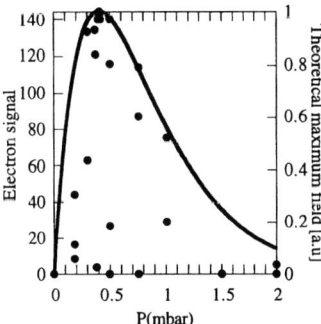

FIGURE 1. Number of electrons in the first channel as a function of the gas pressure, for a moderate laser energy up to 1.5 J. The solid is the amplitude of the longitunal electric field associated with the relativistic plasma wave.

inserted stainless steel filters with various thichnesses in front of some scintillators. The signal was reduced in agreement with the expected one for this given electron energy.

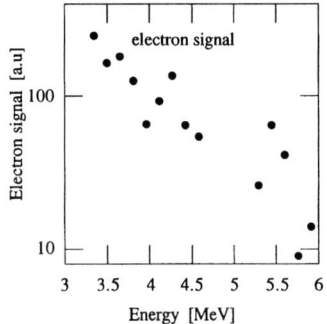

FIGURE 2. Electron spectrum obtained in the resonant condition

LASER BEAT WAVE

In the beat-wave scheme, the accelerating plasma pulses is generated by the beating of two laser waves with slightly different frequencies. In this technique, the beating between two laser beams with slightly different angular frequencies ω_1 and ω_2 excites a longitudinal oscillation in a plasma at the difference frequency

$\delta\omega = \omega_1 - \omega_2$. If the natural oscillation frequency of the electrons in the plasma ω_p is close to the difference frequency $\delta\omega$, the plasma oscillation is excited resonantly and can lead to large electric fields. The phase velocity of the longitudinal field, which is equal to the group velocity of the laser electromagnetic wave in the plasma, is very close to the speed of light, as the plasma density is very low. The first group to demonstrate the generation of such intense relativistic plasma waves (with 0.7 GV/m electric field) at the UCLA in 1993 [6] used two infra-red CO_2 lasers beams. In 1985 the same group has mesured by Thomson scattering the amplitude (1-3 %) of these fast waves [13]. Similar experiments were then made in the INRS [14] in Canada, at RAL [15] in England and at Ecole Polytechnique [16] in France. These last experiments have been done with shorter laser wavelengths ($\lambda \simeq 1\,\mu$m). The longitudinal electric field has been used to accelerate relativistic particles to very high energies. By the fact electron acceleration has been demonstrated with an energy gain of up to 30 MeV with CO_2 lasers ($\lambda \simeq 10\,\mu$m) [13,17] and up to 2 MeV with Nd-Glass lasers ($\lambda \simeq 1\,\mu$m) [16]. In the first case (using two infra-red CO_2 lasers beams) the plasma wave amplitude is limited by the relativistic detuning which occurs when the quiver velocity of the plasma electrons in the wave becomes relativistic [6] . The plasma frequency decreases and drifts away from the exciting beat frequency. The beating then tends to decrease the plasma wave amplitude. In the second case (laser wavelength near $1\,\mu$m), ions begin to move during the growth of the plasma wave thus modifying the resonance condition. The growth time is limited to a few times of the typical ion plasma period. After this delay electron waves decay by modulationnal instability [16] into ion waves and electron waves of lower phase velocities reducing the energy transfer into fast electron plasma waves. Injected electrons at an energy of 3 MeV were observed to be accelerate to 3.7 MeV in agreement with the theoretical value for the electric field of the order of 0.6 GV/m [7]. Such electron spectra obtained in this experiment is presented in Fig.3.

The resonant electron density was 1.115×10^{17} cm^{-3} for which corresponds a plasma wavelength of $100\,\mu$m and a relativistic factor of $\gamma = 94.5$. Also indicated in continuous line is the theoretical electron spectra obtained for a plasma wave amplitude of 2% and a Rayleigh length of 2.4 mm. The future of beat-wave acceleration experiments relies on the possibility of developing short and intense two-frequency laser pulses [18]. This will provide a better coupling between the laser energy and the energy transfer to the fast plasma waves. Indeed in this scheme ions do not have the time to move and modulational instabilities will not destroy the relativistic plasma waves.

SELF MODULATED LASER WAKE FIELD

In the self-modulated laser wakefield regime [2–4] , an intense laser pulse, longer than a few plasma wavelengths is focused on a high-density gas-jet. The laser pulse is modulated by Raman-type instabilities and a very large amplitude plasma wave is generated. This wave can trap electrons and accelerate them along the plasma.

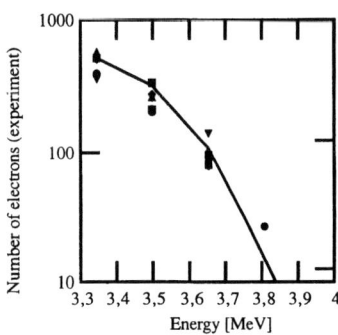

FIGURE 3. Comparison between the best electron spectra and the results of the acceleration model for a peak plasma wave amplitude of $\delta = 2\%$ and a Rayleigh length of 2.4 mm. Each symbol represents the numbers of electrons per channel for one single shot, and the solid line simply joins the predicted values for the number of electrons in each channel. The number of electrons predicted by the model was divided by a factor of 8.

Mori *et al.* [19] discussed the acceleration of electrons to very high energies due to the Raman Forward Scattering instability (RFS). Experimental observations of RFS have been done by Joshi et al. at UCLA who had measured electrons up to 1.4 MeV [20].

Electrons produced in the relativistic wave breaking regime up to 100 MeV have been observed [21] in this type of experiments with multi-TW short pulse lasers. The first experiment was done at RAL in 1994 [8] with 25 J, 1 ps laser beam at $1.054\,\mu$m in underdense plasmas. On this experiment the so-called wavebreaking was demonstrated. It was characterized by the sudden increase in both the number and maximum energies of electrons from the plasma itself as well as by the loss of coherence of the wave as seen from the broadening of the spectrum in the forward direction. The maximum electric field achieved was estimated to be over 100 GV/m. Measurements of the Thomson scattered light on the relativistic plasma waves along the laser axis propagation indicates a self guiding over the whole gas jet in the relativistic regime and a plasma waves amplitude of $40 \pm 20\%$ [22].

A new and very interesting issues is the production of such electron beams with the use of table top TW lasers with higher repetition rates (10Hz). Electrons with energies of up to 10 MeV have been obtained [23], and more recently electrons with energies of up to 70 MeV have been generated in well define homogenous supersonic gas jets [24]. In this last experiment electrons were accelerated in the SMLWF regime for ultra-relativistic laser pulses with $a_0 > 1$, (where a_0 is the normalized vector potential of the laser). For the first time, it was observed that the maximum electron energy increases when the electron density decreases indicating that electrons are accelerated by relativistic plasma waves. The experiment was performed at the LOA with the titanium doped sapphire (Ti:Sa) laser [25] operating at $\lambda_L = 0.82\,\mu m$ in the chirped-pulse amplification (CPA) mode [26] . In this configuration the laser delivered an energy up to 0.6 J (on target) in 35 fs (FWHM) pulses with a linear polarization. The laser beam was focused onto the edge of a gas jet with a f/7.5 off-axis parabola. The laser distribution at full energy at the focal plane was a Gaussian function with a waist $w_0 = 6\,\mu m$ containing 50 % of the total laser energy. This corresponds to typical powers of 20 TW and to on-target intensities I_L of the order of $2 \times 10^{19}\,W/cm^2$. To avoid refraction induced by ionization processes [27,28], the laser beam was focused onto the sharp edge of a 2 (and 4) millimeter diameter laminar plume of helium gas from a pulsed, supersonic gas jet located 1 mm below the focal region. The flat top neutral density profile was perfectly characterized by interferometry [29] . Electron measurement has been done carefully including several null tests were done in order to make sure that the diode signals were really due to electrons. We present in Fig.4 two typical electron spectra obtained at $5 \times 10^{19}\,cm^{-3}$ and $1.5 \times 10^{20}\,cm^{-3}$.

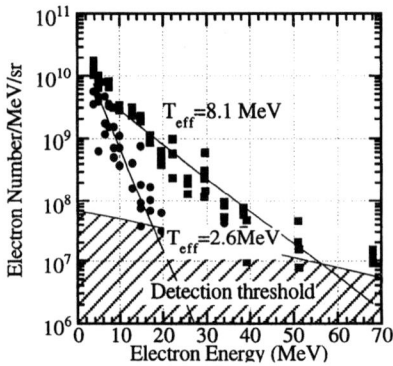

FIGURE 4. Electron spectra measured at $1.5 \times 10^{20}\,cm^{-3}$ (circles) and $5 \times 10^{19}\,cm^{-3}$ (squares). Exponential fit with the deduced effective electron temperature.

The laser parameters were 0.6 J and 35 fs. We observe that the distribution of electrons of energy above 4 MeV is well fitted by an exponential function, characteristic of an effective temperature for the electron beam. These effective temperatures are 8.1 MeV (2.6 MeV) for an electron density of $5 \times 10^{19}\,cm^{-3}$ ($1.5 \times 10^{20}\,cm^{-3}$). We can also deduce a typical value of 54 MeV (15 MeV) for the maximum electron energy. We observe an important decrease of the effective temperature and of the maximum electron energy when increasing the electron density. This is summarized in Fig.5 where we present the maximum electron energy versus the electron density. It decreases from 70 MeV to 15 MeV when the electron density increases from $1.5 \times 10^{19}\,cm^{-3}$ to $1.5 \times 10^{20}\,cm^{-3}$.

First, we can compare the maximum electron energy to the one due to acceleration in relativistic plasma waves with a constant amplitude. This energy is equal

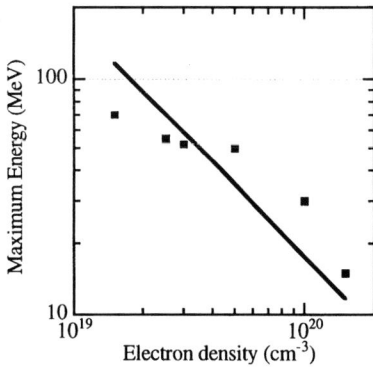

FIGURE 5. Maximum electron energy as a function of the electron density. Squares correspond to experimental values. The continuous line corresponds to theoretical calculation with a normalized electrostatic field $E_z/E_0 = 0.5$ Laser parameters: 0.6 J, 35 fs and 2×10^{19} W/cm^2.

to the product of the electrostatic field by an optimum length. This length is the dephasing length and corresponds to exactly half a wavelength in the wave frame [30] : $W_{max} \approx 4\gamma_p^2(E_z/E_0)mc^2$, where γ_p is the plasma wave Lorentz factor (which is equal to the square root of the critical density to electron density ratio n_c/n_e) and E_z/E_0 is the electrostatic field normalized to $E_0 = cm\omega_p/e$. Presented in Fig.5 is the theoretical value deduced from this equation for a given value of the normalized electrostatic field $E_z/E_0 = 0.5$. Non linear corrections due to the effect of a relativistic pump, to self-channeling [31] and to the reduction of the phase velocity of plasma waves [31] have been neglected. Experimental results are in reasonable agreement with this model. For electron densities greater than 1.5×10^{19} cm^{-3} the maximum electron energy varies as : $E_{max}(MeV) = 0.76n_c/n_e$.

The total charge was also measured with an integrated charge transformer as a function of the electron density, fixing the pulse duration at 35 fs. The result is plotted in Fig.6 as a function of $\omega_p\tau$, from the wakefield regime ($\omega_p\tau \approx 2$)to the self-modulated regime ($\omega_p\tau > 2$).

Near the wakefield regime no electrons were detected, whereas with increasing $\omega_p\tau$ the number of electrons increased. At higher electron densities the number of electrons with energy greater than 3.7 MeV and detected by the spectrometer became smaller than those measured with the charge collector because the radial wake scattered electrons out of the aperture cone of the electron spectrometer. The total charge is about 8 nC in agreement with recent numerical simulations [32], whereas the charge for electrons with energies greater than 3.7 MeV is about 1 nC.

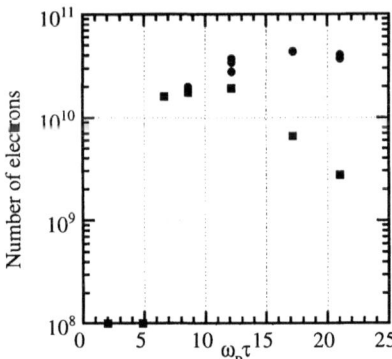

FIGURE 6. Total electron number (circles) and number of electrons with energies greater than 3.7 MeV (squares) as a function of the product $\omega_p\tau$, where ω_p is the plasma frequency and τ is the laser pulse duration (35fs).

PERSPECTIVES

Two majors issues are presented here. The first one is related to the obtention of high particles energies by guiding a laser pulse in very underdense plasmas. The second one is related to the use on the electron beam already produced.

Perspectives: Higher Particles Energies

The key issues in order to increase the energy gain for accelerated particles is the increase of the dephasing length, that is to decrease n_e/n_c, or for a given laser wavelength to decrease the electron density. Electron energies of several hundred MeV to 1 GeV can be obtained by the use of table top TW lasers with the following parameters: 0.1-1 J for the laser energy and pulse duration of the several tens fs. The constrain is that the amplitude of the plasma wave has to be high over a long distance appoximately equal to the dephasing length and typically of 1 to 10 cm. The solution to overcome this problem is the guiding of the laser beam at high intensities. An overview of this aspect has been done by Leemans *et al.* in 1996 with a theoretical approach [33]. Interaction of Terawatt (TW) laser pulses over long distances with a plasma is also of great interest for X-ray lasers [34], harmonic generation [35] and inertial confinement fusion with the fast ignitor scheme [36].

For a gaussian laser pulse, diffraction commonly limits the interaction length to the Rayleigh length $z_R = \pi w_0^2/\lambda_L$. This is why guiding over several Rayleigh lengths is important for the above applications. In the case of laser plasma accelerators, an intense laser pulse propagating in an underdense plasma can excite strong relativistic plasma waves via the ponderomotive force. The maximum energy that electrons can get from the relativistic plasma waves (RPW) is proportional to the product of the RPW amplitude and the dephasing length (the length over which electrons stay in an accelerating arch of the relativistic plasma waves). In order to extract maximum energy from the waves, the electrons must travel in a long and intense RPW. Previous experiments showed that the creation of such a long and intense wave can be achieved through self-focusing [37,22] when the laser pulse power P_L is slightly greater than the critical power for relativistic self-focusing P_C, where $P_C(GW) = 17(\omega_0/\omega_p)^2$, for a gaussian laser pulse. In future experiments, lower density plasmas will tend to be used in order to increase the dephasing length and the extractable energy. Current laser technology hardly provides enough power for self-focusing at lower density (at $n_e = 10^{18}\,cm^{-3}$, $P_C \approx 20\,TW$), thus guiding the pulse in a guiding structure becomes necessary.

Different ways have been explored experimentally recently: tube capillary [38,39], capillary discharge [40] and plasma channel [41–43]. Plasma channels have been obtained by several groups. The first very impressive results are those obtained by Milcherg's group [42,43]. Using an axicon they create a light pipe for moderate intensity laser pulses using 100 ps laser beams. A first pulse prepared a shock driven, axially extended radial electron density profiles which guided a second laser pulse over 70 Rayleigh length. Such channels are not well adapted for high intensity laser guiding because the heavy gas used to preformed the channel is not fully ionized. The group of Downer found a very nice solution to overcome this problem preioinizing helium gas using a pulsed electrical discharge [44]. Nevertheless these experiments have done in gas filled chamber and not optimized for who wants to guide high intensity laser pulse since the laser beam will ionize the surrounding gas before the entrance of the plasma channel. In order to avoid refraction induced by ionization the laser beam must first propagate into vacuum and then enter in a fully ionized plasma channel. Such plasmas channels with vacuum boundaries are created in capillary discharges or in gas jet. Using capillary discharge guiding has been demonstrated by the group of Zigler [40]. Such approach offer the possibility of tapered the electron density profile along the plasma guide which can be of great interest for particle acceleration [45]. Channel formation in long laser pulse interaction with helium gas jets has been extensively studied at LULI. Plasma expansion was governed by a radial thermal wave. Plasma channel formation and guiding using short pulses has been studied by Krushelnick and also by Wagner [37,46]. In these experiments the ponderomotive force expulses radially the electrons, due to the charge separation ions are also expulsed radially generating a channel.

At Luli, the propagation of a pulse with intensities from $10^{17}\,W/cm^2$ to a few $10^{18}\,W/cm^2$ has been studied in a well-defined plasma channel created by a long laser pulse [47,48]. A change of propagation and of the interaction when P_L becomes greater than P_C has been observed. For long pulse (ps) with a laser power lower than the critical power for self-focusing : $P_L/P_C < 1$ ($I_0 = 2 \times 10^{17}\,W/cm^2$), the laser pulse was guided by the preformed plasma channel over 3 Rayleigh lengths (4 mm) and a longitudinal plasma wave was generated by envelope self-modulation of the pulse. For a short pulse (0.3 ps) and $P_L/P_C \gg 1$, the interaction was dominated by self-focusing and Raman instabilities. In Fig.7 interferometry measurement show in the first case a propagation of a single filament through the whole plasma pipe and in the second case a presence of many filaments propagates at large angles.

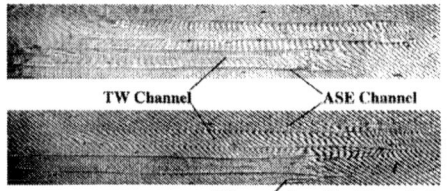

FIGURE 7. Top image: Interferogram of the plasma taken 20 ps after the main pulse in the plasma. Parameters were: $n_e = 1. \times 10^{19}\,cm^{-3}$, $P_L/P_C = 0.8, \tau = 7ps$, Buttom Image: Same but with a delay of 10 ps and for the following parameters: $n_e = 1. \times 10^{19}\,cm^{-3}$, $P_L/P_C = 17, \tau = 0.37\,ps$

Finally a very attractive solution to guide an intense laser beam over a long distance is the use of capillary tube. With such hollow capillary dielectric tubes monomode guiding of $10^{16}\,W/cm^2$ laser pulses over 100 Rayleigh length has been demonstrated [38].

Perspective : Particle Sources Production

Laser plasma accelerator are now available generating electron beams with energy up to several 10 MeV and a total charge of few nC [24].The pulse duration of the electron beam at the output of the plasma is less then the laser pulse duration and can therefore be in the order of 30 fs. Far from the plasma the duration of the electron beam increases due to the broad energy spectrum in the beam. Angular distribution of the electrons beams produced by a laser-plasma accelerator has been measure at RAL by using nuclear activation techniques. The electron yield is observed to increase with the plasma density, and reaches up to 4×10^{11} fast electrons (> 10 MeV) [49]. Electrons are found to be emitted in a cone along the laser axis and the angular spread of the emitted electrons is found to increase with the plasma density. The observed activation yield is an order of magnitude higher than that in solid target experiments with similar laser parameters [49], suggesting potential application in production of activated materials and for isotope production. In order to activate targets a converter is fixed on way of the electron beam to produce Bremmsstrahlung radiation. Such a γ converter is a thick foil (1 mm) of high Z material which absorbed efficiently electrons by Bremmsstrahlung [50]. The γ-flash irradiates and activates the target. Photofission of Uranium has already be obtained by (γ, n) reactions. Using a 1 cm thick uranium target estimation of 4000 prompt neutrons/pulse has been established by Shkolnikov [51].

More recently (γ, n) activation experiments using laser wakefield accelerators has been performed by the group of Leemans in which several radio-isotope have been identified [52]. Focusing a TW laser on the electron beam X-ray pulses can be generated by Thomson scattering with a pulse duration as short as the laser pulse [53]. Such Short X ray pulses (30keV) have been used by the group of Leemans in order to characterize a 50 MeV electron beam produced by a linac [33]. Positrons have also been produced using a laser plasma accelerator [54]. Other applications of such accelerators can be found in chemistry. For example at Michigan the electron charge of the beam was high enough to conduct time resolved investigations of radiation induced chemical events [55]. It is important to note that the laser plasma accelerator offer a unique capability to deliver electron beams physically separated from the source (the laser). This can be useful because the region where the electron beam is produced can be very small and the associated radioactivity can be isolated at lower costs. The second main advantage is the presence on the same site of the perfect synchronized laser and electron beams.

CONCLUSIONS

We have presented a synthesis of experimental results on laser particles acceleration with an underdense plasma. The three considered schemes LWF, LBW and SMLWF have been experimentally demonstrated. A new interested on this field occurs with appearance of Table top TW laser.

ACKNOWLEDGMENTS

This article presents a synthesis number of experimental results obtained in the laser plasma particle context. Due to the large number of very good publications in this new excited field I probably forgot to mention some very good works. I hope that the authors would accept my excuses if I forgot to mention their works. Since I present a lot of results obtained by a fruitful collaboration with many other groups I would like to take the opportunity to knowledge vigorously physicist from the LULI (France), LOA (Laboratoire d'Optique Appliques, France), IC (Imperial College, UK), RAL (Rutherford Appleton Laboratory, UK), UCLA (University of California of Los Angeles USA), LPNHE (Laboratoire de Physique Nuclaire et des Hautes Energies, France), SESI (Solides Irradis), LPGP (Laboratoire de Physique des Gaz et Plasmas, France), CPTH (Centre de Physique Thorique, France). This work hase been partially supported by Ecole Polytechnique, IN2P3-CNRS, SPI-CNRS, and by EU Large Facility Program under Contract No. FMGE CT95 0044.

REFERENCES

1. Tajima, T., and Dawson, J. M., *Phys. Rev. Lett.*, **43**, 267 (1979).

2. Andreev, N. E., Gorbunov, L. M., Kirsanov, V. I., Pogosova, A. A., and Ramaza-shvili, R. R., *JETP Lett*, **55**, 571 (1992).

3. Antonsen, J. T. M., and Mora, P., *Phys. Rev. Lett.*, **69**, 2204–2207 (1992).

4. Sprangle, P., Esarey, E., Krall, J., and Joyce, G., *Phys. Rev. Lett.*, **69**, 2200 (1992).

5. Amiranoff, F., Baton, S., Bernard, D., Cros, B., Descamps, D., Dorchies, F., Jacquet, F., Malka, V., Matthieussent, G., Marques, J. R., Mine, P., Modena, A., Mora, P., Morillo, J., and Najmudin, Z., *Phys. Rev. Lett.*, **5**, 995 (1998).

6. Clayton, C. E., Marsh, K. A., Dyson, A., Everett, M., Lal, A., Leemans, W. P., Williams, R., and Joshi, C., *Phys. Rev. Lett.*, **70**, 37 (1993).

7. Amiranoff, F., Bernard, D., Cros, B., F.Jacquet, G.Matthieussent, Mine, P., Mora, P., Morillo, J., Moulin, F., Specka, A. E., and Stenz, C., *Phys. Rev. Lett.*, **74**, 5220 (1995).

8. Modena, A., Dangor, A., Najmudin, Z., Clayton, C., Marsh, K., Joshi, C., Malka, V., Darrow, C., Neely, D., and Walsh, F., *Nature*, **377**, 606–608 (1995).

9. Umstadter, D., Chen, S. Y., Maksimchuk, A., Mourou, G., and Wagner, R., *Science*, **273**, 472–475 (1996).

10. Marquès, J. R., Geindre, J. P., Amiranoff, F., Audebert, P., Gauthier, J. C., Antonetti, A., and Grillon, G., *Phys. Rev. Lett.*, **76**, 3566 (1996).

11. Siders, C. W., LeBlanc, S. P., Fisher, D., Tajima, T., Downer, M. C., Babine, A., Stepanov, A., and Sergeev, A., *Phys. Rev. Lett.*, **76**, 3570 (1996).

12. Hamster, H., Sullivan, A., Gordon, S., and Falcone, R. W., *Phys. Rev. E*, **49**, 671 (1994).

13. Clayton, C. E., Joshi, C., Darrow, C., and Umstadter, D., *Phys. Rev. Lett.*, **54**, 2343 (1983).

14. Ebrahim, N. A., Lavigne, P., and Aithal, S., *IEEE Trans. Plasma Sci.*, **ns12**, 3539 (1985).

15. Dangor, A. E., *IEEE Trans. Plasma Sci.*, **ps15**, 161 (1987).

16. Amiranoff, F., Laberge, M., Marques, J., Moulin, F., Fabre, E., Cros, B., G.Matthieussent, Benkheiri, P., F.Jacquet, Meyer, J., Mine, P., Stenz, C., and Mora, P., *Phys. Rev. Lett.*, **68**, 3710 (1992).

17. Everett, M., Lal, A., Gordon, D., Clayton, C. E., March, K. A., and Joshi, C., *Nature*, **368**, 527 (1994).

18. Joshi, C., Clayton, C. E., Mori, W. B., Dawson, J. M., and Katsouleas, T., *Comments on Plasma Physics and Controlled Fusion*, **16**, 65 (1994).

19. Mori, W. B., Decker, C. D., Hinkel, D. E., and Katsouleas, T., *Phys. Rev. Lett.*, **72**, 1482–1485 (1994).

20. Joshi, C., Tajima, T., Dawson, J. M., Baldis, H. A., and Ebrahim, N. A., *Phys. Rev. Lett.*, **47**, 1285 (1981).

21. Gordon, D., Tzeng, K. C., Clayton, C. E., Dangor, A. E., Malka, V., Marsh, K. A., Modena, A., Mori, W. B., Muggli, P., Najmudin, Z., and Joshi, C., *Phys. Rev. Lett.*, **80**, 10 (1998).

22. Clayton, C. E., Gordon, D., Marsh, K. A., Joshi, C., Malka, V., Najmudin, Z., Modena, A., Dangor, A. E., Neely, D., and Danson, C., *Phys. Rev. Lett.*, **81**, 100 (1998).

23. Gahn, C., Tsakiris, G. D., Pukhov, A., Meyer-ter-Vehn, J., Pretzler, G., Thirolf, P.,

Habs, D., and Witte, K. J., *Phys. Rev. Lett.*, **83**, 4772–4775 (1999).

24. Malka, V., Faure, J., Marques, J. R., Amiranoff, F., Rousseau, J. P., Ranc, S., Chambaret, J. P., Najmudin, Z., Warton, B., Mora, P., and Solodov, A., *Phys. Plasmas*, **8**, 2605 (2001).

25. Antonetti, A., Blasco, F., Chambaret, J. P., Cheriaux, G., Darpentigny, G., Blanc, C. L., Rousseau, P., Ranc, S., Rey, G., and Salin, F., *applied physics B*, **65**, 197–204 (1997).

26. Strickland, D., and Mourou, G., *Opt. Comm.*, **56**, 219–221 (1985).

27. Monot, P., Auguste, T., Lompr, L. A., Mainfray, G., and Manus, C., *J. Opt. Soc. Am. B*, **9**, 1579 (1992).

28. Malka, V., Wispelaere, E. D., Marqus, J. R., Bonadio, R., Amiranoff, F., Blasco, F., Stenz, C., Mounaix, P., Grillon, G., and Nibbering, E., *Phys. Plasmas*, **3**, 1682 (1996).

29. Malka, V., Coulaud, C., Geindre, J. P., Lopez, V., Najmudin, Z., Neely, D., and Amiranoff, F., *Rev. Sci. Instrum.*, **71**, 2329–2333 (2000).

30. Mora, P., and Amiranoff, F., *J. Appl. Phys.*, **66**, 3476–3481 (1989).

31. Esarey, E., Hafizi, B., Hubbard, R., and Ting, A., *Phys. Rev. Lett.*, **80**, 5552 (1998).

32. Tzeng, K.-C., Mori, W. B., and Katsouleas, T., *Phys. Rev. Lett.*, **79**, 5258 (1997).

33. Leemans, W. P., Widers, C. W., Esarey, E., Andreev, N. E., Shvets, G., and Mori, W. B., *IEEE Trans. Plasma Sci.*, **24**, 331–350 (1996).

34. Burnett, N. H., and Enright, G. D., *IEEE J. Quantum Electron.*, **26**, 1797–1808 (1990).

35. Li, X. F., L'Huillier, A., Ferray, M., Lompre, L. A., and Mainfray, G., *Phys. Rev. A*, **39**, 5751 (1989).

36. Tabak, M., Hammer, J., Glinsky, M. E., Kruer, W. L., Wiks, S. C., Woodworth, J., Campbell, E. M., Perry, M. D., and Mason, R. J., *Phys. Plasmas*, **1**, 1626–1634 (1994).

37. Krushelnick, K., Ting, A., Moore, C. I., Burris, H. R., Esarey, E., Sprangle, P., and Baine, M., *Phys. Rev. Lett.*, **78**, 4047–4050 (1997).

38. Dorchies, F., Marqus, J. R., Cros, B., Matthieussent, G., Courtois, C., Vlikoroussov, T., Audebert, P., Geindre, J. P., Rebibo, S., Hamoniaux, G., and Amiranoff, F., *Phys. Rev. Lett.*, **80**, 720–723 (1998).

39. Mackinnon, M. B. A. J., Gaillard, R., Willi, O., and Offenberger, A. A., *Phys. Rev. E*, **57**, 4899 (1998).

40. Ehrlich, Y., Cohen, C., Zigler, A., Krall, J., Sprangle, P., and Esarey, E., *Phys. Rev. Lett.*, **77**, 4186 (1996).

41. Malka, V., De Wispelaere, E., Amiranoff, F., Baton, S., Bonadio, R., Coulaud, C., Haroutunian, R., Modena, A., Puissant, D., Stenz, C., Hüller, S., and Casanova, M., *Phys. Rev. Lett.*, **79**, 2979–2982 (1997).

42. Durfee III, C. G., Lynch, J., and Milchberg, H. M., *Phys. Rev. Lett.*, **71**, 2409 (1993).

43. Durfee III, C. G., and Milchberg, H. M., *Phys. Rev. E*, **51**, 2368 (1995).

44. Gaul, E. W., Blanc, S. P. L., Rundquist, A. R., Zgadzaj, R., Langhoff, H., and Downer, M. C., *Appl. Phys. Lett.*, **77**, 4112 (2000).

45. Sprangle, P., Hafizi, B., Penano, J. R., Hubbard, R. F., Ting, A., Zigler, A., and Antonsen, T. M., *Phys. Rev. Lett.*, **85**, 5110 (2000).

46. Wagner, R., Chen, S.-Y., Maksimchuk, A., and Umstadter, D., *Phys. Rev. Lett.*, **78**, 3125–3128 (1997).

47. Malka, V., J.Faure, Marques, J. R., Amiranoff, F., Courtois, C., Najmudin, Z., Krushelnick, K., Salvati, M., and Dangor, A. E., *IEEE Trans. Plasma Sci.*, **28**, 1078–1083 (2000).

48. J.Faure, Malka, V., Marques, J. R., Amiranoff, F., Courtois, C., Najmudin, Z., Krushelnick, K., Salvati, M., Dangor, A. E., Solodov, A., Mora, P., Adam, J. C., and Hron, A., *Phys. Plasmas*, **7**, 3009–3017 (2000).

49. Santala, M. I. K., Zepf, M., Watts, I., Beg, F. N., Clark, E., Tatarakis, M., Krushelnick, K., Dangor, A. E., McCanny, T., Spencer, I., Singhal, R. P., Ledingham, K. W. D., Wilks, S. C., Machacek, A. C., Wark, J. S., Allott, R., Clarke, R. J., and Norreys, P. E., *Phys. Rev. Lett.*, **84**, 1459–1462 (2000).

50. Norreys, P., Santala, M., Clark, E., Zepf, M., Watts, I., Beg, F. N., Krushelnick, K., Tatarakis, M., Dangor, A. E., Fang, X., Graham, P., McCanny, T., Singhal, R. P., Ledingham, K. W. D., Creswell, A., Sanderson, D. C. W., Magill, J., Machacek, A., Wark, J. S., Allott, R., Kennedy, B., and Neely, D., *Phys. Plasmas*, **6**, 2150–2156 (1999).

51. Shkolnikov, P. L., Kaplan, A. E., Pukhov, A., and ter Vehn, J. M., *Appl. Phys. Lett.*, **71**, 3471–3473 (1997).

52. Leemans, W. P., Rodgers, D., Catavras, P. E., Geddes, C. G. R., Fubiani, G., Esarey, E., Shadwick, B. A., Donahue, R., and Smith, A., *Phys. Plasmas*, **8**, 2510–2516 (2001).

53. Schoenlein, R. W., Leemans, W. P., Chin, A. H., Volfbeyn, P., Glover, T. E., Balling, P., Zolotorev, M., Kim, K. J., Chattopadahayay, S., and Shang, C. V., *Science*, **274**, 236–238 (1996).

54. Gahn, C., Tsakiris, G. D., Pretzler, G., Witte, K. J., Delfin, C., Wahlstrom, C., and Habs, D., *submitted to Apl. Phys. Lett.* (2001).

55. Saleh, N., Flippo, K., Nemoto, K., Umstadter, D., Crowell, R. A., Jonah, C. D., and Trifunac, A. D., *Rev. Sci. Instrum.*, **71**, 2305–2308 (2000).

Excitation of accelerating plasma waves by counter-propagating laser beams

Gennady Shvets and Nathaniel J. Fisch

Princeton Plasma Physics Laboratory[1]
Princeton NJ 08543

Alexander Pukhov

Max-Planck-Institut für Quantenoptik, D-85748 Garching, Germany

Abstract. Generation of accelerating plasma waves using two counter-propagating laser beams is considered. Colliding-beam accelerator requires two laser pulses: the long pump and the short timing beam. We emphasize the similarities and differences between the conventional laser wakefield accelerator and the colliding-beam accelerator (CBA). The highly-nonlinear nature of the wake excitation is explained using both non-linear optics and plasma physics concepts. Two regimes of CBA are considered: (i) the short-pulse regime, where the timing beam is shorter than the plasma period, and (ii) parametric excitation regime, where the timing beam is longer than the plasma period. Possible future experiments are also outlined.

INTRODUCTION AND MOTIVATION

Plasma is an attractive medium for ultra-high gradient particle acceleration because it can sustain a very high electric field, roughly limited by the cold wave-breaking field $E_{\text{WB}} = mc\omega_p/e \approx \sqrt{n[\text{cm}^{-3}]}\text{V/cm}$, where $\omega_p = \sqrt{4\pi e^2 n/m}$ is the plasma frequency and n is the electron density. To accelerate injected particles to velocities close to the speed of light c, this electric field has to be in a form of a fast longitudinal plasma wave with phase velocity $v_{\text{ph}} \approx c$. The frequency of the fast plasma wave is ω_p, and its wavenumber is $k_p \approx \omega_p/c$. Excitation of such plasma waves can be accomplished by lasers or fast particle beams [1–3].

Below we review the basics of the linear plasma wave excitation in very general terms, without restricting ourselves to the specifics. Let's assume that plasma electrons are subject to the electric field of the fast plasma wave $\vec{E}$, as well as other nonlinear forces $\vec{F}_{\text{NL}}$, for example, the ponderomotive force of one or more

[1] This work was supported by the US DOE Division of High-Energy and Nuclear Physics

laser pulses. The total current $\vec{J} = \vec{J}_p + \vec{J}_2$ which enters Ampere's law $\vec{\nabla} \times \vec{B} = (1/c)\partial_t \vec{E} + (4\pi/c)(\vec{J}_p + \vec{J}_2)$ is intentionally split into two components. The first one, $\vec{J}_p = -en\vec{v}_e$, where $\vec{v}_e$ is the electron fluid velocity, is driven by the electric field $\vec{E}$ and satisfies $\partial_t \vec{J}_p = c^2 n\vec{E}$. The second component $\vec{J}_2$ is driven by the nonlinear ponderomotive force, or could also represent an external current provided by injected electron beam. Taking the time derivative of the Ampere's law yields:

$$\left(\frac{\partial^2}{\partial t^2} + \omega_{p0}^2 \right) \vec{E} + c^2 \nabla \times \nabla \times \vec{E} = -4\pi \frac{\partial \vec{J}_2}{\partial t}, \qquad (1)$$

where the $\nabla \times \nabla \times \vec{E}$ term naturally vanishes in 1D. One can say that the science of making a plasma accelerator is about finding the most effective way of producing the appropriate $J_{2z}(z,t)$. Of course, not every functional form of $J_{2z}(z,t)$ is useful for relativistic particle acceleration. In the rest of this paper we concentrate on using two counter-propagating laser beams to excite $J_{2z}(z - ct)$.

I   COMPARISON OF SINGLE-BEAM AND COLLIDING BEAM ACCELERATORS

The simplest laser-driven plasma accelerator is the plasma beatwave accelerator [1] (PBWA). It employs a pair of co-propagating laser beams with normalized vector-potentials $\vec{a}_{0,1} = e\vec{A}_{0,1}/mc^2$ and frequencies ω_0 and $\omega_1 = \omega_0 - \omega_p$. The nonlinear current J_{2z} is driven by the ponderomotive force of the resulting electromagnetic beatwave according to $\partial_t J_{2z} = en\partial_z(\vec{a}_0 \cdot \vec{a}_1)$. If the two laser-beams are detuned by the plasma frequency ω_p, plasma wave is resonantly excited.

From Eq. (1), to excite a plasma wave, one needs to deposit momentum into the plasma. The source of this momentum is, of course, the laser. However, since the typical laser frequencies $\omega_{0,1} \gg \omega_p$, it is impossible for a laser photon to impart its entire momentum to the plasma. What happens instead is that the frequency of a laser photon is down-shifted by the amount ω_p, depositing the remainder momentum and energy into the plasma. In the case of the PBWA, the higher-frequency photons at ω_0 are scattered into the lower-frequency photons at $\omega_1 = \omega_0 - \omega_p$. Schematically, this process is shown in the top Fig. (1). The phasors of the lasers lie on the $\omega^2 = \omega_p^2 + c^2 k^2$ dispersion curve, and the vector difference of these phasors gives the phasor of the driven plasma wave. The total rate of the momentum transfer to plasma in PBWA is then proportional to the relative momentum transfer per photon $\eta = \omega_p/\omega_0$, times the rate of scattering which is proportional to the beam intensity. Since the relative amount of down-shifting $\eta \ll 1$, high laser intensities are needed to ensure the high overall rate of the momentum transfer. Note that Fig. 1 (top) is also applicable to the laser wakefield accelerator (LWFA) which employs a single ultra-short ($\tau_L \approx 2\omega_p^{-1}$) laser pulse. Broad bandwidth of the pulse implies that it contains a continuum of frequency

Kinematics of Wake Excitation

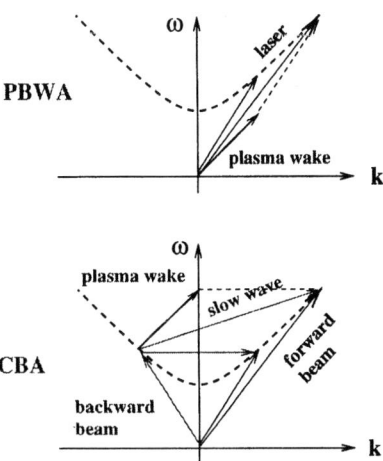

FIGURE 1. PBWA: kinematics of the excitation of the plasma wake by a co-propagating wavepacket consisting of two frequency components differing by $\Delta\omega = \omega_p$. Phase velocity of plasma wake $v_{\text{ph}} \approx vg$, where v_g is the group velocity of the wavepacket; CBA: same, only using an extra counter-propagating laser beam. Nonlinear beating of two slow waves gives rise to fast plasma wake.

pairs differing from each other by ω_p. Because the pulse is short, wake excitation is not resonant, and even larger than in PBWA intensity is needed (typically, close to $10^{18}\text{W}/\text{cm}^2$ to achieve $E/E_{\text{WB}} \sim 0.2$).

In CBA [4,5] we take a very different approach by employing two *counter-propagating* laser beams with differing frequencies: one short and another long ($\tau_p = 2L_p/c$, where L_p is the length of the plasma). When the two beams interact in the plasma, the photons of the higher-frequency beam scatter into the photons of the lower-frequency beam. The crucial difference from the PBWA case is that now approximately twice the total photon momentum is deposited into the plasma: the recoil momentum of scattering a forward moving photon with frequency ω_0 into the backward moving photon with frequency ω_2 is $\hbar\omega_0/c - (-\hbar\omega_2/c) \approx 2\hbar\omega_0/c$. Thus, the laser beams' intensities required to produce a given accelerating field is going to be much smaller for counter-propagating geometry than for the LWFA (or PBWA). More details can be found in Ref. [4].

The bottom drawing in Fig. 1(labeled CBA) illustrates the nonlinear excitation of the fast plasma waves which is significantly more complex than in PBWA (or LWFA). Specifically, we assume that two frequency components, separated by ω_p, are propagating in the forward direction. These two frequency components could either belong to two separate and long laser beams (as in PBWA), or two a single ultra-short laser pulse (as in LWFA). In the latter case, a continuum of such frequency pairs separated by ω_p can be identified. In such pair is shown in Fig. 1. The frequency phasor for the counter-propagating beam is labeled as backward beam. The beating between the different frequency components of the forward beam and the backward beam produce two "slow" plasma waves which are shown as almost-horizontal lines in the drawing. It is the nonlinear mixing of these two slow waves that gives rise to the "fast" plasma wave (labeled as plasma wake). Visually, one can deduce from the drawing that the phase velocity of the fast wave is much larger than that of the slow waves. Mathematically, one can show that the phase velocities of the slow waves roughly scale as $v_{sl} \approx \omega_p/k_0$ while the phase velocity of the fast wake is close to the speed of light. In Section II we derive formulas for the fast wake amplitude and demonstrate that, under some circumstances, it can be orders of magnitude larger than the regular wake produced by only the forward propagating pulse(s).

II   COLLIDING BEAM ACCELERATOR

The following physical problem was simulated using a one-dimensional particle-in-cell (PIC) code VLPL. An ultra-short circularly polarized Gaussian laser pulse with duration $\tau_L = 1.5\omega_p^{-1}$ and normalized vector potential $a_0 = 0.12$, propagating in the positive z direction, collides in a plasma with a long counter-propagating pulse with $a_1 = 0.05$. Plasma density was chosen such that $\omega_p/\omega_0 = 0.05$. The snapshot of the pulse intensity normalized to 2.7×10^{18}W/cm^2 is shown in Fig. 2(a). Two cases, corresponding to the different frequencies of the PB, $\omega_1 = 1.1\omega_0$ and $\omega_1 = 0.9\omega_0$, were simulated. The resulting plasma wakes are shown in Fig. 2(c) and (d), respectively. For comparison, we also plot the wake produced by a single TB *in absence* of the counter-propagating pulse in Fig. 2(b). Since the intensity of the short pulse is chosen non-relativistic, the magnitude of the plasma wake left behind the pulse is much smaller than the limiting (wavebreaking) field according to $E/E_{wb} \sim a_0^2/2$, where $E_{wb} = mc\omega_p/e$. The situation changes dramatically when a counter-propagating beam is added. As Figs. 1(c) and (d) indicate, the addition of the pumping beam increases the electric field of the plasma wake by an order of magnitude. To further illustrate this point, we plotted the regular wake [same as shown in Fig. 2(a)] in Figs. 2(c-d) for comparison. Note that the vertical scales of the Figs. 2(c-d) and Fig. 2(b) differ by a factor 20. Plasma wakes produced as a result of the collision between the counter-propagating beams is referred to as the enhanced wake because it is much larger than the regular wake.

This conclusion about the relative magnitudes of the regular and enhanced wakes is only valid for nonrelativistic laser pulses. It turns out that the magnitude of the enhanced wake $E < (\omega_p/\omega_0)E_{wb}$. This limit is set by the maximum velocity of the plasma electrons which cannot significantly exceed the phase velocity of the beatwave between the short and long laser beam, equal to $v_{sl} = (\omega_0 - \omega_1)/2k_0$. Excitation of the fast (accelerating) plasma wake is a strongly nonlinear process, with the "slow" (short-wavelength) plasma waves generated as intermediaries. Wave-

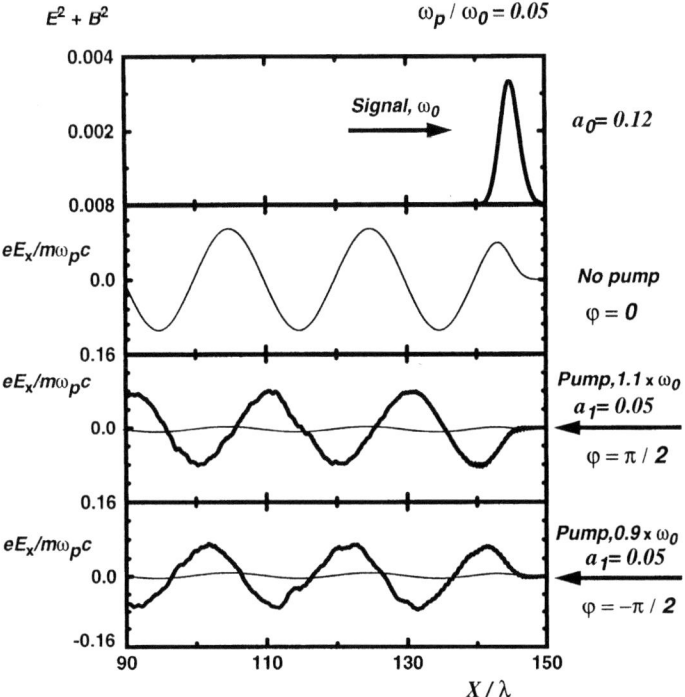

FIGURE 2. Top to bottom: (a) single short laser pulse with $a_0 = 0.12$ and frequency ω_0 propagates from left to right; (b) short pulse generates a weak plasma wake E_x; (c) in the presence of counter-propagating pump with $a_1 = 0.05$ and frequency $\omega_1 = 1.1\omega_0$ the wake is enhanced, and its phase is shifted by $\pi/2$ with respect to the "regular" wake of (b), which is also shown for comparison; (d) Same as (c), only a down-shifted pump with $\omega_1 = 0.9\omega_0$ is used, and the phase shift is $-\pi/2$.

breaking of these intermediaries sets limits the excitation of the fast wave. Both the fast and the slow plasma waves are shown in Fig. 1 (bottom).

A Linear regime: four-wave mixing

The above kinematic illustration in Fig. 1 is, of course, only a cartoon, which does not explain the physical mechanism of the nonlinear mixing between the slow plasma waves. The beating between the slow plasma waves is a novel phenomenon, and we have identified it as a method of driving fast plasma waves. From Eq. (1),

$$\left(\frac{\partial^2}{\partial \zeta^2} + \omega_p^2\right) E_z = -4\pi e \frac{\partial < nv >}{\partial \zeta}, \tag{2}$$

where $\zeta = t - z/c$, and $< nv >= \hat{n}_0 \hat{v}_1^* + \hat{n}_1 \hat{v}_0^* +$ c. c., where $\hat{n}_{0,1}$ and $\hat{v}_{0,1}$ are, correspondingly, fractional density and velocity perturbations in the first and second slow plasma waves. The fast wave, characterized by its amplitude E_z, is then nonlinearly driven by the RHS of Eq. (2). Equation (2) mathematically expresses the nonlinear mixing between the slow plasma waves schematically shown in Fig. 1. Assuming that pulses 0 and 1 are both flat-tops of duration τ_L, the amplitude of the fast wake left behind is

$$\frac{eE_z}{mc\omega_p} = \omega_p \tau_L \frac{(\Delta_0 + \Delta_1)\omega_0^3}{(\Delta_0^2 - \omega_p^2)(\Delta_1^2 - \omega_p^2)}|a_0 a_1 a_2^2|, \tag{3}$$

where $\Delta_0 = \omega_0 - \omega_2$ and $\Delta_1 = \omega_1 - \omega_2$. According to Eq. (3), fast wave generation in the colliding beam accelerator is a four-wave process.

Note that in the particular case of $\Delta_0 + \Delta_1 = 0$ wakefield vanishes. Since $\omega_1 = \omega_0 - \omega_p$, this case corresponds to $\omega_2 = \omega_0 - 0.5\omega_p$. Therefore, the scattering of the photons from beam 0 into beam 2 proceeds at the same rate as the scattering of the beam 2 into beam 1, and the overall momentum deposition into the plasma vanishes. Equation (3) breaks down for $\Delta_1^2 = \omega_p^2$ and $\Delta_0^2 = \omega_p^2$. For example, for $\Delta_1 = 0$ and $\Delta_0 = \omega_p$, the wake amplitude is $eE_z/mc\omega_p = (\omega_0^3 \tau_L^2/2\omega_p)a_0 a_1 a_2^2$.

For a short single-frequency forward-moving pulse, a similar expression for derived in [4] which we present here for completeness:

$$\frac{eE_z}{mc\omega_p} = \frac{\pi\Delta\omega}{8\omega_0}\left(4a_2 a_0 \frac{\omega_0^2}{\omega_p^2}\right)^2 \omega_p^2 \tau_L^2 e^{-\omega_p^2 \tau_L^2/4}\left[e^{-(\omega_p - \Delta\omega)^2 \tau_L^2} + e^{-(\omega_p + \Delta\omega)^2 \tau_L^2} + \frac{2}{3}e^{-\Delta\omega^2 \tau_L^2}\right]$$

$$\tag{4}$$

The most efficient excitation of the accelerating wake requires $\tau_L \approx 2.0\omega_p^{-1}$ and $\Delta\omega = \pm 1.1\omega_p$. For these parameters $eE_z/mc\omega_p \approx 0.6\omega_p/\omega_0 \ (4a_0 a_2 \omega_0^2/\omega_p^2)^2$. The enhanced wake exceeds the regular wake from forward scattering whenever $a_2 > (\omega_p/\omega_0)^{3/2}/4$. For $n_0 = 10^{18}\mathrm{cm}^{-3}$, this corresponds to the pump intensity $I_2 > 2 \cdot 10^{14} \ \mathrm{W/cm^2}$.

B   Nonlinear regime: particle trapping

The above picture of four-wave process resulting in the excitation of a fast wave is only true when all waves in question are linear. Fast plasma wave always remains linear because its amplitude is below the wavebreaking limit. Slow waves (which interfere to drive the fast wave) break much sooner, their breaking limiting the fast wave amplitude. After wavebreaking, particle motion is determined by ponderomotive beatwave force between counter-propagating beams.

The most interesting and easy-to-understand regime corresponds to the single-frequency short pulse of duration $\tau_L < \pi/\omega_p$ which is strong enough to cause wavebreaking. The incidence of wavebreaking is, approximately, determined by the ratio of the bounce frequency $\omega_B = 2\omega_0\sqrt{a_0 a_2}$ and the plasma frequency. In the strongly-nonlinear regime $\omega_B^2 \gg \omega_p^2$ the amplitude of the plasma wave is estimated [5] as

$$\frac{eE_z}{mc\omega_p} = \frac{\langle P_z \rangle}{mc} \sin \omega_p \zeta \approx \mathrm{sign}(\Delta\omega) \left(\frac{\omega_B}{\omega_0}\right) \sin \omega_p \zeta, \tag{5}$$

where $\langle P_z \rangle$ is the average momentum transferred to the plasma by the laser pulse. The physics of this momentum transfer can be visualized by plotting the electron phase space at different times: before the arrival of the short pulse, near the maximum of the short pulse, and right after the wavebreaking (Fig. 3). Numerical

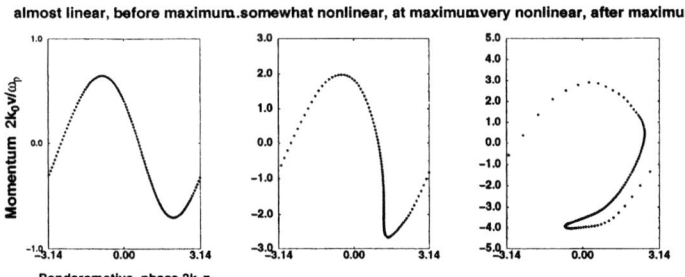

FIGURE 3. Left to right: electron phase space (a) before the arrival of short pulse; (b) near maximum of short pulse; (c) at wavebreaking. Rapid current jolt developing at wavebreaking drives the enhanced wake behind the short pulse.

simulations indicate that the largest momentum gain is achieved for the frequency detuning $\Delta\omega \approx \omega_B$ and pulse duration $\tau_L \approx 2/\omega_B$. For those parameters, plasma electrons execute about half a bounce in the ponderomotive potential, and leave the ponderomotive bucket with average velocity $v_z \approx c\omega_B/\omega_0$. The nonlinear current $J_{2z} = -env_z$ is then inserted into Eq. (1) to yield Eq. (5).

III PARAMETRIC EXCITATION OF PLASMA WAVES BY $2\omega_P$ DETUNING

In the previous section we considered two approaches to excitation of fast plasma waves: one involved two pulses moving in the forward direction and another in the backward direction (beatwave approach), and the other one required a short ($\tau_L \approx 2/\omega_p$) forward-moving pulse and a backward-moving pulse (CBA approach). The beatwave approach is complex for two reasons: (a) three laser pulses are needed, and (b) laser pulses have to be detuned by the plasma frequency. Most laser systems have a fairly small bandwidth (several percent). This forces the plasma density down and reduces the accelerating gradient. CBA can also be challenging because it requires a very short pulse. We have also found from numerical simulations that the optimal operation corresponds to $\Delta\omega = (1.5 - 2.0)\omega_p$. This decreases plasma density even further.

All these limitations, and also the simultaneous availability of Nd:Yag ($\lambda_1 = 1.06\mu m$) and Ti:S ($\lambda_0 = 0.8\mu m$) laser systems in a number of laboratories compelled us to think of other possible techniques of wake excitation. We suggested a novel scheme [6]: parametric excitation of accelerating plasma waves using counter-propagating laser beams detuned by, approximately, $2\omega_p$. Short-pulse duration no longer is required to be comparable to ω_p^{-1}; in fact, it is advantageous to use significantly longer pulses with $\omega_p\tau_L \approx 25$. From experimental standpoint, this could be a fairly attractive regime: if $\omega_0 - \omega_1 = 2\omega_p$, then the desired plasma density $n_p \approx 2.5 \times 10^{19} cm^{-3}$, and the required pulse duration $\tau_L \approx 25\omega_p^{-1}$ corresponds to 160 fs (FWHM). Such plasma and laser parameters are achievable, making the practical implementation of our scheme feasible.

That a plasma wave can be driven unstable by the $2\omega_p$ beatwave was originally proposed by Rosenbluth and Liu [7], who calculate the growth rate of a fast plasma wave $\gamma_{RL} \approx \omega_p a_0 a_1/2$ (co-propagating lasers). This instability is high-order, with growth rate scaling as the square of the pump amplitude. Thus, for pump waves of sub-relativistic intensity, i.e. $a_0, a_1 \ll 1$, this decay instability is too slow to be of great practical interest. We realized that (i) the counter-propagating pump geometry results in a growth rate enhanced by the factor $2\omega_0^2/\omega_p^2$, and (ii) fast (accelerating) plasma waves can be produced in the counter-propagating geometry – a fact overlooked in Ref. [7].

Using the equation of motion for the Lagrangian displacement $\zeta = z - z_0$,

$$\ddot{\xi} + \omega_p^2\xi = ik_0c^2a_0a_1e^{i[\Delta\omega t - 2k_0 z]} + c. \ c., \tag{6}$$

together with the novel *two-wave* ansatz for an electron displacement

$$\xi = A_f \sin\left[k_p z_0 - \omega_p t + \phi_f\right] + A_s \sin\left[k_s z_0 - \omega_p t + \phi_s\right], \tag{7}$$

where A_f (ϕ_f) and A_s (ϕ_s) are the amplitudes (phases) of the fast and slow plasma waves, one can show [6] that both the fast and the slow plasma waves can be driven unstable by a pair of counter-propagating laser beams detuned (approximately) by $2\omega_p$. Of interest to plasma accelerators is, of course, only the fast plasma wave with phase velocity close to the speed of light. We demonstrated that both the fast and the slow waves grow together with the growth rate $\Omega_i = \omega_0^2 a_1 a_0/\omega_p$: much faster then for co-propagating lasers. In fact, the presence of the slow wave is very important since it increases the instability growth rate.

The instability mechanism is easy to understand. Fast plasma wave which varies as $\delta n_f \sim \cos\omega_p(t - z/c)$ modulates the ponderomotive force which oscillates as $f_z \sim \cos(2k_0 z - 2\omega_p t)$ to resonantly drive the slow wave which varies as $\delta n_s \sim \cos(2k_0 - k_p)z - \omega_p t$. In its turn, the slow wave modulates the ponderomotive force, driving the fast wave and completing the feedback loop of the instability. Instability persists until the wavebreaking of the slow wave. Numerical simulations indicate that the amplitude of the fast wave is limited by $E_{max} = mc\omega_p^2/4\omega_0 e$.

Using a one-dimensional time-averaged particle code, we simulated excitation of the fast and slow plasma waves by a short slightly chirped (under-compressed) pulse with the wavelength $\lambda_0 = 0.8\mu m$ which collides with a longer $\lambda_1 = 1\mu m$ pulse in a 10^{19} cm^{-3} plasma. These wavelengths correspond to widely available laser systems (Ti:S and Nd-glass), and the plasma density was chosen to satisfy $\omega_0 = \omega_1 + 2.35\omega_p$. Other laser parameters are as follows: $a_0 = 0.15\exp\left[-\zeta^2/2\tau_L^2\right]$ with $\tau_L = 25$ (160 fs FWHM) and $d\delta\omega/d\zeta = -9.5 \times 10^{-3}\omega_p$ (3% bandwidth). The initial fast plasma wave $\tilde{e}_0 = 10^{-3}$ and $a_1 = 0.0165$ have been assumed. There results of the simulation are shown in Fig. 4, where we observe the excitation of both the fast and the slow plasma waves. Despite the small amplitudes of both

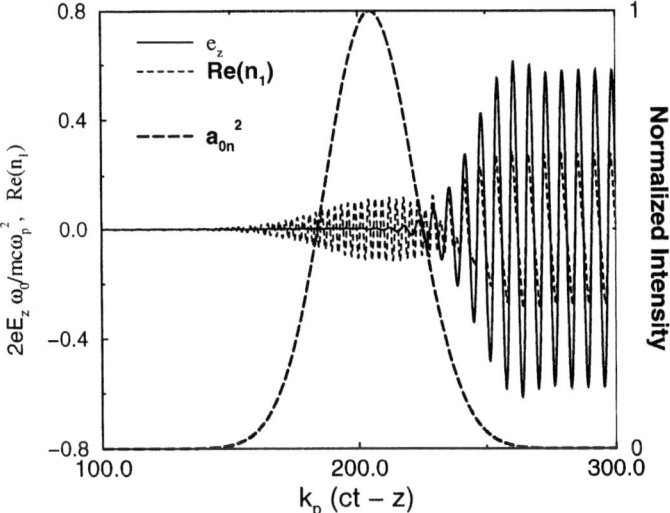

FIGURE 4. Solid line: fast electric field $\tilde{e}_z$, long-dashed line: normalized intensity of short pulse a_{0n}^2, dashed line: density bunching of the slow plasma wave $\mathrm{Re}(\hat{n}_1) = \langle\cos\theta_j\rangle$. Rapidly-varying part part of $\hat{n}_1$ is the driven plasma response inside the laser pulse.

forward and backward pulse, and despite the fact that the duration of the short pulse is too long for the efficient wake generation, we find that a significant fast plasma wave $E_z = 7$ GeV/m is excited. Parameters used in the simulation are fairly standard for Ti:S and Nd-Glass systems.

IV   UTILITY OF COLLIDING BEAM ACCELERATOR

One obvious benefit of the counter-propagating geometry is that very large accelerating wakes (of order 10 GeV/m) can be produced with moderate-intensity lasers ($I \sim 10^{16}$ W/cm^2). Another, less obvious benefit is the ability to control the phase of the accelerating wake. One observes from Fig. 2 that by changing the frequency of the long pulse from $\omega_1 = 1.1\omega_0$ (Fig. 2c) to $\omega_1 = 0.9\omega_0$ (Fig. 2d), the phase of the wake is changed by $\Delta\phi = \pi$. Thus, one can envision a "plasma linac" which consists of independently phase-controlled acceleration sections, separated by drift spaces.

Numerical implementation of the "plasma linac" concept is shown in Fig. 5. Collision of a short "timing beam" (TB) of duration $\tau_L = \omega_p^{-1}$ and normalized vector potential $a_0 = 0.08$ with a long "pumping beam" (PB) $a_1 = 0.012$ is modeled using a 1D version of the Particle-in-Cell (PIC) simulation code VLPL. Fig. 5(a) illustrates the temporal profile of the PB, which moves to the left; Figs. 5(b,c) are the snapshots of the generated plasma wake and the phase space of accelerated electrons, which are continuously injected with initial energy 10 MeV electrons; Fig. 5(d) shows the evolution of the TB as it moves through the plasma. To show how one can control the phase and the magnitude of the resulting plasma wake, we split the PB into two sections: the leading section of duration $\Delta t_1 = 500 \times 2\pi/\omega_0$, where $\Delta\omega = -1.7\omega_p$, and the trailing section $\Delta t_3 = 250 \times 2\pi/\omega_0$, where $\Delta\omega = 1.7\omega_p$. These two pump beam sections are separated by the middle section of duration $\Delta t_2 = \Delta t_3$, where the pump is switched off.

As Figs. 5(a,b) show, the three pump sections map into three spatial acceleration regions, which are different from each other in TB dynamics, magnitude, and *phase* of the plasma wake. In the leading region the pump beam has higher frequency and energy flows into the TB, amplifying it. A strong plasma wake with the peak accelerating gradient of 8 GeV/m is induced. The middle region is void of the pump. Here the TB interacts with the plasma through the usual LWFA mechanism only, producing a weak, < 1GeV/m, accelerating wake. In this region the energy of the injected electrons does not significantly change, as seen from Fig. 5(d). When the trailing (low-frequency) part of the pump collides with the TB, the energy flows from the TB into the PB, Fig. 5(c). Again, a strong plasma wake is induced, Fig. 5(b). This wake, however, is shifted in phase by $\Delta\phi = \pi$ with respect to the leading region. As a result, electrons which gained energy in the leading region are *decelerated* in the trailing region, Fig. 5(d). This shows that both amplitude and phase of the enhanced plasma wake can be controlled by shaping the long low-intensity pump beam.

Plasma linac can be used to prevent phase slippage between ultra-relativistic particles and the wake which has the phase velocity $v_{\rm ph}/c \approx 1 - \omega_p^2/2\omega_0^2$. Since particles are moving slightly faster than the wake crests, they eventually outrun the accelerating phase and move into the decelerating phase of the wake (Fig. (6), left). This occurs after one dephasing length $L_d = \lambda_p^3/\lambda_0^2$. After that, acceleration has to be terminated by terminating the plasma. The next acceleration stage needs to be in phase with the previous one, presenting a serious technical challenge.

In a colliding beam plasma linac shown in Fig. (5) dephasing can be circumvented by taking the length of the leading pump section equal to $2L_d$. Particle phase dynamics is shown in Fig. (6), right. After advancing in phase by $\Delta\phi = \pi$, electron finds itself in the gap between accelerating sections. Accelerating field in the gap is

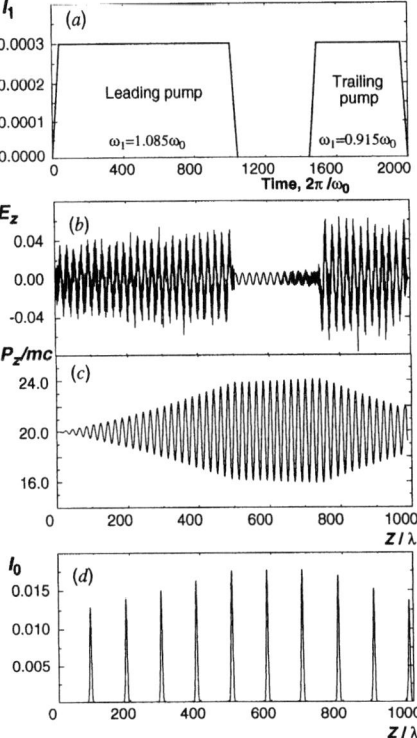

FIGURE 5. Collision between a short timing beam ($a_0 = 0.08$, $\tau_L = \omega_p^{-1}$) and an intermittent pump ($a_1 = 0.012$) in $n_0 = 2.5 \times 10^{18}\text{cm}^{-3}$ plasma ($\omega_0/\omega_p = 20$). 10 MeV electrons are continuously injected into the plasma. (a) Time-dependence of the pumping beam intensity $I_1 = a_1^2$; (b) longitudinal electric field $eE_z/mc\omega_0$; (c) propagation of the TB through the plasma, $I_0 = a_0^2$; (d) phase space of injected electrons.

very small because there is no enhanced wake there. After the gap, electron enters the second accelerating section, where the phase differs from the first section by π. Therefore, electron is in the accelerating phase again. This sequence can be repeated indefinitely, ensuring that electron is never decelerated.

V FUTURE WORK

An important unresolved problem is generation of accelerating plasma waves using the CBA technique in a plasma channel. Plasma channels are important for guiding both long and short laser beams. Moreover, transversely inhomogeneous plasma may impart an unusual structure of the accelerating field with a local minimum on axis. This may result in advantageous transverse focusing properties of the wake, especially in the context of the colliding-beam injector [5].

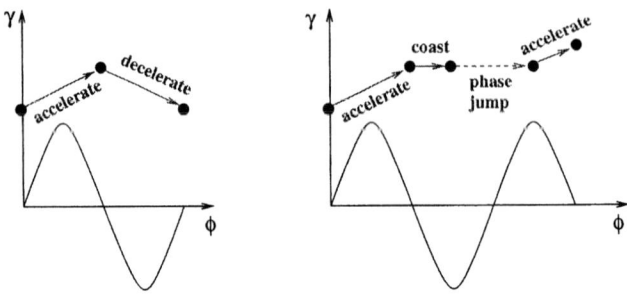

FIGURE 6. Schematic of the phase slippage of electron with respect to the wake in a standard wakefield accelerator (left) and in a "plasma linac" (right).

This work was supported by the DOE Division of High Energy Physics and the Presidential Early Career Award for Scientists and Engineers.

REFERENCES

1. T. Tajima and J. M. Dawson, Phys. Rev. Lett. **43**, 267 (1979).
2. P. Chen, J. M. Dawson, R. W. Huff, and T. Katsouleas, Phys. Rev. Lett. **54**, 693 (1985).
3. E. Esarey, P. Sprangle, J. Krall, and A. Ting, IEEE Trans. Plasma Science **24**, 252 (1996), and references therein.
4. G. Shvets, N. J. Fisch, A. Pukhov, and J. Meyer-ter-Vehn, Phys. Rev. E **60**, 2218 (1999).
5. G. Shvets, N. J. Fisch, and A. Pukhov, IEEE Trans. Plasma Science, **28**, 1194 (2000).
6. G. Shvets and N. J. Fisch, Phys. Rev. Lett. **86**, 3328 (2001).
7. M. N. Rosenbluth, C. S. Liu, Phys. Rev. Lett. **29**, 701 (1972).

Structure Formation, Tearing and Merging of Relativistic Electron Beam Propagating through a Laser Produced Plasma

T. Taguchi*, T. M. Antonsen Jr.†, C. S. Liu†, K. Mima**, Y. Sentoku**, H. Nagatomo** and H. Sakagami‡

*Department of Electrical Engineering, Setsunan University, Neyagawa, Osaka, Japan
†Institute for Plasma Research, University of Maryland, College Park, MD, U.S.A.
**Institute of Laser Engineering, Osaka University, Suita, Osaka, Japan
‡Department of Computer Engineering, Himeji Institute of Technology, Himeji, Hyogo, Japan

Abstract.
Hot electron transport in a high density plasmas has been studied using a two dimensional hybrid code. The results show that the initially cylindrical solid beam evolves into a hollow, annular beam due to the Weibel instability and generates strong magnetic fields on both sides of the annular ring. The annular structure subsequently breaks up into several beamlets due to a mechanism similar to a tearing instability. It is found that the magnetic fields parallel to the direction of beam propagation also grow during the tearing process. These beamlets are attract each other and finally they merged into a single beam which carries a net electric current. The tearing and merging processes have also been analyzed by the hybrid code for a uniform initial condition to determine an asymptotic behaviors of the total number of hot electron beams.

We also have performed a three dimensional particle-in-cell (PIC) simulation to compare with the results of 2D hybrid code. It shows that the spatial variation of cold electron density is much more uniform than that obtained by 2D code.

INTRODUCTION

Ultra-intense, short pulse lasers [1, 2] have been developed in recent years to generate relativistic plasmas. When the irradiation intensity approaches 10^{18} W/cm^2, the electron quiver velocity is close to the speed of light. Near the critical surface, where the plasma frequency equals the laser frequency, the laser generates an intense relativistic electron stream which carries the absorbed laser energy into the overdense plasma. The electric current associated with the electron stream reaches 100 MA or higher in a diameter of a few tens of micrometers when the input laser power is 100 TW or higher. Since the relativistic electron current is much higher than the Alfvén limiting current; $\gamma m_e c^3/e \simeq 17\gamma$ kA for a relativistic electron beam with average energy of $(\gamma - 1)m_e c^2$, the relativistic electron current penetrating into the overdense plasma will be neutralized by a return current induced in the background. Through this process, the absorbed laser energy is transported into the dense plasma. Further, this process has significant consequences for the fast ignition concept [3] of the laser fusion research.

Although the return current carried by thermal background electrons removes the Alfvén current limitation, the resulting bi-Maxwellian electron distribution is susceptible

CP611, *Superstrong Fields in Plasmas:* Second Int'l. Conf., edited by M. Lontano et al.
© 2002 American Institute of Physics 0-7354-0057-1/02/$19.00

to a number of instabilities, including the Weibel instability [4, 5]. The Weibel instability is characterized by the variation of perturbed quantities in the direction transverse to the beam axis. It generates strong magnetic fields transverse to the direction of beam propagation, and the magnetic field causes the beam to beak-up and be deflected.

The electromagnetic dynamics of relativistic electrons generated by ultra-intense lasers has been widely studied by computer simulation [6, 7, 8, 9, 10]. Many of the studies use electromagnetic, particle-in-cell (PIC) codes, which can describe kinetic effects of hot electrons and full electromagnetic processes. However, 2- or 3-D PIC codes require large computer memory because they usually use a few tens of particles per computational grid to reduce statistical fluctuations. PIC codes also consume computer CPU time because they must use small time steps, which are limited by the Courant condition, $c\Delta t/\Delta \leq 1$, where Δt is the time step, Δ is the grid spacing and c is the speed of light. The grid spacing in PIC codes must be the order of the Debye length λ_D [11], then the time step is shorter than $v_{te}/(c\omega_{pe})$, where v_{te} is the thermal velocity of the electrons and ω_{pe} is the electron plasma frequency. On the other hand, the electron MHD (magnetohydrodynamics) approximation (EMHD) is often used to analyze the self-consistent interaction between electrons and strong magnetic fields [12, 13]. The EMHD system describes phenomena that occur on the spatial scale of the order of the collisionless skin depth, c/ω_{pe}, while it describes time scales longer than the electron cyclotron period. However, the conventional EMHD approximation is only applicable to a single electron fluid.

To analyze this phenomena more efficiently, we newly developed a hybrid Darwin code. In this paper, we describe the basic concepts of the hybrid Darwin code and show results which are analyzed by the code for the hot electron transport in a high density plasma. We especially focused on the analysis of beam tearing and merging process associated with a magnetic field generation. Generating the strong magnetic field which results from the Weibel instability, the electron streams, which are initially neutralized, are separated into net current regions, namely net hot electron filaments and net cold electron flows surrounding the hot electron filaments. The initial size of beams is determined by the growth rate of the Weibel instability. These net current filaments are attracted each other by the current force and a pair of filaments merges into a single larger beam. After successive merging process, initial several beamlets merged into a large single beam at a certain parameter.

We also describe here a result from a three dimensional particle-in-cell (PIC) simulation to compare with the 2D hybrid simulation. The simulation result shows that some longitudinal instabilities take place and the return current induced in the background becomes much more uniform than the results of 2D simulation.

HYBRID-DARWIN SIMULATION CODE

In the analysis of the hot electron transport in high density plasmas, the minimum scale length is the order of the collisionless skin depth, c/ω_{pe}, while the total size of electron beam is the order of the beam spot size, which is about $10\,\mu$m. As the result, the ratio of the maximum and the minimum scale length is very large for the hundreds of the critical density, and then the large grid number is necessary for the simulation.

In order to reduce computational time and memory consumption and to simulate large scale plasmas for long time durations, we have developed a new type of simulation code, a hybrid Darwin code [14]. This code has two features. The first is that it uses a hybrid description, i.e., hot electrons and cold electrons are separately described. Hot electrons, whose density is lower, are described as particles, while cold electrons, whose density is higher, are described as a fluid. Since the hot electron density is approximately of the order of the critical density, the density of cold electrons, which is approximately the same as the background density, is always much higher than the hot electron density in overdense plasmas. While the particle treatment retains the kinetic effects of hot electrons, the fluid description of cold electrons enables us to reduce the total amount of memory consumption. The cold electron quantities, such as density or velocity, are only assigned on the computational grids, which are several times fewer in number than the particles. This feature is especially important for the analysis of hot electron dynamics in very high density plasmas. We also treat ions as fluid, so it does not significantly increase memory consumption.

The other feature of our code is that it uses the Darwin approximation to calculate the self-consistent electromagnetic fields [15]. This approximation extends the EMHD model to include space charge effects. In the Darwin approximation, the electric field $\mathbf{E}$ is divided into a divergence-free component $\mathbf{E}_T$, and a curl-free component $\mathbf{E}_L$ such as $\mathbf{E} = \mathbf{E}_T + \mathbf{E}_L$, where $\nabla \cdot \mathbf{E}_T = 0$ and $\nabla \times \mathbf{E}_L = 0$.

The Darwin approximation neglects the displacement current due to the divergence-free component, $\partial \mathbf{E}_T / \partial t$, and then the set of Maxwell's equation are as follows:

$$c^2 \nabla \times \mathbf{B} = \frac{1}{\varepsilon_0} \mathbf{J} + \frac{\partial \mathbf{E}_L}{\partial t}. \tag{1}$$

After taking some mathematical procedure, we obtain the following Helmholtz type equation for $\mathbf{E}_T$.

$$c^2 \nabla^2 \mathbf{E}_T - \left[\left(\omega_{pc}^2 + \omega_{ph}^2 \left\langle \frac{1}{\gamma} \left(I - \frac{\mathbf{v}\mathbf{v}}{c^2} \right) \right\rangle \right) \cdot \mathbf{E} \right]_T = \mathbf{C}_T \tag{2}$$

Here $\omega_{pc}^2 = 4\pi e^2 n_c / m_e$ and $\omega_{ph}^2 = 4\pi e^2 n_h / m_e$ are the local plasma frequencies for cold and hot electrons, respectively, where m_e is the rest mass of an electron, e is the unit charge. The angle bracket $\langle \cdots \rangle$ denotes a locally averaged value for hot electrons. The suffix T in eq.(2) indicates the transverse component, and C is defined as follows:

$$\begin{aligned} \mathbf{C} &= 4\pi e \left(\nabla \cdot (n_c \mathbf{u}_c \mathbf{u}_c) + \nabla \cdot (n_h \langle \mathbf{v}\mathbf{v} \rangle) + \nabla p_c / m_e + \nu_c n_c \mathbf{u}_c \right) \\ &\quad + \left(\omega_{pc}^2 \mathbf{u}_c + \omega_{ph}^2 \langle \mathbf{v}/\gamma \rangle \right) \times \mathbf{B}/c. \end{aligned} \tag{3}$$

Here $\mathbf{u}_c$ is the flow velocity of cold electrons. The quantity p_c is the kinetic pressure of the cold electrons, ν_c is the collisional frequency for cold electrons, and $\mathbf{I}$ is a unit

331

tensor. Transverse component can be extracted using Fourier transformation. On the other hand, the longitudinal component of the electric field, $\mathbf{E}_L$ can be obtained by the scalar potential determined by the Poisson equation.

Since both components of the electric field in the Darwin approximation are calculated directly by either Helmholtz or Poisson equations, it is not necessary to solve the temporal evolution of the electric fields. As a result, high frequency oscillations such as high frequency electromagnetic waves are automatically removed, and the restriction, $c\Delta t/\Delta \lesssim 1$ is not necessary. Since this model includes self-consistent electrostatic fields, the time step is usually shorter than $1/\omega_{pe}$. This time step is longer than the time step of PIC codes ($c\Delta t \leq \Delta \leq v_{tc}/\omega_{pe}$) by the ratio of c/v_{tc}, where v_{tc} is the thermal velocity of cold electrons. Since we also have no constraint that the grid size must be of the order of λ_D, we can take it to be a fraction of c/ω_{pe}, which is usually larger than $\lambda_D = v_{tc}/\omega_{pe}$ in cold plasmas. These advantages enable us to do large scale simulations for the hot electron dynamics in cold overdense plasmas.

TEMPORAL EVOLUTION OF A SOLD CYLINDRICAL BEAM

We have performed two dimensional simulations using the hybrid Darwin code to analyze the evolution of an initially solid cylindrical, hot electron beam propagating in an overdense plasma. In the simulations, vector quantities retain all three components, however, spatial variation is restricted to the two directions (x, y) transverse to the direction of beam propagation. Recently, Honda et al. showed simulation results for such a case using a PIC code [9, 10]. Since they, however, assumed a uniform beam initially, they did not simulate the macroscopic dynamics of beam structure formation that we consider here.

Figures 1–3 show the results. Figures 1(a)–(d) are the hot electron density profile at different times, $\omega_{pe}t =$(a) 39, (b) 67, (c) 77, and (d) 151, where $\omega_{pe}^2 = 4\pi e^2 n_0/m_e$, which is determined by the background ion density n_0. These times are sufficiently short that ions can be taken to be immobile as we have assumed. The initial hot electron beam profile is cylindrically flat-top, and its diameter is 200Δ, where Δ is the grid spacing. The peak hot electron density is set to be $0.1n_0$, while the initial cold electron density is determined to maintain charge neutrality. In the figure, the density is normalized by n_0. The total simulation system is $512\Delta \times 512\Delta$, while we only show the central region of size $120\Delta \times 120\Delta$ where the beam pinches. The normalized collisionless skin depth, $c/(\omega_{pe}\Delta) = 8$ in the simulation. The initial average velocity of the hot electrons, $\langle v_z \rangle$ is $0.885c$ (Lorentz factor, $\gamma_0 = 2.15$), and the initial flow velocity of the cold electrons, u_{cz} is set so as to cancel the initial hot electron current ($n_h\langle v_z \rangle + n_c u_{cz} = 0$).

The dimensionless parameters of the simulation correspond to dimensional parameters as follows. When $n_0 = 10^{22}$ cm^{-3}, $\omega_{pe}^{-1} =0.177$ fsec and $\Delta = 6.64\times10^{-3}$ μm, and hence the diameter of the hot electron beam is about 1.33 μm. Figures (a)–(c) show the results at times corresponding to 6.8, 11.8, 13.7 and 26.8 fsec, respectively. The initial temperatures of both hot and cold electrons are set to be 5 keV.

As shown in the figs. 1(a) and 1(b), the density of the circular edge of the hot electron beam increases, and the beam gradually forms an annulus. This is due to the Weibel

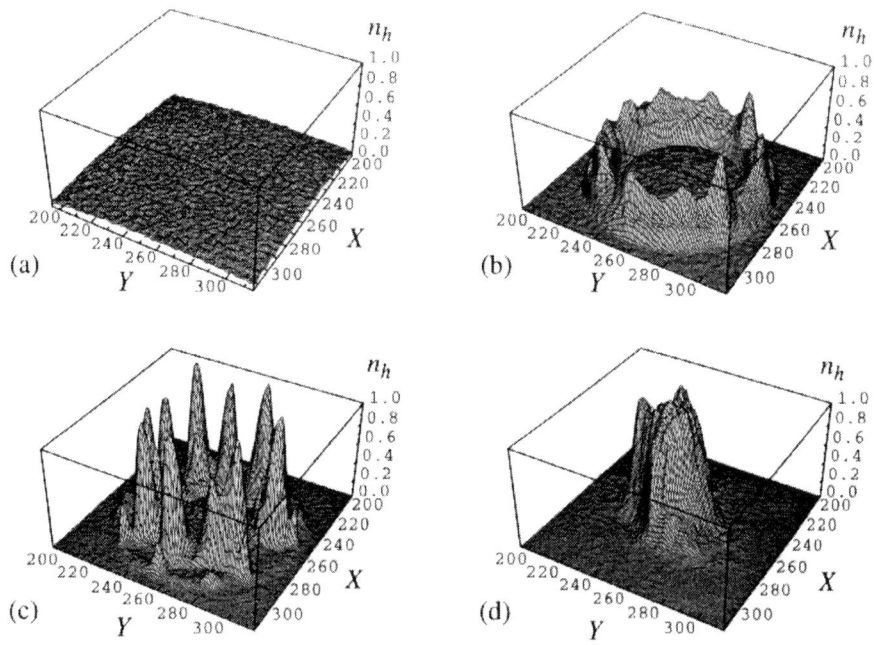

FIGURE 1. Temporal evolution of hot electron density profile, $n_h(x,y)$.

instability, then the thickness of the hollow ring is determined of the most unstable mode of Weibel instability. The initial cylindrical beam is radially inhomogeneous. For this reason the first perturbations to grow have no azimuthal variation. However, as shown in Figs.1(b) and (c), the annular beam gradually pinches, becomes unstable, develops azimuthal structure, and breaks up into several narrow filaments. Since each narrow beam of hot electrons carries a current, the narrow beams attract each other and they eventually coalesce into a single hot electron beam as shown in the Fig.1(d). The process of coalescence is similar to the results of Honda et al. [9, 10].

Figures 2 (a)–(d) show isocontours of z component of the vector potential A_z, which correspond to magnetic field lines in the 2D case. The plots correspond to the same times as in Figs. 1 (a)–(d). Figure 2(a) shows that the magnetic field is generated on both the inside and the outside of the annular beam. Further, the field lines reconnect as the the beam breaks up into several narrow beams, as shown in Figs. 2 (b) and (c). This process is very similar to the tearing instability in magnetized plasmas.

Since ions are immobile and only electrons participate, the process is closely related to whistler mediated magnetic reconnection, which was proposed by Mandt et al. [16]. Whistler mediated reconnection is characterized by the generation of an out-of-plane magnetic field, which is transverse to the original magnetic field generated by the current sheet. Under the current conditions, the generation of the out-of-plane field implies the generation of a magnetic field parallel to the direction of beam propagation.

The arrows in Figs. 2 show the total electron flux vectors, $n_h \langle \mathbf{v}_\perp \rangle + n_c \mathbf{u}_{c\perp}$, which are

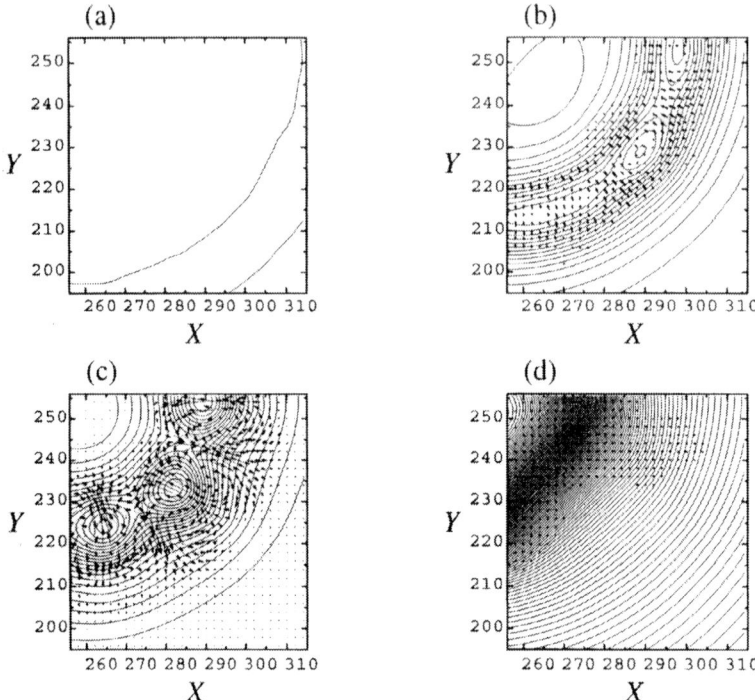

FIGURE 2. Temporal evolution of magnetic field lines. The arrows in (c) shows total electron fluxes $n_c \mathbf{u}_c + n_h \langle \mathbf{v} \rangle$.

proportional to the current density. The vectors clearly indicate the presence of current loops in the plane after the tearing is well developed. The loops generate the out-of-plane magnetic field.

Figure 3 shows the temporal evolution of the maximum strength of the in-plane magnetic field, B_T, and the out-of-plane magnetic field, B_z, respectively. In the figure, the magnetic field is normalized by $B_0 = (4\pi n_0 T_{h0})^{1/2}$, where B_0=31.7 MG for the current parameters. The figure shows that the out-of-plane field grows exponentially after the tearing process becomes significant. The maximum in-plane magnetic field is about 80 MG for the current parameters, while the maximum out-of-plane field reaches at about 28 MG.

There is a major difference between pure whistler mediated reconnection and our results. While whistler reconnection generates a quadrupole out-of-plane field around the X-point [17], our results show a dipole field, which changes sign on both sides of the X-point. The dipole field appears because the hot electron beam has an inward velocity due to the pinch. From our simulation results (Fig.2(b) and (c)), the inward velocity is estimated as about $u_{h\perp} \simeq 0.1c$. On the other hand, the background electrons do not have such a large radial velocity component, although their density, n_c changes as it neutralizes the hot electron density. As a result, the hot electrons and the cold electrons

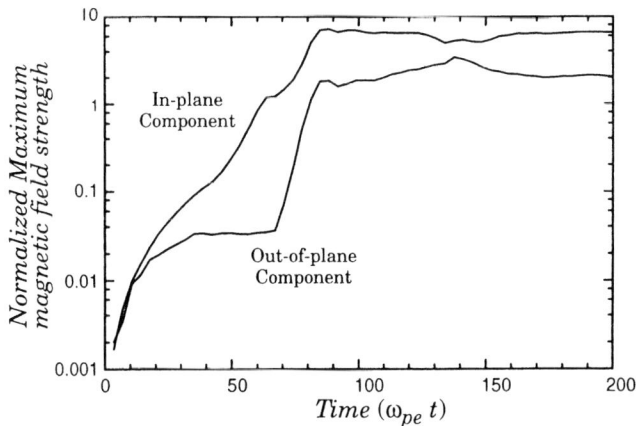

FIGURE 3. Maximum magnetic field strength in the simulation region. Here B_T is the transverse component to the beam propagation, and B_z is the parallel components to it.

have a relative radial flow. When the beam has cylindrical symmetry, the relative flow does not contribute to the generation of an out-of-plane magnetic field. But as the tearing develops, the hot electron flow and the cold electron flow in the plane are also torn, and this produces a current loop with opposite sign on either side of the X-point.

BEAM AREA DISTRIBUTION AND MAGNETIC FIELD SCALING IN A BEAM MERGING PROCESS

As shown in the above section, a cylindrical hot electron beam which is initially neutralized by the cold electron return current breaks up into several beamlets whose size is determined by the growth rate of the Weibel instability. The generated small beamlets are attracts each other, and a pair of closely encountered beamlets merges into a single beam. When the initial area of the beam becomes much wider than the above case, the instability produces a large number of beamlets, and as time goes on, successive merging processes produce various sizes of beams. After a sufficient time has elapsed, the area distribution approaches to an asymptotic function which is determined by the merging rate.

Here we discuss about temporal evolution of a beam area distribution function $g(\sigma, t)$, where σ denotes an area of a beam. This function is normalized as that the total number of beams is defined by $N = \int_0^\infty g(\sigma, t) d\sigma$.

In the merging process, the distribution function, $g(\sigma, t)$, satisfies the following rate equation:

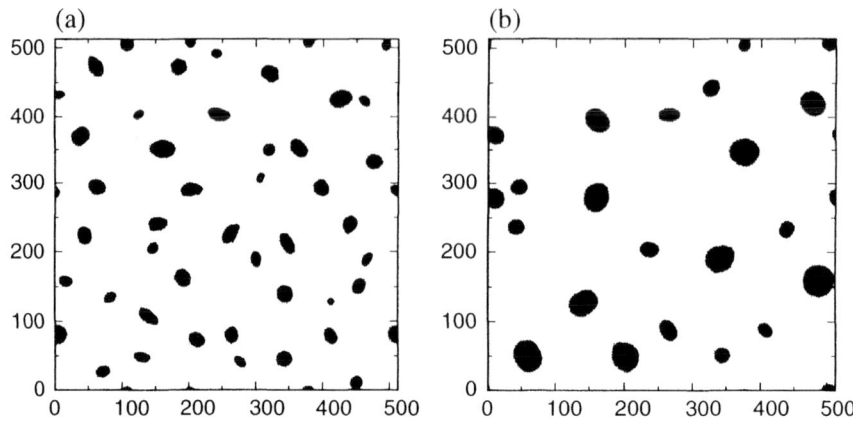

FIGURE 4. Two snapshots of density plot of hot electron density, $n_h(x, y)$.

$$N\frac{dg(\sigma, t)}{dt} = -2g(\sigma, t)\int_0^\infty g(\sigma', t)\Omega\left(\sigma, \sigma'\right) d\sigma'$$

$$+ \int_0^\sigma g(\sigma - \sigma', t)g(\sigma', t)\Omega\left(\sigma - \sigma', \sigma'\right) d\sigma'. \tag{4}$$

Here, $\Omega(\sigma, \sigma')$ is a merging rate of two beams whose beam areas are σ and σ', respectively, where the rate is defined as the number of merging events in a unit time. If we assume the following scaling:

$$\Omega\left(\sigma, \sigma'\right) \propto \frac{1}{(\sigma + \sigma')^\alpha}, \tag{5}$$

then it is found that the total number of beams asymptotically decreases with time as, $N(t) \propto t^{-1/\alpha}$.

If we assume that all beams carry an Alfvén limit current independently of their size, the magnitude of magnetic field surrounding beams, B, is in proportion to an inverse of the beam size, λ, as $B \propto \lambda^{-1}$, then the constant α in eq.(5) can be estimated as $\alpha = 3/2$ [18], and then $N(t) \propto t^{-2/3}$.

We performed a simulation for a wide uniform beam which initially produces much more beamlets than the simulation described in the previous section and investigate the temporal evolution of the total number of beams and the beam area distribution.

Figure 4(a) and (b) shows the density plot of the hot electron density for a different time $\omega_{pe}t =$ (a) 246 and (b) 563. They show that beams at a later time are larger.

Figure 5 shows the temporal evolution of an average beam area and an average hot electron current. From this figure, we can estimate the average area is in proportion to $t^{0.9}$, so that $N(t) \propto t^{-0.9}$, because the total area is conserved in eq.(4). As the result, $\alpha = 1.1$, and then the dependency between the magnitude of magnetic field and beam

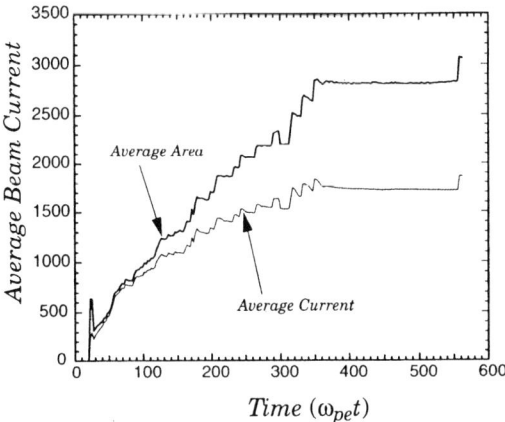

FIGURE 5. Temporal evolution of an average hot electron area and an average hot electron current.

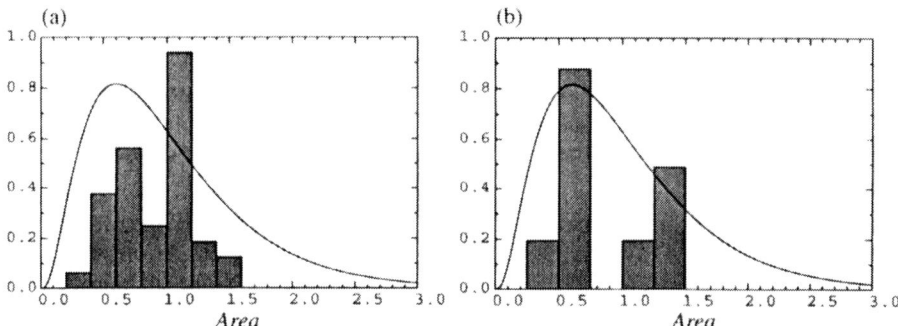

FIGURE 6. Two snapshots of histograms for the beam area.

size is $B \propto \lambda^{-0.1}$. The result indicates that the magnetic field does not depend strongly on the beam size. It means that the generated magnetic field does not change so much after a merging event.

Figure 6(a) and (b) show histograms of the number of beams for the same time of figures 4. The solid curve in the figures is determined by a numerical solution of eq.(4) with $\alpha = 1.1$. The results is seemed to well agree with the simulation results.

COMPARISON WITH A RESULT OF THREE DIMENSINAL PIC SIMULATION

We also perform a three dimensional particle-in-cell (PIC) simulations. The result also shows tearing and merging process when we observe at a certain cross section transverse to the hot electron beam propagation. But there is a major difference. According to the

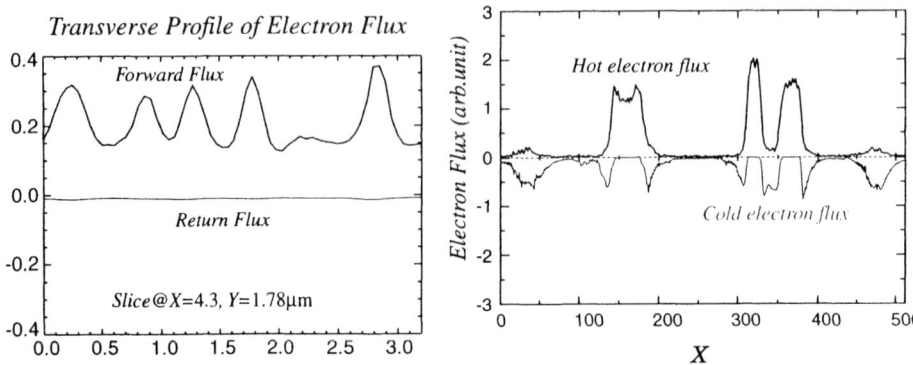

FIGURE 7. Spatial variation of hot electron and cold electron flux calculated by a 3D PIC code (left) and by a 2D hybrid code (right).

3D simulation, the return current induced in the background plasma is almost uniform in the cross section, while the forward propagating energetic hot electrons forms a filaments as shown in 2D simulation (figure 7, left). On the other hand, return currents in 2D simulation change its profile corresponding to the hot electron flow to neutralize the total charge in a whole plane (figure 7, right).

This difference results from the logitudinal instability, such as two stream instability, sausage instability or Kelvin-Helmholz instability. Due to these longitudinal instabilities, cold background electrons suffer mixing process and then they becomes uniform.

CONCLUSION AND DISCUSSION

In conclusion, we have studied the stability of hot electron transport and strong magnetic field generation in overdense plasmas, which is critically important for the fast ignition concept. For this purpose, we have used a newly developed hybrid Darwin code. The results show that the cylindrical hot electron beam, which is initially solid, gradually forms annular structure due to the Weibel instability, and the annular ring breaks up into several narrow beams through a similar process to a tearing instability. The tearing process is accompanied by the generation of an out-of-plane magnetic field. The out-of-plane field in our simulation had a dipole character, which can be explained by the relative velocity between the hot and cold electron flows in the transverse plane. Since the out-of-plane field reaches a strength on the same order of the in-plane field, the three dimensional field lines forms spirals as they surrounds the hot electron beam.

We have also performed a simulation for a wide area beam to investigate asymptotic behavior of the beam area distribution and the asymptotic time dependence of the total number of beams. This study determine a scaling of the magnetic field amplitude with beam size. The scaling study has a lot of meanings. The most important point is to investigate what determines the final net current and how much it flows. This question is crucial especially for the fast ignitor laser fusion program.

Finally we compare the 2D result with a result of 3D PIC simulation. The result of 3D simulation suggests that the longitudinal instability is also important for the long time beam propagation. We will study much more about this point.

We are now developing an integrated code including PIC, hybrid, Fokker-Planck and fluid. Our final goal is to describe all processes which take place in the fast ignition process in laser fusion experiment and to understand all of the processes and to optimize the target design to get a successful operation of the fast ignitor.

This work was supported by JIFT and part of this work is supported by a grant-in-aid for scientific research from the Ministry of Education, Sports Culture, Science and Technology and by the National Science Foundation and the U.S. Department of Energy.

REFERENCES

1. D. Strickland and G. Mourou, Opt. Lett. **20**, 1157 (1995).
2. P. Maine, D. Strickland, P. Bado, M. Pessot and G. Mourou, IEEE J. Quantum Electron. **24**, 398 (1988).
3. M. Tabak, J. Hammer, M. E. Glinsky, W. L. Kruer, S. C. Wilks, J. Woodworth, E. M. Campbell, M. D. Perry, R. J. Mason, Phys. Plasmas **1**, 1626 (1994).
4. E. S. Weibel, Phys. Rev. Lett. **2**, 83 (1959).
5. F. Califano, F. Pegoraro and S. V. Bulanov, Phys. Rev. E **56**, 963 (1997).
6. R. J. Mason and M. Tabak, Phys. Rev. Lett. **80**, 524 (1998).
7. A. Pukhov and J. Meyer-ter-Vehn, Phys. Rev. Lett. **79**, 2686 (1997).
8. Y. Sentoku, K. Mima, S. Kojima, and H. Ruhl, Phys. Plasmas **7**, 689 (2000).
9. M. Honda, J. Meyer-ter-Vehn and A. Pukhov, Phys. Plasmas **7**, 1302 (2000).
10. M. Honda, J. Meyer-ter-Vehn and A. Pukhov, Phys. Rev. Lett. **85**, 2128 (2000).
11. C. K. Birdsall and A. B. Langdon, *Plasma Physics via Computer Simulation*, (MacGraw-Hill, New York, 1985) p.176.
12. A. S. Kingsep, K. V. Chukbar and V. V. Yan'kov, in *Reviews of Plasma Physics*, edited by B. B. Kadomtsev (Consultant Bureau, London, 1990), Vol. 16, p.243.
13. S. V. Bulanov, F. Pegoraro and A. S. Sakharov, Phys. Fluids B **4**, 2499 (1992).
14. T. Taguchi, T. M. Antonsen Jr., C. S. Liu and K. Mima, Phys. Rev. Lett. **86**, 5055 (2001).
15. J. Busnardo-Neto, P. L. Pritchett, A. T. Lin and J. M. Dawson, J. Comp. Phys. **23**, 300 (1977).
16. M. E. Mandt, R. E. Denton and J. F. Drake, Geophys. Res. Lett., **21**, 73 (1994).
17. M. A. Shay and J. F. Drake, Geophys. Res. Lett., **25**, 73 (1998).
18. Y. Sentoku, K. Mima, et al., to be published.

Accelerated Dense Ion Filament Formed by Ultra Intense Laser in Plasma Slab

H. Amitani[a], T. Esirkepov[a,b], S. Bulanov[b,c], K. Nishihara[a],
A. Kuznetsov[b], F. Kamenets[b]

[a]Institute of Laser Engineering, Osaka University
2-6 Yamada-oka, Suita, Osaka 565-0871, Japan
bMoscow Institute of Physics and Technology
Institutskij per. 9, Dolgoprudny, Moscow region 141700, Russia
[c]General Physics Institute RAS, Vavilov Str. 38, Moscow 119991, Russia

Abstract. The ion acceleration is studied with 3D and 2D PIC simulations of the intense laser pulse interaction with the underdense plasma slab. The simulations reveal the forward acceleration of ions at the rear plasma-vacuum interface. A novel mechanism of the ion acceleration is found. Efficient ion acceleration occurs when an intense quasistatic magnetic field is generated at the rear side of the plasma slab. The magnetic field pressure pushes electrons inside the slab shifting them with respect to the plasma ions. As a result large-scale regions of non-compensated positive electric charge are formed. These regions act as an accelerator and a collimator for the ions from the dense plasma filament formed inside the self-focusing channel.

INTRODUCTION

To study the ion acceleration mechanisms in underdense laser plasmas we performed the 2D and 3D Particle-in-Cell simulations. We found that the quasistatic magnetic field plays a key role in the process of the ion acceleration.

Numerous experiments and PIC simulations of a laser-plasma interaction in the range of the laser intensity $10^{18} \div 10^{20} \, \text{W/cm}^2$ and the laser pulse duration <1ps confirm that protons (ions) are accelerated near the both the front and the rear plasma slab edges [1,2]. Taking into account greatness of the ion mass and shortness of the period of laser-plasma interaction, one should invoke a quasi-steady (long-living) electric field to explain the ion acceleration. Such a long-living electric field can be caused by the electric charge separation due to thermal expansion of the electron component into vacuum [3]. This mechanism is effective on a large time-scale, much greater than the length of the 1ps-pulse interaction and for a moderate intensity of the laser radiation. In the case of an ultra short, high intensity laser pulse

CP611, *Superstrong Fields in Plasmas:* Second Int'l. Conf., edited by M. Lontano et al.

its interaction with inhomogeneous plasmas is accompanied by a strong electric charge separation and by the magnetic field generation.

Below we present a novel mechanism for the ion acceleration, when a long-living electric field and corresponding electric charge separation are induced by a 'slowly' varying magnetic field generated by the laser. A strong quasistatic magnetic field is generated in the self-focusing channel of the intense laser pulse by the electric current carried by fast electrons [4]. This magnetic field is associated with the 'slow' vortices in the electron component of the plasma [5,6]. Both inside the self-focusing channel walls and at the plasma vacuum interface on the front and the rare sides of the plasma slab there is a large density gradient. It is known that in inhomogeneous plasma the vortex moves in the direction perpendicular to the plasma density gradient. Thus the magnetized plasma regions are expected to move along the channel walls and along the plasma surface. As the magnitude of the magnetic field is very high, even rather slow motion can induce a substantially intense long-living electric field. In its turn this electric field accelerates and focuses the ions. Our 2D and 3D PIC simulations confirm the validity of this scheme for the ion acceleration at the plasma-vacuum interface at the rear side of the target.

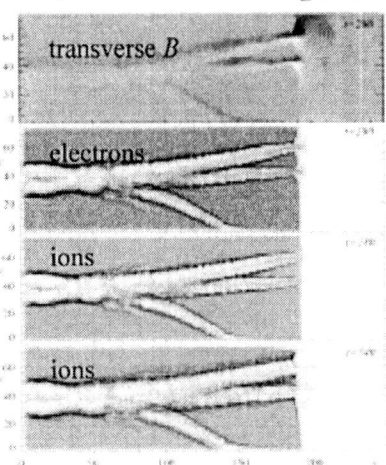

FIGURE 1. 2D case. Filamentation, self-channeling of the laser light, quasistatic magnetic field generation and

SIMULATION RESULTS

We present the results of 2D and 3D simulations of the laser pulse interaction with a finite width plasma slab. We use 2D and 3D Particle-in-Cell code REMP (Relativistic Electro-Magnetic Particle-mesh code), written by one of the authors (T.E.) in 1998-1999, [7]. In the 2D case circularly polarized laser pulse with a dimensionless amplitude

a=eE/(meω_c)=10 (I=5·1019 W/cm²), its size is (x × y) =40λ×12λ, the
plasma density equals n=0.2025n_cr. We perform a series of 2D simulations

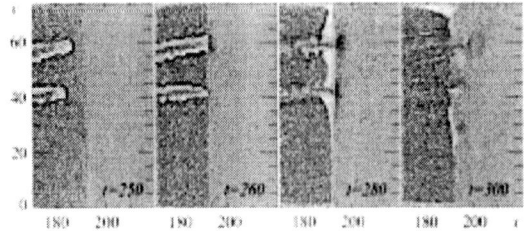 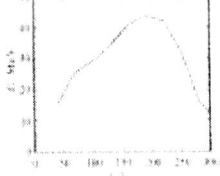

FIGURE 3. 2D case.
Maximal ion energy
vs plasma slab length.

FIGURE 2. 2D case. Electric charge density evolution

with varying plasma slab length from 40λ to 300λ. The total number of
particles is up to 1.5·10⁶. The simulation mesh size is 0.125λ. The boundaries of the computation box absorb both the e.m. radiation and the particles.

In the 3D case the laser pulse is also circularly polarized, its amplitude
a=5 corresponds to the intensity 1.25·10¹⁹ W/cm², the laser pulse size is (x
× y× z) =20λ×6λ×6λ. Plasma slab density is n=0.36n_cr. and the length is
40λ. Corresponding the total numbers of particles is 246·10⁶. The simulation grid has 620×160×160 grid steps with a grid step length equal to 0.1λ.
The boundary conditions along the y- and the z-axes are periodic, the
boundaries along the x-axis are absorbing for the e. m. radiation and particles. The simulations were performed on the 16 processors of NEC SX-5
vector supercomputer at ILE, Osaka University.In the simulations the ion
to electron mass ratio is m_i/m_e=1836, the laser pulse propagates along the
x-axis. In figures the time and space units are normalized on the period
2π/ω and the wavelength λ of the incident radiation, respectively.

In the 2D case in Fig. 1 we see a typical scenario of the laser pulse evolution in underdense plasmas. The laser pulse undergoes self-focusing, defocusing and filamentation. It forms a channel in the plasma. The density
of the plasma inside the channel is less than the unperturbed density. The
density of the channel walls is greater than the critical density. The steep
front is formed in the leading part of the laser pulse and in all the produced
self-focusing filaments. Here the electric double layer is formed [2,6,8].
The double layer is a 'half' of wake field wave; it consists of two layers of
positive and negative longitudinal electric field. On an image of the charge
density the double layer looks like a single positively charged bubble followed by a crescent layer of negative charge [2]. Fast electrons and return

currents induce the quasistatic magnetic field [4]. Inside the channel we see a formation of density filament.

Now we consider the processes, which are seen at the rear plasma-vacuum interface at time, when the leading front of the pulse crosses it. Here the laser pulse produces a cloud of fast electrons, which expands into vacuum. The cloud spreads in the transverse direction, and a considerable portion of electrons returns back to the plasma region in the form of the electron fountain. As a result the circulating electric current generates the dipole magnetic field seen in Fig. 1. The opposite sign magnetic regions expand outwards along the plasma-vacuum interface. The magnetic field pressure pushes the electrons inside the plasma slab in the backward direction with the heavy ions remained at rest [8]. This forms large-scale regions with non-compensated positive electric charge, as it is seen in Fig. 2.

In between these regions we see the density filament. The tip of the filament and the charged layers near the estuary of the channel are accelerated due to the Coulomb explosion. The fast ion filament appears to be well collimated by the transverse component of the electric field produced by positively charged regions shown in Fig.1.

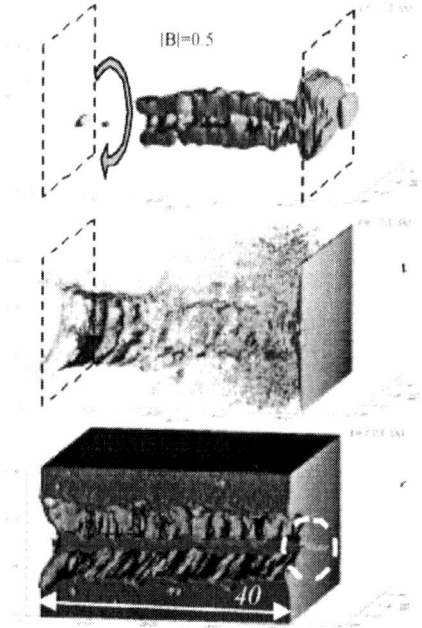

FIGURE 4. (a) magnetic field averaged over laser period, (b) electric charge level $+0.37 n_{pe}$, at t= 75. (c) ion density level 0.75 n_{pe} at t = 125.

The maximal energy of ions depends on the width of the plasma slab, as it seen in Fig. 3. For the parameters under discussion the optimal slab width is ≈200λ, and corresponding maximal energy of ions is 45MeV.

The result of the 3D simulation is presented

343

in Figs. 4, 5, and 6. We slice the 3D simulation box into two parts and plot the only part to reveal the inner structure. We see the same mechanism of the ion acceleration at the rear plasma-vacuum interface as in the 2D case. The laser pulse drills through the plasma slab. The fast electrons with kinetic energy in the range from 50 to 60 MeV induce intense quasistatic magnetic field shown in Fig. 4 (a). The magnitude of the magnetic field is of the order of 100MG×(μm/λ). The magnetic pressure pushes electron component inside the plasma slab, shifting the electrons with respect to the ions, and forms large-scale positively charged regions as we see in Fig. 4(b). The electric field produced by these non-neutral regions accelerates and

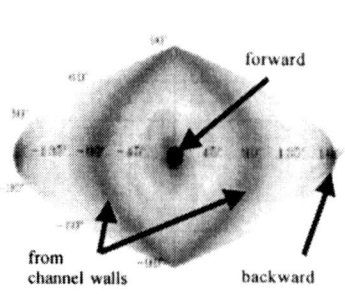

FIGURE 5. Ion energy vs angle at t=225.

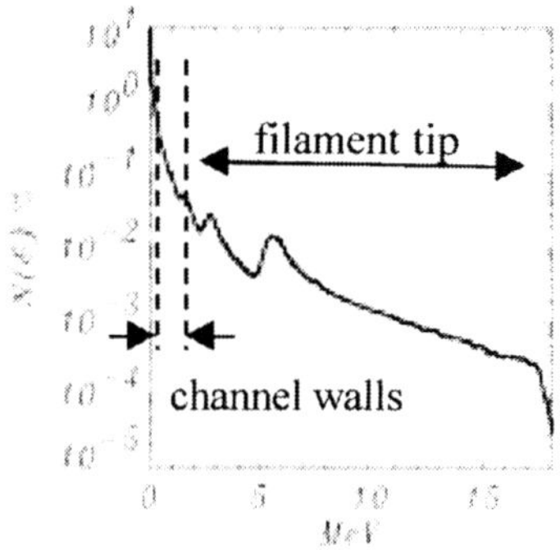

FIGURE 6. Energy spectrum of ions

collimates the ions from the filament. The magnitude of longitudinal electric field is up to 6TV/m×(μm/λ). In Fig. 4(c) we see the accelerated density filament tip and a flexure of the plasma slab edge caused by magnetic field.

Fig. 5 shows the angular distribution of the ion kinetic energy. Each point represents the average kinetic energy of ions going in certain direction in a small solid angle; the darker the point the greater the average energy is. We see three fluxes of ion energy. The flux in the forward direction represents the most energetic ions in the filament tip. It contains relatively small number of ions. The flux containing the greatest number of the ions represents the ions in the channel walls. These ions move almost perpendicularly to the channel axis. Their kinetic energy is up to 2MeV. Energy distribution of the ions is presented in Fig. 6. The diffuse peaks in the energy spectrum correspond to different transverse layers of the plasma. This is an effect of the initial spatial distribution of particles.

CONCLUSION

3D and 2D PIC simulations of intense laser pulse interaction with underdense plasma reveal the forward acceleration of ions at the rear plasma-vacuum interface, the farthest from the pulse entrance. We present a novel mechanism of the ion acceleration. The quasistatic magnetic field generated near the rear plasma-vacuum interface pushes electrons inside the plasma slab. As a result large-scale regions of non-compensated positive charge are formed. These regions act as an accelerator and a collimator for the ions in the density filament formed inside the channel.

ACKNOWLEDGMENTS

We appreciate the help of ILE computer group and Cyber Media Center of Osaka University (Japan). This work was supported by Japan Society for the Promotion of Science, by Russian Ministry of Sciences, and by Russian Foundation for Fundamental Sciences.

REFERENCES

1. Maksimchuk, A., et al., *Phys. Rev. Lett.* **84**, 4108 (2000); Clark, E. L., et al., *Phys. Rev. Lett.* **85**, 1654 (2000); Hatchett, S.P., et al., *Phys. Plasmas* **7**, 2076 (2000); Mackinnon, A.J., et al., *Phys. Rev. Lett.* **86**, 1769 (2001).
2. Esirkepov, T.Zh. et al., *JETP Lett.* **70**, 82 (1999); Bulanov, S.V., et al., *JETP Lett.* **71**, 407 (2000); Sentoku, Y., et al., *Phys. Rev. E* **62**, 7271 (2000).
3. Gurevich, A.V., et al., *Sov. Phys. JETP* **21**, 449 (1966); Gitomer, S.J., et al., *Phys. Fluids* **29**, 2679 (1986).
4. Askar'yan, G.A., et al., *JETP Lett.* **60**, 251 (1994).
5. Bulanov, S.V., et al., *Phys. Rev. Lett.* **76**, 3562 (1996).
6. Bulanov, S.V., et al., "Relativistic Interaction of Laser Pulses with Plasmas", Relativistic Interaction of Laser Pulses with Plasmas, in *Reviews of Plasma Physics*. Volume: 22, edited by V.D. Shafranov, Kluwer Academic / Plenum Publishers, New York, 2001). p. 227.
7. Esirkepov, T.Zh., *Comput. Phys. Comm.* **135**, 144 (2001).
8. Kuznetsov, A.V., et al., *Plasma Phys. Rep.* **27**, 211 (2001).

Observation of Directed Emission and Spectral Narrowing on Xe(L) Hollow Atom Single-(2p) and (2s2p) Double Vacancy Inner-Shell Transitions at 2.8-2.9 Angstroms

A. B. Borisov, K. Boyer, A. Van Tassle, X. Song, F. Frigeni, M. Kado, and C. K. Rhodes

University of Illinois at Chicago, Department of Physics, M/C 273
845 W. Taylor Street, Chicago, IL 60607-7059

Abstract. The results of measurements indicate the achievement of amplification on Xe(L) hollow atom Xe^{34+}, Xe^{35+}, Xe^{36+} and Xe^{37+} transition arrays in the 2.8 – 2.9 Å region involving both single 2p and double 2s2p hole excitations.

I. INTRODUCTION

Ideal conditions for x-ray amplification in the multikilovolt spectral region combine an exceptional set of circumstances. They are summarized by the production of cold, low opacity, spatially directionally organized, and vigorously ($10^{19} - 10^{20}$ W/cm^3) inner-shell state-selectively excited high-Z matter. The results reported below support the conclusion that the alliance [1] of two recently studied phenomena, (α) the direct multiphoton excitation of hollow atoms from clusters [2] with ultraviolet radiation [3] and (β) a nonlinear mode of confined propagation in plasmas resulting from relativistic/charge-displacement self-channeling [4,5], can successfully produce the union of this demanding set of requirements. The two reasons for this outcome are cooperation and control. The <u>cooperative</u> action of these two processes is easily understood; the former <u>locally</u> furnishes the high state-selective power density required, while the latter efficiently organizes the excited volume <u>spatially</u> into a long narrow directional structure. Moreover, the ability to <u>control independently</u> the action of these two phenomena enables the <u>optimization</u> of the configuration of the amplifying medium. Importantly, based on both experimental findings and a corresponding theoretical analysis, a procedure enabling the identification of these optimum conditions has also been developed [5].

Experimental evidence [1-4, 6-10] and corresponding theoretical analyses [5, 11-14] have led to the conclusion that the hollow atom Xe(L) emission at $\lambda \cong 2.9$ Å generated by 248 nm excitation of Xe clusters [2] in a self-trapped plasma channel is a model system that closely represents the ideal conditions sought for multikilovolt x-ray amplification. Specifically, the basis for this conclusion is comprised of (a) a detailed examination of Xe(L) spectral data [3,6,8], (b) theoretical analyses of the

CP611, *Superstrong Fields in Plasmas:* Second Int'l. Conf., edited by M. Lontano et al.
© 2002 American Institute of Physics 0-7354-0057-1/02/$19.00

mechanisms of cluster excitation [1, 11-13] and channeled propagation [1, 5, 14], and (c) calibrated measurements of the Xe(L) energy yield [9].

II. EXPERIMENTAL RESULTS

A. Apparatus

The experimental arrangement of the cluster target system and associated x-ray diagnostics is illustrated in Fig. (1). The gaseous xenon cluster target [1,2] was provided by a pulsed valve having an aperture of 1.5 mm which was operated at a maximum backing pressure of ~ 125 psia, a method that produced an average Xe density of ~ $3-6 \times 10^{19}$ cm^{-3}. The incident ultraviolet radiation was communicated to the cluster medium with an f/3 off-axis parabolic optic. The x-ray spectra were recorded in third order diffraction [6] with two identical von Hámos spectrographs equipped with mica crystals and Kodak RAR 2492 film. The plasma channels were observed on a single-pulse basis with an x-ray pinhole camera that viewed the Xe(M) radiation [1].

B. Xe(L) Hollow Atom Spectra

Characteristic Xe(L) spectra recorded without channel formation from Xe cluster targets under comparable conditions with irradiation at 248 nm and 800 nm are shown in Fig. (2). The large enhancement (~3000-fold) in the Xe(L) hollow atom emission (~2.9Å) generated with Xe clusters undergoing ultraviolet (248 nm) excitation over that observed with infrared (800 nm) excitation has a solid experimental basis [2]. The comparative spectra shown in Fig. (2) clearly demonstrate that the interaction with the clusters is grossly altered by changing the wavelength from 800 nm to 248 nm. Moreover, this observation now has a corresponding theoretical interpretation [12, 13]. The fundamental physical reason for the behavior illustrated by the data in Fig. (2) is that the relatively shorter optical period of the ultraviolet wave causes the electron-cluster interactions to be far more localized, controlled, and effective.

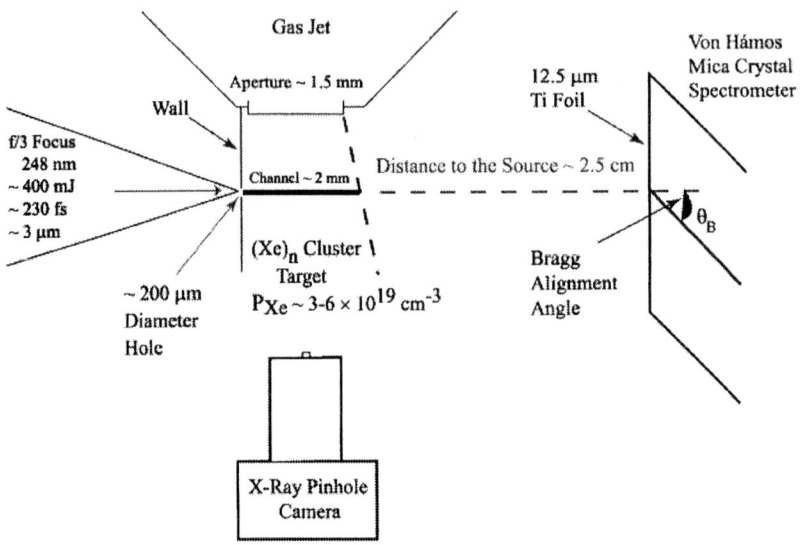

FIGURE 1. Experimental configuration used for observation of amplification of Xe(L) radiation in self-trapped channels. The x-ray pinhole camera was equipped with a ~ 10 μm thick Be foil enabling the morphology of the channel to be visualized by the Xe(M) emission (~ 1 keV). The wall defining the entrance plane having the 200 μm aperture was fabricated from ~ 100 μm thick steel and the incident 248 nm pulse was focused with an f/3 off-axis parabolic optic to a spot size of ~ 3 μm. The entrance of the von Hámos spectrograph viewing the forward directed emission was protected with a Ti foil of 12.5 μm thickness. The angle θ_B was normally arranged to be appropriate for the Xe^{34+} component at 2.88 Å. An identical von Hámos spectrograph, equipped with Muscovite mica from the same cut, was also used to record the transversely emitted spontaneous emission.

C. Channeled Propagation

The ability to produce narrow self-trapped channels in which the propagating intensity can approach ~ 10^{20} W/cm^2 is crucial for the production of amplified x-rays. These channels [1, 4-6, 9, 10, 13] control an exceptional power density. Evaluation of the coupling between the channeled ultraviolet radiation and the clusters [5, 7] indicates an atom-specific power of ~ 1W/atom, a value considerably greater than that of a vigorous thermonuclear environment. The presence of the channels grossly alters the morphology of the Xe(L) spectra in both the transverse and longitudinal directions with respect to the axis of the channel.

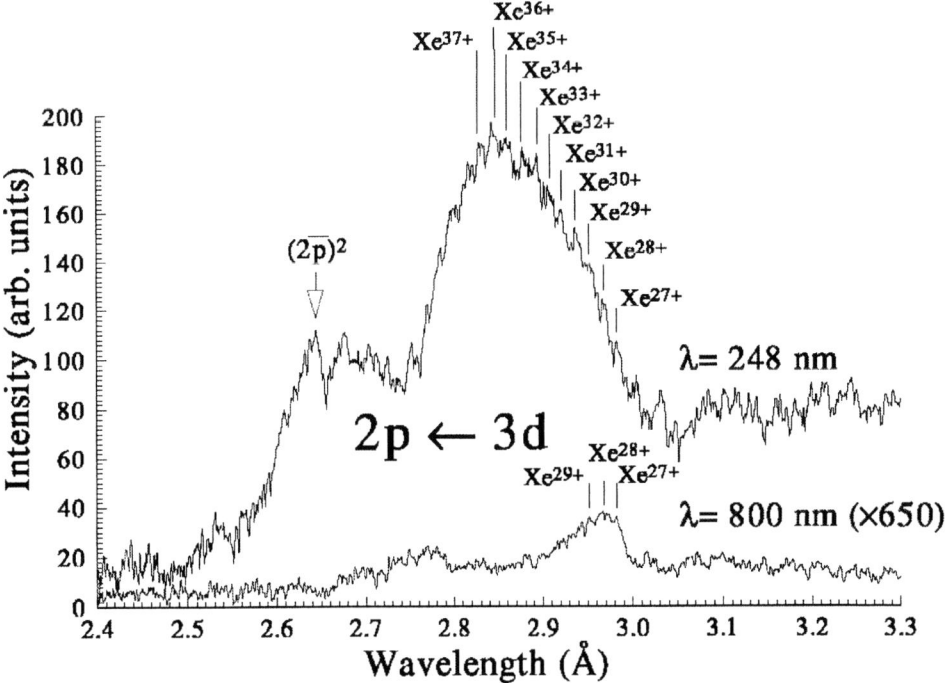

FIGURE 2. Comparative Xe(L) spectra observed without channel formation by irradiation of Xe clusters with wavelengths of 248 and 800 nm. The positions of the identified Xe^{q+} charge states are indicated. The ordinate of the 800 nm data (lower curve) is increased by a factor of 650 in order to enhance the visibility of the spectrum. The peak intensity (per pulse) of Xe(L) emission is approximately 3000-fold greater with the 248 nm excitation. Figure taken from Ref. [2].

D. Preliminary Xe(L) Spectra Produced from Self-Trapped Channels

A Xe(L) spectrum recorded in the forward (axial) direction is shown in Fig. (3). It differs grossly from the corresponding spectrum illustrated in Fig. (2) for which no channel was present. The assignment of the Xe^{34+} transition array is derived from calculations of the transition wavelength [15]. A correlated spectral hole-burning in transversely recorded spectra is simultaneously recorded; specifically, the Xe^{34+} intensity is significantly reduced when channels are present. Similar results have been observed on single-vacancy transitions on the Xe^{35+} and Xe^{36+} arrays.

Comparable results to those shown in Fig. (3) have also been obtained at ~ 2.8 Å on a transition identified as Xe^{37+} in which the excited level is a 2s2p double vacancy [8]. These results actually suggest that double vacancy production may be the dominant excitation in the central core of the channel.

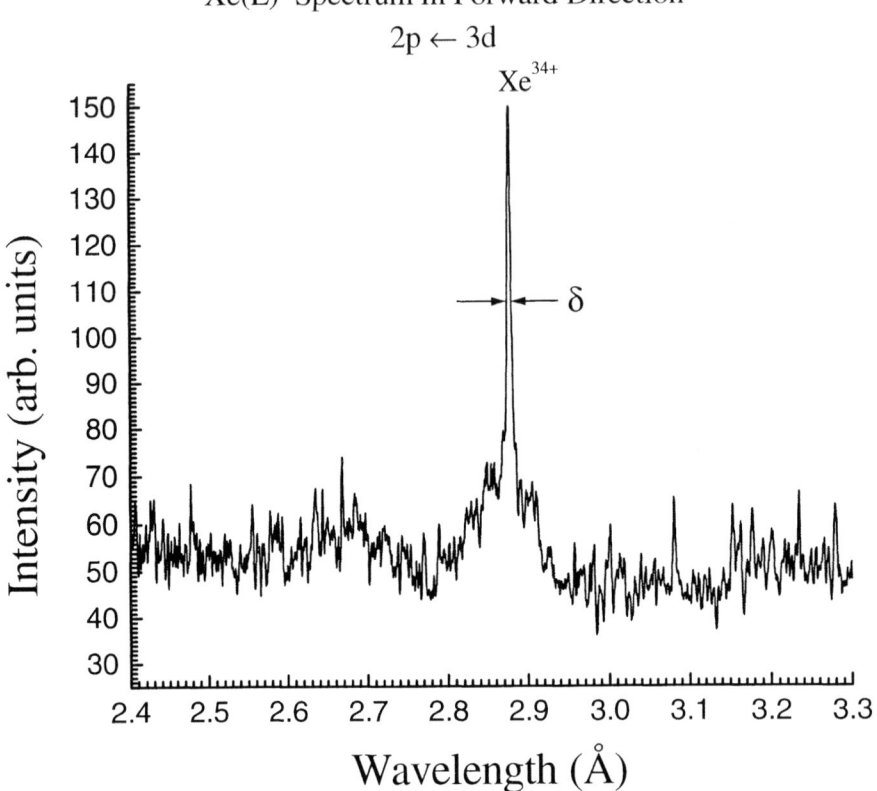

FIGURE 3. Xe(L) spectrum recorded with a mica von Hámos spectrometer in the forward (axial) direction from a channel. The width (FWHM) of the narrow Xe^{34+} feature is $\delta \sim 8.4$ eV.

III. CONCLUSIONS

It is crucial to note that the cooperative alliance [11] formed by the two physical processes described above has a fundamental physical basis that is the core of the concept; it is not an accidental arrangement. Herein lies the trick that enables the amplification to be successfully achieved. These two unusual forms of radiatively excited systems, the hollow atoms produced from the clusters and the self-trapped plasma channels, have a basic underlying structural relationship. They can be considered as two examples of a single new class of highly excited ordered matter [16]. Specifically, they respectively represent 3D and 2D systems whose corresponding excitation energies are derived from the symmetric displacement of electronic charge. From this structural similarity, it can be shown [1] that the system comprised of the plasma channel and the atomic cluster medium can be arranged so that essentially identical conditions are required to excite both of these forms of highly energetic matter. This outcome enables the formation of the confined channels under circumstances for which multikilovolt emission from hollow atoms perforce is simultaneously and optimally generated within them. The preliminary experimental findings reported above are consistent with the achievement of this consummate condition with the example of Xe(L) emission at $\lambda \sim 2.9$ Å.

ACKNOWLEDGMENTS

Support for this research was provided by the Army Research Office (Contracts DAAH04-94-G-0089 and DAAG55-97-1-0310) and the Sandia National Laboratories (Contracts DEAC04-94AL85000 and BF3611).

REFERENCES

1. A. B. Borisov, A. McPherson, B. D. Thompson, K. Boyer, and C. K. Rhodes, "Ultrahigh Power Compression for X-Ray Amplification: Multiphoton Cluster Excitation Combined with Non-Linear Channeled Propagation," *J. Phys. B* 28, 2143 (1995).
2. A. McPherson, B. D. Thompson, A. B. Borisov, K. Boyer, and C. K. Rhodes, "Multiphoton-Induced X-Ray Emission at 4–5 keV from Xe Atoms with Multiple Core Vacancies," *Nature* 370, 631 (1994).
3. K. Kondo, A. B. Borisov, C. Jordan, A. McPherson, W. A. Schroeder, K. Boyer, and C. K. Rhodes, "Wavelength Dependence of Multiphoton-Induced Xe(M) and Xe(L) Emissions from Xe Clusters," *J. Phys. B* 30, 2707 (1997).
4. A. B. Borisov, A. V. Borovskiy, V. V. Korobkin, A. M. Prokhorov, O. B. Shiryaev, X. M. Shi, T. S. Luk, A. McPherson, J. C. Solem, K. Boyer, and C. K. Rhodes, "Observation of Relativistic and Charge-Displacement Self-Channeling of Intense Subpicosecond Ultraviolet (248 nm) Radiation in Plasmas," *Phys. Rev. Lett.* 68, 2309 (1992).
5. A. B. Borisov, J. W. Longworth, K. Boyer, and C. K. Rhodes, "Stable Relativistic/Charge-Displacement Channels in Ultrahigh Power Density ($\sim 10^{21}$ W/cm^3) Plasmas," *Proc. Natl. Acad. Sci. USA* 95, 7854 (1998).
6. A. B. Borisov, A. McPherson, K. Boyer, and C. K. Rhodes, "Z-λ Imaging of Xe(M) and Xe(L) Emissions from Channeled Propagation of Intense Femtosecond 248 nm Pulses in a Xe Cluster Target," *J. Phys. B* 29, L113 (1996).
7. A. McPherson, A. B. Borisov, K. Boyer, and C. K. Rhodes, Competition between Multiphoton Cluster Excitation and Plasma Wave Raman Scattering at 248 nm," *J. Phys. B* 29, L291 (1996).
8. A. B. Borisov, A. McPherson, K. Boyer, and C. K. Rhodes, "Intensity Dependence of the Multiphoton-Induced Xe(L) Spectrum Produced by Subpicosecond 248 nm Excitation of Xe Clusters," *J. Phys. B* 29, L43 (1996).
9. A. McPherson, J. Cobble, A. B. Borisov, B. D. Thompson, F. Omenetto, K. Boyer, and C. K. Rhodes, "Evidence of Enhanced Multiphoton (248 nm) Coupling from Single-Pulse Energy Measurements of Xe(L) Emission Induced from Xe Clusters," *J. Phys. B* 30, L767 (1997).
10. A. B. Borisov, X. Shi, V. B. Karpov, V. V. Korobkin, J. C. Solem, O. B. Shiryaev, A. McPherson, K. Boyer, and C. K. Rhodes, "Stable Self-Channeling of Intense Ultraviolet Pulses in Underdense Plasma Producing Channels Exceeding 100 Rayleigh Lengths," *JOSA B* 11, 1941 (1994).
11. K. Boyer and C. K. Rhodes, "Atomic Inner-Shell Excitation Induced by Coherent Motion of Outer-Shell Electrons," *Phys. Rev. Lett.* 54, 1490 (1985).
12. W. Andreas Schroeder, F. G. Omenetto, A. B. Borisov, J. W. Longworth, A. McPherson, C. Jordan, K. Boyer, K. Kondo and C. K. Rhodes, "Pump Laser Wavelength-Dependent Control of the Efficiency of Kilovolt X-Ray Emission from Atomic Clusters," *J. Phys. B* 31, 5031 (1998).
13. W. A. Schroeder, T. R. Nelson, A. B. Borisov, J. W. Longworth, K. Boyer, and C. K. Rhodes, "An Efficient, Selective Collisional Ejection Mechanism for Inner-Shell Population Inversion in Laser-Driven Plasmas," *J. Phys. B* 34, 297 (2001).
14. A. B. Borisov, A. V. Borovskiy, O. B. Shiryaev, V. V. Korobkin, A. M. Prokhorov, J. C. Solem, T. S. Luk, K. Boyer, and C. K. Rhodes, "Relativistic and Charge-Displacement Self-Channeling of Intense Ultrashort Laser Pulses in Plasmas," *Phys. Rev. A* 45, 5830 (1992).

15. R. D. Cowan, *The Theory of Atomic Structure and Spectra* (University of California Press, Berkeley, CA., 1982).
16. G. Marowsky and Ch. K. Rhodes, "Hohle Atome-Eine Neue Form von Hochangeregter Materie," *Neue Zürcher Zeitung*, Nr. 254, 1. November 1995, S. 42.

kHz Femtosecond Laser-Plasma hard X-ray and fast Ion Source

A. Thoss, G. Korn, M.C. Richardson[1], M. Faubel[2], H. Stiel, U. Voigt,
C.W. Siders[1], T. Elsaesser

Max-Born-Institut, Berlin, Germany
[1] also at School of Optics & CREOL, UCF, Orlando, USA
[2]Max-Planck-Institut für Strömungsforschung, Göttingen,
Tel: 407 823 6819 Fax: 407 823 3570 email:mcr@creol.ucf.edu

Abstract. We describe the first demonstration of a new stable, kHz femtosecond laser-plasma source of hard x-ray continuum and K_α emission using a thin liquid metallic jet target. kHz femtosecond x-ray sources will find many applications in time-resolved x-ray diffraction and microscopy studies. As high intensity lasers become more compact and operate at increasingly high repetition-rates, they require a target configuration that is both repeatable from shot-to-shot and is debris-free. We have solved this requirement with the use of a fine (10-30 μm diameter) liquid metal jet target that provides a pristine, unperturbed filament surface at rates > 100 kHz. A number of liquid metal targets are considered. We will show hard x-ray spectra recorded from liquid Ga targets that show the generation of the 9.3 keV and 10.3 keV, K_α and K_β lines superimposed on a multi-keV Bremsstrahlung continuum. This source was generated by a 50fs duration, 1 kHz, 2W, high intensity Ti:Sapphire laser. We will discuss the extension of this source to higher powers and higher repetition rates, providing harder x-ray emission, with the incorporation of pulse-shaping and other techniques to enhance the x-ray conversion efficiency.

Using the same liquid target technology, we have also demonstrated the generation of forward-going sub-MeV protons from a 10 μm liquid water target at 1 kHz repetition rates. kHz sources of high energy ions will find many applications in time-resolved particle interaction studies, as well as lead to the efficient generation of short-lived isotopes for use in nuclear medicine and other applications. The protons were detected with CR-39 track detectors both in the forward and backward directions up to energies of ~ 500 keV. As the intensity of compact high repetition-rate lasers sources increase, we can expect improvements in the energy, conversion efficiency and directionality to occur.

The impact of these developments on a number of fields will be discussed. As compact, high repetition-rate femtosecond laser technology reaches towards focused intensities ~ 10^{19} W/cm², many new applications of high repetition rate hard x-ray and MeV ion sources will become practical.

CP611, *Superstrong Fields in Plasmas:* Second Int'l. Conf., edited by M. Lontano et al.
© 2002 American Institute of Physics 0-7354-0057-1/02/$19.00

INTRODUCTION

The many new and exciting discoveries that have recently been made in investigations of the interaction of intense femtosecond laser pulses with matter, have all been made in regimes governed by the performance parameters of the lasers used in these investigations, generally the complex, low-repetition-rate, or single-shot, multi-stage CPA, oscillator amplifier systems required to reach the focused intensities needed (~ 10^{17} W/cm^2). In this regime, apart from its precise alignment, little sophistication need be attached to the target system, since there is in general ample time for its manual replacement between shots. Much of the potential of these new discoveries however, may well only have practical significance when the development of high power, short pulse laser technology takes this regime into the domain of compact high repetition rate facilities. These facilities will be the sources of relativistic electrons, ultra-short duration x-rays and collimated MegaelectronVolt ions that will potentially open applications in time-resolved x-ray diffraction, x-ray radiography, perhaps proton tomography and proton cancer therapy. High repetition rate, kHz, femtosecond Ti:Sapphire lasers are now approaching this regime. As a consequence there is now the need to incorporate a target geometry that (i) can refresh the target material at a rate sufficient to provide a virgin target surface to each laser pulse, (ii) maintain an open, preferably ~4π, access to the target, and (iii) mitigates target debris to an extent that associated optics and diagnostic instrumentation do not get damaged or destroyed by collateral particulate debris emanating from the target.

In this paper we present two investigations that operate in this regime. For the first time we have combined a high repetition rate kHz high intensity 50 fs laser with a liquid microscopic liquid jet targets. In one case, we utilize a liquid Ga jet as a metal target to create strong femtosecond K_α emission in the 10 keV region[1]. This source will provide an idea source for time-resolved Bragg x-ray diffractions studies of structural changes in solid-state and biological media[2]. In a second experiment, we have used a micron-scale liquid water jet, and have detected semi-collimated high energy MeV protons emanating from the rear side of the water jet. Extrapolations of these experiments to higher intensities, will, based on previous low repetition-rate experiments[3-7], provide higher fluxes of higher energy protons, up to several 10's of MeV in highly collimated beams. Although medical applications of these sources may be years away, pump-probe investigations of the interactions of these particles with biological and solid state materials will be immediately possible.

KHZ 9.25 KEV X-RAY SOURCE FROM A LIQUID GALLIUM JET TARGET

For ultrashort x-ray sources to be more widely used, they must become compact and exhibit high average power. Moreover they must operate in a continuous regime with minimal laser adjustment and target replenishment. Developments in both laser and target technology are making this a reality. The development of high repetition rate (kHz), table-top femtosecond Ti:Sapphire lasers is now reaching the point where amplified systems can provide the focused intensities required[8-10]. Amplified milliJoule laser pulses of several tens of femtoseconds duration focused to spot sizes of a few microns, produce intensities in excess of 10^{16} W/cm^2. The plasmas created by these lasers need target configurations that are continuous in operation, presenting each laser pulse with a virgin target surface that remains precisely aligned within the short Rayleigh range of the high numerical aperture focusing optics employed (usually a few microns). Rotating disc, or drum targets do not easily meet these requirements. Reeled tape or wire targets may be better, but all these approaches suffer from the additional drawback in that they present a target surface that is many times greater in spatial extent than the focused spot size. This not only limits the solid angle of the useful x-ray emission emanating from the plasma, but, more seriously, leads to the generation of large amounts of target material projectiles and debris that are ruinous to the long-term life of the associated laser and x-ray optics. The severity of this problem has lead to the development of innovative approaches that bring the target size down to the size of the focus volume, ultimately to targets having a mass equal only to that of the number of x-ray radiating atoms required. These new approaches, based on 10-80 micron diameter liquid droplet[11] or cryogenic gas jet technology[12], have demonstrated debris-free operation in excess of 10^7 shots, even for optics located only a few centimeters from the source. Moreover, these targets present almost 4π–illumination from the source, and their limited spatial extent ensures a stable, localized, point x-ray source, an important requirement for many applications. They have however, been limited to only few liquids (water, alcohol, CuN solution, ethylene glycol) or gases (N$_2$, Xe, Ar)[13]. This is a serious drawback for hard x-ray generation at high laser intensities where high Z materials have higher conversion efficiency.

Here we report the use of a liquid metal jet as a target for a laser plasma. For kilohertz operation, the liquid metal must, have a flow velocity high enough to ensure that the jet is replenished and restored after each laser pulse. This is easily achieved, for separate hydrodynamic studies[14] show that for flow velocities of between 10 and 100 m s^{-1}, will allow repetition rates up to 100 kHz. In addition the jet's small lateral size, with diameters as small as 10 μm allows for near perfect matching to the spatial extension of the laser focus. The narrow spatial confinement of the radiation source provided by this geometry could be favorable for the generation of very short x-ray bursts in the 100fs time range.

The experiments reported here concentrated on laser plasmas produced from stable Ga jets. In these experiments a jet diameter d ~ 30 μm, was used. Experimentally, stable jet lengths of the order of 3.6 mm were determined with a simple HeNe laser scattering diagnostic, agreeing well with the expectations of theory of liquid jet stability[15]. The liquid Ga jet was located vertically within a vacuum chamber, at the focus of a f = 7 cm lens (Fig. 1). This focused the output of the high power, 1 kHz repetition-rate, Ti:Sapphire laser system onto the Ga jet to a spot diameter of ~10 μm at a point approximately 2 mm from the nozzle. The CPA-laser, which consists of a mode-locked Ti:sapphire oscillator, a regenerative amplifier[1] and a double-pass booster amplifier provided compressed 50 fs pulses at a center wavelength of 780 nm (spectral bandwidth 25 nm) with a maximum average power of 3 W, corresponding to a single pulse energies of 1.8 mJ. The effective contrast ratio of the amplified pulse was ~ 10^4. The intensity on target was 3×10^{16} W/cm^2.

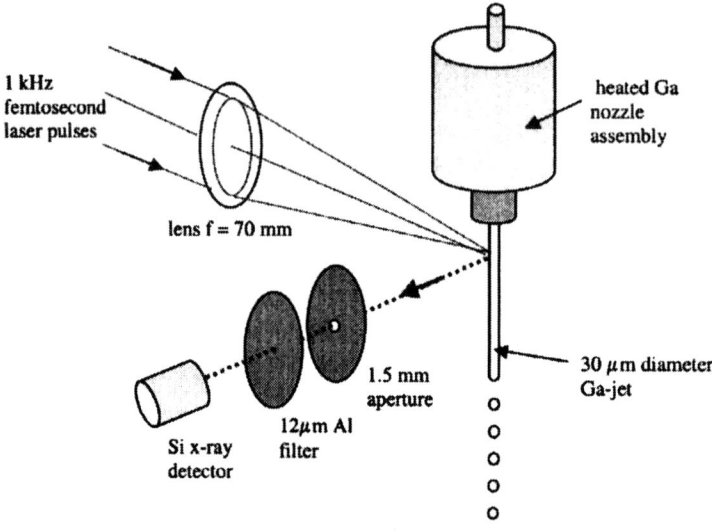

FIGURE 1. Experimental arrangement of the liquid Ga jet laser plasma source.

The spectrum of the x-ray emission emanating from the plasma was determined with the use of a Si-photodiode-based energy-dispersive, x-ray detector[16]. It was placed 300 mm from the liquid jet source. To avoid pile-up effects a 1.5 mm diameter pinhole was positioned 3 cm from the detector. A 12 μm Al-foil placed directly at the detector entrance acted as a cut-off filter for energies below 2.5 keV. The detector output was connected via an amplifier to the input of a 1024 channel multi-channel amplifier. An X-ray spectrum from the plasma generated from the Ga liquid jet is shown in Fig. 2. It consists of a broad continuum and two narrow features at 9.3 keV and 10.3 keV. These two peaks correspond to the characteristic K_α and - K_β lines of Ga. The K_α feature has two components, at 9.22 keV and 9.25 keV, corresponding to the $K_{\alpha 1}$ and $K_{\alpha 2}$ lines respectively. The broad continuum is predominantly free-free collisional (Bremsstrahlung) emission from the plasma, the fall-off at the low energy end of the spectrum being due to the low energy limit of the detector's sensitivity. The characteristic K_α and K_β lines originate from the collision of the high energy electrons with cold target material as they propagate forward through the jet, or from those electrons that stream outward from the plasma interaction, toward the laser, and are captured by the Coulomb field they create, and stream back into cold jet material in the vicinity of the plasma. The energy dispersive detector also gives an estimate of the absolute photon yield from the plasma. This was measured to be ~ 4 x 10^4

photons/sr/pulse. On the assumption of isotropic emission into 4π steradians, this would indicate a total emission power of x-ray's above the Al edge at 2.5 keV, of ~ to 5×10^8 photons /sec. Doubtlessly, the efficiency of this conversion can be improved, and as higher-powered lasers become available, the hard x-ray emission available from this type of source will increase. Moreover the use of other materials will provide a broad spectral range of x-ray emission, either from the plasma continuum emission, or from the characteristic K_α and K_β. Given it's relatively debris-free nature, this ultra-short laser plasma source will therefore find wide applications in x-ray diffraction studies of solid-state and biological materials, and other time-resolved studies, (radiography, fluorescence spectroscopy, etc).

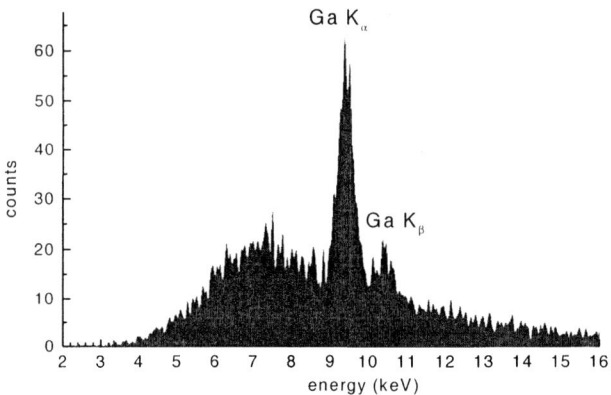

FIGURE 2. Spectral emission of femtosecond laser plasma from a Ga jet target.

HIGH-ENERGY PROTON GENERATION WITH A 1 KHZ HIGH INTENSITY FEMTOSECOND LASER

Many recent experimental investigations of laser-plasmas produced by high-intensity, ultra-short pulse lasers from solid targets have shown the generation of MeV protons, predominantly from the rear-side of the target[5,6]. At sufficiently high laser intensities, $> 10^{18}$ W/cm^2, these protons display distinct collimation and can contain a sizeable fraction of the initial laser energy. In fact, the intensity scaling of both the proton energy and their conversion efficiency from laser light appears to be more than linear, prompting suggestions that at higher laser intensities, such collimated beams of multi-MeV protons might be used for proton tomography, proton therapy for cancer treatment and other similar applications. Up to now, these studies have been made with low repetition-rate laser systems, either sub-picosecond Ti:Sapphire / Nd:glass hybrid laser systems, or ~100 fs Ti:Sapphire lasers. In order to exploit the potential of these high energy particles there is a need to take this exciting field towards operation at higher repetition-rates, with continuous operation, utilizing compact, operator-free laser systems, that is, towards a regime in which the laser-target system is simply regarded as a continuous source of pulsed, collimated MeV protons. Here we describe experiments that approach this regime, with a novel target geometry capable of providing a stable target surface to the laser beam. By using a femtosecond laser system operating at 1 kHz repetition frequency, we take, to the best of our knowledge for the first time, these experiments on high-energy proton generation into a new operating domain. This regime, we believe, will open a broad range of new scientific avenues and applications.

The type of targets we are using in this investigation will also contribute to the understanding of the mechanisms responsible for this intense proton generation. In the majority of previous studies, the proton emission has been shown to originate from thin layers of water vapor or hydrocarbons deposited on the backside of the target, (generally resulting from contaminants within the evacuated vacuum targets). In the experiments describe here, we use as a target, a microscopic stream of water, some 10-30 μm in diameter, a target already rich in hydrogen atoms, and flowing sufficiently fast that it is unlikely to accumulate contaminants from the vacuum chamber. These experiments in addition present a different target geometry to the focused laser beam. Whereas nearly all previous investigations have used massive planar targets, in the present case we use a thin cylindrical target having a diameter similar in size to the laser focal size.. This clearly defined microscopic target geometry has possible implications for the interpretation of the generation mechanisms of the protons themselves and the subsequent directionality of their emission.

The general configuration of the experiments reported here is shown in Fig. 3. The jet, of the Faubel design[14] , comprised basically of a specially designed orifice, of diameter 10 –30 μm, through which water is ejected from a pressurized reservoir which provides target material sufficient for several hours of operation. Proton emission from the target was detected with 50 mm square CR-39 track detectors. Two detectors were located ~ 3 cm from the target, one monitoring the proton emission emanating from the rear of the target, subtending a total solid angle of ~40 degrees in the forward direction of the laser beam. The second detector monitored protons accelerated from the front surface of the target on one side of the off-axis parabola, detecting particles coming from target at angles of 10° to 90° from the backward direction of the laser beam. When developed for 6 hrs in 6.25 mol solution of NaOH , at a temperature of 70° C, CR-39 is sensitive to protons having a minimum energy of 100 keV. The resulting circular tracks in the surface of the CR-39 are ~1 –8 μm diameter, and were analyzed with a high resolution microscope.

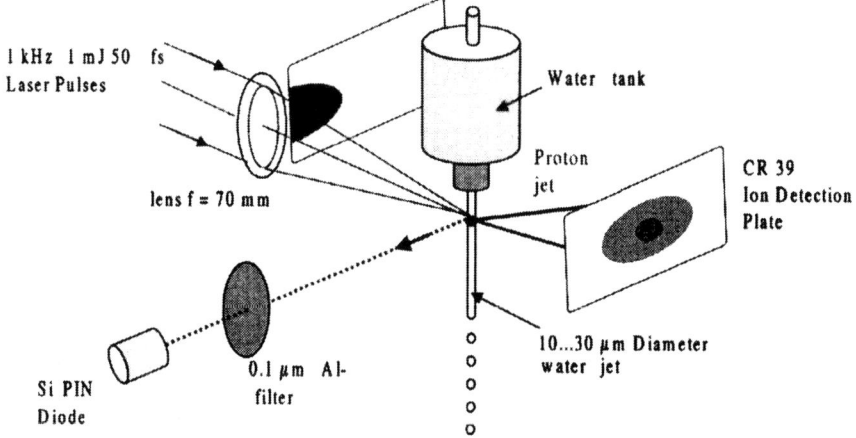

FIGURE 3. Experimental set-up for the detection of protons.

The angular distribution of the protons from the rearside of the target in the plane normal to the jet direction had a circular symmetry and is shown in Fig 4. This data was taken for a burst of 10^4 laser shots over 10 secs. The horizontal distribution of the protons >100 keV emanating from the target has a (FWHM) angle of 40°. The total number of protons coming from the rearside of the target per shot was estimated to be 3×10^3.

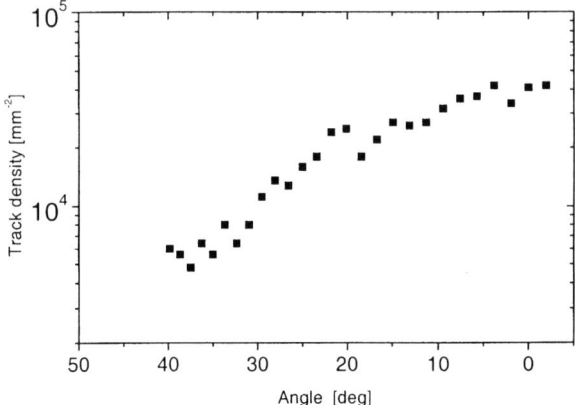

FIGURE 4. Horizontal angular distribution of proton tracks on CR 39 detection plate counted up to the center of a circular distribution. Exposure time was 10 s (10^4 shots).

Spectrometric measurements were made of the energy distribution of the protons using various thicknesses of mylar film. Thin strips of these films were laid across these detectors to measure both the energy spectrum of the protons, and its angular dependence. A typical analysis of the energy distribution of the protons in the forward direction is shown in Fig 5.   Layers of 2 μm and 4 μm of mylar selected protons having energies in excess of ~ 300 keV and ~ 500 keV. Protons of higher energy were not detected with thicker layers of mylar absorbers.   The scaling of the maximum proton energy measured in these experiments agrees well with previous measurements made with planar 'non-hydrogenic' targets and larger low repetition-rate laser systems.

The conversion efficiency of laser energy to protons in these experiments is fairly low, 10^{-5} %. However we can expect this efficiency to rise, as the laser intensity is increased. With the continued improvements to high repetition-rate femtosecond laser systems, intensities approaching 10^{19} W/cm^2, can be expected in the near future. In this regime, proton generation will be much more efficient, and their maximum energy will be in the multiple- MeV range. Whether these particle fluxes will be sufficient to stimulate interest in their use for proton tomography or proton cancer therapy may not be sure. However they will, without doubt be enough to initiate studies of the impact of short bursts of protons on solid-state and biological matter.   The availability of synchronised femtosecond optical pulses, either at the laser wavelength, or at other wavelengths in IR, visible or UV range, derived from this wavelength by nonlinear optical methods, will provide many ways for time-resolved pump-probe determinations of reversible and non reversible structural changes. Further on, one can envisage these methods being augmented with time-resolved x-ray diffraction studies facilitated by a high-repetition-rate femtosecond x-ray source of the type described here.

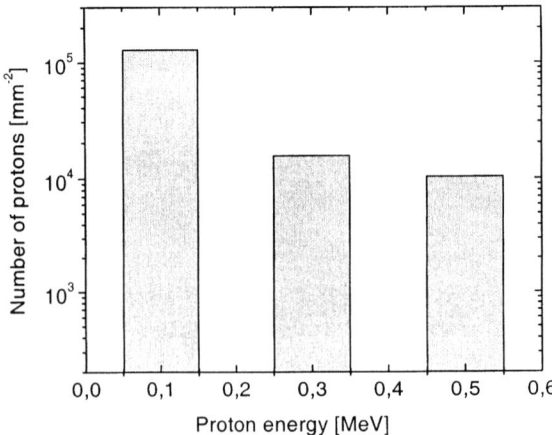

FIGURE 5. Proton energy distribution counted on CR 39 with no mylar, 1 layer and 2 layers of 2 μm mylar.

In summary, we have described briefly two important outcomes of the marriage of high intensity, high-repetition-rate femtosecond laser technology with a new, microscopic liquid jet target design. We expect to see rapid development this interaction regime in the future. At MBI, we are currently extending the performance of the laser system to higher powers, and adapting the liquid jet technology for other target materials.

The authors acknowledge useful discussions with Drs Peter Fews, Peter Norreys and Christoph Rose-Petruck, the provision of materials by Dr Steven Jacobs, and the strong support of Dr Nikolai Zhavoronkov in the development and operation of the laser system.

REFERENCES

1. G. Korn, A. Thoss, M. Faubel, H. Stiel, U. Voigt, M. Richardson, T. Elsaesser, in *Conference on Lasers and Electro-Optics (CLEO/US) 2001*, OSA Techn. Digest (Optical Society of America, Washington, DC 2001), paper CTuC4, p.113.

2. D. von der Linde et al., Laser Part. Beams **19**, 15 (2001), and A. Rousse, C. Rischel, J.C. Gauthier, Rev. Mod. Phys. **73**, 17 (2001).

3. Maksimchuk, S. Gu, K. Flippo, D. Umstadter, Phys. Rev. Lett. **84**, 4108 (2000).

4. E.L. Clark, et al., Phys. Rev. Lett. **85**, 1654 (2000).

5. S.P. Hatchett et al., Phys. Plasmas **7**, 2076 (2000).

6. Y. Murakami, Y. Kitagawa, Y. Sentoku, M. Mori, R. Kodama, K.A. Tanaka, K. Mima, T. Yamanaka, Phys. Plasm. **8**, 4138 (2001).

7. Zepf et al., Phys. Plasma **8**, 2323 (2001).

8. V. Bagnoud, F. Salin, Appl. Phys. B **70** (Suppl), S165 (2000).

9. J. Squier, G. Korn, G. Mourou, Opt. Lett. **18**, 625, (1993).

10. O. Albert, H. Wang, D. Liu, Z. Chang, G. Mourou, Opt. Lett. **25**, 1125 (2000).

11. M. Richardson, D. Torres, C. DePriest, F. Jin, G. Shimkaveg, Opt. Comm. **145**,109 (1998).

12. M. Berglund, L. Rymell, H.M. Hertz, T. Wilhein, Rev. Scient. Instr. **69**, 2361 (1998).

13. G. Korn, A. Thoß, M. Faubel, H. Stiel, U. Vogt, M. Richardson, T. Elsaesser, subm. to Opt. Lett. (2001) and references therein.

14. M. Richardson, A. Thoß, G. Korn, M. Faubel, T. Elsaesser [to be published].

15. M.Faubel, in *"Photoionization and Photodetachment"*, Ed. C.Y. Ng, World Scientific, Singapore, p.634-690 (2000).

16. Amptek Inc., Bedford, MA, the signal from the XR-100CR detector was amplified with a PX2CR and recorded with a MCA8000A.

Fast coronal ignition and the problem of energy transport

P. Mulser*, H. Ruhl[†], S. Hain* and D. Bauer*

*Theoretical Quantum Electronics (TQE), Darmstadt University of Technology, Hochschulstr. 4A, 64289 Darmstadt, Germany
[†]Max-Born-Institut, Max-Born-Str. 2a, 12489 Berlin, Germany

Abstract. An alternative scheme of fast ignition is proposed and the key issue of dense matter heating, common to all fast ignition schemes, is discussed.

1. Introduction

Since Fast Ignition (FI) was proposed in 1994 to ignite deuterium/tritium (DT) pellets [1] it attracted vivid interest owing to the new prospects it opens in the field of Inertial Confinement Fusion (ICF) with lasers and heavy ion beams [2]. In contrast to standard central ignition FI consists in heating a portion of precompressed DT pellet to temperatures above 8 keV during the time of typically 20 ps by a laser beam of at least several 10^{19} Wcm^{-2} intensity. During the last seven years the concept of FI has evolved theoretically, and at the same time, has stimulated preparatory experiments in leading high power laser laboratories (LLNL, ILE, RAL, LULI, MPQ). Currently we may distinguish three FI approaches, fundamentally differing from each other: Fast Beam Ignition (FBI) by intense laser-generated particle beams [1, 3]; Cone-guided Fast Ignition (CFI) of a dense region of the pellet by guiding the laser beam along the inner wall of a cone consisting of a high-Z element [4]; finally, Fast Coronal Ignition (FCI) [5] which, together with the problem of energy transfer to compressed matter, is the subject of this paper. The FI concept offers the advantage of separating pellet ignition from its compression phase and additional degrees of freedom in pellet design.

2. Holeboring and FI exclude each other

We consider first the penetration of the laser beam in cold matter. Basic estimates [5] show that the laser beam acts like an impermeable piston driven by the light pressure $p_L = (1+R)I/c$ (I laser intensity, R reflection coefficient, c speed of light). In front of the critical surface a strong bow shock forms which runs with speed v_0 into the precompressed cold pellet (our assumption here) of density ρ_0 (Fig.1(a)). The momentum balance reads with the compression κ as follows,

$$p_L = \rho_0 v_0^2 (1 - \kappa^{-1}). \tag{1}$$

In an ideal classical gas $\kappa = (\gamma+1)M^2/[(\gamma-1)M^2+2]$, γ adiabatic coefficient, M Mach number. For $\gamma = 5/3$ the maximum achievable compression of $\kappa = 4$ results. The density ρ_P

CP611, *Superstrong Fields in Plasmas:* Second Int'l. Conf., edited by M. Lontano et al.
© 2002 American Institute of Physics 0-7354-0057-1/02/$19.00

at the vertex of the piston is higher owing to stagnation and is determined from Bernoulli's equation in differential form $\nabla v^2/2 + \nabla p/\rho = 0$. Under the most reasonable assumption that the shocked matter is isothermal it can be integrated along the streamline on the axis,

$$\rho_P = \kappa \rho_0 \exp(v_1^2/2s^2), \quad v_1 = (1 - \kappa^{-1})v_0, \quad s^2 = p_S/\rho_S. \tag{2}$$

(see Fig.1(a)). As we shall see Eqs. (1) and (2) have far-reaching consequences.

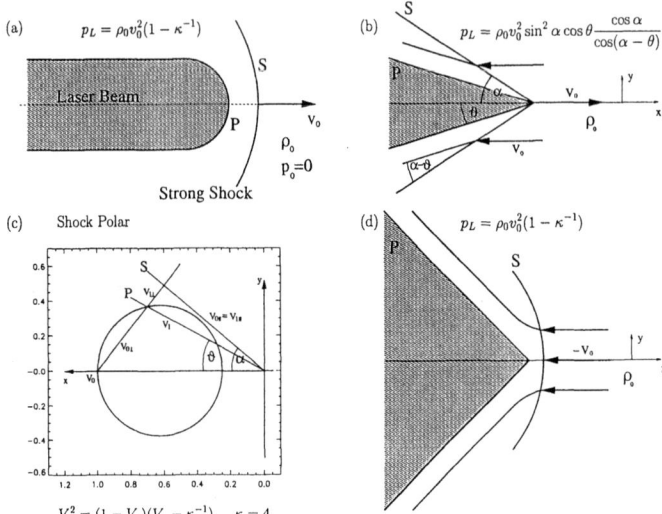

Fig. 1. Holeboring in cold matter: alternative models for matter displacement. (a) Laser acts as impermeable piston, p_L light pressure. ρ_0 undisturbed precompressed pellet density, v_0 holeboring speed. Oblique shock in (b) is determined from the shock polar (c). Beyond limiting angle ϑ shock detaches from cone (d).

It is important to compare them with expressions from alternative models. Consider a cone-shaped piston of narrow aperture angle θ propagating at speed v_0. From the vertex of the piston P a co-axial conical shock originates of aperture angle α (Fig.1(b)). Under the assumption of negligible pressure p_0 the possible states behind the shock cone (region 1) can be determined from the shock polar, which in Fig.1(c) is given for $\kappa = 4$. When normalized to v_0 it reads $V_y^2 = (1 - V_x)(V_x - \kappa^{-1})$. From the continuous transition of the tangential component through the shock front, i.e., $v_{0\parallel} = v_{1\parallel}$ (see for instance any textbook of gasdynamics)., one derives

$$p_L = \rho_0 v_0 v_1 \sin^2 \alpha \cos \theta = \rho_0 v_0^2 \sin^2 \alpha \cos \theta \frac{\cos \alpha}{\cos(\alpha - \theta)}. \tag{3}$$

For angles θ such that the straight line starting from the origin does not have any point in common with the shock polar, the shock S forms at finite distance from the cone with finite curvature on the axis (Fig. 1(c)), and Eq.(1) applies again. For a strong shock, equivalent to $p_0 = 0$, the limiting angle is given by $\tan \theta = (\kappa - 1)/2\kappa^{1/2}$. For $\kappa = 4$ results $\tan \theta = 3/4$ and

$\theta = 37°$. The corresponding angle α follows from Fig. 1, i.e., $\alpha = 64°$, and the angular factor in Eq.(3) assumes the value 0.32. For the cone and shock angles of $\theta = 28°$ and $\alpha = 40°$ indicated in the Figure the angular factor is 0.29. The factor decreases monotonously with decreasing aperture angle of the cone and, at given p_L, v_0 increases. For the strong shock of Fig.1(a) and (c) the factor becomes $1 - 1/4 = 0.75$ from Eq. (1).

Alternatively, for a piston of curvature R moving through a homogeneous fluid a potential flow establishes and the pressure at the vertex becomes the Bernoulli stagnation pressure, $p_L = \rho_0 v_0^2/2 + p_0$. The efficiency of holeboring is evaluated for a density profile $\rho_0 = h(x/L)^\alpha$. From $p_L = f\rho_0 v_0^2$ the depth x/L as a function of t is given by

$$\frac{x}{L} = \left(\frac{\alpha+2}{2L}t\right)^{\frac{2}{\alpha+2}} \left(\frac{p_L}{fh}\right)^{\frac{1}{\alpha+2}}. \tag{4}$$

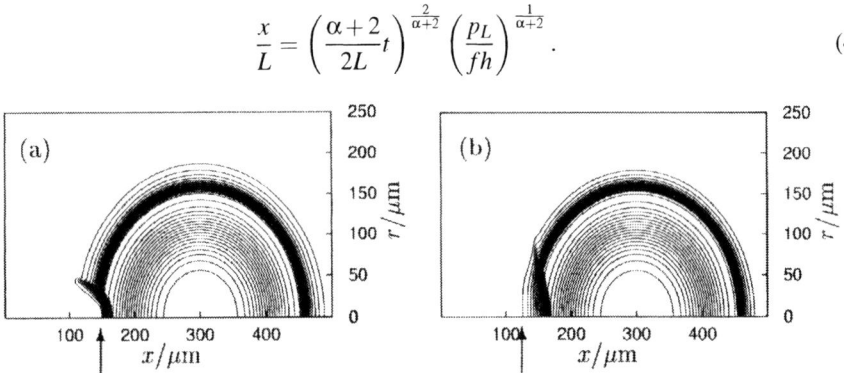

Fig. 2: Doughnut-like pellet is the result of laser pulse compression of a 5 mg DT pellet along a low adiabate; maximum density $\rho_{max} = 350$ gcm^{-3}. It is used in all subsequent FCI studies. The position of the laser beam front is indicated by ↑. (a): holeboring by a beam of $I = 10^{21}$ Wcm^{-2} after 20 ps in the absence of thermal transport. The holeboring effect in the cold matter is clearly seen. (b): $I = 10^{21}$ Wcm^{-2} with energy deposition at $\rho = 4.6$ gcm^{-3} (outer contour) and diffusive heat conduction; no indication of any kind of hole formation (shock is driven by the heat wave).

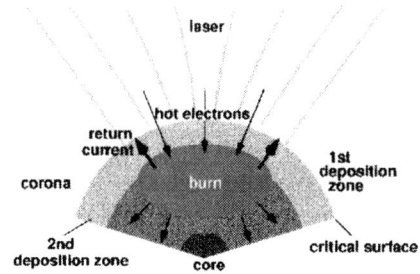

Fig. 3: Scheme of Fast Coronal Ignition (FCI).

For $I = 10^{21}$ Wcm^{-2}, $R = 0$, $h = 400$ gcm^{-3}, $t = 20$ ps, $f = 0.5$, $L = 0.01$ cm, $\alpha = 2$ follows in cold matter $x/L = 0.40$, $\rho/h = 0.16$, $\rho = 65$ gcm^{-3}. For $\alpha = 3$ the result is not much different: $x/L = 0.53$, $\rho/h = 0.15$, $\rho = 60$ gcm^{-3}. The conclusion is that holeboring works at superhigh intensities in cold matter. At an intensity of $I = 10^{19}$ Wcm^{-2} holeboring stops after 20 ps at $\rho = 21$ gcm^{-3} and $\rho = 24$ gcm^{-3}, respectively. Numerical simulations and estimates show that for ignition to work the electron temperature in the low density deposition zone, typically 4 gcm^{-3}, must reach 100 keV. The corresponding pressure is $p_0 = 4 \times 10^{16} \rho$ [cgs] and subtracts from p_L in Eqs.(1), (3). Since $p_L - p_0 \simeq f\rho_0 v_0^2$ holeboring at $I = 10^{21}$ Wcm^{-2} stops at the latest when the density $\rho = 5$ gcm^{-3} is reached. This estimate is confirmed by the simulation of Fig.2. The unavoidable conclusion is that holeboring and fast ignition exclude each other: When there is holeboring no ignition occurs, and when there is ignition holeboring is absent. Fast ignition based on electron energy transport has to start from the corona (see Fig.3), or, in CFI, from the top of the cone.

3. Fast coronal ignition (FCI)

Unless a process is found by which almost all laser beam energy is converted into beams of highly energetic heavy particles or electrons, ignition of the pellet can only start from a low density region in the corona. In Fig.3 the scheme of FCI is illustrated. In the underdense corona the laser energy is converted into kinetic electron and magnetic field energy (first deposition zone). From there the energy is transmitted to a region of typically $1 - 5$ gcm^{-3} by fast electron stopping, electron thermal conduction and return current heating to initiate a selfsustained burn wave (second deposition zone). To study the FCI scheme numerical simulations are performed on a 5 g DT shell which is precompressed along a low adiabate by a ns laser pulse until it stagnates in the center and a maximum density of 350 gcm^{-3} is reached on a concentric shell. The mass concentration assumes a doughnut-like shape, as shown in Fig.2. This pellet is used in all following simulations. Fig. 4 shows successful FCI with flux-limited Spitzer heat conduction (a), and without limiter (b). Deposition of the laser energy occurs at $\rho = 4.6$ gcm^{-3} (outer contours in the two pictures at earliest times) for 20 ps. The evolving shock is driven by the heat wave and the thermonuclear deflagration wave and is not a consequence of holeboring. Owing to Spitzer flux limitation most of the energy supplied (75 kJ) to the pellet in (a) is dispersed in the corona. Nevertheless the overall burn

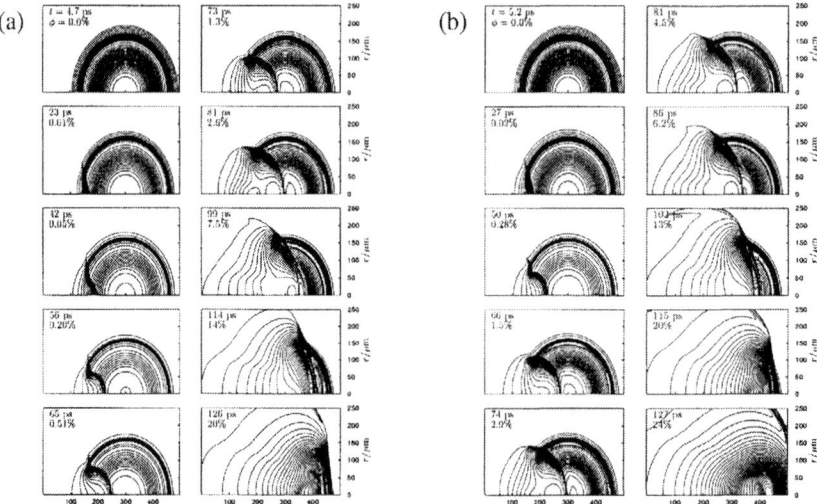

Fig. 4: Fast ignition and burn of pellet from Fig.2. (a): Flux-limited Spitzer heat flow. Laser intensity $I = 10^{21}$ Wcm^{-2}; 75 kJ energy deposited at $\rho = 4.6$ gcm^{-3} at position of outer contour in first picture ($t = 4.7$ ps). (b): No flux limited Spitzer heat flow. Laser Intensity $I = 4 \times 10^{20}$ Wcm^{-2} deposited during 20 ps; deposition energy 30 kJ. "Free ignition intensity" is $I = 2 \times 10^{20}$ Wcm^{-2}, "free ignition energy" is 15 kJ. Most of the energy in (a) is dispersed in the corona. High burn efficiency: 25 %.

efficiency is as high as 25 %. In order to find out how much energy is needed for successful FCI, so-called "free ignition energy" and "free ignition intensity" correspondingly, in (b) the same simulation is done without flux limitation. In this way a free ignition intensity of $I_F = 4 \times 10^{20}$ Wcm^{-2} and a free ignition energy amounting to $E_F = 15$ kJ are found for our (not optimized) pellet.

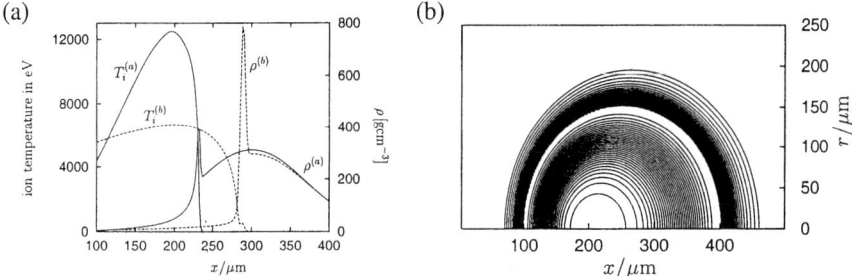

Fig. 5: (a): Ion temperature and mass density profiles on the axis $r = 0$ from 2D burn simulations; parameters: pellet mass 1.7 mg, peak density $\rho_{DT} = 311$ gcm^{-3}, characteristic length of density profile $L = 100 \, \mu$m, pulse duration $\tau = 20$ ps (a),(b), free ignition intensity $I = 1.2 \times 10^{20}$ Wcm^{-2} (a), and $I = 0.8 \times 10^{20}$ Wcm^{-2} (b). The profiles are taken after $t = 44$ ps (a) and $t = 104$ ps (b). While case (a) shows a propagating burn front (moderate shock compression $\kappa < 2$), in case (b), dashed lines, the compression front decouples from the heat front and ignition fails (strong shock, $\kappa > 2$). (b): Fast ignition ($I = 10^{21}$ Wcm^{-2}) of FCI pellet from Fig.2 with artificially imposed asymmetry. No burn depletion is observed relative to Fig.4.

An essential condition of a burn wave which is self-sustaining can be deduced from Fig. 5(a): The shock in front of the heat wave must not decouple from the heat front. The influence of asymmetry on FI is studied with the asymmetric pellet of Fig. 5(b), and an earlier result [5] is reconfirmed: Considerable deviations from spherical symmetry show almost no burn depletion as soon as ignition has started. It should be noted that pellets of Gaussian density structure are more easily ignited than doughnut-like ones (lower E_F in [5]).

The criterion $\rho R = 0.1 - 0.3$ for ignition does not apply to FCI since this condition is purely static whereas in Fig.4 the energy diffusing from the hot spot is not lost and the heat wave is supported by thermonuclear energy release. In contrast to central spark ignition FCI starts very slowly. In Fig.4 only 1% of DT is burned 50 ps after the end of the laser pulse.

4. Energy transport and deposition

FI in all its variants (FBI, CFI, FCI) stands and falls with the possibility to efficiently transport the energy from the 1st (low density) to the 2nd (high density) deposition zone. In order to study this question with the most advanced means actually at hand laser light coupling at $I\lambda^2 = 5 \times 10^{19}$ Wcm$^{-2}\mu$m^2 is simulated in a collisionless 3D PIC in a 33 times overdense target [6]. The simulation shows that most of the absorbed energy (30–40%) goes into fast electrons in the range 1–40 MeV and dc magnetic fields. The electron flow pinches into coalescing filaments. In the strong B-field the forward electrons (I_z) diverge laterally ($I_\perp$) and finally go over into a hot return current. This is the reason for their short penetration in forward direction ($\approx 2 \, \mu$m). Lateral current and energy flows ($I_\perp, q_\perp$ in Fig. 6) exceed several times the longitudinal quantities $|I_z|$, q_z. Of particular interest is the energy density distribution ε compared with the particle density n (Fig. 6(c)) since $\varepsilon/n \sim p/\rho$ is a measure for local energy deposition, or equivalently, for diffusive transport. From Fig. 6 it must be concluded that in the absence of collisions the energetic electrons heat the low density corona surrounding the compressed pellet as a consequence of **magnetic insulation**: No well localized hot spot is formed.

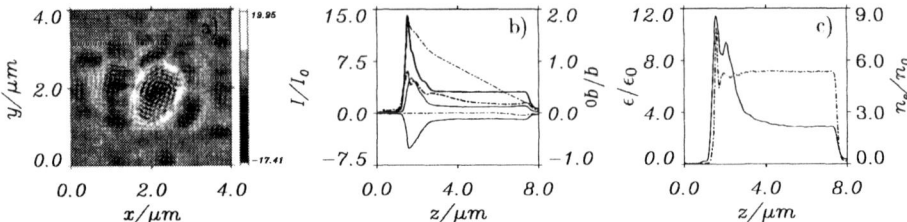

Fig. 6: 3D PIC simulation (6.9×10^9 cells) of energy transport and deposition at 179 fs for a laser of $I\lambda^2 = 5 \times 10^{19}$ Wcm$^{-2}\mu$m^2 incident along the z-axis onto a 33 times overdense plasma. The hot electron stream is unstable and decays into filaments which coalesce: (a) current and return current filaments taken at $z = 2\,\mu$m. Plot (b) shows the total direct and return currents I_z (thin solid curves) and $I_\perp$ (bold) parallel and transverse to the incident beam as well as the energy fluxes q_z and $q_\perp$. $q_\perp$ is highest, the return energy flux $q_z < 0$ nearly vanishes. In plot (c) the line energy density $\varepsilon(z)$ and the line particle density n_e are shown. There is almost no energy exchange between the hot and the cold background electrons. Normalization constants: $I_0 = 10^5$ A, $q_0 = 10^{12}$ W, $\varepsilon_0 = 10^4$ Jm^{-1}, $n_0 = 10^{15}$ cm^{-1}.

Summary and outlook

If energy transport from the first to the second deposition zone works properly FCI represents a promising scheme of ICF with lasers and heavy ion beams. According to present understanding of superintense laser-matter interaction a considerable fraction of the absorbed energy (30–40%; up to 70% at long times) is transferred to jets of fast electrons. It is an open question whether the fast electrons can drive a return current able to heat dense enough coronal matter or whether direct diffusive transport is strong enough to do the job. There are also indications from PIC simulations that appreciable energy is transported by magnetic fields which decay in the interior of the pellet [6]. In any case an efficient deposition mechanism is needed because FCI seems to work above a certain density threshold which for realistic intensities lies above 1 gcm^{-3}. Only the next generation of simulations on a ps timescale (and high energy experiments) will show whether collisionality will be efficient enough to prevent magnetic insulation and/or to merge into diffusive electron transport. Actually, in dense matter the energy transport is terra incognita, a widely unexplored territory.

REFERENCES

1. M. Tabak et al., Phys. Plasmas **1**, 1626 (1994).
2. J. Lindl, Phys. Plasmas **2**, 3933 (1995).
3. M. Roth et al., Phys. Rev. Lett. **86**, 436 (2001).
4. M. Zepf et al, preprint under preparation; M. Zepf, E.L. Clark et al., Phys. Plasmas **8**, 2323 (2001).
5. S. Hain and P. Mulser, Phys. Rev. Lett. **86**, 1015 (2001).
6. H. Ruhl, submitted for publication.

Generation and Astrophysical Applications of Relativistic Pair Plasmas with Ultra-Intense Lasers

Edison Liang

Rice University, Houston, TX 77005-1892

Abstract. We discuss the production of relativistic electron-positron plasmas by the interaction of ultra-intense lasers with high-Z targets. We review the potential astrophysical applications of such laser experiments, including gamma-ray bursts, blazar jets and black holes. We propose future experiments involving collisions of such pair fireballs and the study of collisionless shocks in pair plamsas.

INTRODUCTION

The development of Peta-Watt class ultra-intense lasers has enabled the study of new regimes of plasmas, including relativistic plasmas, electron-positron plasmas, and plasmas with up to Giga-gauss magnetic fields. Current and pending Peta-Watt lasers can be focussed to intensities such that $I\lambda^2 >> 10^{19}$, where I is laser intensty in W/cm^2 and λ is laser wavelength in microns. At such intensities most of the laser energy is converted into superthermal electrons [1] with a quasi-Maxwellian temperature of $T_{hot} = mc^2[(1 + I\lambda^2/1.4.10^{18})^{1/2} - 1]$ where m is electron rest mass and c is light speed. Hence $T_{hot} > mc^2$ for $I\lambda^2 > 1.4.10^{18}$. Such relativistic electrons are capable of pair production when they collide with the ions and with each other. Also such electrons will radiate bremsstrahlung gamma-rays which in turn are capable of pair production.

During the past few years we have investigated, both analytically and with particle-in-cell (PIC) simulations, the kinematics and microphysics of pair plasma generation by ultra-intense lasers and their subsequent evolution [2]. We have also explored their potential astrophysical applications, especially issues related to scaling. Nonthermal electron-positron pair plasmas are known to be abundant in many astrophysical environments, from pulsars to blazars. In the last few years, the discovery that gamma-ray bursts are at cosmological distances also led to the conclusion that these events involve ultra-relativistic plasmas which are likely composed of pairs. The plasma physics of such relativistic electron-positron pairs is poorly understood, especially in the collisionless regime. Hence the laboratory experiments will provide critical new insight into these exotic astrophysical phenomena.

CP611, *Superstrong Fields in Plasmas:* Second Int'l. Conf., edited by M. Lontano et al.

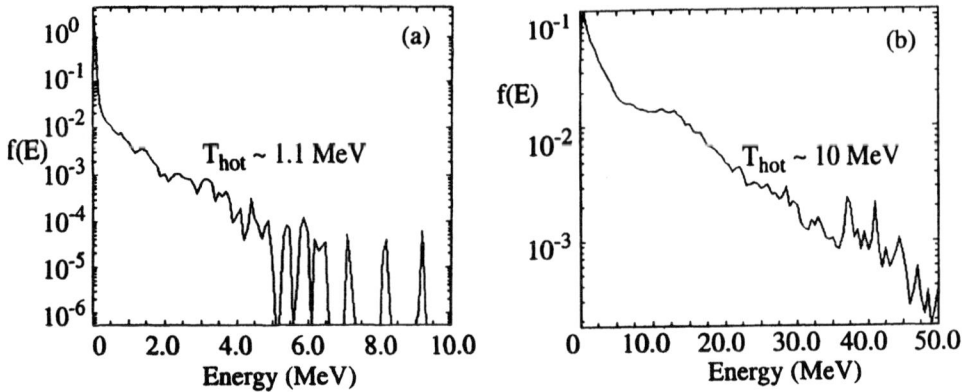

FIGURE 1. PIC simulations of superthermal electron energy spectra generated by gold targets hit by ultra-intense lasers (a) $I = 8.6.10^{18}$. (b) $I = 1.4.10^{20}$, from [2].

POSITRON PRODUCTION WITH ULTRA-INTENSE LASERS

When the superthermal electrons interact with high-Z ions they will pair produce via the Bethe-Heitler cross-section [3]. In most laser laboratory regimes this process should dominate over leptonic and photonic pair production processes. Using PIC simulations we have generated superthermal electron spectra for a range of laser intensities for Au targets (Fig.1) [2]. By convolving such electron spectra with the Bethe-Heitler cross section we obtain the pair creation rate per target electron Γ, which is plotted as function of incident laser intensity in Fig.2. We see that the rate rises steeply until I reaches around 4.10^{19} after which it rises more slowly due to the log dependence of the Bethe-Heitler cross section on electron energy. Hence to produce the maximum number of pairs for a given laser pulse energy, it is better to keep the intensity at around 10^{20} but make the laser pulse as long as possible, instead of compressing the pulse to get to higher intensity. As an example, Fig.2 shows that for $I = 10^{20}$ and a pulse duration of 10 ps, $\Gamma t = 2.10^{-3}$. This prediction is in fact confirmed to first order by the pioneering experiments of Cowan et al [4] using the Peta-Watt laser at LLNL (Fig.3). Fig.4 shows PIC simulations of the predicted electron and positron spectra, which compare favorably, at least qualitatively, to the experimental data.

DOUBLE-SIDED ILLUMINATION AND PAIR FIREBALL

In order to produce a dense pair fireball, we propose to use double-sided laser illumination. The laser pondermotive pressure from both lasers will confine the

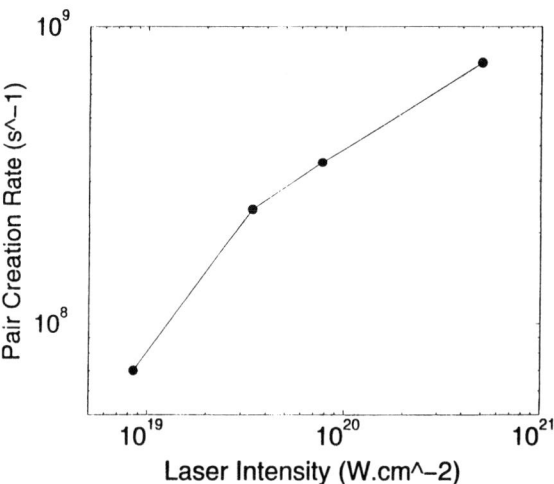

FIGURE 2. Pair creation rate per target electron as a function of incident laser intensity, from [2].

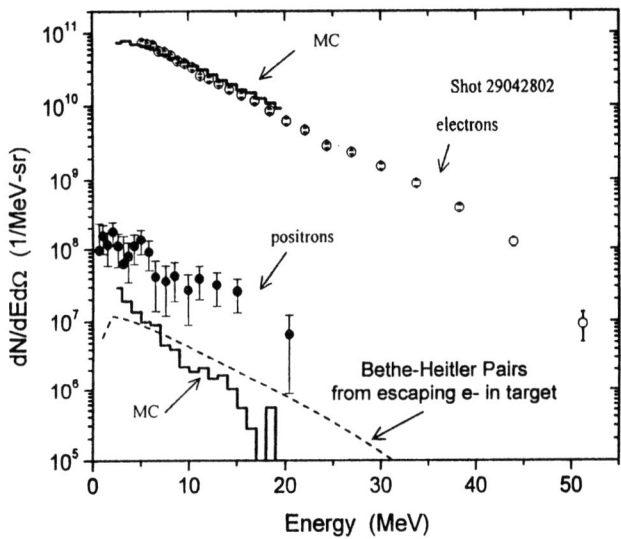

FIGURE 3. Experimental data of superthermal electron and positron energy spectra obtained with the LLNL Peta-Watt laser, from [4].

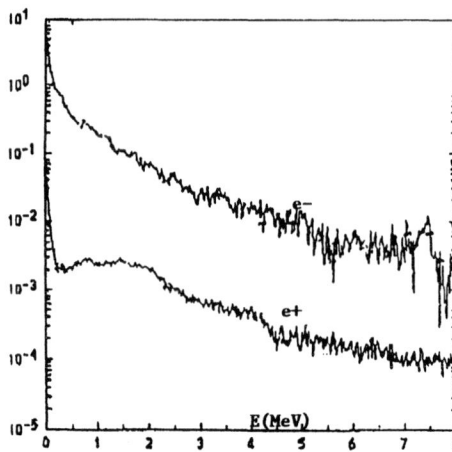

FIGURE 4. PIC simulation of superthermal electron and positron energy spectra relevant to the experiment of Fig.3, from Wilks unpublishied.

pairs and superthermal electrons, and drive them to interact with the target ions multiple times, thus optimizing the number of generations of pairs that can be produced for a given laser pulse energy. PIC simulations of such double-sided laser-target interactions are currently underway.

After the laser pulse is turned off, both the pair plasma and the electron-ion plasma will explode. However, since the the pairs are much lighter than the ions, they will stream out ahead of the ion plasma. After a while we should achieve an almost pure relativistic pair fireball with a small ion core. This expanding fireball can in principle reach bulk Lorentz factors of 10's and its energy distribution resembles a Maxwellian. To our knowledge, no other laboratory schemes of pair production can generated similar regimes of pair plasmas.

The plasma physics of such pair fireballs is of fundamental interest. But we can also visualize its tremendous potential to simulate a variety of astrophysical phenomena, from blazar jets to gamma-ray bursts.

COLLIDING FIREBALLS AND GAMMA-RAY BURSTS

As an example we have explored the application of pair fireballs to the study of gamma-ray bursts. The currently popular paradigm of gamma-ray bursts is that these are ultra-relativistic jets of plasmoids ejected by a rapidly rotating black hole during formation at the collapsing core of a massive star. The gamma-rays are emitted by so-called "internal shocks" which are created by the collision of plasmoids moving at slightly different Lorentz factors, due to the unsteadiness of

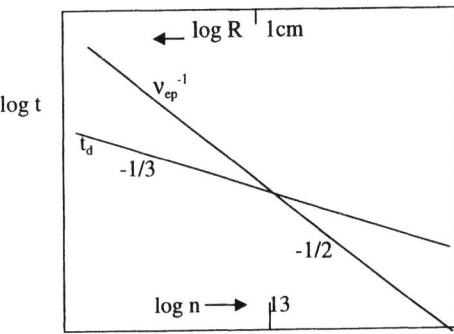

FIGURE 5. Comparison of fireball dynamical time and inverse electron plasma frequency as functions of the fireball size (top scale) and density (bottom scale)

the central accretion process, or interactions with the surrounding material [5]. These collisions occur at moderate relative Lorentz factors (say a few to 10's) even though the overall bulk Lorentz factor of the plasmoids may be much higher (100's or even 1000's). Due to the low density, the kinetic energy of the colliding plasmoids are most likely converted to internal and radiation energy by collisionless processes. This regime is likely accessible with the laser generated fireballs, except that the space and time scales are greatly compressed, and we have to study in detail the scaling properties to make sure that the relevant physics we learn in the laser experiments are indeed applicable to the astrophysical environments.

Hence we propose a colliding fireballs experiment. Such an experiment will involve two adjacent fireballs generated by four Peta-Watt class lasers. Technologically this is quite feasible with the conversion of one quad of the NIF lasers. We have begun to explore the plasma microphysics of such colliding pair fireballs, both theoretically and with PIC simulations. For example, Fig.5 shows the plasma frequency compared to the dynamical time of the expanding fireball. We see that the plasma

frequency drops below the inverse dynamical time when the pair density drops below $10^{13} cm^{-3}$, or when the fireball is larger than about 1 cm. Hence to have a chance of achieving a collisionless shock, the fireballs must collide before their densities drop below about $10^{13} cm^{-3}$. This requires that the fireballs be formed no more than a cm apart. An alternative is to add an external magnetic field so that the electron gyroradius becomes small compared to the fireball size. Of course detailed PIC simulations will be needed to see if a collisionless shock will indeed occur.

FUTURE PLANS

In addition to the construction of four Peta-Watt class lasers, the proposed colliding relativistic pair fireball experiments will require new types of advanced diagnostics. This will include space-and-time-resolved x-and-gamma-ray spectroscospy, diagnostics of pair density, velocity and energy spectra, and measurement of in-situ magnetic fields. All of these will be extremely challenging.

On the simulation side, clearly we will also need new tools. The current plan calls for linking three different types of codes for end-to-end simulations. The first is the PIC codes used to simulate the laser-particle interactions. The second is the high energy particle interaction codes to take into account all of the pair related processes. The third is the codes that simulate the radiation output from the plasma. For the first we will continue to use ZOHAR developed at LLNL. For the second we plan to use FLUKA developed at CERN. For the third we will use the Monte Carlo radiation code developed in-house at Rice University.

This work is partially supported by LLNL Contract No. B510243 and NASA grant NAG5-3824.

REFERENCES

1. Wilks, S.C., Kruer, W.L., Tabak, M. and Langdon, A.B., Phys. Rev. Lett. 69, 1383 (1992).
2. Liang, E., Wilks, S.C., and Tabak, M., Phys. Rev. Lett. 81, 4887 (1998).
3. Heitler, W., Quantum Theory of Radiation (Oxford, London, 1954).
4. Cowan. T. et. al, Part. Beams 17, 773 (1999).
5. Meszaros, P. and Rees, M.J., ApJ 530, 292 (2000)

Laboratory Astrophysics Using High Intensity Particle and Photon Beams

Pisin Chen

Stanford Linear Accelerator Center
Stanford University, Stanford, CA 94309

Abstract.
   Recent years have seen growing interests in the laboratory investigation of astrophysical phenomena that can be addressed by ultra intense electromagnetic fields. Such
fields are becoming available through the advancement of technologies in lasers as well
as high-energy particle beams. History has shown that the symbiosis between direct
observations and laboratory studies is instrumental in the progress of astrophysics. We
believe that a dedicated facility for laboratory astrophysics using high intensity particle and photon beams can help to significantly advance our knowledge in some critical
aspects of astrophysical phenomena.

INTRODUCTION

History has shown that the symbiosis between direct observations and laboratory
studies is instrumental in the progress of astrophysics. One prominent example was
the seminal theoretical investigation by H. Bethe in the 1930s, and the detailed
experimental measurement by W. Fowler in the 1950s, of the nuclear fusion reaction
that eventually helped to determine the rate of stellar evolution, and thus the age
of the Sun.

Current frontier astrophysical phenomena typically involve one or more of the
following conditions:
• Very high intensity, high temperature processes, such as gamma ray bursts
(GRBs) and supernovae;
• Extremely high energy events, such as ultra high energy cosmic rays (UHECRs)
and blazars;
• Super strong field environments, such as that in the vicinity of balck holes and
neutron stars.
Insights into the underlying fundamental physical mechanisms and processes that
generate these phenomena require controlled laboratory experiments.

Furthermore, the complexity of such systems renders the eventual understanding
far beyond the reach of any fully theoretical treatment. Progress can only be made
with a joint effort of carefully designed experiments coupled with sophisticated

CP611, *Superstrong Fields in Plasmas:* Second Int'l. Conf., edited by M. Lontano et al.
© 2002 American Institute of Physics 0-7354-0057-1/02/$19.00

astrophysical simulations. Laboratory experiments can explore the most complex aspects of the problem as well as verify the validity of simulations designed for environments far from accessible in terrestrial conditions.

High energy, high intensity electron and positron beams, such as that at SLAC, can be made ultra-short ($\sim$ 50 fsec), high energy ($\sim$ 30 GeV/e), ultra-high intensity ($> 10^{20}$ W/cm^2), and can be operated at high repetition rate (10 Hz). In addition, such particle beams can be efficiently converted to high fluence photon beams (tunable from x-ray to gamma-ray) by either colliding with laser pulses or channeling through an undulator or a crystal. As high electromagnetic fields and event rates are essential to the investigations of various astrophysical phenomena, high energy charged particle beams as well as the converted high fluence photon beams provide a unique tool for such purposes.

UNIVERSE AS A LABORATORY

Our Universe is a vast laboratory which produces physical phenomena in their most extreme conditions far beyond the reach of terrestrial settings. For our purpose in this article, we review a few salient features in the current frontier as briefly mentioned in the Introduction, with the understanding that this is only a minute sampling of what the Universe can offer.

A Very High Intensity, High Temperature Processes

Gamma ray burst (GRB) events are one of the most violent releases of energy in the Universe, second perhaps only to the Big Bang itself. Within a brief time of $1 - 100$ seconds an energy up to $10^{52}\Omega_\gamma/4\pi$ erg (Ω_γ is the solid angle of the GRB emission), is released as (predominantly) gamma rays in the range of hundreds of keV. Such energy is equivalent to a substantial fraction of the restmass of a typical star (for $\Omega_\gamma = 4\pi$). It is now commonly believed that such prodigious sources of energy are originated at cosmological distances. This conclusion was derived from the optical observations simultaneous to the flashes of γ-rays of the long bursts ($t_b \sim 10 - 100\,\mathrm{s}$), where the red-shifts of the optical spectra indicate that $z \sim \mathcal{O}(1)$.

B Extremely High Energy Events

One of the unresolved mysteries in current astrophysics has been the composition and the origin of the observed ultra high energy cosmic rays (UHECRs) with energy beyond 10^{20}eV. Although it still awaits further statistics, the events observed so far appear isotropic in their arrival direction. The lack of obvious local source suggests that the origin of UHECR may be cosmological.

376

C   Super Strong Field Environments

In the vicinity of compact objects such as neutron star and black holes, the electromagnetic as well as gravitational fields are believed to be extremely intense. For example the magnetic fields around a neutron star is approaching the Schwinger critical field strength, i.e., $\sim 4.4 \times 10^{13}$G, while the gravity near the event horizon of a black hole is so strong that general relativity has to be invoked in order to properly describe its dynamics.

Furthermore, under such super strong fields quantum effects play essential roles. For example under the Schwinger critical field condition (where electric field with comparable strength also exists) the QED vacuum becomes unstable. Black holes, on the other hand, can provide a fertile test bed for the eventual understanding of quantum gravity, for example, via Hawking radiation.

LABORATORY STUDIES OF THE UNIVERSE

It happens that many aspects of the above mentioned extreme astrophysical phenomena, though cannot be actually reproduced in the earth-bound laboratory per se, can be investigated by using the very high intensity photon and particle beams with the state-of-the-art laser and accelerator technologies. Many of the above-mentioned astro-phenomena involves extremely high energy particles interacting with high density plasma and super strong fields. Some of these conditions can in principle be simulated in the laboratory setting.

In addition, as direct astrophysical observations cannot be done in a cotrolled fashion, due to the very nature of astrophysics, laboratory experiments can help to characterize or calibrate these observations. Furthermore, the very complex astrophysical environments mentioned above renders fully theoretical treatment impossible, and large scale computer simulations are indispensable. Yet limited by the CPUs and other constraints, even computer simulations require approximations and assumptions. Laboratory experiments can help to bench-mark the simulation codes and provide their validation.

In terms of the applications to laboratory astrophysics, there are essentially two categories of experiments.
(A) Experiments that looks for insights into underlying fundamental physical mechanisms and processes of astrophysical phenomena such as the nature of GRB, UHECR, etc;
(B) Experiments that characterize/calibrate direct observations or validate/benchmark astrophysical simulation codes.

Current technologies can produce high energy particle beams and laser beams with intensities at or above 10^{22} Watt/cm. Such high intensity of electromagnetic energy can in general couple very well with high density plasma, as evidenced by the well-studied advanced accelerator concepts of Laser Wakefield Accelerator [1] and the Plasma Wakefield Accelerator [2]. For laboratory astrophysics purposes,

TABLE 1. Parameters for a Laboratory Astrophysics Facility

Particle Beam Parameters	
Particle Type	e^+, e^-
Beam Energy [GeV]	28
Number of Particles per bunch	2×10^{10}
Beam Size σ_r [μm]	10
Bunch Length σ_z [μm]	12
Normalized Emittance ϵ_n [m]	3×10^{-5}
Bunch Density [cm^{-3}]	10^{18}
Peak Power [Watt/cm^2]	10^{21}
Photon Beam Parameters	
Conversion Laser Wavelength [μm]	1
Laser Energy [Joule]	0.1
Pulse Length [femto-sec]	100
Laser Peak Power [W/cm^2]	10^{19}
Undulator-Induced Photon Wavelength [Å]	1.5
Undulator-Induced Photon Number	10^9
Compton-converted Photon Energy [GeV]	tunable up to 15

however, the utilities of laser and the particle beam are well complimentary to each other. Laser photons, whose energy is typically at the eV range, can couple more effectively with solid material, and is most suitable in the study of supernova explosion dynamics. On the other hand, particle beams, whose energy can be tens of GeV per particle, tend to be more penetrating and thus more suitable for the studies of high energy astrophysical processes. It is thus high desirable that a dedicated facility can be created where a combination of both particle and photon beams can be invoked.

We note that the Stanford Linear Accelerator Center (SLAC), with its sizable accelerator infrastructure, can provide a unique leverage in operating such a facility for the astrophysical community. For example, by adding a "chicane" to the existing three-kilometer LINAC, the 30 GeV electron as well as positron beams can be compressed to about 12 μm in the rms bunch length, or < 100 femto-second in total pulse length, comparable to state-of-the-art laser pulse length. Such electron or positron beam can be focused to a few microns in radius at its focus. It can also radiate high fluence x-ray photons by propagating through a undulator. Furthermore, if high energy gamma rays are desirable for certain astrophysical studies, such as the GRBs, they can be produced through Compton scattering of particle beam against table-top high power lasers. The following table summarizes the possible beam parameters that can in principle be attained in such a facility.

It is hoped that different combinations of these beams can provide a wide spectrum of researches in astrophysics.

378

COSMIC PLASMA WAKEFIELD ACCELERATION

As one example of laboratory astrophysics experiments, we discuss the nature of cosmic acceleration that is responsible for the observed ultra high energy cosmic rays. We will give a brief introduction to a new theory based on the plasma wakefields excited by Alfven shocks propagating in a highly relativistic plasma [3].

While the landing locations of UHECRs appear to be random, there exist several "doublet" and "triplet" events, where consecutive events within a short (e.g., a year) time interval land within a narrow solid angle. This further suggests that, though radon, the sources that produce these UHECRs can in principle generate more than one such particles. One well-known challenge, if the UHECRs are actually high energy protons, is that high energy cosmic protons can hardly survive through the collisions with the cosmic microwave background (CMB) photons when the proton-photon center-of-mass energy exceeds the mass of the Δ hyperon. This leads to a cut-off in the cosmic ray energy spectrum, the Greisen-Zespien-Kuzmin (GZK) cut-off, at around 5×10^{19}eV, unless the proton was initially produced within a radius less than 100 Mparsec.

So far the theories that attempt to explain the UHECR can be largely categorized into the "top-down" and the "bottom-up" scenarios. The top-down scenario invokes novel astrophysical processes such as topological defects for the generation of super-heavy exotic particles that survive from primordial time in the early universe till now. One strong argument for the top-down scenario is, not being protons, the survival of such exotic particles are not subjected to the GZK cut-off. This, however, is only a retreat that resorts to the lack of knowledge on the part of these exotic particles. Furthermore, processes such as the topological defect may not be entirely local and random, but have global structures instead. If so, then the arrival locations of UHECRs cannot be isotropic.

The bottom-up scenario assumes the carriers of the UHECRs are the known, ordinary particles that have been accelerated by some means to the extremely high energy. To circumvent the GZK-limit, T. Weiler proposes an alternative. If instead of protons, neutrinos are the energy carrier, then the much weaker interaction ensures that they can travel across a much larger cosmic distance. These ultra-high energy neutrinos would eventually have to be annihilated by the cosmic background neutrinos and turn into Z-bosons, or the "Z-burst". If this happens within the GZK radius, the proton that is produced from the Z-decay will reach the earth. For such a scenario to work, it requires that the original particle that eventually produce the ultra high energy neutrino be accelerated to an energy which is one to two orders of magnitude more energetic than that of the final particle that arrives at the earth. Even if the GZK-limit can be circumvented through the Weiler process, the challenge for the bottom-up scenarios remain severe.

The existing paradigm for cosmic acceleration, namely the Fermi acceleration, as well as its variants, such as the relativistic shock acceleration, cannot succeed in producing the necessary energy. These acceleration mechanisms are essentially "collisional" processes. Namely, they rely on either the deflection by magnetic field domains or the scattering by particles in the shock medium. In either case the particle would suffer severe energy losses through synchrotron radiation or inelastic collisions. These menachisms are thus very hard to produce particles beyond 10^{18}eV. There is also the Eddington acceleration mechanism which relies on the photon-electron interaction, with Compton scattering as its underlying physical process. However, as the Klein-Nishina cross section is inversely proportional to the particle initial energy, the Eddington mechanism also suffers a diminishing return at extreme high energies.

The new proposal for cosmic acceleration based on plasma wakefields, on the other hand, is a "collisionless" process. The plasma wakefield is essentially a space-charge field induced in the plasma by a propagating high intensity electromagnetic field. Such a electric field is parallel to the direction of wave propagation, and is therefore Lorentz invariant, as is also the case for all conventional accelerators. A test particle can ride on such a plasma collective field and gain energy from it without actually colliding with any plasma particle. The Lorentz invariance of the field, augmented by the extremely high acceleration gradient that plasma wakefields can typically provide, are the keys that opens the door for particles, such as protons, to gain energies very efficiently within one mean free path.

We note that the environments necessary for such an astrophysical plasma wake-field acceleration are cosmically abundant. In addition to the GRBs, the active galactic nuclei (AGNs) as well as black holes can also provide the conditions. As a concrete example, however, we shall invoke GRBs as our setting for the UHECR production. It happens that with the extremely violent release of energy, a GRB can provide the following two crucial ingredients necessary for an efficient cosmic acceleration: (1) The GRB can excite intense fields that can in turn induce huge accelerating gradient; and (2) This accelerating field propagates at near the speed of light so that sufficiently large energy boost can take place during the interaction.

For condition (1) to occur, we note that upon the coalescence of neutron stars, violent collisional energy can be released in the forms such as acoustic shocks and very importantly transverse (Alfven) shocks. As these intense waves propagate through the leptosphere which is exploding due to the immense photon pressure with sufficiently high but not highly relativistic flow ($\Gamma \sim \mathcal{O}(1)$), these form strong localized shocks, with the longitudinal shocks induced from the acoustic energy and the transverse (or Alfvenic) shocks from the transverse (or magnetic) energy. Although it remains to be investigated if the acoustic shocks can be efficiently converted into the above described required accelerating gradient. it can be demenstrated [3] that the transverse shocks thus created can be an efficient accelerating agent that is suitable for ultra high energy particle acceleration.

Regarding condition (2), while the above-mentioned Eddington acceleration is not sufficient in accelerating particles to ultra high energies, it is nevertheless effective in accelerating large quantities of particles up to sufficiently relativistic energies. The instant release of 10^{52}ergs of photons at energies around 0.5MeV promptly accelerate a large number of electrons and positrons (and through these hadrons) as soon as intense photon flux emanates from the opaque surface of the

GRB. This gives rise to a relativistic flow of electron-proton plasma with Lorentz factor Γ 10^5 immediately (within 1m) upon the entry to the GRB plasmosphere. Within this surface lies the leptosphere which is opaque even to high energy gamma rays.

The Alfvenic shocks riding on this relativistic plasma is propagating at near the speed of light and behave as an authentic EM wave. As the corresponding dimensionless vector potential, or the strength parameter, a_0 10^{11}, and the plasma density in such an environment is around $10^{20}/cm^3$, one finds the plasma wakefield so induced as strong as $G = eE$ $10^{12}eV/cm$. The "dephasing length", is

$$L_{dp} = 2\Gamma^2\lambda_p \approx 10^{10}cm, \tag{1}$$

whereas the mean-free-path for a proton at energy $10^{20}eV$ in such a medium happens to be comparable. Therefore within one proton mean-free-path some lucky protons can be accelerated up to beyond $10^{22}eV$, sufficient for further cascades through through neutrinos required by the Weiler mechanism for cosmic transport.

A POSSIBLE EXPERIMENT ON COSMIC ACCELERATION

The key assumptions in the cosmic plasma wakefield acceleration mechanism introduced above can in principle be tested in the laboratory setting. Specifically, one should be able to verify
• The generation of near-luminous waves in relativistic plasma flows;
• The generation of high gradient nonlinear plasma wakefields;
• The acceleration of trapped e^+/e^- to higher energies through the plasma wakefield;
• The energy spectrum as a result of the stochastic acceleration process.

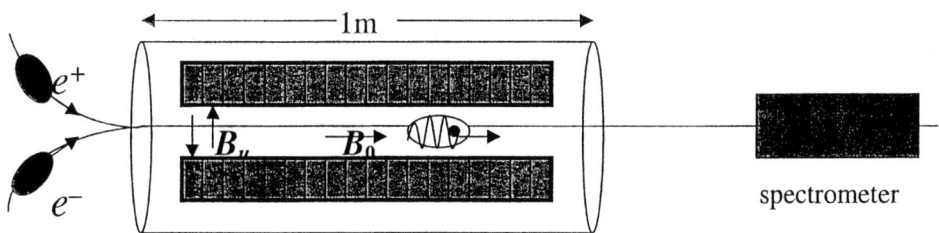

Figure 1. Schematic diagram for cosmic acceleration experiment

To achieve these gaols, we envision the creation of a relativistic plasma by converging a high energy, 30 GeV, SLAC positron beam with a lower energy, 50 MeV, electron beam with comparable densities at n_p $10^{16}/\mathrm{cm}^3$. We then send this "relativistic neutral plasma" through a 1-meter-long axial solenoidal magnetic field with $B_0 = 10$ Tesla, which is superimposed with a transverse undulator magnetic field with strength $B_u = 1$ Tesla. The interaction of the relativistic plasma with such a configuration of magnetic fields should excite an Alfven wave within the plasma with a velocity of v_A/c 10^{-2}. The corresponding acceleration gradient is estimated to be G 3 GeV/m. Figure 1 shows the schematic drawing of the conceptual design of such an experiment.

SUMMARY

We have presented the arguments for the need of laboratory investigation in astrophysics. We suggest that high intensity particle and photon beams are powerful tools for this purpose, and the existing accelerator infrastructure at SLAC should be ideal for establishing such a facility. We gave an explicit example, namely the cosmic plasma wakefield acceleration, that can be possibly performed at such facility. It should be emphasized, however, that this is just one sampling of a wide spectrum of possibilities with this facility. For example, the relativistic e^+e^- plasma can also be a great value in studying other astrophysical processes such as the jets from AGNs.

It is exciting to realize that, entering the new century, using the Universe as a laboratory as well as using laboratory to study the Universe, is now becoming possible. The joint efforts in both approaches should help to significantly advance the ultimate understanding of our cosmos.

REFERENCES

1. T. Tajima and J. M. Dawson *Phys. Rev. Lett.* **43**, 267 (1979).
2. P. Chen, J. M. Dawson, R. Huff, and T. Katsouleas *Phys. Rev. Lett.* **54**, 693 (1985).
3. P. Chen, T. Tajima, and Y. Takahashi, in preparation, 2001.

5. LASERS FOR ULTRAHIGH INTENSITY PHYSICS

Status And Future Developments Of Ultrahigh Intensity Lasers At JAERI

K. Yamakawa, Y. Akahane, M. Aoyama, Y. Fukuda, N. Inoue, J. Ma, and H. Ueda

Advanced Photon Research Center, KANSAI Research Establishment
Japan Atomic Energy Research Institute
8-1 Umemidai, Kizu, Kyoto 619-0215, Japan

Abstract. We review progress in the generation of multiterawatt optical pulses in the 10-fs range. As an example, the design, performance and characterization of a Ti:sapphire laser system based on chirped-pulse amplification at the Advanced Photon Research Center, Japan Atomic Energy Research Institute is described. This system produces 20-fs pulses with peak and average powers of 100-TW and 20-W, respectively at a 10-Hz repetition rate and is currently being applied to perform high-field atomic ionization experiments at a relativistic intensity regime. We also discuss extension of the system that is presently upgrading to the petawatt power level..

INTRODUCTION

Recent advances in femtosecond laser sources are making intensities approaching ~ 10^{21} W/cm^2 available for the study of nonlinear relativistic optics [1, 2]. At such intensities the electron velocity in the laser field becomes relativistic and exhibits highly nonlinear motion, thus making it possible to investigate entirely new classes of physical effects. Potential applications of these lasers include the generation of ultrafast x-ray radiation [3-7], ultrahigh-order harmonic generation [8-18], photo-ionization pumped x-ray lasers [19, 20], optical field ionization x-ray lasers [21-24], laser wakefield particle acceleration [25-29], laser induced nuclear photophysics [30, 31], laboratory-based astrophysics [32, 33] and fast ignitor fusion [34, 35].

The technique of chirped pulse amplification (CPA) has opened new avenues for the production of very high-energy ultrashort duration pulses without optical damage to amplifiers and optical components [36, 37]. The combination of CPA and ultrabroad-band solid-state laser materials has made it possible to produce terawatt and even multiterawatt femtosecond pulses with ever increasing average powers [38-41]. The CPA technique consists of four basic components: (1) a short pulse oscillator, (2) a pulse expander, (3) an amplifier, and (4) a pulse compressor. In ultrafast CPA, a short pulse is generated by a mode-locked laser oscillator and is temporally stretched by an antiparallel grating pair pulse expander [42]. The low energy and long duration chirped pulse is then amplified to a high energy commensurate with the saturation fluence of the solid state laser amplifiers. The amplified pulse is then compressed to a transform-limited short pulse of high peak power with a parallel grating pair compressor [43].

Over the past 12 years, peak powers from terawatt CPA systems have steadily increased from less than a terawatt initially to a present record of 1500-TW (1.5-PW) [44]. At the same time, pulse durations have steadily decreased from greater than a

CP611, *Superstrong Fields in Plasmas:* Second Int'l. Conf., edited by M. Lontano et al.
© 2002 American Institute of Physics 0-7354-0057-1/02/$19.00

picosecond to a current record 16-fs [45]. Extremely modest amounts of energy can now be used to achieve multiterawatt peak powers. At the 20-fs pulse duration, for example, only ~ 20-mJ of energy is required to achieve a peak power of 1-TW. Because much less energy is required to reach the same peak power, the size of the laser system can be significantly reduced and the repetition rate of the system can be very high. For instance, the size of the inertial-confinement-fusion Nd:glass laser system such as the NOVA laser at the Lawrence Livermore National Laboratory (LLNL) [46] requires a 200 meter long building to produce 100-TW peak powers in 1-ns duration and the system cannot be operated at more than 1 shot per hour. On the other hand, our 100-TW Ti:sapphire laser system occupies an area of only ~ 15-m^2 and produces the same peak power at 10 shots per second. This capability is especially significant because it allows reliable, high repetition rate, ultrahigh peak power lasers to become realistic laboratory tools for investigations requiring ultrahigh intensities. In addition the high average power is also desired to produce high fluxes of energetic particles or x-rays and to allow signal averaging techniques to be applied to relativistic laser/matter investigations.

In this paper we review the evolution of CPA into the 10-fs range. As an example, the design, performance and characterization of a compact three-stage Ti:sapphire CPA laser system at the Japan Atomic Energy Research Institute is described. This system is designed to produce sub-20-fs pulses with peak and average powers of 100-TW and 20-W, respectively at a 10-Hz repetition rate. We also discuss extension of this system to the petawatt power level and potential applications in the nonlinear relativistic intensity regimes.

ULTRAFAST CPA TECHNIQUES

The progression of the CPA systems with respect to pulse duration, is illustrated in Fig. 1. The CPA technique has been demonstrated with a variety of laser materials such as Nd:glass [44, 47-54], alexandrite [55, 56], Ti:sapphire (Ti:Al$_2$O$_3$) [38-41, 45, 57-62] and Cr:LiSAlF [63, 64]. These materials all have relatively large saturation fluences of the order of joules per square centimeter, relatively long upper state lifetimes and broad bandwidths. While the first generation of CPA systems were based on Nd:glass amplifiers and generated high energy picosecond pulses, the relatively narrow bandwidth of Nd:glass has limited amplified pulse duration to a few 100's of femtoseconds. To date, pulses as short as 450-fs with over a petawatt peak power have been generated by using a large scale, single-shot-per-hour, inertial-confinement-fusion, Nd:glass laser [44]. While Nd:glass amplifiers have good energy storage and can easily be scaled to large volumes, they are in general limited to low repetition rates and low average power operation because of the poor thermal characteristics of laser glasses. Nevertheless, a terawatt laser with a repetition rate of 1-Hz has been built using a flashlamp-pumped Nd:glass slab power amplifier [65].

Using larger gain bandwidth materials such as Ti:sapphire [66] and Cr:LiSAF [67], however, permits the amplification of sub-100 femtosecond pulses from the Kerr-lens mode-locked oscillators [68-74]. In particular, Ti:sapphire has several desirable characteristics including a high saturation fluences (~ 0.9-J/cm^2), a high thermal conductivity (46-W/mK at 300 K) and a high damage threshold (> 5-J/cm^2) for producing high-peak and high-average power pulses. Its gain bandwidth of ~ 230-nm at Full width at half maximum (FWHM) could in principle support transform limited pulses of ~ 3-fs. Recently pulses shorter than two optical cycles have been generated directly from Kerr-lens mode-locked Ti:sapphire oscillators by using prism pairs and double chirped mirrors in combination with and without a semiconductor saturable absorber mirror [75, 76]. As for the amplification system, Sartania et al. have demonstrated the generation of 5-fs, 0.5-mJ pulses at a 1-kHz repetition rate using the technique of hollow fiber based pulse compression and ultrabroadband chirped mirrors [77]. Although Ti:sapphire amplifier systems for the generation of pulses with duration of around 20-fs have been demonstrated [78, 79], the amplification of 20-fs pulses to energies greater than one joule has only recently been accomplished [41, 62]. The difficulty lies in the control of two major effects: high-order phase distortion in the amplification chain and gain narrowing in the amplifying media.

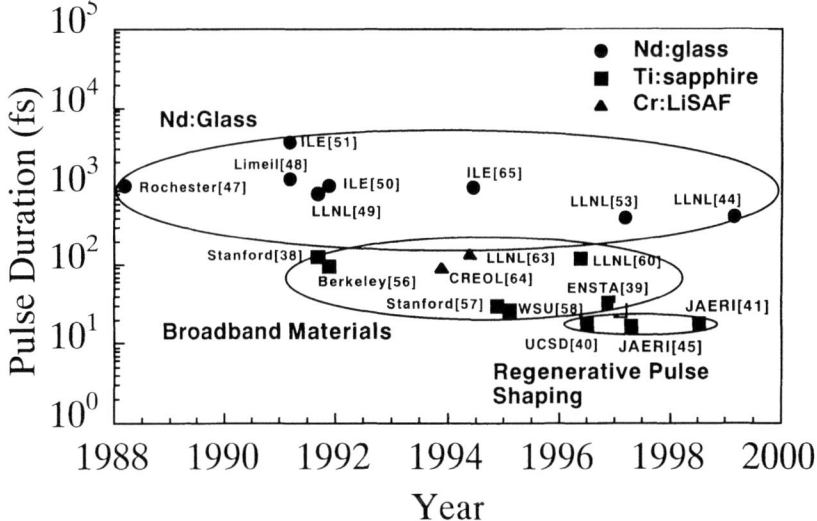

FIGURE 1. Representative history of the evolution of terawatt CPA pulse duration.

Terawatt level pulses with durations of 100-fs – 1-ps have been produced with the elimination second- and third-order dispersions by a number of laser systems. For ultrashort pulse systems (≤ 20-fs), however, the fourth-order-dispersion must be eliminated. This concern has been addressed by a number of groups, which have proposed and demonstrated dispersive optical systems that are capable of controlling dispersion up to fourth order [80-83]. For example, tests of the system described by Lemoff and Barty [81] indicate that broadening during the amplification and recompression of a 10-fs Gaussian pulse would be limited to less than 1-fs. With such a system, amplification of 10-fs optical pulses is therefore limited primarily by gain narrowing during amplification and the bandwidth of the optical components in the amplification chain.

Recent progress in ultrafast CPA systems has also utilized regenerative pulse shaping to counter gain narrowing [85, 86]. By including a frequency dependent filter to eliminate gain narrowing, regenerative pulse shaping allows the production of very short duration and high energy amplified pulses. With this technique the pulse duration of the amplified compressed pulses has been reduced by a factor of 2, reaching now 16-fs at a 10-TW level [45].

DESIGN, PERFORMANCE AND CHARACTERIZATION OF A 100-TW, SUB-20-FS, 10-HZ TI:SAPPHIRE LASER SYSTEM

As an example, the design, performance and characterization of a compact three-stage Ti:sapphire CPA laser system at the Advanced Photon Research Center (APRC), Japan Atomic Energy Research Institute is described in this section. The laser system is the front end of a 4-stage amplification system which is planned to eventually produce peak powers on the order of one petawatt (20-J in 20-fs). A schematic of the laser system is shown in Fig. 2.

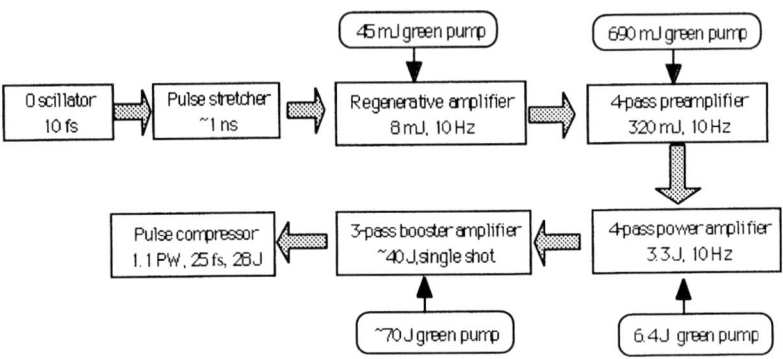

FIGURE 2. Schematic of the petawatt Ti:sapphire laser system.

Briefly, the system is seeded with pulses from an all-solid-state mirror-dispersion-controlled Ti: sapphire oscillator capable of producing 10-fs pulses. These pulses are temporally stretched by a factor of 100,000 in an all-reflective, cylindrical mirror-based pulse expander. Then, stretched pulses are amplified with three amplifier stages. The Ti: sapphire amplifier chain consists of a regenerative amplifier, and two multi-pass amplifiers. These amplifiers are pumped by a fraction of the 532 nm radiation at 10 Hz from Q-switched Nd: YAG lasers. The regenerative amplifier is a stable TEM00 cavity and the resonator is 1.8-m long and uses two cavity mirrors [61]. The amplifier output energy was adjusted to be approximately 8-mJ. Two, 3-μm thick etalons (Melles Griot Corp.) are used for regenerative pulse shaping in the regenerative amplifier. When the stretched pulse was amplified in the regenerative amplifier without a spectral filter, the spectrum narrowed to 28-nm as shown in Fig. 3. However, by using the etalons to produce a frequency dependent attenuation and selectively amplifying the wings of the spectrum, in the spectrum of the amplified pulse was broadened to 82-nm FWHM (Fig. 3). In addition, it should be noted that modified filters also permitted the simultaneous amplification of two colors. Pulses centered at 765-nm and 855-nm have been amplified simultaneously without unwanted amplification at the peak of the gain profile. Such pulses should lead to compact sources of high power, tunable mid-infrared light via difference frequency mixing. It should also be possible to produce energetic, multi-wavelength, femtosecond pulses with this technique.

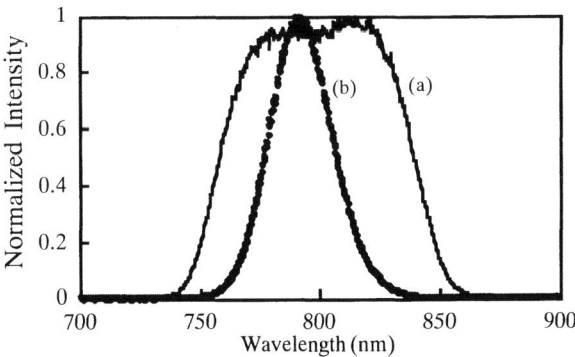

FIGURE 3. The spectra for amplified pulses from the regenerative amplifier (a) with and (b) without the thin solid etalons.

Further amplification is accomplished in a 4-pass pre- and power-amplifiers. The power-amplifier uses a water cooled 40-mm diameter 25-mm long Ti:sapphire crystal (Union Carbide Corporation) with anti-reflection coatings on both faces and is pumped with a custom built Nd:YAG laser which is capable of producing ~ 7 J of 532-nm radiation at 10-Hz. With 6.4-J of pump light incident upon the crystal the amplifier has produced 3.3-J of 800-nm radiation. This amplifier provided a total saturated gain of 10 and a small signal gain of ~ 5. Under this condition, this amplifier has reached 90% of the theoretical maximum conversion efficiency of 532-nm pump light to 800-nm radiation.

The output of the power amplifier passes through relay imaging optics with a magnification ($M = 3.3$). This provides a spatially uniform beam for the pulse compression gratings and collimates the beam diameter to an approximately 50-mm. The vacuum pulse compressor consists of two parallel, gold-coated, 1200-grooves/mm, ruled gratings (Richardson Grating Labs.), which had a measured diffraction efficiency of ~ 91%. A fraction of the compressor output was sent to a single-shot autocorrelator. The FWHM of the measured pulse duration is 18.7-fs. The duration of the transform limit, as calculated from the measured, amplified spectrum after the compressor is 17-fs. The high degree of agreement suggests that the compressed pulses are nearly transform limited. The transmission of the compressor, including the multilayer dielectric- and gold-coated turning optics, was ~ 57%, yielding a compressed output pulse energy of 1.9-J, which implies a peak power for the laser pulse in excess of 100-TW.

In high-field physics experiments, such a low-intensity pedestal (pre- and post-pulses) and/or ASE would create a low density plasma in advance of the main laser pulse and thus significantly alter the physics of the laser/matter interaction. Therefore detailed characterization and control of the temporal shape and phase of the laser pulse is crucial to the study of high-intensity laser-matter experiments. In order to fully characterize the pulse duration and phase of 20-fs laser pulses, second harmonic generation (SHG) frequency-resolved optical gating (FROG) technique [87] was used. Figure 4 shows the pulse intensity and phase in time retrieved from the SHG FROG trace. The compressed pulse accompanied with small pre- and post-pulses resulting that the predominant phase distortion is quartic.

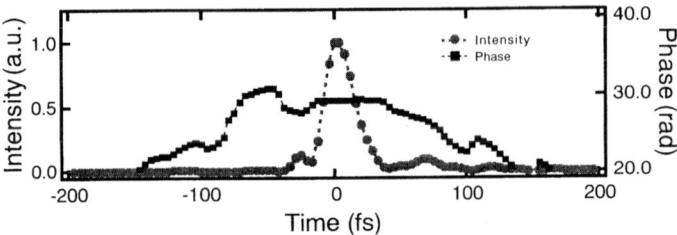

FIGURE 4. Retrieved intensities and phases of the SHG FROG trace for the compressed pulse.

A high dynamic range cross-correlation signal of the compressed pulses is shown in Figure 5. Each point of cross-correlation trace with a time resolution of 670-fs corresponds to an average of ten laser shots. The detection limit of this apparatus is approximately 10^{-8}. The measured contrast is of the order of 10^{-6} limited by ASE mainly coming from the regenerative amplifiers. ASE can be easily suppressed by two orders of magnitude by using a solid-state saturable absorber with a preamplifier before the pulse expander [88].

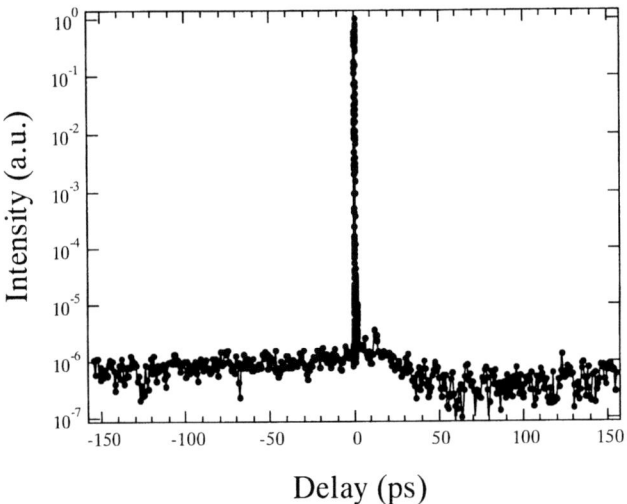

FIGURE 5. High dynamic range cross-correlation trace of a compressed pulse.

Since the CPA laser systems are now approaching focused laser intensities of 10^{20} - 10^{21} W/cm^2, the quality of the beam corresponding to diffraction limited spot size is crucial. The spatial beam quality was determined by focusing the attenuated output with a 3-m focal-length spherical mirror and measuring the spot size at the focus with a high dynamic range 16bit CCD camera. Under the full power operations of the laser system, the spatial quality of the beam produced a spot which was 2- and 2.5-times that of a diffraction limited gaussian beam in vertical- and horizontal-planes, respectively. We have determined that approximately 80% of the focused energy lies within the spot. With an f/3 off-axis parabolic mirror, focused intensities of ~ 3 x 10^{20} W/cm^2 should therefore be possible with this beam quality.

HIGH FIELD ATOMIC IONIZATION

Among the numerous possible applications of 100 TW-class, ultrafast laser systems are studies of relativistic laser/mater interactions. We have installed two target chambers for optical field ionization and hard x-ray generation experiments as shown in Fig. 6. The beam switchyard can be used to switch one experiment and another easily and quickly, so that we run both type of experiments very efficiently. In order to clarify the ionization mechanisms of an atom including the ionization rate [89] at relativistic intensities, we have studied tunneling ionization of Argon using our laser system with intensities over 10^{19} W/cm^2. Using an off-axis parabolic mirror of focal length of 161 mm, we obtain the focal spot diameter of 6.8 μm at full width at half maximum (FWHM), which corresponds to the estimated peak intensity of 1.1 x 10^{19} W/cm^2. The uncertainly in the absolute value of the intensity is approximately factor of two. Argon gas is introduced into the vacuum chamber by means of a precision leak valve. The ion spectra are obtained with a 1-m time-of-flight spectrometer, which has a few hundred volt/cm field potential. Dual micro channel plates (MCP) are used to detect the argon ions.

FIGURE 6. Two targeting systems for high filed ionization and hard x-ray generation experiments.

At the highest laser intensities the first eight ion charge states (Ar$^+$ to Ar^{8+}) were clearly observed and highly-charged argon ions up to Ar^{16+} were also detected. In order to clarify the ionization mechanism of L-shell states of Argon as a function of laser intensity, we used a half wave plate and polarizer to vary the intensity of the pulse. The results of the ion-yield data for Ar^{9+} and Ar^{10+} as a function of laser intensity are shown in Fig. 7. The theoretical curves calculated by the simple ADK model [90] are also shown in this figure. It seems that the experimental data behave similar to ADK within our experimental dynamic range, but we could not conclude at the moment without more precise measurements and comparison to another models. The detailed understanding of the ionization dynamics of the single atoms in this study will be a basis for the future investigations of more complex systems such as molecules or clusters in the relativistic intensity regime.

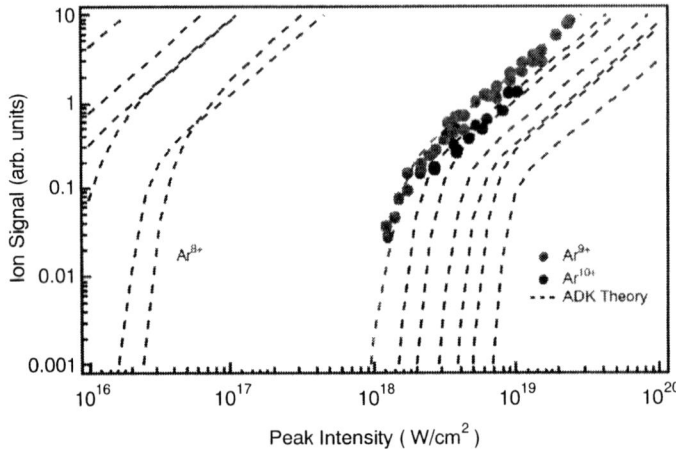

FIGURE 7. Argon ion yield data compared to sequential ADK model as a function of incident laser intensity for a 20 fs 800 nm linearly polarized laser pulse.

TOWARD A PETAWATT

The potential scalability of ultrafast CPA architectures described in this paper to higher energies is an important issue considering peak powers up to the petawatt level. To scale the system to peak powers above 100-TW requires larger size gain media, higher energy pump lasers and larger diameter gratings. In this section, we discuss these issues including of suppression of parasitic oscillation across the large-aperture Ti:sapphire amplifier disk, optimization of the spectrum of the amplified pulse at the output of the amplifier and compensation of high-order dispersion in the laser chain.

Ti:sapphire disks with up to 100-mm in diameter are available with current growth technologies. There are major problems with ASE (amplified spontaneous emission) and PO (parastic oscillation) across the aperture in the large Ti:sapphire crystal. PO will reduce the stored energy in the amplifier, thus the amplifier efficiency also decreases. A technique for suppressing these PO modes based on index matching the crystal edges with an absorbing doped polymer thermoplastic was developed and demonstrated for large aperture Ti:sapphire disk amplifiers having high refractive index (n=1.76) [91]. The estimated PO gain on the disk faces to the beam diameter of the pump laser is shown in Fig.8. Fresnel reflection at the interface between the Ti:sapphire crystal and the thermoplastic is estimated to be ~0.048 %. Therefore, PO should not occur across the input face until the transverse gain reaches ~2,100, corresponding to a pump fluence of ~5.6 J / cm². The booster amplifier has been designed to achieve efficient energy extraction without PO, and the amplified energy has been expected to be as much as ~40-J for a ~70-J green pump.

Furthermore, the pulse stretcher and compressor for the petawatt laser system are also considered to produce sub-30 fs laser pulses. These components are the Offner triplet stretcher and the Tracy type compressor based on the mixed grating scheme [83]. In order to compensate for the phase distortion of the materials up to fourth order in the laser chain, we have chosen a 1,200 groove / mm ruled grating in the stretcher and 1,480 groove / mm holographic gratings in the compressor. Although the fourth-order dispersion compensation can be accomplished here, the residual fifth-order dispersion would broaden the initial 20-fs pulse to 25-fs duration. The compressor will consist of four gold-coated holographic gratings in vacuum. The sizes of the gratings are 220-mm x 165-mm for the first and last gratings and 420-mm x 210-mm for the second and third gratings, respectively. The diffraction efficiency of these gratings was expected to be greater than 92% over the 100-nm bandwidth (centered at 800-nm), and thus the overall efficiency should be greater than 70%. Based on the efficiency of the gratings, the energy of the compressed pulse is estimated to be > 28-J. Thus, the peak power for laser pulse is expected to be > 1.1-PW.

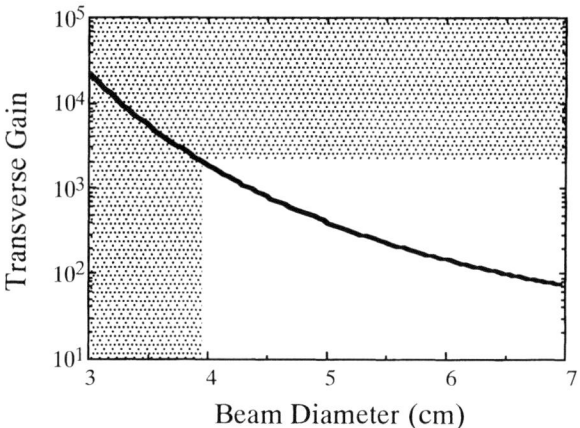

FIGURE 8. Calculated transverse gain as a function of the beam diameter for 70-J pump energy. Shaded region is above parastic oscillation threshold.

CONCLUSION

The ultrafast CPA architectures that have been described in this article open to a new route for the production of the optical pulses with duration on the order of ten femtoseconds to be amplified to multiterawatt peak powers. With a compact three-stage Ti:sapphire CPA system, near diffraction limited and spectrally limited sub-20-fs duration pulses with a peak power in excess of 100-TW have been produced. Since this laser system operates reliably at a 10-Hz repetition rate, it should allow the use sampling and averaging techniques even in the ultrahigh intensity regime (> 10^{20} W/cm^2). We have presented a systematic study of optical field ionization of Argon (n = 2) in the relativistic laser field. Ar^{16+} has been observed at laser intensity of over 1 x 10^{19} W/cm^2. Inner shell (L-shell) photoionization rate for Ar^{9+} and Ar^{10+} as a function of laser intensity has been measured over 3 orders of magnitude. The study presented here should lead to better understanding of relativistic plasmas of atoms, molecules and clusters in all of their complex interactions. We are also currently upgrading the system to the petawatt power level. Based on design studies which include the suppression of parasitic oscillation across the large-aperture Ti:sapphire amplifier disk, the optimization of the amplified pulse spectrum at the output of the amplifier and the compensation of high-order dispersion in the laser chain with a mixed grating scheme, scaling of 20-fs CPA technology to the petawatt peak power level seems quite feasible. These lasers have opened the door to investigations of entirely new classes of physical effects and applications and should have a major impact on science and technology.

REFERENCES

1. G. Mourou, C. P. J. Barty, and M. D. Perry, Phys. Today, vol. 51, pp. 22-28, 1998.
2. D. Umstadter, S.-Y. Chen, A. Maksimchuk, G. Mourou, and R. Wagner, Science vol. 273, pp. 472-475, 1996
3. J. D. Kmetec, C. L. Gordon, III, J. J. Macklin, B. E. Lemoff, G. S. Brown, and S. E. Harris, Phys. Rev. Lett., vol.68, pp.1527-1530, 1992.
4. J. D. Kmetec, IEEE J. Quantum Electron., vol. QE-28, pp. 2382-2387, 1992.
5. Z. Jiang, J. C. Kieffer, J. P. Matte, M. Chaker, O. Peyrusse, D. Gilles, G. Korn, A. Maksimchuk, S. Coe, and G. Mourou, Phys. Plasma, vol. 2, pp. 1702-1711, 1995.
6. J. Workman, A. Maksimchuk, X. Liu, U. Ellenberger, J. S. Coe, C. Y. Chien, and D. Umstadter, Phys. Rev. Lett., vol. 75, pp. 2324-2327, 1995.
7. J. F. Pellrtier, M. Chaker, and J. C. Kieffer, Opt. Lett., vol. 21, pp. 1040-1042, 1996.
8. A. L'Huillier, and P. Balcou, Phys. Rev. Lett., vol. 70, pp. 774-777, 1993.
9. J. J. Macklin, J. D. Kmetec, and C. L. Gordon III, Phys. Rev. Lett., vol. 70, pp. 766-769, 1993.
10. J. Zhou, J. Peatross, M. M. Murnane, and H. C. Kapteyn, Phys. Rev. Lett., vol. 76, pp. 752-755, 1996.
11. C. Spielmann, N. H. Burnett, S. Sartania, R. Koppitsch, M. Schnurer, C. Kan, M. Lenzner, P. Wobrauschek, and F. Krausz, Science, vol. 278, pp. 661-663, 1997.
12. M. Schnurer, C. Spielmann, P. Wobrauschek, C. Streli, N. H. Burnett, C. Kan, K. Ferencz, R. Koppitsch, Z. Cheng, and F. Krausz, Phys. Rev. Lett., vol. 80, pp. 3236-3238, 1998.
13. T. Sekikawa, T. Ohno, T. Yamazaki, Y. Nabekawa, and S. Watanabe, Phys. Rev. Lett., vol. 83, pp. 2564-2567, 1999.
14. Y. Tamaki, J. Itatani, Y. Nagata, M. Obara, and K. Midorikawa, Phys. Rev. Lett., vol. 82, pp. 1422-1425, 1999.
15. P. A. Norreys, M. Zepf, S. Moustaizis, A. P. Fews, J. Zhang, P. Lee, M. Bakarezos, C. N. Danson, A, Dyson, P. Gibbon, P. Loukakos, D. Neely, F. N. Walsh, J. S. Wark, and A. E. Dangor, Phys. Rev. Lett., vol. 76, pp. 1832-1835, 1996.
16. R. Lichters, J. Meyer-ter-Vehn, and A. Pukhov, Phys. Plasmas., vol. 3, pp. 3425-3437, 1996.
17. P. Gibbon, IEEE J. Quantum Electron., vol. 33, pp. 1915-1924, 1997.
18. D. von der Linde, Appl. Phys. B, vol. 68, pp. 315-319, 1999.
19. H. C. Kapteyn, Appl. Opt., vol. 31, pp. 4931-4939, 1992.
20. S. J. Moon and D. C. Eder, Phys. Rev. A, vol. 57, pp. 1391-1394, 1998.
21. N. H. Burnett and P. B. Corkum, J. Opt. Soc. Am. B, vol. 6, pp. 1195-1199, 1989.
22. P. Armendt, D. Eder, and S. Wilks, Phys. Rev. Lett., vol. 66, pp. 2589-2592, 1991.
23. Y. Nagata, K. Midorikawa, S. Kubodera, M. Obara, H. Tashiro, and K. Toyoda, Phys. Rev. Lett., vol. 71, pp. 3774-3777, 1993.
24. B. E. Lemoff. G. Y. Yin, C. L. Gordon III, C. P. J. Barty, and S. E. Harris, Phys. Rev. Lett., vol. 74, pp. 1574-1577, 1995.
25. T. Tajima and J. M. Dawson, Phys. Rev. Lett., vol. 43, pp. 267-270, 1979.
26. P. Sprangle, E. Esarey, A. Ting, and G. Joyce, Appl. Phys. Lett., vol. 53, pp. 2146-2148, 1988.
27. K. Nakajima, D. Fisher, T. Kawakubo, H. Nakanishi, A. Ogata, Y. Kato, Y. Kitagawa, R. Kodama, K. Mima, H. Shiraga, K. Suzuki, K. Yamakawa, T. Zhang, Y. Sakawa, T. Shoji, N. Yugami, M. Downer, and T. Tajima, Phys. Rev. Lett., vol. 74, pp. 4428-4431, 1995.
28. A. Modena, Z. Najmudin, A. E. Dangor, C. E. Clayton, K. A. Marsh, C. Joshi, V. Malka, C. B. Darrow, C. Danson, D. Neely, and F. N. Walsh, Nature, Vol. 377, pp. 606-608, 1995.
29. R. Wagner, S.-Y. Chen, A. Maksimchuk, and D. Umstadter, Phys. Rev. Lett., vol. 78, pp. 3125-3128, 1997.

30. T. E. Cowan, A. W. Hunt, T. W. Phillips, S. C. Wilks, M. D. Perry, C. Brown, W. Fountain, S. Hatchett, J. Johnson, M. H. Key, T. Parnell. D. M. Pennington, R. A. Snavely, and Y. Takahashi, Phys. Rev. Lett., vol. 84, pp. 903-906, 2000.

31. K. W. D. Ledingham, I. Spencer, T. McCanny, R. P. Singhal, M. I. K. Santala, E. Clark, I. Watts, F. N. beg, M. Zepf, K. Krushelnick, M. Tatarakis, A. E. Dangor, P. A. Norreys, R. Allott, D. Neely, R. J. Clark, A. C. Machacek, J. S. Wark, A. J. Cresswell, D. C. W. Sanderson, and J. Magill, Phys. Rev. Lett., vol. 84, pp. 899-902, 2000.

32. R. Sauerbrey, Phys. Plasmas, vol. 3, pp. 4712-4716, 1996.

33. H. Takabe, AIP Conference Proceedings of the International Conference on Superstrong Fields in Plasmas, No.426, pp. 560-570, Edited by: M. Lontano, G. Mourou, F. Pegoraro, and E. Sindoni, AIP Press, 1998.

34. M. Tabak, J. Hammer, M. E. Glinsky, W. L. Kruer, S. C. Wilks, J. Woodworth, E. M. Campbell, M. D. Perry, and R. J. Mason, Phys. Plasmas, vol. 1, pp. 1626-1634, 1994.

35. C. Deutsch, H. Furukawa, K. Mima, M. Murakami, and K. Nishihara, Phys. Rev. Lett., vol. 77, pp. 2483-2486, 1996.

36. D. Strickland and G. Mourou, Opt. Commun., vol. 56, pp. 219-221, 1985.

37. M. D. Perry and G. Mourou, Science, vol. 64, pp. 917-924, 1994.

38. J. D. Kmetec, J. J. Macklin and J. F. Young, Opt. Lett., vol. 16, pp. 1001-1003, 1991.

39. J. P. Chambaret, C. Le Blanc, A. Antonetti, G. Cheriaux, P. F. Curley, G. Darpentigny, and F. Salin, Opt. Lett. vol. 21, pp. 1921-1923, 1996.

40. C. P. J. Barty, T. Guo, C. Le Blanc, F. Raksi, C. Rose-Petruck, J. Squier, K. R. Wilson, V. V. Yakovlev, and K. Yamakawa, Opt. Lett., vol. 21, pp. 668-670, 1996.

41. K. Yamakawa, M. Aoyama, S. Matsuoka, T. Kase, Y. Akahane and H. Takuma, Opt. Lett., vol. 23, pp. 1468-1470, 1998.

42. O. E. Martinez, IEEE J. Quantum Electron., vol. QE-23, pp. 59-64, 1987.

43. E. B. Treacy, IEEE J. Quantum Electron., vol. 5, pp. 454-458, 1969.

44. M. D. Perry, D. Pennington, B. C. Stuart, G. Tietbohl, J. A. Britten, C. Brown, S. Harman, B. Golick, M. Kartz, J. Miller, H. T. Powell, M. Vergino, and V. Yanovsky, Opt. Lett. Vol. 24, pp. 160-162 , 1999.

45. K. Yamakawa, M Aoyama, S. Matsuoka, H. Takuma, C. P. J. Barty and D. Fittinghoff, Opt. Lett., vol. 23, pp. 525-527, 1998.

46. C. Bibeau, D. R. Speck, R. B. Ehrlich, C. W. Laumann, D. T. Kyrazis, M. A. Henesian, J. K. Lawson, M. D. Perry, P. J. Wegner, and T. L. Weiland, Applied Optics, vol. 31, pp. 5799-5809, 1992.

47. P. Maine, D. Strickland, P. Bado, M. Pessot, and G. Mourou, IEEE J. Quantum Electron., vol. QE-24, pp. 398-403, 1988.

48. C. Sauteret, D. Husson, G. Thiell, S. Seznec, G. Gray, A. Migus, and G. Mourou, Opt. Lett., vol. 16, pp. 238-240, 1991.

49. F. G. Patterson, R. Gonzales and M. D. Perry, Opt. Lett., vol. 16, pp. 1107-1109, 1991.

50. K. Yamakawa, H. Shiraga, Y. Kato and C. P. J. Barty, Opt. Lett., vol. 16, pp. 1593-1595, 1991.

51. K. Yamakawa, C. P. J. Barty, H. Shiraga and Y. Kato, IEEE J. Quantum Electron., vol. QE-27, pp. 288-294, 1991.

52. C. Rouyer, E. Mazataud, I. Allais, A. Pierre, S. Seznec, C. Sauteret, G. Mourou, and A. Migus, Opt. Lett., vol. 18, pp. 214-216, 1993.

53. B. C. Stuart, M. D. Perry, J. Miller, G. Tietbohl, S. Herman, J. A. Britten, C. Brown, D. Pennington, V. Yanovsky, and K. Wharton, Opt. Lett., vol. 22, pp. 242-244, 1997.

54. C. N. Danson, J. Collier, D. Neely, L. J. Barzanti, A. Damerell, C. B. Edwards, M. H. R.Hutchinson, M. H. Key, P. A. Norreys, D. A. Pepler, I. N. Ross, P. F. Taday, W. T. Toner, M. Trentelman, F. N. Walsh, T. B. Winstone, and R. W. W. Wyatt, Journal of Modern Optics, vol. 45, pp. 1653-1669, 1998.

55. M. Pessot, J. Squier, P. Bado, G. Mourou, and D. J. Harter, IEEE J. Quantum Electron., vol. QE-25, pp. 61-66, 1989.

56. M. Pessot, J. Squier, P. Bado, G. Mourou, and D. J. Harter, Opt. Lett., vol. 14, pp. 797-799, 1989.

57. C. P. J. Barty, C. L. Gordon III, and B. E. Lemoff, Opt. Lett., vol. 19, pp. 1442-1444, 1994.

58. J. Zhou, C.-P. Huang, M. M. Murnane, and H. C. Kapteyn, Opt. Lett., vol. 20, pp. 64-66, 1995.

59. A. Sullivan, H. Hamster, H. C. Kapteyn, S. Gordon, W. White, H. Nathel, R. J. Blair, and R. W. Falcone, "Multiterawatt, 100-fs laser," Opt. Lett., vol. 16, pp. 1406-1408, 1991.

60. A. Sullivan, J. Bonlie, D. F. Price, and W. E. White, Opt. Lett., vol. 21, pp. 603-605, 1996.

61. K. Yamakawa, A. Magana, P. H. Chiu, and J. D. Kmetec, IEEE J. Quantum Electron., vol. QE-30, pp. 2698-2706, 1994.

62. B. Walker, C. Toth, D. N. Fittinghoff, T. Guo, D.-E. Kim, C. Rose-Petruck, J. A. Squier, K. Yamakawa, K. R. Wilson, and C. P. J. Barty, Opt. Exp., vol. 5, pp. 196-202, 1999.

63. T. Ditmire and M. D. Perry, Opt. Lett., vol. 18, pp. 426-428, 1993.

64. P. Beaud, M. Richardson, E. J. Miesak, and B. H. T. Chai, Opt. Lett., vol. 18, pp. 1550-1552, 1993.

65. K. Yamakawa, H. Sugio, H. Daido, M. Nakatsuka, Y. Kato, and S. Nakai, Opt. Commun., vol. 112, pp. 37-42, 1994.

66. P. F. Moulton, J. Opt. Soc. Am. B, vol. 3, pp. 125-33, 1986.

67. S. A. Payne, L. L. Chase, L. K. Smith, W. L. Kway, and H. M. Newkrik, J. Appl. Phys., vol. 66, pp. 1051-1054, 1989.

68. D. E. Spence, P. N. Kean, and W. Sibbett, Opt. Lett., vol. 16, pp. 42-44, 1991.

69. B. E. Lemoff and C. P. J. Barty, Opt. Lett., vol. 17, pp. 1367-1369, 1992.

70. M. T. Asaki, C.-P. Huang, D. Garvey, J. Zhou, H. C. Kapteyn, and M. M. Murnane, Opt. Lett., vol. 18, pp. 977-479, 1993.

71. A. Stingl, Ch. Spielmann, F. Krausz and R. Szipocs, Opt. Lett. vol. 19, pp. 204-206, 1994.

72. N. H. Rizvi, P. M. W. French, and J. R. Taylor, Opt. Lett, vol. 17, pp. 1605-1607, 1992.

73. J. M. Evans, D. E. Spence, W. Sibbett, B. H. T. Chai, and A. Miller, Opt. Lett., vol. 17, pp. 1447-1449, 1992.

74. I. T. Sorokina, E. Sorokin, E. Wintner, A. Cassanho, H. P. Jenssen, and R. Szipocs, Appl. Phys. B, vol. 65, pp. 245-253, 1997.

75. U. Morgner, F. X. Kartner, S. H. Cho, Y. Chen, H. A. Haus, J. G. Fujimoto, E. P. Ippen, V. Scheuer, G. Angelow, and T. Tschudi, Opt. Lett., vol. 24, pp. 411-413, 1999.

76. D. H. Sutter, G. Steinmeyer, L. Gallmann, N. Matuschek, F. Morier-Genoud, U. Keller, V. Scheuer, G. Angelow, and T. Tschudi, Opt. Lett., vol. 24, pp. 631-633, 1999.

77. S. Sartania, Z. Cheng, M. Lenzner, G. Tempea, C. Spielmann, F. Krausz, and K. Ferencz, Opt. Lett., vol. 22, pp. 1562-1564, 1997.

78. C. G. Durfee, III, S. Backus, M. M. Murnane, and H. C. Kapteyn, IEEE J. Select. Topics Quantum Electron., vol. 4, pp. 395-406, 1998.

79. Y. Nabekawa, Y. Kuramoto, T. Togashi, T. Sekikawa, and S. Watanabe, Opt. Lett., vol. 23, pp. 1384-1386, 1998.

80. W. E. White, F. G. Patterson, R. L. Combs, D. F. Price, and R. L. Shepherd, Opt. Lett., vol. 18, pp. 1343-1345, 1993.

81. B. E. Lemoff and C. P. J. Barty, Opt. Lett., Opt. Lett., vol. 18, pp. 1651-1653, 1993.

82. S. Kane and J. Squier, J. Opt. Soc. Am. B, vol. 14, pp. 1237-1244, 1997.

83. J. Squier, C. P. J. Barty, F. Salin, C. Le Blanc, and S. Kane, Appl. Opt., vol. 37, pp. 1638-1641, 1998.

84. L. M. Frantz and J. S. Nodvik, J. Appl. Phys., vol. 34, pp. 2346-2349, 1963.

85. C. P. J. Barty, G. Korn, F. Raksi, C. Rose-Petruck, J. Squier, A.-C. Tien, K. R. Wilson, V. V. Yakovlev, and K. Yamakawa, Opt. Lett. vol. 21, pp. 219-221, 1996.

86. K. Yamakawa, T. Guo, G. Korn, C. Le Blanc, F. Raksi, C. Rose-Petruck, J. Squier, K. R. Wilson, V. Yakovlev, and C. P. J. Barty, Proceedings of the SPIE, "Generation, Amplification, and Measurement of Ultrashort Laser Pulses III," vol. 2701, pp. 198-208, 1996.

87. R. Trebino, K. W. DeLong, D. N. Fittinghoff, J. N. Sweetser, M. A. Krumbugel, B. A. Richman, and D. J. Kane, Rev. Sci. Instrum., vol. 68, pp. 3277-3295, 1997.

88. J. Itatani, J. Faure, M. Nantel, G. Mourou, and S. Watanabe, Opt. Commun., vol. 148, pp. 70-74, 1998.

89. S. Augst, D. D. Meyerhofer, D. Strickland, and S. L. Chin, J. Opt. Soc. Amer. B, vol. 8, pp. 858-867, 1991.

90. M. V. Ammosov, N. B. Delone, and V. P. Krainov, Zh. Eksp. Teor. Fiz., vol. 91, pp. 2008-2013, 1986. [Sov. Phys. JETP, vol. 64, pp. 1191-1194, 1986.]

91. F. G. Patterson, J. Bonlie, D. Price, and B. White, Opt. Lett. Vol. 24, pp. 963-965, 1999.

Strong-Field Physics with Mid-Infrared Lasers

I.V. Pogorelsky

Accelerator Test Facility, BNL, Upton, NY 11973, USA

Abstract. Mid-infrared gas laser technology promises to become a unique tool for research in strong-field relativistic physics. The degree to which physics is relativistic is determined by a ponderomotive potential. At a given intensity, a 10 μm wavelength CO_2 laser reaches a 100 times higher ponderomotive potential than the 1 μm wavelength solid state lasers. Thus, we can expect a proportional increase in the throughput of such processes as laser acceleration, x-ray production, etc. These arguments have been confirmed in proof-of-principle Thomson scattering and laser acceleration experiments conducted at BNL and UCLA where the first terawatt-class CO_2 lasers are in operation. Further more, proposals for the 100 TW, 100 fs CO_2 lasers based on frequency-chirped pulse amplification have been conceived. Such lasers can produce physical effects equivalent to a hypothetical multi-petawatt solid state laser. Ultra-fast mid-infrared lasers will open new routes to the next generation electron and ion accelerators, ultra-bright monochromatic femtosecond x-ray and gamma sources, allow to attempt the study of Hawking-Unruh radiation, and explore relativistic aspects of laser-matter interactions. We review the present status and experiments with terawatt-class CO_2 lasers, sub-petawatt projects, and prospective applications in strong-field science.

INTRODUCTION

Four years ago at the the first meeting of this series we talked about emerging terawatt (TW) CO_2 laser technology and what 10 μm beams promise for strong physics applications [1]. Since then, CO_2 laser interacting with relativistic electron beams produced acceleration and x-ray radiation effects to unmatched quality and intensity [2,3]. Two CO_2 lasers that already attained or approach TW level are in operation on the US East and West coasts [4,5], and advanced ideas about petawatt-class CO_2 lasers are conceived. This paper provides a "status report" on this still low profile but promising laser technology.

Why are we interested in CO_2 lasers? There are at least two reasons. Both of them are based on the ten times longer wavelength of the CO_2 lasers to compare with the solid state lasers.

The first attractive point is that CO_2 lasers allow a viable compromise between conventional RF linear accelerators that reached perfection in the beam quality and

CP611, *Superstrong Fields in Plasmas:* Second Int'l. Conf., edited by M. Lontano et al.
© 2002 American Institute of Physics 0-7354-0057-1/02/$19.00

optical laser drivers that far prevail in acceleration gradients but do not demonstrate decent beam quality so far.

The degree to which physics is relativistic is determined by a ponderomotive potential

$$W_{osc} = e^2 E_L^2 / 2m\omega^2 , \tag{1}$$

where e and m are correspondingly the electron charge and mass and ω is the laser frequency $\omega = 2\pi c/\lambda$. At a given intensity, the 10-μm CO_2 laser reaches 100 times higher ponderomotive potential than the 1-μm solid-state lasers. This implies a possibility of the proportional increase in throughput of such processes as laser acceleration [6], x-ray production via Thomson scattering [7], etc.

At first glance, these arguments can be dismissed by exercising tighter focusing of the short-wavelength beams:

$$w_0 = \frac{2}{\pi} \lambda M^2 F\#, \tag{2}$$

where w_0 is the Gaussial beam radius at the level of $1/e^2$, M^2 is the beam quality factor equal to 1 for ideal Gaussian beam, and $F\#$ is a ratio of the lens focal length to the initial beam diameter, $f/2W_0$. However, such statement has very limited relevance.

Let us look first into conditions for laser acceleration. Until now, electrons in laser acceleration experiments have been spread over a range of energies with just a few particles observed near the maximum of the accelerating field. Next goal in laser accelerator development is to demonstrate practically meaningful monoenergetic acceleration that resembles qualities of conventional accelerators. What is actually needed to achieve this goal? Accelerating field normally exists in form of a sinusoidal (or shaped close to it) relativistic wave. This may be laser or plasma wave. In order that co-propagating electrons are accelerated monoenergetically they shall occupy a small portion of the wave period and shall be focused to a spot small to compare with radial scale of the wave. Low emittance e-beams are typically focused to 10-100 μm size that is accessible with a CO_2 laser.

Considering volumetric interactions such as Thomson scattering in stationary plasma or field ionization, ten times tighter focus of the 1-μm laser results in the 1000 times smaller interaction volume proportional to $w_0^2 \times z_0$, where z_0 is Rayleigh distance $z_0 = \pi w_0^2 / \lambda$. This leads to the observation that 1 TW CO_2 laser may produce a process yield equivalent to the 100 TW solid state laser.

STATUS OF TWps-CO$_2$ LASER TECHNOLOGY

Above considerations justify efforts in development of mid-IR laser technology. Unfortunately these efforts are still scarce and are not consistent with a promise that this technology provides. There are just two research facilities in the US, UCLA and BNL ATF, that promote this technology.

The UCLA Neptune terawatt CO_2 laser system (see Fig.1) includes master oscillator; optical switch controlled by YAG laser to produce a picosecond CO_2 pulse; regenerative amplifier followed by booster amplifier shown in Fig.1 that is recycled Los-Alamos Antarcs laser built in 1980's. The system produces dual wavelength radiation used for next-generation laser beatwave acceleration experiment that is presently in preparation.

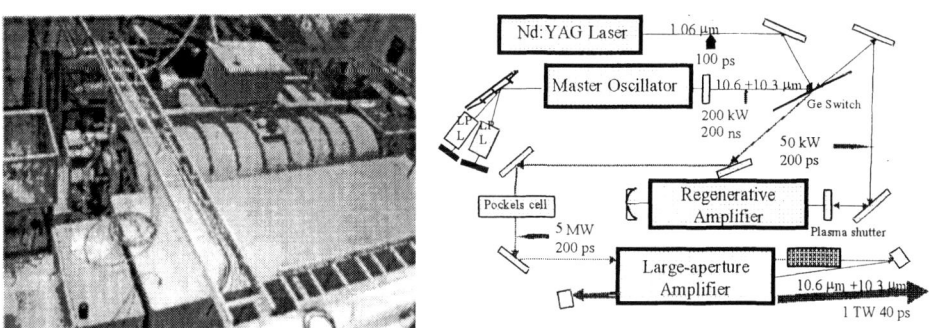

FIGURE 1. Optical diagram of the UCLA Neptune laser and picture of the large-aperture amplifier.

The main complication in building ultra-fast CO_2 lasers is deep modulation of their gain spectrum by rotational structure. This bandwidth limitation can be alleviated by pressure broadening or using multi-isotope mixtures [5]. Both these approaches are not practical for the big-volume Neptune amplifier designed for 3-atm pressure. However, UCLA researchers found an intricate solution to this problem. Focusing laser pulse in a gas cell they observed frequency chirp due to gas ionization [4]. When the pulse is returned back into the amplifier, the chirped tail is filtered out by a relatively narrow individual rotational line. In combination with gain saturation and power broadening, spectral filtering allows to compress the laser pulse from 200 ps to 40 ps. Similar to pressure broadening, power broadening allows to build a bridge between rotational lines. In principle, as soon as the vibrational band is smeared out into a quasi-continuum as short as 1 ps pulses can be amplified directly.

BNL ATF offers another example of a picosecond CO_2 laser with the TW capability named PITER I, where we capitalize on the pressure broadening effect. To this end, one of a kind 10-atm, big-volume booster amplifier has been constructed. The laser action is excited by the 1 MV x-ray preionized discharge. Voltage is applied to electrodes shown in Fig. 2 where the amplifier is opened for maintenance. The 10 J output is extracted through a 10 cm diameter window.

Shown in Fig. 2 the combination of the principle elements of PITER I looks similar to the Neptune laser. Mode-locked solid state laser helps to generate 10 μm picosecond pulse by turning on a semiconductor optical switch. Regenerative preamplifier, combined with 4 additional passes through the same active medium, increases the power to 1 GW. Presently, ATF is still in the process of upgrading its CO_2 laser system to the TW level. 10 atm booster amplifier is already installed and is in operation. However, initial elements still need to be upgraded to deliver a proper 1 ps pulse to the booster amplifier. Meantime, even operating at the present 200 ps pulse duration and 30 GW peak power the ATF laser still enables cutting edge experiments as we discuss in the next Section.

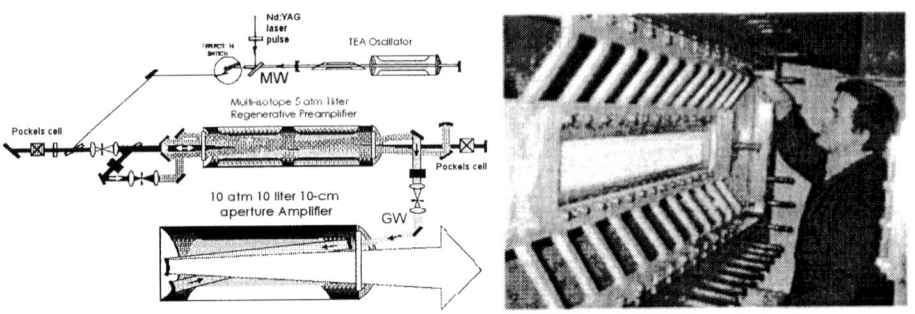

FIGURE 2. Optical diagram of the BNL PITER I CO_2 laser and picture of the booster amplifier.

CURRENT AND PROSPECTIVE PROOF-OF-PRINCIPLE EXPERIMENTS

Many researchers understand the convenience of using long-wavelength lasers in combination with low-emittance relativistic beams for proof-of-principle particle acceleration experiments. As a result, nearly all non-plasma laser acceleration experimental efforts in the US are concentrated now at the ATF - a user's facility operating on a regular basis for high energy physics studies. For this mission ATF is equipped with a high-brightness 70-MeV linac synchronized to high power laser pulses.

A variety of laser acceleration schemes are presently under test or in preparation at the ATF. After testing the inverse Cherenkov scheme based on direct electron acceleration by radially polarized laser field in unionized gas [8] the ATF proceeded to processes based on second order interaction where laser accelerates electrons affected by external electric or magnetic field. In inverse free electron laser (IFEL) linearly polarized laser beam is phased with a planar wiggling of electrons in magnetic undulator producing additional accelerating force in the direction of the local propagation [9]. Similar to this, in LACARA (laser driven electron cyclotron autoresonance accelerator) circularly polarized laser field enhances spiral motion of electrons propagating in a superconducting solenoid [10].

Several schemes based on the first order direct interaction of the Gaussian or Bessel laser focus with e-beam are under consideration. They are based on combinations of axicon or spherical focusing of annular-shaped beams.

Concluding the list, ATF approaches technical capabilities to conduct grating linac experiment proposed at early days of ATF more than a decade ago [11].

Staged Electron Laser Accelerator (STELLA)

Each of the listed above accelerators could be a subject of a separate paper. We review here just one that illustrates advantages of long-wavelength radiation for non-plasma laser accelerators. This experiment has produced the results that advanced accelerator community characterizes as a step towards laser accelerators of the next-generation [12].

In the case of the direct acceleration of relativistic electrons in the laser beam a condition for monoenergetic acceleration requires that electron beam shall be focused much tighter than the laser and grouped into microbunches exactly to the laser period. We can add to it that such bunch train shall be much shorter than the laser pulse envelope. There is probably only one possible way to produce such microbunch train: to use the same laser.

The idea is simple. If we achieve sinusoidal energy modulation in the laser wave and allow electrons to drift or transmit them through dispersive compressor, a good portion of them group together into a small microbunch and this is exactly periodical to the laser wavelength. Modeling of this process that assumes practically achievable electron beam parameters and CO_2 laser shows sub-micron microbunches (see Fig. 3). Such short microbunches can be accelerated monoenergetically in the 10-μm laser beam.

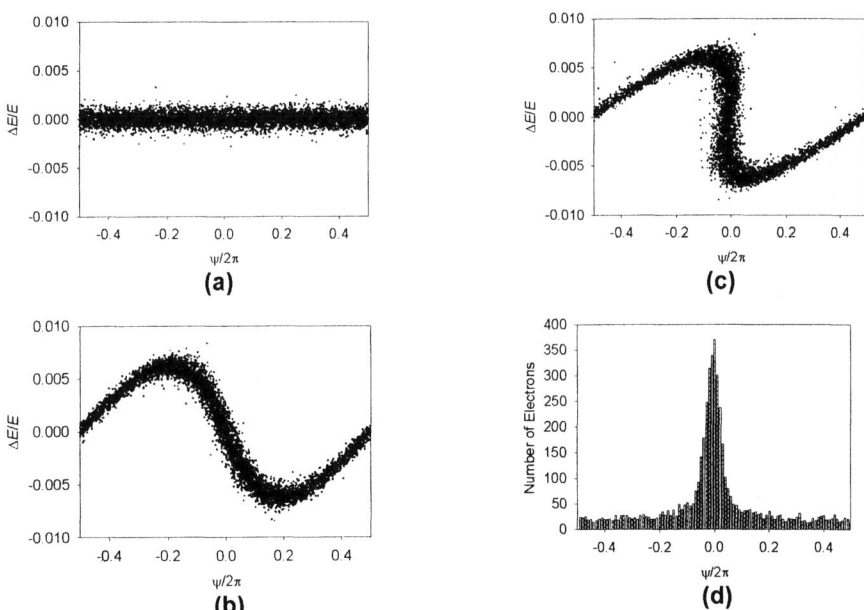

FIGURE 3. Simulations of electron beam bunching at $\Delta E/E=1.2\%$ energy modulation in IFEL wiggler: a) initial uniform energy distribution; b) energy modulation at the wiggler exit; c) energy distribution at the entrance to the ICA cell; d) longitudinal density distribution in which 50% of the electrons are bunched into FWHM=0.63 μm.

A concept of monoenergetic laser accelerator evolves into two-stage scheme where the first stage serves as a buncher while the second produces monoenergetic acceleration. This is an idea of the first staged electron laser acceleration experiment abbreviated to STELLA. As is shown in Fig. 4, a single laser beam is split between two 30 cm long, 10-period IFEL wigglers.

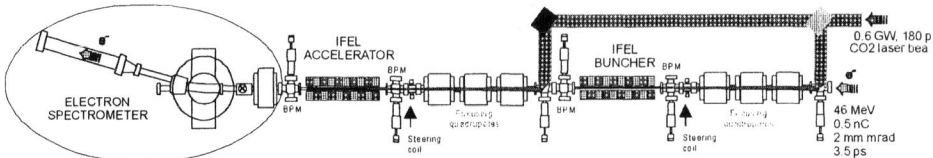

FIGURE 4. Principle diagram of STELLA experiment

Extent of phase control over the two-stage acceleration process is illustrated by the set of spectra shown in Fig. 5. This sequence of spectra is obtained when the optical delay of the accelerating laser beam changes half wavelength from the maximum acceleration to ultimate deceleration of electrons.

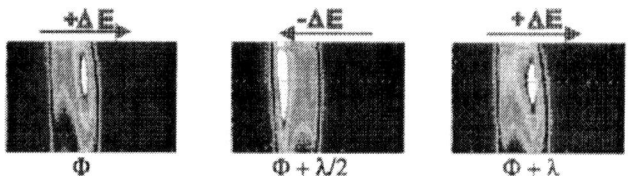

FIGURE 5. Phase control over microbunch acceleration in STELLA experiment

Note that no active phase stabilization has been applied. The demonstrated phase control is possible due to a relatively long wavelength of the CO_2 laser to compare with naturally occurring thermal drifts and vibrations in the optics and accelerator components. The longer wavelength also relaxes requirements to the microbunch duration. Thus, STELLA gives an example how a mid-IR laser facilitates the task of demonstrating the next generation laser accelerator.

Prospective Next-Generation LWFA

There are good prospects for setting at the ATF the next-generation LWFA experiment. The CO_2 laser is a meaningful candidate for this application due to the quadratic wavelength scaling of the ponderomotive potential (see Eq.1) that drives a plasma wake. This ultimately allows the attainment of a high acceleration gradient even at a low plasma density

$$E_{acc}^{max}[GV/m] = 2.8 \times 10^4 \left(\lambda/w_0\right)^2 P_L[TW]/\lambda_p[\mu m].$$ (3)

Reasonably low plasma density ($\sim 10^{16}$ cm^{-3}) is desirable because it allows to accelerate proportionally higher bunch charge,

$$N_e < n_e \left(c/\omega_p\right)^3 = 4 \times 10^6 \lambda_p [\mu m] \qquad (4)$$

at a small electron energy spread and emittance. To validate this claim, the optimum parameter space for the next-generation LWFA has been verified by simulations [13] made for 1 ps CO_2 laser that is under development at the ATF. Simulations demonstrate that, in order to achieve beam quality comparable to conventional accelerators, the injected electron bunch shall be much shorter than has been achieved so far using conventional RF techniques. For example, 200 fs minimum bunch duration demonstrated so far in FR linacs, being just 10% of the resonance plasma wavelength (λ_p=800 μm) at the axis of the parabolic plasma channel, is still not sufficient to ensure an adequate beam quality control.

A possible solution could be using a plasma wake as a bunch compressor. Simulations show that injecting the 5 MeV electron bunch at the negative slope of the accelerating field we introduce the energy modulation that compresses the bunch 10 times. This approach to monoenergetic LWFA is illustrated by the scheme in Fig. 6.

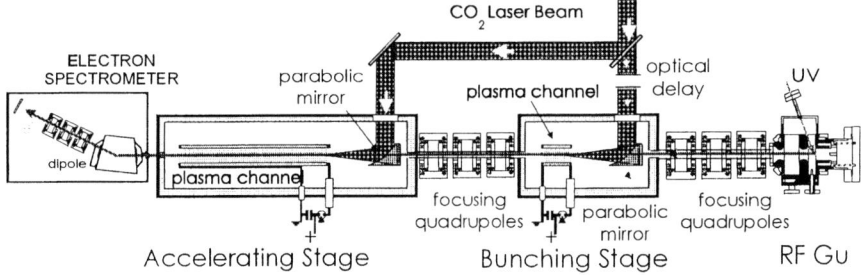

FIGURE 6. Diagram of prospective two-stage monoenergetic LWFA.

The 200 fs 5 MeV electron bunch produced by a photocathode RF gun is focused into the bunching LWFA stage where it is compressed to 20 fs (see Fig. 7).

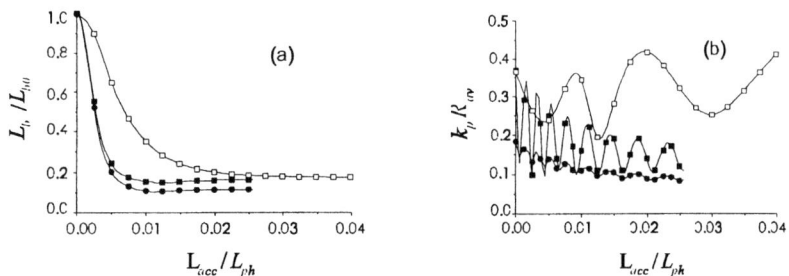

FIGURE 7. Bunch compression in channel guided LWFA;
(a) longitudinal bunch compression, (b) transverse bunch dynamics;
open squares –ϕ_{max}=0.04, R_{av}^0 =50 μm, solid squares - ϕ_{max}=0.19, R_{av}^0 =50 μm, circles -
ϕ_{max}=0.19, R_{av}^0 =25 μm.

Simulations shown in Fig.7 assume the following initial e-beam parameters: energy 5 MeV, energy spread 1.5%, geometric emittance 0.6 mm.mrad, bunch length 200 fs, and the rms radius R_{av}^0 =25–50 μm. We see that a good control over the bunch compression process is possible at a small initial e-beam radius R_{av}^0 =25 μm and relatively strong wakefield potential ϕ_{max}=0.19. Injecting the compressed bunch into the accelerating stage we observe also reduction in the relative energy spread from 5 to 1-2% and preservation of the normalized emittance.

It is important to see if these properties could be maintained when a similar configuration is used in a multi-stage scheme. Simulations [14] demonstrate a steady increase of the electron energy and a good control over the e-beam quality. Such performance requires a precise control of the optimum phases for injection and extraction of bunches from the accelerating stages.

We conclude this paragraph with a statement that analytical and computer calculations prove a possibility of the multi-GeV monoenergetic electron accelerator driven by a picosecond CO_2 laser and utilizing a conventional electron injector.

High-Brightness Relativistic Thomson X-Ray Source

Another application of a high-power CO_2 laser that we analyze here is x-ray generation via Thomson scattering.

A laser beam interacting with a counter-propagating relativistic electron beam behaves like a wiggler of an extremely short period. Relevant expressions for the wavelength, angular divergence, intensity, and brightness of the produced x-rays follow:

$$\lambda_x = \lambda / 4\gamma^2 , \tag{5}$$

$$\theta_0 = 1/\gamma , \tag{6}$$

$$N_X[photon/pulse] = 6.7 \times 10^{11} E_L^{eff} [J]Q[nC]\lambda[\mu m]/r_L^2[\mu m], \tag{7}$$

$$B[photon/mm^2 mrad^2 \sec] = N_x \gamma^2 / 2(\pi r_b)^2 \tau_b . \tag{8}$$

With a 70 MeV e-beam and a CO_2 laser, as short as 1 Å x-rays can be produced at the divergence and spectral bandwidth comparable with conventional synchrotron sources. Thus, much more compact and economical accelerator can be used. X-ray pulse duration is close to the electron bunch length, which can be on the femtosecond scale.

Let us consider what is the optimum choice for a laser and an accelerator when designing a high-brightness laser synchrotron source for a particular x-ray wavelength. As long as λ_x is considered as an invariant then choosing the CO_2 laser, with its wavelength 10 times longer than a solid state laser, requires a 3 times more energetic e-beam. This immediately improves angular divergence of the produced x-rays. X-ray yield will rise 10 times proportional to λ. This stems from the facts that number of x-ray photons is proportional to the number of delivered laser photons, which is proportional to λ at the fixed laser energy. When combining these factors together, we source opens the prospect for up to 100 times increase in the brightness of the produced x-rays to compare with the 1-μm laser.

Using these considerations as design criteria, we assembled Thomson source in the ATF electron beamline as is shown in Fig. 8. Electron beam and CO_2 laser pulses are focused at the interaction point in a head-on collision. Laser beam is backscattered into x-rays. Dipole magnet separates electron beam from x-rays that pass a foil window to a Si detector.

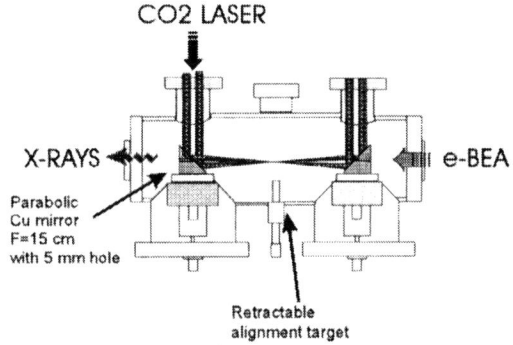

FIGURE 8. Interaction cell of the ATF Thomson x-ray source

Using 15 GW laser we obtained the highest photon yield ever demonstrated via laser Thomson scattering on relativistic electron beams [15]. On the diagram in Fig.9 you see earlier results obtained in LBNL and NRL. Lambda-proportional x-ray yield and counter-propagation configuration are the factors that explain a high position of the ATF source in this competition. We move now to the next stage where 1 TW CO_2 laser will be used in the same configuration and actually in the same interaction cell. We expect to observe strong harmonics and a noticeable shift in the fundamental peak energy (see Fig.10) due to relativistic mass shift of electron [15]. This is an illustration of one of our introductory statements that CO_2 laser allows attaining relativistic physics at rather moderate power.

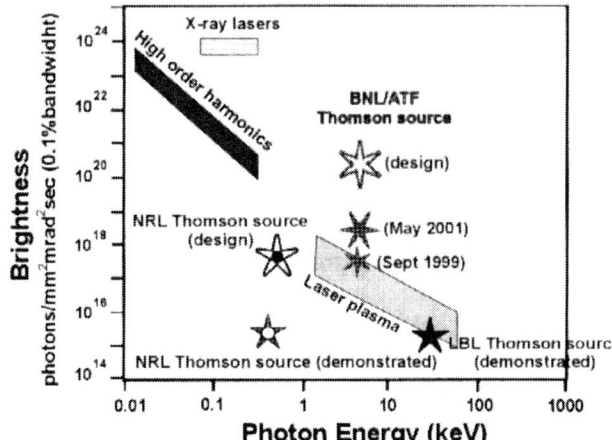

FIGURE 9. ATF Thomson scattering source. Demonstrated and design parameters

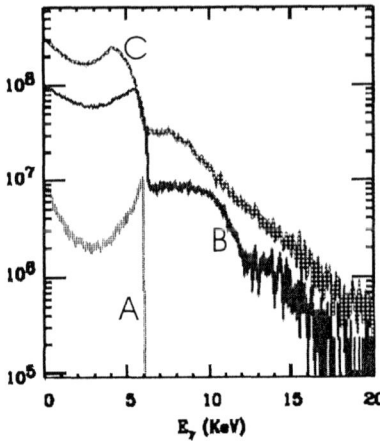

FIGURE 10. Thomson scattering spectra simulated for conditions of the recent ATF test done with the 30 GW CO_2 laser (A) and projects for 0.3 TW (B) and 1 TW laser power (C).

CONCLUSIONS

The BNL ATF is engaged in a string of the next-generation laser acceleration experiments including STELLA II, LACARA, and LWFA in a plasma channel. Simultaneously, UCLA will attempt next-generation beatwave experiment. All these experiments use terawatt CO_2 laser and strive to demonstrate quasi-monoenergetic electron acceleration at a gradient above conventional accelerators and over extended interaction distance. Another application of the terawatt CO_2 lasers - demonstration of nonlinear Thomson scattering and realization of ultra-bright femtosecond Thomson x-ray source may provide the first time opportunity to initiate studies in a parameter space approaching LCLS regime. Success in these experiments will establish a respectful position of CO_2 lasers within a family of ultra-fast lasers.

A reasonably confidential forecast towards 100 TW CO_2 lasers and beyond can be based on the already existing laser modules. For example, energy up to 1 kJ can be potentially extracted from the 30-cm aperture Neptune amplifier [16]. As we discussed, power broadening at 10^{11}-10^{12} W/cm^2 provides sufficient bandwidth for the 1-ps pulse amplification. Thus, it is just a matter of modifying the front end of the Neptune laser system to produce a proper seed pulse. The ATF laser designed to operate at 10 J, 1 ps is an example of a system that may serve for this purpose. Such input would be sufficient to reach 1 PW peak power from the Neptune amplifier after 4 passes with gradual beam expansion to the full aperture of the amplifier in a saturated regime.

There are also potentials to shorten CO_2 laser pulse down to 100 fs. The most straightforward way is to build 7 THz broad continuum in the gain spectrum by mixing all possible isotopes of oxygen and carbon. Such continuum can support direct amplification of the 100 fs pulse. Another possibility to reach 100 fs pulse duration starting with 1-2 ps pulse is due to pulse chirping via gas ionization followed by compression in the material with a negative dispersion (e.g., conventional ZnSe IR window) [17].

Combination of the energy boost above 100 J with prospective pulse compression to 100 fs may allow CO_2 laser to approach the multi-petawatt level. The 1 PW 10 μm radiation focused to diffraction limit with F#=2 optics produces field with a=120 that makes possible efficient realization of a number of exotic highly relativistic processes such as GeV ion and electron acceleration via Coulomb explosion or direct ponderomotive expulsion from the laser focus, study of Unruch radiation, etc.

ACKNOWLEDGEMENTS

The author wishes to thank W.D. Kimura, S. Tochitsky, and K. Kusche for providing graphic material and technical help in preparing this paper. The reviewed here results shall be credited to several research teams including: STELLA collaboration (see participants in ref. [2]), Japan-US collaboration on polarized positron source [3,15], N. Andreev and S. Kuznetsov [13,14]. Special thanks to Optoel Co. [5] for help in developing the ATF CO_2 laser.

The work is supported by the US Department of Energy.

REFERENCES

1. Pogorelsky, I.V., "Sperstrong Fields in Plasmas", *AIP Conference Proceedings*, **426**, 415 (1998)

2. Kimura, W. D., van Steenbergen, A., Babzien, M., Ben-Zvi, I., Campbell, L.P., Cline, D.B., Dilley, C.E., Gallardo, J. C., Gottschalk, S.C., He, P., Kusche, K.P., Liu, Y., Pantell, R.H., Pogorelsky, I.V., Quimby, D.C., Skaritka, J., Steinhauer, L.C. , and Yakimenko, V., *Phys. Rev. Lett.*, **86**, 4041 (2001)

3. Pogorelsky, I.V., Ben-Zvi, I., Hirose, T., Kashiwagi, S., Yakimenko, V., Kusche, K., Siddons, P., Skaritka, J., Kumita, T., Tsunemi, A., Omori, T., Urakawa, J., Washio, M., Yokoya, K., Okugi, T., Liu, Y., He, P. and Cline, D., *Phys. Rev. Special Topics - Accelerators and Beams*, **3**, 090702 (2000)

4. Tochitsky, S. Yu., Narang, R., Filip, C., Clayton, C.E., Mash, K.A. and Joshi, C., *Optics Lett.* **24**, 1717 (1999)

5. Pogorelsky, I.V., Ben-Zvi, I., Babzien, M., Kusche, K., Skaritka, J., Meshkovsky, I., Dublov, A., Lekomtsev, V., Pavlishin, I., Boloshin, Yu. and Deineko, G.," Laser Optics '98, St. Petersburg, June22-26 1998, "Superstrong Laser Fields and Applications", *Proc. of SPIE* **3683**, 15 (1999)

6. Pogorelsky, I.V., *Nucl. Instrum. and Methods in Phys. Res.* A, **410**, 524 (1998)

7. Pogorelsky, I.V., *Nucl. Instrum. and Methods in Phys. Res.* A, **411**, 172 (1998)

8. Fontana, J.R. and Pantell, R.H., *J. Appl. Phys.* **54**, 4285 (1983); Kimura, W.D., Kim, G. H., Romea, R. D., Steinhauer, L. C., Pogorelsky, I.V., Kusche, K.P., Fernow, R.C., Wang, X., and Liu. Y., *Phys. Rev. Lett.* **74**, 546 (1995)

9. Palmer, R.B, *J. Appl. Phys.* **43**, 3014 (1972)

10. Hirshfield, J.L. and Wang, C., *Phys. Rev.* E **61**, 7252 (2000)

11. Palmer, R.B., Proc. "Laser Acceleration of Particles", AIP **91**, 179 (1982); Fernow, R.C. and Claus, J., "Advanced Accelerator Concepts", Port Jefferson, NY, 1992, AIP **279**, 212 (1993)

12. Colby, E.R., and Gai, W., "Advanced Accelerator Concepts", Santa Fe, NM, 2000, AIP **569**, 47

(2001)

13. Andreev, N.E., Kuznetsov, S.V., and Pogorelsky, I.V., **3**, 021301 (2000)

14. Pogorelsky, I.V. Andreev, N.E. and Kuznetsov, S.V., Proceedings of LASERS' 99, Quebec, Canada, December 13-16, 1999, STS Press, McLean, 313 (2000)

15. Kumita, T., Kamiya, Y., Hirose, T., Pogorelsky, I.V., Ben-Zvi, I., Kusche, K., Siddons, P., Yakimenko, V., Omori, T., Yokoya, K., Urakaw, J., Kashiwagi, Wasio, M., Zhou, F., and Cline, D., to be published in *Phys. Rev. Special Topics - Accelerators and Beams*, Conference Edition "Laser-Beam Interactions 2001"

16. Tochitsky, S. Yu., Filip, C., Narang, R., Clayton, C.E., Mash, K.A., and Joshi, C., Proceedings of LASERS' 2000, Albuquerque, NM, December 4-8, 2000, STS Press, McLean, 417 (2001)

17. Corkum, P.B., *IEEE J. Quantum Electron.* **QE-21**, 216 (1985)

Compression of High Power Lasers in Plasma

Nathaniel J. Fisch*, Vladimir M. Malkin* and Gennady Shvets*

*Princeton University, Princeton NJ 08540, USA

Abstract.
   While achievable laser intensities have grown remarkably during recent years, mainly due to the method of chirped pulse amplification, to attain very much higher powers would demand suitable material gratings for handling very high power and very high total energy. However, plasma is an ideal medium for processing very high power and very high total energy, making feasible, in principle, much higher laser intensities than might otherwise be contemplated. The idea in plasma is to store energy in a long pump pulse which is quickly depleted by a short counterpropagating pulse. This counter-propagating wave effect has already been employed in Raman amplifiers using gases or plasmas. At very high power, there are nonlinear effects in plasma that enter which make such methods particularly suitable for high power pulse compression.

INTRODUCTION

Laboratory laser intensities have grown remarkably during recent years due to the method of chirped pulse amplification (CPA) [1, 2]. In CPA, optical gratings are used to strech a short pulse, separating it into its frequency components. A broadband optical amplifier is then employed to amplify the low-powr stretched signal. Complementary gratings then reconstitute the original, but now highly amplified, signal. This method has been extraordinarily successful, but it does require a final material grating subject to fluence limits. Currently, gratings for $1\,\mu$ light are imagined to be limited to the range of several J/cm^2. The method also requires uniform amplification over a larger bandwidth.

For example, using CPA technology, a 100 fs pulse can be expanded to 1 ns, pumped to 1 J/cm^2 and then recompressed to 100 fs, giving output intensities of 10 TW/cm^2. Using 10^3 cm^2 gratings, 1 kJ of laser power can then be delivered in 100 fs for an output power of 10 PW. With a vacuum focus to say a spot size $30\,\mu \times 30\,\mu$, *i.e.*, focusing by a factor of 10^8, intensities of 10^{21} W/cm^2 can be attained. However, the use of 10^3 cm^2 gratings withstanding fluences of 1 J/cm ^{2}, while feasible, is expensive and technologically challenging. Using CPA to attain much higher power will require gratings that will eventually be too large to produce. Moreover, for high power at wavelengths shorter than a micron, it would be necessary also to develop lasers, amplifiers, and gratings operating efficiently at short wavelengths. Whereas at one micron, fluences of several J/cm^2 can be contemplated, at shorter wavelengths, only much lower fluences can be imagined.

Yet plasma is an ideal medium to form a grating capable of processing very high power and very high total energy. We have in mind compression of powers to exawatts per square cm or fluences to kilojoules per square cm, prior to the vacuum focus. For energy applications, pulse compression does not need high fidelity within each frequency range. It turns out that the limiting effects in plasma are the nonlinear effects

CP611, *Superstrong Fields in Plasmas:* Second Int'l. Conf., edited by M. Lontano et al.

associated with nearly relativistic electron velocities in the wave fields; hence, at higher frequency, the plasma is even more capable of processing high power, since the velocities in a constant power laser scale inversely with frequency. Thus, plasma is ideal for applications for delivering simply the highest possible power.

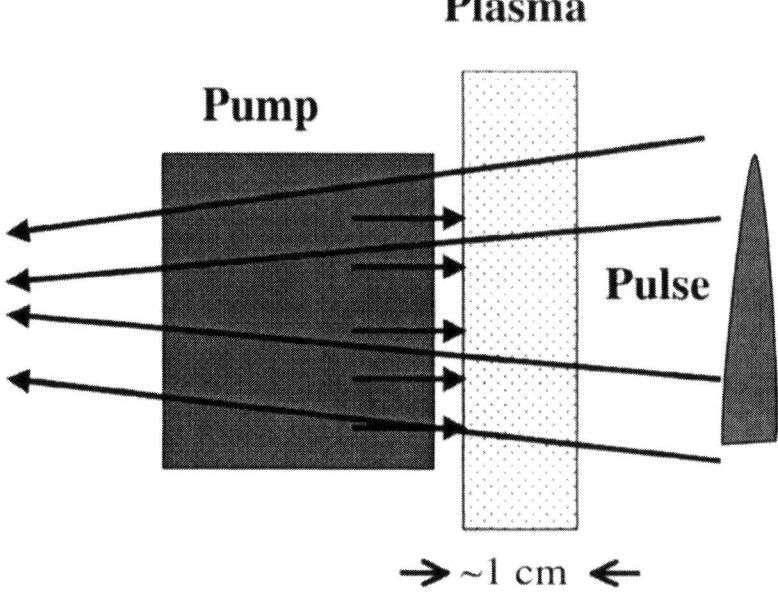

FIGURE 1. Schematic of the counter-propagating geometry: pump laser, about twice the 1 cm plasma length, travels to the right; a counter-propagating seed pulse interacts with the pump, and is timed to enter plasma just as pump leaves plasma; and the pulse is depicted as focusing to point beyond plasma.

The basic geometry for the use of plasma is shown in Fig. 1. A long pump laser loses its energy to a short counter-propagating short pulse, with the plasma serving as a coupling medium between the two lasers. Pulse compression occurs, since most of the energy of the long pulse then resides in the short pulse. Several effects in different plasma regimes can be contemplated to produce this backscattering compression. However, in addition to an efficient backscattering of pump energy into the short conterpropagating wave, several issues must be resolved successfully before the compression effect can be considered useful. For the compression effect to occur altogether, it important that the short pulse not only deplete the long pump pulse, but that it remain short as it propagates through the long pump. In fact, there are a number of pulse lengthening effects that must be avoided. Also, as can be seen from Fig. 1., the pump must propagate through transparent plasma to the seed pulse, without encountering deterioration. The pulse must not only be stable during amplification, but it also must retain its focus as it leaves the plasma. It turns out that the stability of the pulse is the major limitation on the plasma length. Thus, as soon as the pulse experiences instability, the plasma must be terminated, and the amplified pulse, however much it has been amplified, must be extracted from the plasma.

This paper briefly reviews some of the important issues in achieving useful pulse compression effects.

REGIMES OF PULSE COMPRESSION

Plasmas have, in fact, long been contemplated as media suitable for compressor/amplifiers. In particular, the compression of laser light through Raman backscatter has been suggested in gases, liquids and plasmas via the counter-propagating wave geometry depicted in Fig. 1. An early review of pulse compression of excimer lasers in gases is given by [4]. The advantages of using plasma were recognized by Capjack et. al [5]. Raman compression in gas mixtures, from tens of nanoseconds to tens of picoseconds, has been achieved at about 25% efficiency for energies in the range of 0.1 to 10 J [6, 7].

In the counter-propagating geometry, the energy is stored in a long low-intensity pump pulse, or possibly a train of low-intensity pump pulses. The short counterpropagating pulse, or "pumped pulse", can achieve intensities far higher than the pump pulse, so long as it remains short. Of course, for very high powers, Raman media other than plasmas will not be practicable. The early work on compression in plasma was focused on low-power regimes, where the Raman backscattering took place in the stationary regime, namely where the plasma wave was highly collisionally damped. The issues that plagued this early work in plasma were the stringent requirements on plasma homogeneity (because of narrow-bandwidth amplification) and because the amplification lengths were long, in part due to the collisional damping that reduced the efficiency of the interaction.

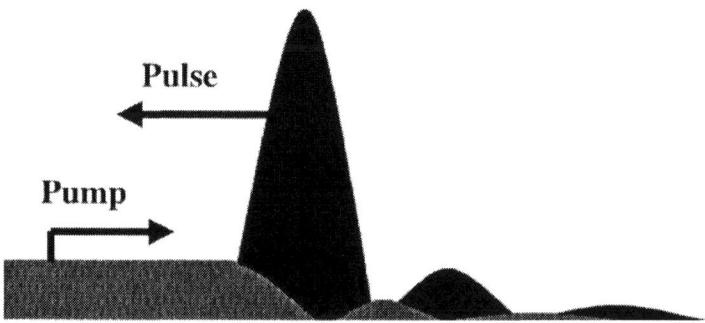

FIGURE 2. Depletion of pump by short pulse in time-asymptotic limit in resonant Raman backscattering regime.

However, at very high power, there are nonlinear effects that enter which may make for better compression. At high power the wave backscattering tends to occur faster, while the large velocities of electrons oscillating in the wave fields tend to reduce collisonal damping. There are several mechanisms by which the pump power might be coupled into the counterpropagating pulse. One coupling mechanism involves a so-called "superradiant" or Compton scattering, where the nonlinear interaction of the plasma electrons with the lasers dominates the plasma restoring motion due to charge imbalance [8]. A second mechanism is resonant backward Raman scattering (see Fig. 2), which can occur fast enough that the amplification process outruns deleterious processes associated with the ultraintense pulse [9]. A third mechanism involves coupling at an ionization front [10].

In the Compton scattering regime, the counter-propagating waves are coupled to each other by electrons that execute nonlinear motion in the beats of the counter-propagating waves. The restoring force arising from the displacement of electrons from positive charge centers is essentially negligible. In this regime, the pump laser is of somewhat higher frequency than the pumped laser; the pulse length of the pumped pulse is less than a plasma length, c/ω_p; and the laser intensities satify $ab > \omega_p^2/4\omega^2$, where a and b are the pump and pumped pulse normalized vector potentials. In this regime, a short weak pulse can be very significantly amplified, but the pump depletion tends not to be complete. Of course, as the amplified pulse intensity grows, the pump depletion becomes greater. For small pump depletion, under Compton scattering, the pulse energy grows like $z^{3/2}$, where z is the distance into the plasma. The available pump energy for depletion grows like z, so the depletion fraction grows like $\sqrt{z}$. The numerical example quoted in Ref. [8] gives about 40% pump depletion. At issue, however, is whether the short pulse can grow to large enough amplitude to deplete the pump before the pulse itself succumbs to modulational [11, 12] or other instabilities. The advantage of this regime, however, may be that exact resonance is not required.

On the other hand, in the Raman scattering regime, where resonance is required, it can be shown that the pulse grows so fast that it can outrun the deleterious instabilities. As shown in Ref. [9], within several growth times of the deleterious instabilities, the pulse grows to amplitudes far exceeding the instability thresholds, $i.\ e.$, to overcritical powers. Pump depletion quickly ensues, so that the efficiencies are limited in principle only by the so called "Manley-Rowe" relations; in other words, since a pump photon is converted to a counter-propagating pulse photon downshifted by the plasma frequency, the fraction of pump energy that will be left in the plasma wave is ω_p/ω. Typically, this fraction would be about 1/10. The remaining 90% of the pump power can, in principle, be converted entirely to the backscattered short pulse.

In this resonant Raman scattering regime, there are many laser wavelengths in the pump pulse and at least several laser wavelengths in the short seed pulse. Hence, the solution to these equations may be envisioned as the interaction between slowly-varying wave envelopes. This interaction between the pump and pulse wave envelopes is shown schematically in Fig. 2. Here the right-going pump is depleted as it encounters a left-going short pulse. Note that both the pump and the pumped pulse envelopes assume a fluctuation in amplitude. Note too that there is virtually no pump energy going off to the right after the encounter with the pulse. This is the well-known "π-pulse" solution, which is a self-similar solution for the three-wave coupled equations of a constant pump encountering a very short counterpropagating downshifted seed pulse [9]. Moreover, it is an attractor solution, in the sense that many initially prepared waves will evolve to the same asymptotic state. What happens is that the pump amplifies the leading edge of the seed pulse, which consumes all the pump energy, with ω_p/ω of the pump energy going into the Langmuir wave. Since the pump is effectively depleted by the leading edge of the seed, the trailing edge of the seed is shadowed, resulting in a reshaping of the seed pulse so that it narrows as it grows. Also, after the pump depletes, the amplified pumped pulse, together with the generated Langmuir waves, regenerate the pump wave via the same 3-wave interaction. The regenerated pump, moving to the right, subsequently decays into the another Langmuir phonon and counterpropagating pulse photon, with the process repeating until no pump energy remains. This repetition results in the the characteristic amplitude variation of the pump and seed pulse in the π-pulse solution.

Both the Compton scattering regime and the resonant Raman backscattering regime require, at least in the most simple implementation, that the pump traverse the plasma before encountering the pulse. The pump length is then optimally twice the plasma length, with the pump front leaving the plasma just as the seed pulse enters the plasma, and with the pump tail entering the plasma just as the amplified seed leaves the plasma slab. Thus, compressing a 50 ps pump requires a plasma length of about 1 cm. In the case of resonant Raman backscattering, with say ω_p/ω about 0.1, the plasma electron density must be about 10^{19}cm^{-3}. If the pump were at 10^{14} W/cm^2, the compression could give output pulses of 10^{17} W/cm^2. The Compton scattering regime might be accessed at much lower plasma density, depending upon the pump and pulse intensities.

In traversing the plasma, which could be noisy, the pump is subject to collisionless instabilities, such as Raman backscattering and forward scattering. One way to avoid the issues in pump traversal of the full plasma is to inject the pump somewhat from the side of the plasma. However, that can be quite complicated, since the interaction region is moving axially. In the case of ionization-front scattering, this issue is avoided, since the pump propagates through a neutral gas. The ionization front is formed by the high power pulse, as it traverses the same gas. Once the plasma is formed, the coupling can occur by means of the resonant backward Raman backward scattering effect. However, to avoid ionization in cases of typical interest, it may be necessary to employ lower pump power densities, say about 10^{12} W/cm^2. The lower pump power density would imply that the interaction length to achieve the output power of the above example would have to be on the order of a meter rather than a centimeter. However, in a gas, the longer interaction length is still quite practicable.

The compression effect is likely best achieved through the careful preparation, alignment, and timing of a counterpropagating seed pulse. However, more advantageous scenarios might also be imagined. For example, it would be to great advantage if the timing could be achieved automatically, i. e., if the backscattered wave arose spontaneously at just the right spot and time in the plasma, namely just at the point of the pump front as it is about to exit plasma. In principle, backscattering off of stationary plasma oscillations near the plasma exit could accomplish this (see Fig. 3). The seed is then the plasma oscillation, which has nearly vanishing group velocity, and hence does not need to be timed precisely. The backscattered pulse is then generated by the same resonant backward Raman scattering process off of a prepared plasma wave, acoustic turbulence, or any other slow moving plasma perturbation. While this method, in principle, is the most simple to implement, it is an open question whether the simplified method can result in a backward-propagating seed signal with the correct parameters to enter into a useful

Plasma

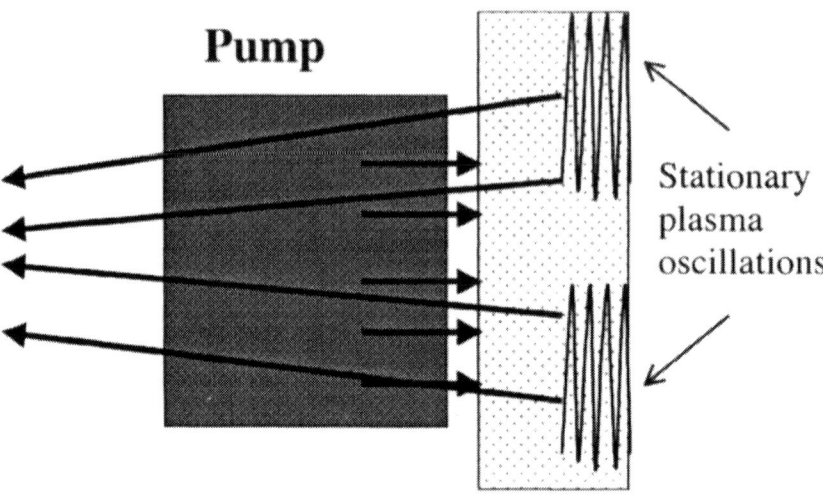

Pump

Stationary
plasma
oscillations

FIGURE 3. No seed pulse is employed. Instead seed is essentially stationary plasma wave or acoustic wave at far end of plasma.

compression regime.

COMPRESSION BY RESONANT RAMAN BACKSCATTERING

The beauty of pulse compression via the resonant Raman backscattering effect is that the effect is both well-studied and inherently one of the simplest plasma effects. The output pulse reaches very high intensities, but only in the sense that it leads to extreme pulse intensities on target. However, within the plasma, the non-focused power is not particularly high, in the sense that the plasma effects are fundamentally not extreme. The coupling occurs between essentially low-intensity counter-propagating light waves via essentially low-amplitude longitudinal cold plasma oscillations. The motion of electrons in the cold plasma wave is coherent, essentially sinusoidal motion, with no overtaking or wave-breaking. The motion of electrons in the lasers are also non-relativistic. Only as the pulse is amplified to its maximum value, which is essentially at the point that these effects are no longer so simple, do the electron trajectories begin to become complicated. At that point, other effects indeed enter, and that is roughly the boundary of the regime which can be considered to be the resonant Raman backscattering regime. And at that point the plasma is terminated and the pulse is extracted.

Hence, in the resonant Raman backscattering regime, there should be only a limited number of surprises in our understanding of the plasma effects. The short time scales involved mean that only electron dynamics, rather than ion dynamics, enter importantly.

This further limits the possibilities of unintended effects that could hinder the useful effects. Yet in this regime the time scales are not so short that the waves cannot be simply described as envelope pulses with a central carrier frequency. In compressing from tens of picoseconds to tens of femtoseconds, there will still be many laser wavelengths in the pulse wavepacket. Thus, there should be high confidence in using relatively simple analytical tools to describe this interaction.

The resonant Raman backscatter equations can thus be written as

$$a_t + ca_z = \omega_p f b, \tag{1}$$

$$b_t - cb_z = -\omega_p f^* a, \tag{2}$$

$$f_t = -\omega ab^*/2, \tag{3}$$

Here a and b are vector-potential envelopes of the pump and pulse, respectively, in units of $m_e c^2/e \approx 5 * 10^5 \mathrm{V}$, and f is the envelope of the Langmuir wave electrostatic field $\vec{E} = E\vec{e}_z$ in units of $m_e c\omega_p/e = c\sqrt{4\pi m_e n_e} \approx \sqrt{n_e[\mathrm{cm}^{-3}]}\,\mathrm{V}/\mathrm{cm}$, defined by formulas

$$(A_x + iA_y)e/m_e c^2 = ae^{i(k_a z - \omega_a t)} + be^{i(k_b z - \omega_b t)}, \tag{4}$$

$$Ee/m_e c\omega_p = fe^{i[(k_a - k_b)z - \omega_p t]} + c.c., \tag{5}$$

where A_x and A_y are components of the real vector-potential $\vec{A}$ in the plane transverse to the propagation direction z; for the pump propagating in the positive and the seed-pulse in the negative direction, $k_a = \sqrt{\omega_a^2 - \omega_p^2}/c$ and $k_b = -\sqrt{\omega_b^2 - \omega_p^2}/c$; ω_p, ω_b, and ω_a are the plasma, laser-seed and laser-pump frequencies, and subscripts t and z denote time and space derivatives. The pulse duration is larger than ω_p^{-1}. Both lasers are circularly polarized. Self-nonlinearities of lasers and Langmuir wave are neglected. Plasma ions are assumed to be immobile. The Langmuir wave group velocity is neglected in comparison with the speed of light. For $\omega_b \gg \omega_p$, one may assume, as it is done above, $\omega_a \approx \omega_b = \omega$ and $k_a \approx -k_b \approx \omega/c$ in all the equation coefficients.

An initially weak seed b will be amplified by an undepleted pump field; in the constant pump or linear regime, exact solutions exist which show that while the seed front moves with the velocity of light c, the envelope maximum only moves with $c/2$ [13]. Hence, in this regime the counter-propagating wave is stretched. It is only in the nonlinear regime, i.e., the pump-depletion regime, that the pulse is compressed. In this regime, the pulse front effectively shadows the pulse maximum and tail, so that the maximum catches up to the front. Also, in this regime, all short enough and intense enough initial seeds will asymptotically reach the π-pulse solution [9]. That a self-similar solution, depending only on the ratio z/t, exists can be seen from a scaling of the equations. Essentially, the Raman depletion of the pump by the seed takes place in a distance that varies inversely with the seed pulse amplitude. This is also the effective width of the seed pulse, since what is further than this distance behind the seed front is effectively shadowed. However, at complete pump depletion, the pulse energy must grow linearly with distance (or time) traveled, because that is all the available energy in the pump. Thus, the amplitude and energy of the seed pulse must grow linearly with distance, while the width contracts inversely with distance. Thus we expect a self-contracting self-similar asymptotic solution.

The π-pulse solution has about 50% of the energy in the first lobe (see Fig. 2), and the rest of the pulse energy appears in succeeding lobes. If there is a process that disrupts the Langmuir wave, say Langmuir wavebreaking or collisional damping of the Langmuir wave, then as much as 80% of the pulse energy can be captured in the first lobe, since the regeneration of the pump will be incomplete. However, the effects that disrupt the Langmuir wave should not be so disruptive that the desired amplification process is also disrupted. Hence, the wavebreaking should only be arranged near threshold. Alternatively, the collisional damping length of the Langmuir wave should be no shorter than the width of the first lobe.

The effects that first disrupt the process as described are the modulational and near-forward Raman scattering instabilities of the pulse, once it has grown to amplitudes much greater than the pump amplitudes. However, within a few growth times of these instabilities, the pulse can reach very high levels of unfocused power within the plasma [9]. Table 1 gives examples of the output parameters that can be expected under various parameters for the resonant Raman pulse compression.

TABLE 1. Examples of resonant Raman backscatter pump and output pulse parameters, for different laser wavelength, near the threshold of Langmuir wave breaking.

Laser wavelength (μm)	1/40	1/4	1	10
Pump duration (ps)	1.25	12.5	50	500
Pump intensity (W/cm^2)	1.6×10^{17}	1.6×10^{15}	10^{14}	10^{12}
Pump vector-potential (a_0)	0.006	0.006	0.006	0.006
Laser-to-plasma frequency ratio	12	12	12	12
Concentration of plasma (cm^{-3})	1.1×10^{22}	1.1×10^{20}	7×10^{18}	7×10^{16}
linear e-folding length (cm)	0.00043	0.0043	0.013	0.13
Total amplification length (cm)	0.018	0.18	0.7	7
Output pulse duration (fs)	1	10	40	400
Output pulse fluence (kJ/cm^2)	160	16	4	0.4
Output pulse intensity (W/cm^2)	1.6×10^{20}	1.6×10^{18}	10^{17}	10^{15}

Note in particular that at shorter wavelength, the plasma is even more accomodating in processing high power. By way of comparison, if they can be built at all, material gratings will withstand much less fluence at short wavelength. Hence, if compression is to be done at very short wavelengths, the practical issue is the suitable high-energy low-power pump laser worthy of compression.

Even at 1 μ, however, the fluences and unfocused powers are quite remarkable compared to what might be achieved with mateial gratings prior to the vacuum focus. One important issue is the extent to which this power can be focused on target by retaining the integrity of the seed pulse phase fronts prior to entering the plasma. Another important issue is the ability of th pump to traverse the full plasma length without undergoing premature Raman backscattering off plasma noise, prior to encountering the pulse. This second issue is addressed in the next section.

DETUNED RAMAN BACKSCATTERING

The issue of pump stability was, in fact, recognized in early studies of Raman compressors in gases and other Raman media. It was recognized that not only were instabilities of the amplified seed pulse worrisome, but so were the possibilities for both the forward scattering and the backscattering of the pump (see Fig. 4). In gases, the forward scattering is slightly larger than the backward scattering.

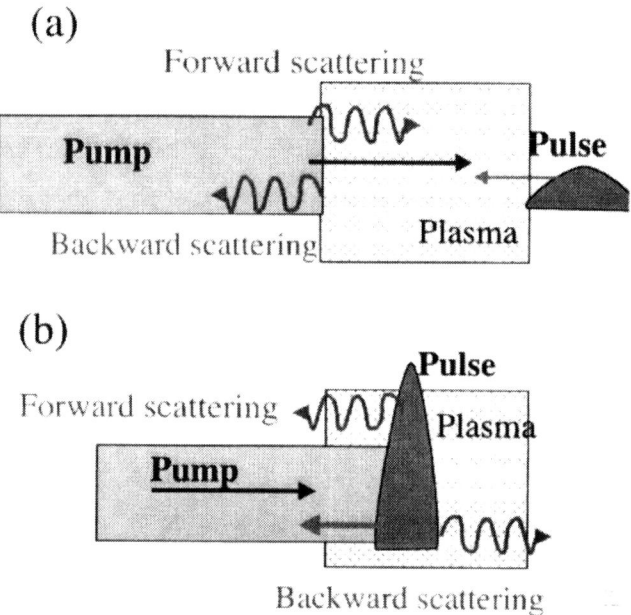

FIGURE 4. Instability of pump and pulse to forward and backward Raman backscattering instabilities. (a) pump is susceptible as it traverses full plasma to meet pulse. (b) pulse is similarly unstable as it reaches amplitude larger than pump.

In Fig. 4a, we schematically depict the forward and backward scattering instabilities associated with the pump in a Raman media, which is taken here to be plasma. Although the pump is of smaller amplitude than the pulse, it has to propagate the full plasma length before reaching the pulse. Hence, instabilities associated with the pump can be serious even at lower amplitude. Note that both the backward scattered wave and the forward scattered wave can draw pump energy for essentially the plasma length.

In Fig. 4b, we schematically depict the forward and backward scattering instabilities associate with the pulse in a Raman media, which is also taken here to be the plasma. The backward scattering of the pulse in gases would be less important than the forward scatering, because the backscattered light quickly goes through the short pulse, whereas the forward scattered light propagates with the short pulse. In plasmas, this issue is balanced by the fact that the backward scattering occurs with a larger growth rate.

In contemplating gas-based Raman compressors, it was also recogized that these issues could be addressed by detuning the resonant interaction. The Raman medium could be made to have a Raman gradient, where the Raman frequency would change with axial position (see Fig. 5). Thus, forward scattered light co-propagating with a pump (or pulse) would be born in resonance with the pump, but as both pump and scattered signal co-propagate some distance, the Raman media no longer accomodates a resonant interaction between the two waves. Moreover, it was also recognized [14], that the pump entering the plasma could be chirped too, so that the front of the pump would be at one frequency, while the back of the pump could be at a higher or lower frequency. Thus, should the front of the pump be backscattered off of some noise, then as the unwanted backscattered energy propagates through the pump, it will encounter pump power with which it is not resonant, which will halt the resonant backscatter. Now, if both the Raman gradient and the pump are graded or chirped, as depicted in Fig. 5b, then the chirps add. So for example, it would be possible to use the Raman gradient to detune the forward scattering effect, and then use a chirped pump to compensate the Raman gradient, so that the backscattering is not detuned at all.

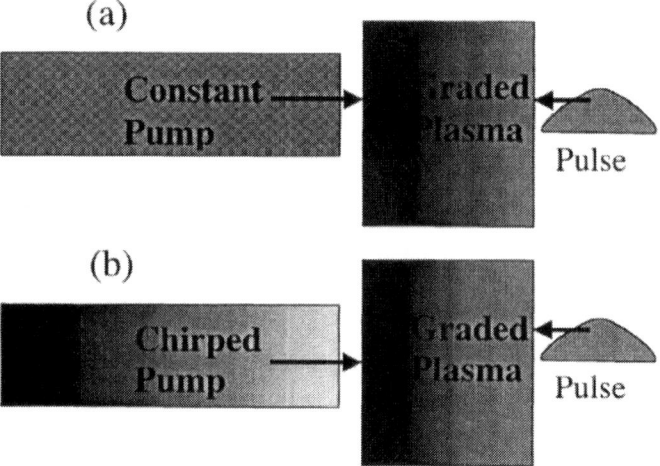

FIGURE 5. Schematic implementation of detuning. (a) Pump is susceptible to Raman instabilities as it traverses full plasma to meet pulse, but graded plasma density produces a plasma frequency gradient so that a forward or backward scattering resonance cannot be satisfied over appreciable distances. (b) pump is chirped to compensate the Raman frequency gradient. As seen by a counterpropagating wave, the detuning caused by density gradient is compensated by encountering a gradient in the pump laser frequency.

In gases, these effects might be usefully used therefore to handle unwanted forward scattering effects, which are the more worrisome effects, while retaining effective backscattering. In plasma, the backscattering is the more serious worry, but it cannot be detuned altogether, because it is also the effect that amplifies the useful pulse. However, quite remarkably, at high power, there is an important nonlinear filtering effect that occurs in plasma [15]. This filtering effect is that the resonance detuning affects differently the amplification of the desired signal through backscattering of the pump and the amplification of noise, also through backscattering of the pump.

On the one hand, the amplification of the desired signal occurs only linearly with distance in the pump-depletion or π-pulse regime, whereas the amplifcation of noise occurs exponentially in time or distance. Hence, one would expect that eventually the plasma length for amplification might be limited by this effect at any noise level. On the other hand, although the pumped pulse grows only linearly with distance, it also contracts as grows, rendering its bandwidth larger. In the pump-depletion regime, as the short pulse becomes even shorter and acquires larger bandwidth, it can tolerate detuning equal to its bandwidth. Thus, it will continue to be amplified, even as a pulse arising from noise, initially growing exponentially with distance, will eventually be saturated due to detuning.

The fact that the Raman interaction in the pump-depletion regime tolerates large detuning effects is quite important. It can be used not only to stabilize the pump, even as the signal amplification is unaffected, but it can also be used to destabilize other unwanted resonant instabilities even as the signal amplification is unaffected (see, for example, Ref. [16]).

OUTSTANDING ISSUES AND CONCLUSIONS

Plasma is the natural medium for processing high power and high fluence lasers. There are, in fact, several backscatter effects, in somewhat different regimes, that may lead to compression of lasers at high power in plasma. Each of the regimes and each of the mechanisms imagined here has its own set of outstanding issues that must be resolved both theoretically and experimentally, before one can have confidence that the favorable compression regimes discussed here can in fact be accessed. Among these issues is the extent to which the pulse can retain its focus as it emerges from the plasma.

Moreover, the practical realization of these effects depends upon the robustness of these effects to deviations from the ideal conditions contemplated. However, until one method is shown to work, there is some comfort to be derived from the fact that a number of rather different mechanisms might be useful in producing compression at high power.

Particular attention was paid to the resonant Raman backscatter effect. It is encouraging that this is an essentially simple effect in the sense that electron motion is quite coherent even as high power pulse compression is achieved. Significantly, since the effect relies on resonance, it is possible to introduce detuning in such a way that the amplification of unwanted noise can be suppressed even as the useful amplification persists.

It should be recognized that while there have been many related experiments, none have realized compression at high power. Recently, backscattered amplification was reported in a Li-F recombining plasma [17], but a π-pulse solution, or even any compression solution, has yet to be demonstrated in plasma.

Yet the compression of lasers in plasma is likely to be a key component to the eventual realization of much higher laser intensities than are presently achieved. As applications emerge particularly for submicron wavelengths, the compression effect in plasmas will become even more central to the development of suitable high-power lasers, since alternative means of pulse compression are not suited to short wavelength.

As a first step to improving present technology, a laser system might employ a plasma component as well as conventional CPA components. A high saturation-fluence amplifier, but possibly with relatively narrow gain bandwidth, might be used to produce a suitable laser pump for use in plasma-based compression.

ACKNOWLEDGMENTS

This work is supported by DOE contract DE-FG030-98DP00210.

REFERENCES

1. Mourou, G. A., Barty, C. P. J., and Perry, M. D., Phys. Today, **51**, 22 (1998).
2. Perry, M. D., Pennington, D., Stuart, B. C., *et al.*, Optics Lett., **24**, 160 (1999).
3. Maier, M., Kaiser, W., and Giordmaine, J. A., Phys. Rev. Lett., **17**, 1275 (1966); Phys. Rev., **177**, 580 (1969).
4. Murray, J. R., Goldhar, J., Eimerl, D., and Szoke, A., IEEE Journal of Quantum Electronics, **QE-15**, 342 (1979).
5. Capjack, C. E., James, C. R., and McMullin, J. N., Journal of Applied Physics, **53**, 4046 (1982).
6. Nishioka, H., Kimura, K., Ueda, K., and Takuma, H., IEEE Journal of Quantum Electronics **29**, 2251 (1993).
7. Takahashi, E., Matsumoto, Y., Matsushima, I., Okuda, I., Owadano, Y., and Kuwahara, K., Fusion Engineering and Design **44**, 133 (1999).
8. Shvets, G., Fisch, N. J., Pukhov, A., and Meyer-ter-Vehn, J., Phys. Rev. Lett., **81**, 4879 (1998).
9. Malkin, V. M., Shvets, G., and Fisch, N. J., Phys. Rev. Lett., **82**, 4448 (1999).
10. Malkin, V. M., and Fisch, N. J., Phys. Plasmas **8**, 4698 (2001).
11. Litvak, A. G., Zh. Eksp. Teor. Fiz. **57**, 629 (1969) [Sov. Phys. JETP **30**, 344 (1970)].
12. Max, C., Arons, J., and Langdon, A. B., Phys. Rev. Lett. **33**, 209 (1974).
13. Bobroff, D. L., and Haus, H. A., J. Appl. Phys., **38**, 390 (1967
14. Caird, J. A., IEEE Journal of Quantum Electronics, **QE-16**, 489 (1980).
15. Malkin, V. M., Shvets, G., and Fisch, N. J., Phys. Rev. Lett., **84**, 1208 (2000).
16. Malkin, V. M., Tsidulko, Y., and Fisch, N. J., Phys. Rev. Lett. **85**, 4068 (2000).
17. Ping, Y., Geltner, I., Fisch, N. J., Shvets, G., and Suckewer, S., Phys. Rev. E: Rapid Comm. **62**, R4532 (2000).

6. APPLICATIONS OF SUPERSTRONG PULSES TO HIGH ENERGY PHYSICS

Superstrong Field Science

T. Tajima* and G. Mourou[†]

*Lawrence Livermore National Lab., Univ. of California, Livermore, CA 94550
[†]Center for Ultrafast Optical Science, University of Michigan, Ann Arbor, MI 48109

Abstract. Over the past fifteen years we have seen a surge in our ability to produce high intensities, five to six orders of magnitude higher than was possible before. At these intensities, particles, electrons and protons, acquire kinetic energy in the mega-electron-volt range through interaction with intense laser fields. This opens a new age for the laser, the age of nonlinear relativistic optics coupling even with nuclear physics. We suggest a path to reach an extremely high-intensity level 10^{26-28} W/cm^2 in the coming decade, much beyond the current and near future intensity regime 10^{23} W/cm^2, taking advantage of the megajoule laser facilities. Such a laser at extreme high intensity could accelerate particles to frontiers of high energy, tera-electron-volt and peta-electron-volt, and would become a tool of fundamental physics encompassing particle physics, gravitational physics, nonlinear field theory, ultrahigh-pressure physics, astrophysics, and cosmology. Such a laser intensity may also be very beneficial to an alternative, more direct approach of fast ignition in laser fusion. We suggest a new possibility to explore this.

INTRODUCTION

Over the past fifteen years, we have seen a revolution in laser intensities [1]. This revolution stemmed from the technique of chirped pulse amplification (CPA), combined with recent progress in short-pulse generation and superior-energy-storage materials like Ti:sapphire, Nd:glass, and Yb:glass. The success of this technique was due to its general concept, which fits small, university-type, tabletop-size systems as well as large existing laser chains built for laser fusion in national laboratories like CEA-Limeil in France, Lawrence Livermore National Lab. (LLNL), Los Alamos National Lab., and Naval Research Lab. in the U.S.; Rutherford in the UK, Max Born Institute in Germany, and the Institute of Laser Engineering in Osaka, Japan. A record peak power of petawatt (PW = 10^{15} W) has been produced at LLNL. CPA lasers have given access to a regime of intensities that was not accessible before, opening up a fundamentally new physical domain [2]. After a rapid increase in the 1960s with the invention of lasers, followed by the demonstration of Q-switching and mode-locking, the power of lasers stagnated due to the inability to amplify ultrashort pulses without causing unwanted nonlinear effects in the optical components. This difficulty was removed with the introduction of the technique of chirped pulse amplification, which took the power of tabletop lasers from the gigawatt to the terawatt — a jump of three to four orders of magnitude. This technique was first used with conventional laser amplifiers and more recently extended to Optical Parametric Chirped Pulse Amplifiers (OPCPA) [3]. A number of laboratories are presently equipped with CPA ultrashort-pulsed terawatt lasers such as Laboratoire Optique Applique in France, University of Lund in Sweden, Max-Planck Institute in

CP611, *Superstrong Fields in Plasmas:* Second Int'l. Conf., edited by M. Lontano et al.
© 2002 American Institute of Physics 0-7354-0057-1/02/$19.00

Garching, Jena University, and the Japan Atomic Energy Research Institute in Kansai. The CPA-enabled short-pulse generation has advanced to the single-cycle regime [4]. The peak power in the 1990s reached 100 TW, demonstrated at JAERI Kansai. More recently, deformable mirrors have been incorporated into CPA, making it possible with low f# parabola to focus the laser power on a 1 μm spot size [5]. Present systems deliver focused intensities in the 10^{20} W/cm^2 range. In the near future, CPA systems will be able to produce intensities of the order of 10^{22} W/cm^2. As indicated in Fig. 1, we will see a leveling off of laser intensity for tabletop-size systems at 10^{23} W/cm^2. This limit [1] is imposed, as we will explain later, by the saturation fluence - energy per unit area — of the amplifying medium and the damage threshold of the optical elements. In a well-conceived CPA system the saturation fluence is of the order but less than the damage threshold. Once this limit is reached, we will have accomplished a leap in intensities of eight orders of magnitude and the only way to increase the focused intensity further will be by increasing the beam size, leaving the following questions to be answered: How could we go higher in intensity? And how much higher?

Because the highest intensities will rely on the largest pump available, we explore if it could be technically feasible to build a large scale CPA (OPCPA) pumped by a megajoule system of the type of the NIF (National Ignition Facility) in the U.S. and the LMJ (Laser Megajoule) in France. Power in the zetta (10^{21}) watt range could be produced, yielding a focused intensity of 10^{28} W/cm^2. These intensities well beyond the currently accessible will open up a new physical regime. The trend in intensity increase is represented in Fig. 1.

En route to "zettawatt"

Although a zettawatt system could be built using Yb:glass, with the advantages of being relatively compact due to the high F_{sat} of this material and being diode pumpable, much development work needs to be accomplished to reach this intensity level with this material. The proposed systems described below have been stimulated by the construction , both in France and in the U.S, of lasers delivering a few megajoules of energy as well as the availability of large telescope technology (10m diameter) and deformable mirrors.

Let us recall that the NIF and LMJ systems will deliver 2 MJ at 350 nm, i.e., the third harmonic of 1060 nm in 3 ns. The energy at 530 nm, for a long pulse of 10-20 ns, could be as high as 5 MJ. This energy could be used to pump a CPA-type system to produce of the order of 1 MJ of energy in 10 fs on a spot size of 1 μm [5]. Such a system will have a power of the order 10^{20} W or 100 exawatts with an intensity of 10^{28} W/cm^2. To produce these phenomenal characteristics, we have two alternatives. The first one would be to use this large pump energy to drive an OPCPA system. This elegant technique [3] has been demonstrated to the joule level at 800 fs level a few years ago by I. Ross *et al.* from Rutherford [6] and is being implemented in the same laboratory to the 10 PW, 30 fs level. It has the advantage that it could use KDP as the nonlinear medium. KDP has been produced in large dimensions at a relatively low cost for the NIF and LMJ programs. The working fluence on the crystal will be 1 J/cm^2, leading to a

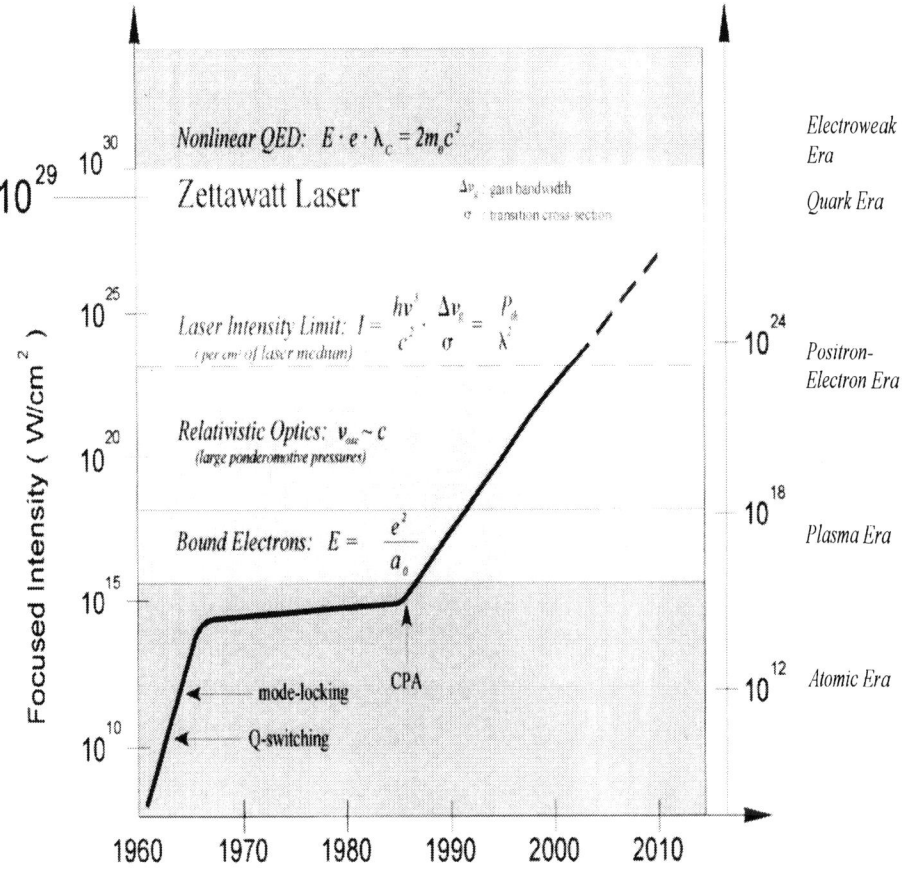

FIG. 1. The leap in laser intensity in time and new frontiers that access.

beam diameter of around $10\,m^2$ to accommodate the whole pump energy. This technique has the potential to produce a 10-fs pulse or shorter, but needs to be demonstrated at least at the joule level. The second approach, though more conservative, will be to drive a CPA Ti:sapphire, a well-established technology. The amplifier will be composed of a large matrix of Ti:sapphire rods. The size of the matrix will be dictated by the saturation fluence $F_{sat} = 1\,J/cm^2$, corresponding to a $10\,m$ diameter beam. The amplifier could be composed of 2500, $20 \times 20\,cm$ pieces, each $2\,cm$ in length. Each piece will have to be segmented in order to avoid the transverse amplified spontaneous emission. Ti:sapphire of $20 \times 20\,cm$ has already been grown [10]. In both schemes the beams can be focused by a large parabola. The dielectric coating used for the parabola will have a damage threshold of $1\,J/cm^2$ (for short pulses) [7, 8], imposing a parabola size of $10\,m$ in diameter. This parabola will have the same diameter as the Keck telescope. The phase front will be interferometrically controlled by an active matrix of deformable mirrors. The grating compressor could be made out of meter-size gratings assembled interferometrically in a matrix geometry. The size of the grating for a megajoule short-pulse system will be also of the order of $10\,m$ in diameter, dictated by a damage threshold of $1\,J/cm^2$ (such a damage threshold has just been reported [9]). Each large grating will be composed of 100 of $1\,m^2$-size gratings.

Note that although large, the number of optical components involved will be small, compared to a MJ system calling for 4500 laser slabs, 800 large KDP frequency converter crystals, and 500 gratings of meter-size. An alternative method using a plasma has been suggested [10], but will lead to the same output energy limited by the pump. In addition the beam must go through a plasma, which is highly undesirable.

A Stepping stone: The exawatt laser system

If a zettawatt laser, although feasible, could seem too grandiose at this time, an exawatt system on the other hand, which would produce $10\,kJ$ in $10\,fs$, i.e., $10^{25}\,W/cm^2$, could be readily constructed. Only one percent or $30\,kJ$ of the NIF/LMJ energy would be necessary. The beam size will be of the order of one meter in diameter. The amplifying method will be composed of a matrix of 25 Ti:sapphire $20 \times 20\,cm^2$ crystals [10] and two gratings of meter-size. The segmented telescope will have a one-meter aperture. The wave front will be corrected by a large deformable mirror.

In the following we explore a few examples of applications of such intense lasers that may enable new ways to investigate fundamental physics.

HIGH-FIELD SCIENCE IN EXTREME FIELDS

The major signpost of contemporary high-field science is the entry into the relativistic regime, which is characterized by the quivering momentum of electrons in the laser fields reaching the speed of light times the electron rest mass. This field is reached when the laser intensity is on the order of $10^{18}\,W/cm^2$ for a typical optical frequency of the laser we mentioned above. When the laser field is much less than this, free electrons behave harmonically to the field's optical oscillations. Although these oscillations of electrons are still important enough to couple with various collective motions of free electrons, there is small orbital nonlinearity associated with the quivering motion in this regime. For bound electrons in atoms or molecules the quivering motion in the laser fields is once again, by and large, a perturbation on their orbital motion. This perturbation is enough to cause various interesting phenomena in the non-relativistic regime. There is such a wealth of phenomena known in this regime that it is hard to list them briefly, but they include the multi-photon process of ionization, that of transitions, Raman and Brillouin scatterings, and various optical nonlinearities [11] arising from the material's response. In the relativistic regime with the intensity above $10^{18}\,\text{W/cm}^2$, in addition to the above phenomena, there emerge new classes of effects largely arising from the relativistic nonlinearities of electrons in the high field.

The field intensity in this regime means that the electron momentum in the light is typically $eE_0\omega_0$, which becomes of the order of and exceeds m_0c, where ω_0 is the laser angular frequency, E_0 the laser electric field, and m_0 the electron rest mass. When the electron momentum exceeds m_0c, the electronic orbit ceases to be harmonic and linear. It becomes a figure-8 motion, including higher-harmonic components. The photon pressure is exerted individually on an electron by the electron-photon collision through the Thomson cross-section

$$\sigma_T = \frac{8\pi}{3} \cdot \frac{e^4}{m_0^2 c^4} \approx 7 \cdot 10^{-25}\,\text{cm}^2. \tag{1}$$

When the flux of laser at the intensity entering the relativistic regime is shone on an electron (with Lorentz factor γ), this causes a force on it

$$F = \frac{\sigma_T}{\gamma}\frac{E_0^2}{4\pi}. \tag{2}$$

When the laser intensity is $I = 10^{26}\,\text{W/cm}^2$, the force acting on an electron is $10^{-2} \sim 10^{-1}\,\text{erg/cm}$. Or the acceleration acting on an electron (originally) at rest is

$$a_e = \frac{f}{m_0} \sim 10^{25-26}\,\text{cm/s}^2, \tag{3}$$

where m_0 is the rest mass. [In contrast to this, the Schwinger acceleration is

$$a_S = 2 \cdot 10^{31}\,\text{cm/s}^2, \tag{4}$$

at which an electron gains energy by m_0c^2 over the Compton length $\lambda_C = \hbar/m_0c$, and the pair creation becomes prevalent.] A large flux of photons bombarding an electron causes such high acceleration through collisions between photons and electrons (Compton collisions). Such acceleration may be called the Eddington acceleration, as Eddington introduced the stellar luminosity at which the gravitational pull is balanced by this photon collisional acceleration (the Eddington luminosity).

The Eddington acceleration may be likened to the acceleration of water molecules near the surface of water in a lake when there is a breeze passing over the water surface. The collisional viscosity created by collisions between water molecules and wind molecules gives rise to a water flow. However, we also observe that when the breeze gets stronger or becomes a gale, the water surface is no longer smooth and acquires ripples or waves. Such ripples facilitate an increase in the effective viscosity of water for the wind, so that the wind momentum is anomalously effectively transported to water molecules with a much faster rate through the Kelvin- Helmholtz instability at the interface of the two fluids. What happens in the case of strong photon flux ("wind") in a plasma? Just like the strong wind on the water surface, the strong photon flux is capable of creating plasma waves, which in turn causes enhanced viscosity and thus anomalous momentum transport from photons to electrons of the plasma. This latter process is through a collective interaction. The plasma wakefield excitation [12] is typical of this, in which plasma waves generated in this process are accentuated, and new processes of collective interaction emerge. In this ultraintense regime, electrons may be accelerated not only through the electrostatic field that is set up by the ponderomotive force of the laser, but also directly by the ponderomotive force itself to very high energies. If we apply this laser at the resonance absorption at high densities near the compressed laser fusion target, on the other hand, much of the laser energy may be converted into relatively low-energy copious electrons, which could constitute a new alternative to the fast ignition fusion. The acceleration of heavier particles (protons and other nuclei) to relativistic energies will become possible, too. Either by direct baryon acceleration by this, or other process, (it will take a variety of experimental realizations such as a target irradiation, cluster irradiation, converging imploding shells, etc.), we will access the nuclear regime of matter reminiscent of the early epoch of the Big Bang .

The production of extremely high energy or copious gamma rays will happened. The extreme high photon pressure (which already exceeds Gbar in the presently available intense lasers) may be finally directly utilized to directly compress matter in this ultrarelativistic regime, because baryons, too, become relativistic. If so, unprecedented densities of matter may be created. The combination of extreme intensity lasers and high-energy particle beams that can be created by the conventional high-energy physics accelerator will further multiply our ability to expand our frontier horizon. This will be the merging point of high-energy physics and high-field science. In the following, we list several examples of exciting new frontiers of fundamental physics that may be explored by this regime of intensities. In many of the applications we discuss, in this high-intensity regime the interaction length between the laser and matter is expected to be extended beyond the Rayleigh length, as the relativistic mass effect of electrons sets the self-focusing threshold at $10^{10}(\omega/\omega_p)^2$ W [13].

Particle acceleration

The pulse or self-modulation of a photon wavepacket with sufficient intensity induces a longitudinal electric field (in the x-direction, as part of plasma oscillations) as

$$E_x \sim \sqrt{\frac{n_e}{n_{18}}} \, a_0^2 \qquad (\text{GeV/cm}), \qquad (a_0 \leq 1), \qquad (5)$$

428

where $n_{18} = 10^{18}/cm^3$ and n_e is the electron density, $a_0 = eE_0/m_0\omega_0c$ the normalized vector potential of the laser, which is sometimes called the quivering velocity normalized to c (or quivering momentum normalized to m_0c). The energy gain over the interaction length l_x is

$$\Delta\varepsilon \sim \sqrt{\frac{n_e}{n_{18}}} a_0^2 l_x \,(\text{GeV}). \tag{6}$$

When a_0 exceeds unity (ultrarelativistic; $I \geq 10^{18}\,\text{W/cm}^2$), the pressure of the photon wavepacket becomes so large that nearly all electrons are evacuated from the laser packet [14]. The photons plow through the plasma with electrons piling up in front of the pulse ("snow plow" of electrons), yielding the snow plow acceleration [15] with momentum gain of electrons

$$p_x = \frac{1}{4}\frac{E_0^2}{n_ec} = \frac{E_c^2}{n_em_0c^2}a_0^2, \tag{7}$$

where $E_c = m_0\omega_0c/e$. Here the laser pulse is assumed to have fully interacted with the plasma. If the pulse quickly diffracts before the full interaction, say over the Rayleigh length, p_x is simply proportional to E_0.

Now let us imagine, for the moment, that the laser has only half the period (unipolar) [16] or subcyclic [17]. In this case the energy (or momentum) gain is

$$\Delta\varepsilon \sim m_0c^2a_0^2. \tag{8}$$

Since the Lawson-Woodward theorem [18] prohibits any overall acceleration for fully oscillatory (i.e., usual) electromagnetic waves in vacuum in infinite space, the above energy gain is compensated for by decelerating phase. There are, however, many instances that break the theorem requirements. For example, it may be possible in this extreme relativistic regime that electrons are accelerated to very high energy, immediately reaching the speed of light and becoming in phase with the photon over a sufficiently long distance, so that by the time they become dephased, the EM wave may decay away for some reason, such as by radiative decay or pump depletion to the acceleration. If this happens, the above energy gain (or a portion of it) may be preserved. In extreme high-intensity regimes such effects will become significant. If the laser spreads over the Rayleigh length, the energy gain is proportional to a_0. The transverse momentum gain (p_y) is always proportional to a_0. In a broad general way we can say that the interaction between the laser and electrons becomes more coherent, as the laser intensity increases because the laser and electrons move more coherently over greater distance. This is the signature of the acceleration by the ponderomotive potential of photon fields in ultrarelativistic intensity regimes.

Thus it is possible to see electrons at energies of up to $\sim 100\,\text{TeV}$ at the laser intensity of $10^{26}\,\text{W/cm}^2$ and even up to $\sim 10\,\text{PeV}$ at $10^{28}\,\text{W/cm}^2$. The accelerating gradient is $200\,\text{TeV/cm}$ and $2\,\text{PeV/cm}$, respectively. Note that such energies ($100\,\text{TeV}$ and $10\,\text{PeV}$) if collided, correspond to $10^{19}\,\text{eV}$ and $10^{23}\,\text{eV}$ for fixed target experiments. These energies rival or exceed those of the highest energy cosmic rays, which are observed up to $3 \times 10^{20}\,\text{eV}$. Of course, it is not easy to attain correspondingly high luminosity for collisions at such high energies. Even though the exploration of particle physics at pb may not be within reach, we may use such particles in pursuit of (other) fundamental physics at the energy frontier. We might recall that Anderson first discovered mesons in cosmic rays, followed by more detailed studies of those particles in cyclotrons and other accelerators. Perhaps the present way-out parameters in the energy frontier may herald some new phenomena. One such example may be the test of Lorentz invariance [19] in extreme high energies. For such a test, unlike the detection of new particles with pb cross-section, the luminosity requirement may be much relaxed.

When we irradiate an extremely relativistic laser pulse ($I \sim 10^{26}\,\mathrm{W/cm^2}$) on a thin film ($\gtrsim 1\,\mu\mathrm{m}$) of a metal at a tight spot of $(1\,\mu\mathrm{m})^2$ followed by a microhole in a metallic slab over more than one cm, we hypothesize that the laser pulse picks up metallic electrons from the film and continues to propagate through the microhole, as it remains focused. If this proves to be the case, the amount of electrons that are to be accelerated by this pulse is in the ballpark of

$$N_e \sim n_e A l_b \sim 3 \cdot 10^{10}, \tag{9}$$

where we assumed $n_e \sim 10^{24}\,\mathrm{cm^{-3}}$, area $A = (1\,\mu\mathrm{m})^2$, the bunch length $l_b \sim 3 \cdot 10^{-6}\,\mathrm{cm}$. In such a large pickup, the pump depletion due to the energy transfer to electrons is significant. In fact, it may play a fundamental role in this acceleration to turn the electromagnetic energy into particle kinetic energy without returning to the decelerating phase, providing one way to break the Lawson-Woodward theorem's constraint, as we mentioned earlier. If we take 1/10 of the above electrons to gain 10 TeV and if we focus electrons (and positrons) down to $10^{-6}\,\mathrm{cm}$ (or $10^{-7}\,\mathrm{cm}$) at focus (the collision point), the luminosity of the colliding events is of the order of

$$L = 10^{31} f\,/\mathrm{cm^2/s}, (\mathrm{or}\ 10^{33} f), \tag{10}$$

where f is the collision repetition rate. Nakajima has considered a similar but more daring luminosity scenario [20].

Although we have no room to enumerate the x-ray generation from those high fields and subsequent high energy electrons, the quantity and quality of emitted x-rays are extraordinary in this regime. This will be discussed again in the future.

Fast ignition fusion

One special case of laser energy conversion into electrons is through the resonance absorption at the critical density. The concept of fast ignition in laser-driven inertial fusion [21] calls for laser beam of $\sim 10\,\mathrm{psec}$ duration at the intensity exceeding $10^{20}\,\mathrm{W/cm^2}$ to be absorbed at the critical density ($\sim 10^{21} - 10^{22}/\mathrm{cm^3}$), creating a beam of electrons in the several MeV range. The idea is to separate the roles of laser into two functions: one to compress the fuel with least amount of entropy increase so that the fusion fuel is compressed to a highest density with least amount of laser energy, and the other is to heat the fuel to the thermonuclear ignition temperature ($\sim 10\,\mathrm{KeV}$) when the main compression is achieved. The latter step may be carried out according to Tabak et al. by the appropriate range of energetic (several MeV) electrons that are transported from the crust of the target ($< 10^{22}/\mathrm{cm^3}$) to the surface of the fuel (at $\sim 10^{26}/\mathrm{cm^3}$) at the pulse duration of ~ 10 ps. In order for electrons to trigger the fusion ignition, the condition

$$\rho r \lesssim 0.5 \qquad \left(\frac{g}{cm^2}\right) \qquad (11)$$

has to be fulfilled [21]. Here ρ is the density of the compressed fuel and r the electron range and thus approximately the size of the fuel at compression. The laser pulse length is therefore given as

$$\tau = 40 \left(\frac{100 \; gcm^{-3}}{\rho}\right) ps \qquad (12)$$

which yields the pulse length between 10-20 ps for the compressed fuel density of 200-300 g/cm^3. The laser energy required for this drive is estimated from simulation [22] to be

$$E_{laser} = 80 \left(\frac{100 \; gcm^{-3}}{\rho}\right)^{1.8} kJ \qquad (13)$$

yielding the required laser energy of about 50 kJ, while 10-20 kJ of electron energy needs to be delivered to the spot. Although this idea is potentially capable of reducing the necessary laser energy by almost an order of magnitude or increasing the fusion gain by an order of magnitude at the same laser energy, there remains a considerable uncertainty in the efficiency of energy conversion from short-pulse (10ps) laser to electron beam to the compressed fuel and in the stability and reliability of electron beam. For example, laser and electron beam have to propagate through the over dense plasma, through which filamentation and kinking instabilities are found to arise. In order to cope with this problem, the ignitor laser is further split into the hole boring one and energy deliverer.

We suggest that an alternative method of fast ignition by much shorter-pulse laser may be possible. In this, although the total amount of energy necessary to be delivered is unchanged ($50 \sim 100$ kJ), we shorten the pulse length to 10fs so that the local laser intensity reaches of the order of 10^{25} W/cm^2. Since the resonance frequency reduces inversely proportional to $\sqrt{n_e}$, the resonance density becomes in the order of 10^{25}/cm^3, a very close proximity of the fully compressed fuel. This way we may be avoiding the difficult and long energy transport of electron beam from the density region of $< 10^{22}$/cm^2 to 10^{26}/cm^3. With the recent success of improved target design with a conic aperture for fast ignitor laser beam access [23], our ultrafast approach may be further bolstered as a direct energy delivery vehicle. It remains to be seen, however, how much fraction of the laser energy is consumed due to the pump depletion [24] while it interacts with the surrounding plasma. It also needs to be investigated what kind of electron energy spectrum is generated in the ultra-intense laser beam. The mission of the electron energy conversion in fast ignition is orthogonal to the previous Subsec. 3A of super-high energy electron generation. The production of a small fraction of extremely high-energy electrons is tolerated, as long as the majority of energy is in several MeV electrons.

Baryon acceleration

Many thought it difficult to accelerate protons and heavier particles by light, as massless light propagates at the speed of light, while protons are massive and nonrelativistic — until last year, when the Petawatt Laser experiment [25] and other experiments [26, 27, 28] showed that protons have been accelerated much beyond a megaelectron-volt. The observed transverse emittance is about 0.5 mm mrad, while the longitudinal one is about MeV-psec [29]. These early experiments already rival or even surpass those of the conventional ion sources in some of crucial parameters. The main mechanism of laser proton acceleration in the above experiments is due to the space charge set up by energetic electrons that are driven forward away from the back surface of the target slab. The energy of protons is thus dependent on that of electrons. Bulanov and others showed [31] in simulation that at a laser intensity of $I = 10^{23}$ W/cm^2, protons are accelerated beyond a giga-electron-volt. If this process of proton acceleration scales with the intensity (as the electron energy does), we may be able to see 100 GeV protons and 10 TeV at $I = 10^{26}$ and 10^{28} W/cm^2, respectively. However, it may also be possible that this process is now due directly to the photon pressure beyond the intensity regime of $I = 10^{24}$ W/cm^2. The energy expected through this mechanism is about the same as that through the space charge mechanism.

It is not clear how much energy will be in protons. In the Petawatt Laser experiment, about 10% of laser energy (300 J) was converted into proton energy — 30 J (beyond 1 MeV) [25]. If we take this conversion efficiency in the extreme relativistic laser intensity, then more than 1 kJ of proton energy is expected for the case of intensity 10^{26} W/cm^2. If we further take a flat energy spectrum, approximately 10^{11} protons are accelerated beyond 10 GeV in this intensity regime. If we can generate the solitary accelerating structure, such energy may exceed 100 GeV. If we can converge these in a colliding pair of beams at focus, the luminosity of colliding hadrons is $\mathcal{L} = 10^{34} f$, if we focus on 10 nm. The expected number of nuclear events is on the order of 10^9 per shot.

Such an intense, relativistic, compact proton source has a number of fascinating applications. It may be applicable to ion radiography, fast ignition of fusion, etc., among many others. An additional application is pion (or muon) and neutrino beam generation. With sufficiently relativistic proton energies the emittance of created pions can be sufficiently small. If so, they may be promptly accelerated pions to sufficiently high energies before the space charge effect expands the beam emittance and before they die out. In this the prompt acceleration and its compactness are important, both of which are the forte of the laser accelerations in an application. Protons beyond a certain energy (several hundred MeV) in matter induce through the nuclear strong interaction the creation of pions, which in turn decay into muons and neutrinos in a matter of 20 ns, if nonrelativistic, propagating mere 6 m at most: $\pi \rightarrow \mu + \nu$. With the conventional accelerating gradient of, say, 20MeV/m over this distance of 6m, we can increase the pion energy by 120MeV, which will increase the lifetime of pions but not by an order of magnitude. On the other hand, the laser acceleration with its far-greater gradient would increase the energy and lifetime of pions far more than this (and also reduce the emission cone angle). This would contribute to a further smaller emittance, which in turn contributes to higher energy, lower emittance muons and neutrinos.

Nonlinear QED and horizon physics

At the intensity 10^{28} W/cm^2, the electric field is only an order of magnitude less than the Schwinger field as discussed below. At this field, fluctuations in vacuum are polarized by laser to yield copious pairs of real electron and positron. In collider physics a similar phenomenon happens when the so-called Υ parameter reaches unity. In reality, even below the Schwinger field, the exponential tail of these fluctuations begins to cause copious pair productions.

Though pair production has been demonstrated at SLAC (E144-experiment) [32] by the interaction of γ-ray with an intense laser at intensities in the range of 10^{18} W/cm^2, the direct production of pairs by a high-intensity laser from vacuum remains elusive. The rule of thumb for threshold of pair production derives from the simple argument that it is the field necessary for a virtual electron to gain an energy $2m_0c^2$ during its lifetime δt, imposed by the Heisenberg uncertainty principle $\delta t = \hbar/m_0c^2$, the energy gain length, $c\delta t$ is the Compton length λ_c. Hence, the breakdown field E_S, the Schwinger field, is $E_S = m_0c^2/e\lambda_c$ where $\lambda_c = 0.386$ pm $E_S = 2 \cdot 10^{16}$ V/cm. (The laser field E_S is related to the laser intensity I_ℓ by $E_S^2 = Z_0 I_S$ where Z_0 is the vacuum impedance. For $Z_0 = 377\,\Omega$, we find a value of $I_S = 10^{30}$ W/cm^2).

This approach gives an estimate for the threshold for pair creation and do not provide the number of pairs that could be created for a given intensity. The probability of spontaneous production of pair creation per unit time per unit volume by Schwinger [33] is

$$w = \frac{1}{\pi^2} \frac{\alpha}{\delta t} \frac{1}{\lambda_c^3} \left(\frac{E}{E_S}\right)^2 \sum_{n=1}^{\infty} \frac{1}{n^2} \exp\left(-n\pi \frac{E_S}{E}\right), \tag{14}$$

where α is the fine structure constant. The number of pairs N for a given laser field is

$$N = V\tau_p w \tag{15}$$

where V is the focal volume. For $V = 10^{-12}$ cm^3 and $\tau_p = 10$ fs, we find the generation of 10^{24} pairs at the Schwinger intensity I_S. It is interesting to note that the intensity to create a single pair is still the gargantuan intensity of 10^{27} W/cm^2. However, as indicated earlier (in Subsec. 3E), the presence of matter such as nuclei or electrons reduces this field by a considerable amount.

The interaction of intense laser with high-energy electrons will enhance some of the parameters even further. For example, counterstreaming electron beam and laser can produce copious polarized (high-quality) positrons. This can serve as a polarized positron source for one thing. In an extreme field regime, on the other hand, the laser field is enhanced by the Lorentz factor of the electron beam, so that the effective field from the electron frame may far exceed the Schwinger value. If this field is exceeded by much more than these orders of magnitude, direct production of other particles such as muons out of "vacuum" may be observed.

In addition to the test of nonlinear fields, we are able to explore what may be called 'horizon physics.' According to Einstein's equivalence principle, a particle that is accelerated feels gravity in the opposite direction of the acceleration. The acceleration due to the electric field of the laser at this intensity is huge: $a_e \sim 10^{30}$ and 10^{31} cm/s^2, at

$I = 10^{26}$ and 10^{28} W/cm^2, respectively. An observer at rest (or in an inertial frame of reference) sees the horizon at infinity if there is no gravitation. On the other hand, an observer near a black hole sees the horizon at a finite distance where the gravitation diverges. Equivalently, an observer who is being accelerated (feeling immense equivalent gravity) now also sees the horizon at a finite distance. Any particle ("observer" — a wave function) that has a finite extent has one side of its wave function leaking out of the horizon. The Unruh radiation is emitted when this happens [34]. Unruh radiation is a sister to the Hawking radiation [35]. The Unruh temperature

$$kT_U = \frac{\hbar a_e}{2\pi c} \tag{16}$$

which is about 10^4 eV and 10^5 eV, for $I = 10^{26}$ and 10^{28} W/cm^2, respectively. The radiative power of Unruh radiation increases in proportion to a_0^3 (or $I^{3/2}$), in contrast to the Larmor radiation power of a_0^2 [36]

$$P_U = \frac{12}{\pi} \frac{r_e \hbar}{c} a_0^3 \omega_0^2, \tag{17}$$

where r_e is the electron classical radius. Since in the extreme relativistic regime the radiation is dominated by Larmor radiation, the Unruh signal has to compete with the Larmor with the ratio of powers

$$\frac{P_U}{P_L} = \frac{\hbar \omega_0}{m_0 c^2} a_0. \tag{18}$$

At this regime, the Unruh is down only by a few orders of magnitude, and it has been suggested to circumvent the noise to observe the Unruh signal by exploiting polarization, etc. [36]. This P_U yields 10^4 eV/s and 10^7 eV/s, for $I = 10^{26}$ and 10^{28} W/cm^2, respectively.

The shrinkage of the distance to the horizon by the violent acceleration allows us to probe other aspects of gravitational physics. For example, Arkani-Hamed $et~al.$ [37] suggested that extra dimensions of the quantum gravity may have manifestations in a relatively low extra dimension (n) in this four-dimensional world. The distance over which this may be manifested is

$$r_n \sim 10^{30/n - 17} ~ \text{cm.} \tag{19}$$

The distance to the horizon created by the intense laser acceleration of an electron

$$d = \frac{c^2}{a_e} = \frac{\lambda}{2\pi} \frac{1}{a_0}, \tag{20}$$

could exceed the above distance Eq. (19) if n is less than or equal to 4.

CONCLUSIONS

Zetta- and exawatt-lasers will allow us a glimpse into some of the most energetic and enigmatic phenomena of astrophysics such as GRB's, and their associated effects on EHECR, the final energy frontier in the Universe ('Extreme Universe' in the highest energy and from the cosmological distance). We marvel at such fascinating possibilities of exploring fundamental physics that we have just barely scratched the surface so far on this subject. It is also noted that this laser could bring many frontiers of contemporary physics, i.e. particle physics, nuclear physics, gravitational physics, nonlinear field theory, ultrahigh pressure physics, relativistic plasma and atomic physics, fusion science, astrophysics, and cosmology together. Because of such significant potential scientific impacts, though it amounts to nontrivial efforts and developments, it seems worthy of further and more serious consideration of this extreme high-field science. We believe that the contemporary scientific development is timely.

ACKNOWLEDGMENTS

We appreciated the comment by Profs. N. Fisch, M. Key, and P. Mulser on fast ignition. This work was supported in part by the U.S. Dept. of Energy Contracts DE-FG03-96ER-54346, W-7405-Eng. 48, and in part by the National Science

REFERENCES

1. Strickland, D. , and Mourou, G. A., *Opt. Comm.*, **56**, 219 (1985).
2. Mourou, G. A., Barty, C. P. J., and Perry, M. D., *Physics Today*, January, 1998.
3. Dubeis, A., *et al.*, *Opt. Comm.*, **88**, 437 (1992).
4. Morgner, U., Kätner, F. X., Cho, S. H., Chen, Y., Haus, A. H., Fujimoto, J. G., and Ippen, E. P., *Opt. Lett.*, **24**, 411 (1999).
5. Druon, F., Cheriaux, G., Faure, J., Nees, J., Nantel, M., Maksimchuk, A., and Mourou, G. A., *Opt. Lett.*, **23**, 1043-1045 (1998); Albert, O., Wang, H., Liu, D., Chang, Z., and Mourou, G. A., *Opt. Lett.*, **125**, 1125 (2000).
6. Ross, I.N., *et al..*, *Opt. Comm.*, **144**, 125 (1997).
7. Du, D., Liu, X., Korn, G., Squier, J., and Mourou, G. A., *Appl. Phys. Lett.*, **64**, 3071 (1994).
8. Stuart, B., Feit, M., Rubenckik, A., Shore, B., and Perry, M., *Phys. Rev. Lett.*, **74**, 2248 (1995).
9. Migus, A., Private Communications.
10. M. Felt, Private Communications.
11. Bloembergen, N., *Nonlinear Optics*, Addison-Wesley, Reading, 1965.
12. Tajima, T., and Dawson, J. M., *Phys. Rev. Lett.*, **43**, 267 (1979).
13. Sprangle, P., Tang, C. M., and Esarey, E., *IEEE Trans. Plasma Sci.*, **15**, 145 (1987); Barnes, D. C., Kurki-Suonio, T., and Tajima, T., *ibid.* **15**, 154 (1987).
14. Ashour-Abdalla, M., LeBoeuf, J. N.,Tajima, T. , Dawson, J. M., and Kennel,C. F., *Phys. Rev. A* , **23**, 1906 (1981).
15. Tajima, T., *Laser Part. Beams*, **3**, 351 (1985).
16. Scheid, W.and Hora, H., *Laser Part. Beams*, **7**, 315 (1989).
17. Rau, B., Tajima,T., and Hojo, H., *Phys. Rev. Lett.*, **78**, 3310 (1997).
18. Lawson, J.D., *IEEE Trans. Nucl. Sci.*, **NS-26**, 4217 (1979); Woodward, P.M., *J. IEE* , **93**, Part III A, 1554 (1947).
19. Sato, H., and Tati, T., *Prog. Theor. Phys.*, **47**, 1788 (1972).

20. Nakajima, K., this Proceedings, (2001).
21. Tabak, M., *et al.*, *Phys. Plasmas*, **1**, 1626 (1994).
22. Atzeni, Jpn. J. Appl. Phys. (1995).
23. Kodama, R., *et al.*, *Nature*, **412**, 798 (2001).
24. Horton, W., and Tajima, T., *Phys. Rev. A*, **34**, 4110 (1986).
25. Key, M. *et al.*, in *First Int'l. Conf. Inertial Fusion Sci. Appl.*, Bordeaux, France 1999.
26. Maksimchuk, A. *et al.*, *Phys. Rev. Lett.*, **84**, 4108 (2000).
27. Clark, E. L., *et al.*, *Phys. Rev. Lett.*, **85**, 1654 (2000).
28. Snavely, R.A., *et al.*, *Phys. Rev. Lett.*, (2000).
29. Roth, M., *et al.*, this Proceedings.
30. Esirkepov, T. Zh., Liseikina, T.V., Califano, F., Naumova, N. M., Vshirkov, V. A., Pegoraro, F., Bulanov, S. V., *JETP Lett.*, **70**, 82 (1999); also Pukhov, A. *et al.*, in Max-Planck, Y. Ueshima *et al.*, in JAERI, S. Wilks *et al.*, in LLNL, among other groups, have shown similar results.
31. Koga, J. K. *et al.*, this Proceedings.
32. Bula, C., *et al.*, *Phys. Rev. Lett.*, **76**, 3116 (1996).
33. Schwinger, J., *Phys. Rev.*, **82**, 664 (1951).
34. Unruh, W., *Phys. Rev. D.*, **14**, 870 (1976).
35. Hawking, S. W., *Nature*, London, **248**, 30 (1974).
36. Chen, P.S., and Tajima, T., *Phys. Rev. Lett.*, **83**, 256 (1999).
37. Arkani-Hamed, N., Dimopoulos, S., and Dvali, G., *Phys. Rev. Lett.*, **84**, 586 (2000).

Interaction of intense laser pulses with particle beams

D. Habs[*], Georg Pretzler[†], Ulrich Schramm[*] and Tobias Schätz[*]

[*]LMU München, 85748 Garching, Germany
[†]LMU München, 85748 Garching, Germany and MPQ Garching, 85748 Garching, Germany

Abstract. We treat some aspects of the interaction of intense laser pulses with ion and electron beams. First, the recently demonstrated crystalline ion beams are discussed concerning their possibility of creating coherent radiation up to the MeV regime. Second, the possibilities of pulsed lasers to guide, focus, and bunch state-of-the art electron beams.

CRYSTALLINE CHARGED PARTICLE BEAMS AND LASERS

Recently crystalline ion beams have been realized in a table top storage ring PALLAS by laser cooling [Schätz et al.(2001), Schramm et al.(2001)]. It appears possible to reach crystalline beams also in high energy storage rings by introducing special transverse laser cooling schemes [Schramm(2001)]. Crystalline beams do not only represent the most brilliant beams, but many new fascinating properties occur after the phase transition. E.g., an extremely small emittance growth is observed [Schätz et al.(2001)]. Here we want to discuss the interaction of crystalline ion beams with laser light.

With the proper Doppler tuning, counter-propagating laser light of energy $\hbar\omega$ excites an internal resonant transition of the ions at the much larger energy $2\gamma\hbar\omega$, where γ is the Lorentz Factor of the ion beam. When the photons are reemitted, they are boosted by the same factor again up to an energy of $4\gamma^2\hbar\omega$ in the laboratory frame. In a crystalline beam, this emission can occur Mössbauer-like, transferring the recoil momentum to the total crystal, resulting in a very energy-sharp emission. If we excite the ions of the beam by a laser pulse, we obtain inversion and amplified spontaneous emission may occur. This emission is enhanced because all ions emit into a narrow opening core with angle $2/\gamma$ and because all ions in a beam crystal are very accurately aligned like pearls on a string in the direction of emission. Although the number of emitting ions is limited to about 10^{11} rather intense coherent light with photon energies up to the MeV range may be obtained. In this transformation of laser light the increased energy is obtained from the accelerating cavities of the ion beam.

In the more long term future it seems possible to obtain also crystalline electron beams. One possible way would be to invert present electron cooling schemes and to produce crystalline electron ion beams by sympathetic cooling with crystalline ion beams. In this way electron beams with orders of magnitude improved normalized emittance compared to present day values of $0,5\pi\,\mathrm{mm}\,\mathrm{mrad}$ could be reached. Crystalline electrons have the fascinating property that the spontaneous emission of synchrotron radiation is very strongly suppressed, due to the destructive interference of emission between

CP611, *Superstrong Fields in Plasmas:* Second Int'l. Conf., edited by M. Lontano et al.

different electrons. [Jackson(1975), Primack and Blümel(1999)].

In the following we want to discuss the possibilities to transport and focus electron beams by donut-shaped laser beams. While the study will show that electron beams with present day electron emittances require rather large laser intensities, the much reduces emittances of crystalline electron beams open up interesting perspectives with moderate laser systems.

PREPARATION OF AN ELECTRON-FOCUSING LASER FOCUS

When a gaussian laser beam (wavelength λ_0, propagation direction x, linearly polarized along z such that the E-field amplitude is given by $E_z(r) = E_1 \exp\left(-(r/w_1)^2\right)$, with $r = \sqrt{y^2 + z^2}$) is focused by a perfect lens of focal length f, the field amplitude distribution in the focus is given by

$$E_z(r) = E_0 \exp\left(-(r/w_0)^2\right) \qquad \text{with} \quad E_0 = E_1 \frac{w_1}{w_0}, \quad w_0 = \frac{f\lambda_0}{w_1\pi}. \tag{1}$$

Apart from this transverse field, a longitudinal field is also present in the focus which is necessary for the Maxwell equation $\nabla \vec{E} = 0$ to be fulfilled. In general, this leads to

$$E_x(x,y,z) = \frac{i}{k}\nabla_\perp E_\perp(x,y,z) . \tag{2}$$

For the gaussian above, we get a field amplitude of

$$E_x(y,z) = E_0 \frac{\lambda_0 z}{\pi w_0^2} \exp\left(-(r/w_0)^2\right) \tag{3}$$

in the focus. When more complicated field amplitude and phase distributions are present at the focusing lens, the field at any point $\vec{r}_f$ close to the focus can always be recovered by (numerically) solving the Huygens-Fresnel diffraction integral

$$\vec{E}(\vec{r}_f) = \frac{1}{\lambda_0} \oint_W \vec{E}_{amp}(\vec{r}_w) \exp\left(i(kR(\vec{r}_w,\vec{r}_f) + \varphi_0(\vec{r}_w))\right) \frac{\exp\left(-ikR(\vec{r}_w,\vec{r}_f)\right)}{R(\vec{r}_w,\vec{r}_f)} \, dW \tag{4}$$

over the field distribution in a plane W close to the focusing lens, with $R(\vec{r}_w,\vec{r}_f)$ the geometrical distance between the points $\vec{r}_w$ and $\vec{r}_f$ on the lens and in the focus, respectively, $\vec{E}_{amp}(\vec{r}_w)$ the amplitude distribution and $\varphi_0(\vec{r}_w)$ the phase front deviation from a perfect sphere. The integral (4) must be solved for each vector component separately. Assuming spherical wave fronts and field vectors parallel to these phase fronts and setting the coordinate origin into the geometric focus, z-polarized light gives

$$\vec{E}_{amp,z}(\vec{r}_w) = \vec{E}_{amp}(\vec{r}_w) \frac{-x_w}{\sqrt{x_w^2 + y_w^2}} \tag{5}$$

$$\vec{E}_{amp,x}(\vec{r}_w) = \vec{E}_{amp}(\vec{r}_w) \frac{z_w}{\sqrt{x_w^2 + y_w^2}} . \tag{6}$$

438

In the following, we present focal field distributions which were obtained by numerically solving eq. (4) for distinct near field field distributions, and empirical formula were derived for describing these fields near the beam axis. A fully analytical treatment will be given elsewhere.

A laser field distribution as given in eqs. (1), (3), together with the analogous magnetic fields, will always act as a symmetric defocusing lens for electrons (the symmetry is achieved by the combination of fields, see [Quesnel and Mora(1998)]), no matter how the laser beam is polarized. This global action of the laser on the electrons (without considering the fast oscillations) is described by the ponderomotive potential

$$U_p(r,t) = mc^2 \frac{a(r,t)^2}{4\bar{\gamma}} = \frac{e^2 I(r,t)\lambda_0^2}{8\pi^2 \varepsilon_0 \bar{\gamma} m_0 c^3} , \qquad (7)$$

where $a(r,t) = (eE(r,t))/(\omega m_0 c)$ is the dimensionless local light amplitude and $\bar{\gamma}$ is the relativistic factor of the electron averaged over the fast oscillations.

Intuitively, one could guess that for creating a focusing lens for electrons, a donut-shaped laser focus is required. However, as in the case of a pure gaussian, the interplay of the transverse and longitudinal electric and magnetic fields is crucial for the overall action of the laser on electrons and therefore, this interplay must be studied carefully. Different methods are possible for creating an *intensity depression* on the laser axis, for example:

(1) Amplitude modulation in an equivalent plane: A beam stop (hard or soft) is placed on axis in an intermediate focus and is demagnified and imaged into the focus. This will give the desired amplitude shape, but is difficult to do in practice since high-power radiation has to be used which causes serious damage problems with the beam stop.

(2) Using a donut-shaped oscillator mode: While it is rather easy to achieve donut modes from standard laser oscillators, this is difficult if not impossible with short-pulse devices, e.g. KLM-oscillators. Amplification and transport are also technically complex, especially when different polarization directions are involved (see beyond).

(3) Introducing a cork-screw-type phase variation: A pure gaussian (or any other nicely shaped) laser beam is used as the input, and a proper phase plate introduces a phase shift of $\Delta\varphi = \arctan(y/z)$ around the beam axis. The resulting focus electric fields are shown in Fig. 1 and show a nicely ring-shaped transverse field, and a strong longitudinal component on axis.

Up to $r = w_0/2$, the transverse component is well described by

$$E_z(r) \approx E_0 \frac{r}{w_1} \exp\left(-\frac{r^2}{w_1^2}\right) , \qquad (8)$$

with $r = \sqrt{y^2 + z^2}$, $w_1 = w_0 \sqrt[4]{2}$, and E_0, w_0 parameters of the gaussian without circular phase (see Fig. 1). The longitudinal component in the center (up to $r = w_0/2$) is

$$E_x(y,z) \approx E_0 \frac{\lambda_0}{2\pi w_1} \left(1 - \frac{z^2}{w_2^2}\right) \exp\left(-\frac{r^2}{w_1^2}\right) , \qquad (9)$$

439

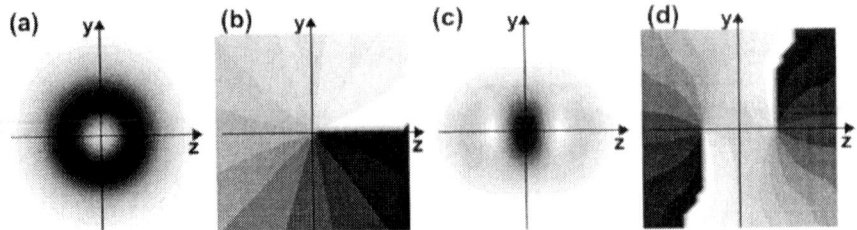

FIGURE 1. Qualitative field structure in the focus of a gaussian laser beam with cork-screw-type phase variation: transverse (a) field amplitude and (b) phase, and longitudinal (c) field amplitude and (d) phase.

with $w_2 = w_0/\sqrt[4]{2}$. The magnetic fields have analogous structure; the longitudinal B_z-component is rotated by $\pi/2$ and has a phase shift of $\pi/2$ on axis compared to the E_z-component.

(4) **Phases close to an oscillator donut mode:** An oscillator donut mode may often be a superposition of two Hermite-gaussian modes (e.g. TEM$_{01}$ and TEM$_{10}$, [Siegman(1986)]). The phase structure of such a mode can be approximated in the near field by introducing an optical component which rotates the field vector in four quarters of the beam cross section by different multiples of $\pi/2$ (see example in Fig. 2(a)).

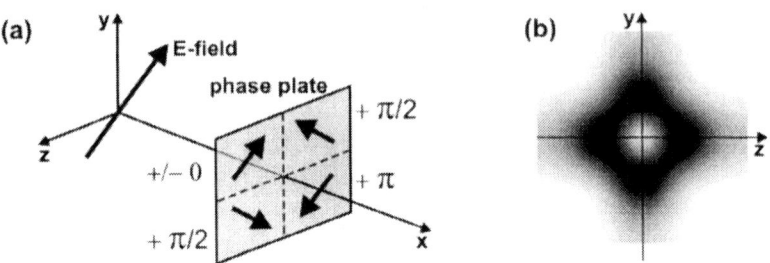

FIGURE 2. **(a)** In the near field (around the focusing lens), a special phase plate can introduce rotation of the light polarization vector by different multiples of $\pi/2$ in each quarter of a gaussian beam. **(b)** In the far field, the transverse light amplitude is distributed in a donut-like structure (same scaling as in Fig. 1).

Four different combinations of polarization rotations were investigated, yielding all the same transverse field amplitude distribution shown in Fig. 2(b), which is roughly donut-shaped and is described in the center by eq. (8).

The four cases are schematically shown in the top row of Fig. 3. Note that the polarization rotation induced in the near field is preserved in the focus. The respective longitudinal field amplitudes are shown in the bottom row and are (up to $r < w_0$) described by

$$E_{x,1}(y,z) \approx 2E_0 \frac{\lambda_0}{2\pi w_1} \left(1 - \frac{r^2}{w_1{}^2}\right) \exp\left(-\frac{r^2}{w_1{}^2}\right) \tag{10}$$

440

$$E_{x,2}(y,z) \approx 2E_0 \frac{\lambda_0}{2\pi w_0} \frac{z^2 - y^2}{w_1{}^2} \exp\left(-\frac{r^2}{w_1{}^2}\right) \qquad (11)$$

$$E_{x,3}(y,z) \approx 2E_0 \frac{\lambda_0}{2\pi w_0} \frac{2zy}{w_1{}^2} \exp\left(-\frac{r^2}{w_1{}^2}\right) \qquad (12)$$

$$E_{x,4}(y,z) \approx 2E_0 \frac{\lambda_0}{2\pi w_1} \frac{2zy\,(z^2 - y^2)}{w_1{}^4} \exp\left(-\frac{r^2}{w_1{}^2}\right), \qquad (13)$$

with $r = \sqrt{y^2 + z^2}$ and $w_1 = w_0 \sqrt[4]{2}$ as above. The magnetic fields are similar in structure and inverted in phase; a case-1 electric field is accompanied by a case-4 magnetic field and vice versa, as are cases 2 and 3.

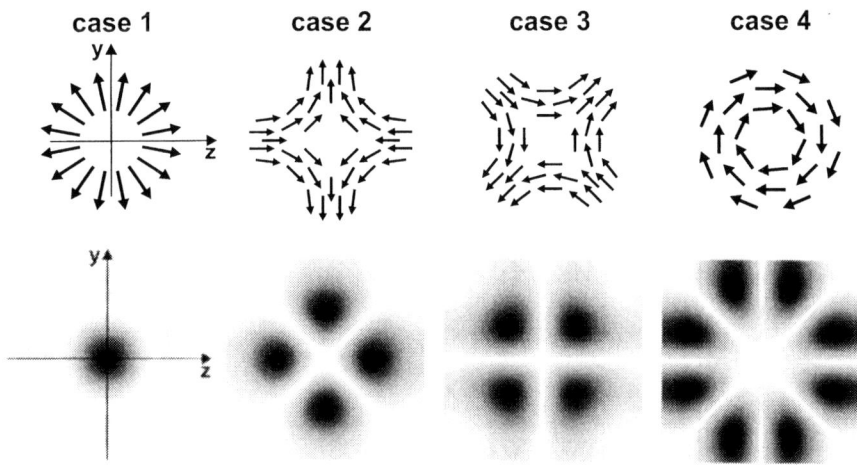

FIGURE 3. **top** Electric field vectors in the *transverse* focal electric field. This distribution is very similar to that induced in the near field, only smoothed (compare case 2 and Fig. 2(a)). **bottom** Electric field amplitudes in the *longitudinal* focal electric field.

The options (3) and (4) allow to use a conventional ultrashort laser amplifier chain which creates something like a gaussian beam profile and plane wave fronts and creating a donut-like focal spot simply by inserting proper phase-plates into the beam before the final focusing optics. While there are different possibilities to create rather similar profiles of the transverse fields, the longitudinal field component can be tailored differently as shown in Fig. 4.

ACTION OF THE FOCAL FIELDS ON RELATIVISTIC ELECTRONS

For the full numerical treatment of laser-electron interaction, an electron is traced in small time-steps, the fields are calculated at each electron position (using eq. (4)) and the relativistic equations of motion are solved for the electron at each position. This method includes the correct phase distributions (also the Guoy-terms which are different

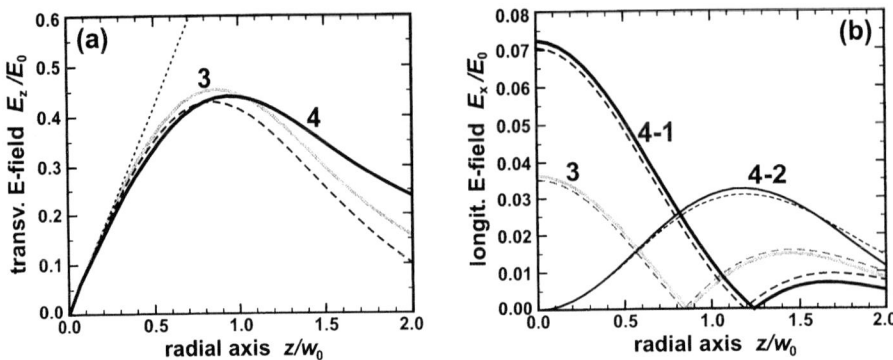

FIGURE 4. Amplitude distribution of the (a) *transverse* and (b) *longitudinal* focal electric field for different cases: 3 circular phase plate, 4 4-quarter phase plate, 4-1, 4-2 case 1 and case 2 (see Fig. 3). The formal approximations given in the the text are also shown (dashed lines).

for the different field components) and allows the treatment of marginal effects before and behind the focus as well. However, such full calculations with sufficient time steps are rather time-consuming.

For the general overview of the phenomena given here, we used the calculated field distributions in the central focal plane and assumed them unchanged within the Rayleigh length. In the following, we investigate several model cases with respect to general properties of the laser-beam – electron interaction.

The most general result of the numerical calculations is that the global interaction is correctly described by the ponderomotive potential which is defined as in eq. (7). Near the center, where the transverse fields may be approximated by $E(r) \approx E_0 r / w_1$ (see eq. (8)), we end up with a harmonic focusing potential

$$U_p(r) \approx \frac{(eE_0\lambda_0)^2}{\bar{\gamma}m_0c^2} \frac{r^2}{(4\pi w_1)^2} . \tag{14}$$

Using the identities for the laser intensity I_0 on axis

$$I_0 = \frac{\varepsilon_0 c}{2} E_0^2 \quad \text{and} \quad I_0 = \frac{2}{\pi} \frac{P}{w_0^2} \quad \rightarrow \quad E_0^2 = \frac{4}{\varepsilon_0 c} \frac{P}{w_0^2 \pi} , \tag{15}$$

with the laser power P, we get the relation

$$U_p(r) \approx \frac{m_0c^2}{\gamma} \frac{P}{P_0} \left(\frac{\lambda_0 r}{w_0^2}\right)^2 \quad \text{with} \quad P_0 = 4\sqrt{2}\pi^3 \frac{\varepsilon_0 m_0^2 c^5}{e^2} \approx 0.12 \text{ TW} . \tag{16}$$

An example for the motion of electrons in a long laser pulse is shown in Fig. 5, and there is an evident analogy to the dynamics in a multi-element quadrupol array as used in accelerators: First, there is a fast low-amplitude oscillation in the laser field which overtakes the particle (like the oscillation in the fields of the single quadrupol elements) and second, a large-scale betatron-type oscillation with wavelengths

$$\lambda_{osc} = 2\gamma^2\lambda_0 \tag{17}$$

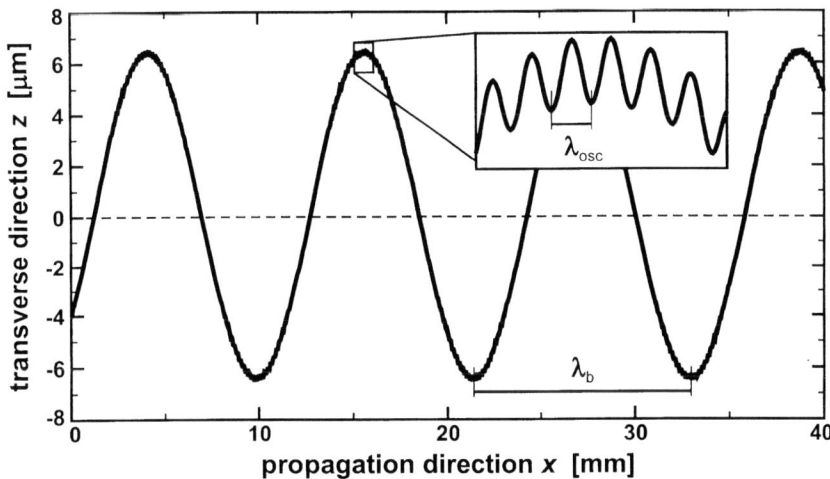

FIGURE 5. Electron trajectory (only in xz-plane; xy-plane is analogous) in a donut-shaped laser focus. Incident electron: $\gamma = 10$. Laser parameters: cw power $P = 1.3\,\text{TW}$, gaussian, circular phase shift (cf. Fig. 1), focused such that $w_0 = 25\,\mu m$ without phase modification, infinitely long focus for demonstration. The fast oscillation in the laser field (λ_{osc}) and the slow betatron-like oscillation in the ponderomotive potential (λ_b) are clearly discernible.

$$\lambda_b = 2\pi c \sqrt{\frac{\gamma m}{(dU_p/dr)/r}} = \sqrt{2\frac{P_0}{P}}\,\pi\gamma\,\frac{w_0^2}{\lambda_0}\,. \tag{18}$$

GUIDING OF ELECTRONS WITHIN A LASER FOCUS

An electron beam can be guided inside the donut focus when the ring-shaped pondero-motive potential barrier around the beam is higher than its transverse energy. Assuming that the potential is harmonically increasing up to $r = w_0/2$ and considering the xz-plane only, an electron with initial radial position z_0 and initial angle to the axis of φ_0 must fulfill the condition

$$U_p(z_0) + \frac{\gamma m_0(\varphi_0 c)^2}{2} \leq U_p(w_0/2)\,. \tag{19}$$

Inserting the normalized electron beam emittance $\varepsilon = z_0\varphi_0\gamma$, one ends up with the condition

$$\frac{z_0^2}{s^2} + \frac{s^2}{z_0^2} \leq \frac{w_0^2}{4s^2} \qquad \text{with} \quad s^2 = \sqrt{\frac{P_0}{P}\frac{w_0^4\varepsilon^2}{2\lambda_0^2}}\,. \tag{20}$$

The left side is smallest for $s = z_0$, reducing the condition to $2 \leq w_0^2/(4s^2)$, leading to the final condition for the laser power:

$$P \geq 32\frac{\varepsilon^2}{\lambda_0^2}P_0\,. \tag{21}$$

For example, with $\epsilon = 0.5\,\text{mm}\,\text{mrad}$ and $\lambda_0 = 1\,\mu\text{m}$, a laser power of $P \approx 1\,\text{TW}$ is needed for guiding an electron beam within the laser ring focus, and we see that neither the electron energy nor the focusing geometry play a role. With a given electron beam radius z_0, this leads to an optimum laser focus radius of $w_0 = \sqrt{8} z_0$. Assuming that electron guiding works reliably within the double Rayleigh length, the guiding length is

$$L_g = 2L_R = 16\pi \frac{z_0^2}{\lambda_0} . \tag{22}$$

With an initial electron beam radius of $z_0 = 100\,\mu\text{m}$, and laser wavelength of $\lambda_0 = 1\,\mu\text{m}$ we get $L_g \approx 50\,\text{cm}$, provided the laser pulse is long enough and the electrons are fast enough to have such a long overlap.

FOCUSING: THIN LENS

When short light pulses and not too much relativistic electrons are used, interaction is so short that the radial position r of the electron remains essentially unchanged during interaction and only its direction is modified. An example is shown in Fig. 6 and resembles the situation with a thin lens in optics.

The situation is described by the equation of motion in the radial direction

$$m_0 \gamma \frac{\partial^2 r}{\partial t^2} = \frac{\partial U_p(r,t)}{\partial r} . \tag{23}$$

We assume that (a) initially, the beam is parallel to the axis $((\partial r/\partial t)(t=0) = 0)$, (b) the radial position r is unchanged during interaction, and (c) the ponderomotive potential (and the laser pulse intensity) as seen by the electron is gaussian in time

$$U_p(r,t) = U_p(r) \exp\left(-\frac{t^2}{(\Delta t)^2}\right) \qquad \text{with} \quad \Delta t = 2\gamma^2 \Delta\tau , \tag{24}$$

where $\Delta\tau$ is the gaussian laser intensity pulse duration in the lab frame. With these assumptions, Integration of eq. (23) yields an electron angle to the axis after interaction of

$$\varphi_r \approx \frac{1}{c}\left(\frac{\partial r}{\partial t}\right)_{end} = \frac{1}{8\sqrt{\pi^3}} \frac{\Delta\tau\, e^2 \lambda_0^2}{m_0^2 c^3} \frac{\mathrm{d}\, E^2(r)}{\mathrm{d}r} . \tag{25}$$

Note that the focusing angle φ_r does not depend on the electron energy, because for larger γ, the effects of weaker focusing force and the longer interaction distance will compensate exactly. For going on, the radial donut field $E(r)$ of eq. (8) is used in linear approximation, and the total laser pulse energy for a Gaussian is expressed as $E_L = P\sqrt{\pi}\Delta\tau$ (with the laser peak power P). Thus we end up with

$$\varphi_r \approx E_L \frac{4c}{P_0} \frac{\lambda_0^2}{w_0^4} r \qquad \text{and} \quad f = \frac{1}{E_L} \frac{P_0}{4c} \frac{w_0^4}{\lambda_0^2} . \tag{26}$$

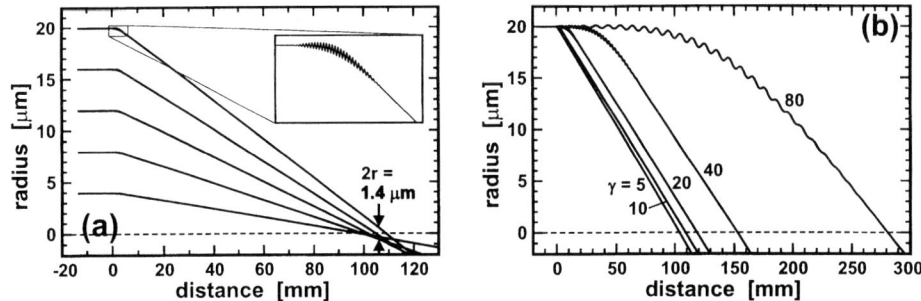

FIGURE 6. Focusing electrons with a donut laser pulse: **(a)** focus quality for $\gamma = 10$ electrons using a 4 TW, 40 fs laser pulse focused to $w_0 = 100\,\mu\text{m}$ with the donut mode shown in Fig. 1. The aberrations are from the non-linearity of the electric field increase (see Fig. 4(a)). **(b)** same laser pulse, but electrons with different energies.

The equations (26) are valid when the focus length f is much larger than the interaction length $c\Delta t$ (thin-lens approximation). Focusing will also work without this approximation, i.e. when the radial position r changes significantly during interaction (thick lens analogy, see Fig. 6(b)), but integrating eq. (23) is more complicated.

In the next step, we drop the condition that the input electrons are parallel and use a realistic electron beam with a normalized emittance of $\varepsilon = \gamma z_0 \varphi_0$. We set the criterion that focusing is acceptably good (i.e. the deviation of the field $E(r)$ from linearity, see Fig. 4) for $z_0 \leq w_0/5$, which requires the electron angles to be distributed within

$$|\varphi_0| \leq 5\frac{\varepsilon}{w_0\gamma}. \tag{27}$$

Focusing will only work if $|\varphi_r|/|\varphi_0| > 1$, leading to the criterion

$$E_L\gamma\frac{\lambda_0^2}{w_0^2} > \frac{25\varepsilon}{4c}P_0, \tag{28}$$

On the other hand, the focusing interaction L_f should be within the double Rayleigh length $2L_R$:

$$L_f = 2\gamma^2 c\Delta\tau \leq 2\pi\frac{w_0^2}{\lambda_0} = 2L_R, \tag{29}$$

which leads, together with eq. (28) to a criterion for possible ranges of w_0:

$$\frac{E_L\gamma\lambda_0^2 4c}{25\varepsilon P_0} \geq w_0^2 \geq \frac{c\Delta\tau\gamma^2\lambda_0}{\pi}. \tag{30}$$

Setting all $\geq$ to = gives an estimation for the threshold of laser and electron parameters which will allow focusing at all

$$\frac{P\lambda_0}{\gamma\varepsilon} > \frac{25}{4\sqrt{\pi^3}}P_0 \approx P_0 \approx 0.12\,\text{TW} \quad \text{with} \quad w_0 = \sqrt{\frac{c\Delta\tau\gamma^2\lambda_0}{\pi}}. \tag{31}$$

For example, with a laser wavelength of $\lambda_0 = 1\,\mu$m, electrons with $\gamma = 10$ and $\varepsilon = 0.5\,$mm mrad require a minimum laser power of $P > 0.6\,$TW for being focused, and Like in the case above, these criterion are valid for the thin-lens case.

When working with a real electron beam, only that portion of the beam will be focused which overlaps with the laser in the focus, i.e. a rather short part of the beam only.

BUNCHING OF HIGHLY RELATIVISTIC ELECTRONS

The longitudinal fields in the focus are not useful for effective direct electron acceleration because the Guoy phase shift (of $\sim 0.75\pi$ within twice the Rayleigh length) makes the light phase fronts travel through the focus faster than the speed of light. Therefore, even very fast electrons ($E \gg 1\,$GeV) run out of phase and feel both acceleration and deceleration within the interaction length.

However, the longitudinal E-field may be used for inducing slight energy variations for different electron groups and therefore add a density periodicity of the laser wavelength to an electron beam. From numerics, we find that within four times the Rayleigh length, energy variations of very roughly

$$|\Delta E_{el}| \leq \sqrt{\frac{P}{P_0}} \; [MeV]\,, \tag{32}$$

may be crated for electrons travelling on axis. This means energy variations of $|\Delta E_{el}| \leq 2\,$MeV and $64\,$MeV for a $P = 1\,$TW and a $P = 1\,$PW laser, respectively.

The process is delicate and depends a lot on the laser beam properties before and behind the focus and therefore also on the beam quality and focusing conditions of the laser beam. We will further investigate this situation in more detail.

REFERENCES

[Schätz et al.(2001)] Schätz, T., Schramm, U., and Habs, D., Nature (London) **412**, 717 (2001).
[Schramm et al.(2001)] Schramm, U., Schätz, T., and Habs, D., Phys. Rev. Lett. **87**, 184801 (2001).
[Schramm(2001)] Schramm, U., *Crystalline ion beams*, Habilitationsschrift, LMU München, 2001.
[Jackson(1975)] Jackson, J., *Classical Electrodynamics*, John Wiley & Sons, New York, 1975, second edn.
[Primack and Blümel(1999)] Primack, H., and Blümel, R., Phys. Rev. E **60**, 957 (1999).
[Quesnel and Mora(1998)] Quesnel, B., and Mora, P., Phys. Rev. E **58**, 3719–3732 (1998).
[Siegman(1986)] Siegman, A., *Lasers*, University Science Books, Mill Valley, CA, 1986.

Super-High Energy Particle Acceleration by Super-Strong Laser pulses in Plasmas

Kazuhisa Nakajima

High Energy Accelerator Research Organization, Oho 1-1, Tsukuba, Ibaraki, 305-0801, Japan

Abstract. The recent tremendous progress of ultraintense lasers creates novel concepts of high energy particle generation and acceleration due to super-strong field interactions with matter. In a state of the art of ultra-intense lasers, the highest laser intensities reach a TeV range in terms of the ponderomotive energy exerted on matter. This implies that a forefront of high energy particle physics phenomena may be revealed with super-strong laser-matter interactions. As an accelerator-oriented application, the recent status of laser wakefield accelerator (LWFA) experiments proceeded at JAERI-APR are presented to accomplish high energy gains of the order of GeV with high quality electron beam injection. A new concept of super high energy particle acceleration mechanism based on super-strong laser-plasma interactions is presented, named as "Dirac accelerator". When this super high energy particle acceleration mechanism takes place in the dense matter, super intense interactions of high energy particles will be produced.

INTRODUCTION

A novel particle acceleration concept was proposed by Tajima and Dawson [1], which utilizes plasma waves excited by intense laser beam interactions with plasmas for particle acceleration, known as laser-plasma accelerators. In particular recently there has been a great experimental progress on the laser wakefield acceleration of electrons since the first ultrahigh gradient acceleration experiment made by Nakajima et al. [2]. Recent world-wide experiments have successfully demonstrated that the self-modulated LWFA mechanism is capable of generating ultrahigh accelerating gradient of ~ 100 GeV/m. In the self-modulated LWFA, however, the maximum energy gain has been limited at most to 100 MeV with energy spread of ~ 100 % because of dephasing and wavebreaking effects in highly dense plasmas where thermal plasma electrons are accelerated. The first high energy gain acceleration exceeding 200 MeV has been observed with the injection of an electron beam at an energy matched to the wakefield phase velocity in a fairly underdense plasma [3]. Hence the second-generation research has dealt with the injection of ultrashort electron bunches into a correct acceleration phase of laser wakefields and the optical guiding of ultraintense ultrashort laser pulses in underdense plasmas in

CP611, *Superstrong Fields in Plasmas:* Second Int'l. Conf., edited by M. Lontano et al.

order to accomplish high energy gains of more than 1 GeV and high quality beam acceleration with a small energy spread. Here the design parameters of GeV laser wakefield accelerator are presented from the points of view on the ultrashort pulse beam injection and the optical guiding. The recent status of the laser acceleration research at JAERI-APR are reported.

The ponderomotive motion of charged particles in the field of non uniform electromagnetic wave is under investigation for a long time, being initiated by the P. L. Kapitza and P. A. M. Dirac [4]. We propose a new regime of laser ponderomotive acceleration of particles named as "Dirac accelerator", which is realized when the group velocity of high-power laser pulse is less than the vacuum speed of light. We intend to show that in this regime, the energy gain of particles is determined by the group velocity and does not depend on laser intensity. The laser intensity determines the probability for the particle to be reflected from the leading edge of the pulse, and, thus, the number of particles captured in the acceleration. Here we present some examples of applications of Dirac accelerator concept to superhigh energy particle accelerators with GeV-TeV range energies. We discuss vision of particle colliders with extremely high luminosities at the TeV range, based on super sutrong laser-matter interactions for exploring the super high energy particle physics.

PARTICLE INTERACTION WITH STRONG ELECTROMAGNETIC FIELDS

The peak amplitude of the transverse electric field of a linearly polarized laser pulse is given by

$$E_L[\text{TV/m}] \simeq 2.7 \times 10^{-9} I^{1/2}[\text{W/cm}^2] \cong 3.2 a_0/\lambda[\mu\text{m}], \tag{1}$$

where I is the laser intensity, λ is the laser wavelength, and a_0 is the laser strength parameter defined by $a_0 \equiv eA_0/m_e c^2$ in terms of the peak amplitude of the laser vetor potential A_0 and the electron rest energy $m_e c^2$. Using the laser peak intensity $I = cE_L^2/8\pi = ck^2 A_0^2/8\pi$, the laser strength parameter is given by

$$a_0 = (2e^2\lambda^2 I/\pi m_e^2 c^5)^{1/2} \cong 0.85 \times 10^{-9}\lambda[\mu\text{m}]I^{1/2}[\text{W/cm}^2]. \tag{2}$$

Physically a_0 is equal to the normalized momentum of the electron quiver motion in the laser field.

In the strong laser field, an electron absorbs energy and momentum from the wave to cause mass shift from m_e to $m_e\gamma_L$, where $\gamma_L = (1 + a_0^2/2)^{1/2}$. The effective potential of the electron inside the laser field for $a_0 << 1$ is

$$U_{eff} = m_e c^2 (1 + a_0^2/2)^{1/2} \approx m_e c^2 + m_e c^2 a_0^2/4 \tag{3}$$

In the nonrelativistic regime, the ponderomotive force or the field gradient force can be defined as

$$\mathbf{F} = -\nabla U_{eff} \approx -m_e c^2 a_0^2/4 \tag{4}$$

The laser pulse propagating in underdense plasmas expels plasma electrons exerted by the ponderomotive force to excite plasma waves. In the highly relativistic regime for $a_0 \gg 1$, if the initial mometum of a free electron is smaller than the quiver momentum, i.e. $\gamma\beta < a_0/\sqrt{2}$, the electron is reflected from the laser pulse. This condition gives $\gamma < \gamma_L$. The reflection of particles from the laser pulse moving in plasmas results in their acceleration.

GEV LASER WAKEFIELD ACCELERATOR DEVELOPMENTS

As an intense laser pulse propagates through an underdense plasma, the ponderomotive force expels electrons from the region of the laser pulse. This effect excites a large amplitude plasma wave (wakefield) with phase velocity approximately equal to the group velocity of laser pulse, given by $v_p = c(1 - \omega_p^2/\omega^2)^{1/2}$, where ω is the laser frequency. The maximum axial wakefield occurs at the plasma wavelength, $\lambda_p[\mu m] \simeq 0.57\tau$ in a plasma with the resonant electron density, $n_0[\text{cm}^{-3}] \simeq 3.5 \times 10^{21}/\tau^2$ in terms of a FWHM pulse duration τ [fs]. When a Gaussian driving laser pulse with the peak power P [TW] is focused on the spot size r_0 [μm], the maximum axial wakefield yields

$$(eE_z)_{max}[\text{GeV/m}] \simeq 8.6 \times 10^4 P\lambda^2/(\tau r_0^2 \gamma_L), \tag{5}$$

where $\gamma_L = (1 + a_0^2/2)^{1/2}$ takes account of nonlinear relativistic effects, and $a_0 = 6.8\lambda P^{1/2}/r_0$ is the laser strength parameter for the linear polarization [5].

Several effects limit the energy gain in a single-stage of laser-plasma accelerators; laser diffraction, electron dephasing, pump depletion and laser-plasma instabilities. In order to achieve the acceleration energy gains of higher than 1 GeV in a single stage with a cm-scale length, it is necessary to extend the acceleration length limited by diffraction effects of laser beams. We develop the channel-guided LWFA in which both the driving laser pulses and particle beams can be guided through Z-pinch capillary discharge plasmas with a cm-scale length. The parameters to test electron acceleration of GeV energies are shown in Table 1. The designs of the LWFA are based on availability of the 10 Hz table-top ultrashort, ultrahigh peak power Ti:Sapphire laser pulses with duration of 20 fs and energy of 2J developed at JAERI-APR [6].

In order to produce a high quality electron beam with low momentum spread and small pulse-to-pulse energy stability, it is required that femtosecond electron bunches should be injected with the energy higher than trapping threshold and femtosecond synchronization with respect to a wakefield accelerating phase space, which is typically less than 100 fs in a longitudinal scale and 10 μm in a transverse size. For this purpose, we have developed the laser acceleration test facility at JAERI-APR, which can deliver a high quality electron beam consisting of a

TABLE 1. Parameters of the GeV channel-guided laser wakefield accelerators.

Energy gain [GeV]	0.5	1	10
Pulse duratuon τ [fs]	20	50	100
Peak power [TW]	100	40	20
Spot radius [μm]	30	20	10
Laser strength parameter	1.8	1.7	2.4
Plasma density [10^{18} cm^{-3}]	8.8	1.4	0.35
Accelerating gradient [GeV/cm]	1.9	0.7	0.55
Diffraction length [cm]	1.1	0.5	0.12
Dephasing length [cm]	0.4	5.5	56
Channel length [cm]	no	1.5	20

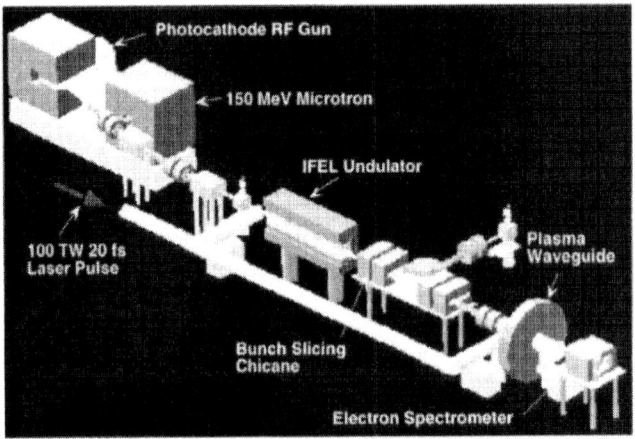

FIGURE 1. A schematic of the Laser Acceleration Test Facility at JAERI-APR.

photocathode RF gun and a compact race-track microtron [7], and a femtosecond bunch generation system shown in Fig. 1. A copper photocathode of the RF gun is illuminated by an UV light of 263 nm with an incident angle of 68°, delivered from a compact all solid-state Nd:YLF laser system. This laser can generate the output pulse energy of 200 μJ at 263 nm with fluctuation of 0.3 % and the pulse width of approximately 6 ps FWHM. The microtron accelerates a low emittance beam injected from photocathode RF gun at 4 MeV to 150 MeV after 25 turns. The performance of the beam injector has been measured for the beam intensity, the beam emittance and the bunch length by observing the synchrotron radiation and the optical transition radiation from the electron beam. The beam injector generated a 150 MeV single electron bunch at 10Hz with a charge of 95 pC, that was the transmission efficiency of 80 % after 25 turns. The normalized beam emittance was as low as < 5πmm·mrad and the bunch width was as short as < 10 ps in FWHM.

We have conceived a method capable of generating a femtosecond electron pulse

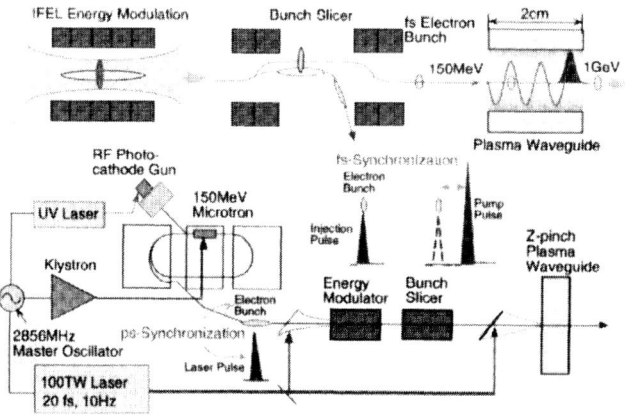

FIGURE 2. A schematic of the femtosecond bunch slicing and synchronization system.

injected into a correct wakefield phase within a few femtoseconds. Generation of femtosecond electron pulses with femtosecond synchronization is based on slicing a bunch through a process of energy modulation created in the interaction of electrons with a femtosecond laser pulse split from a main pump pulse. This technique of slicing a bunch of energy modulated electrons has been demonstrated to produce femtosecond synchrotron radiation pulses in a electron storage ring at the Advanced Light Source of Lawrence Berkeley National Laboratory [8].

We apply this energy modulation technique to production of an ultrashort slice of a few 10 femtoseconds duration from an electron bunch of a few picoseconds duration delivered by the microtron. The mechanism of energy modulation is based on the inverse free electron laser, which generates the efficient energy exchange between electrons and laser fields in an undulator when the laser wavelength λ_L satisfies the resonance condition of free electron lasers, given by $\lambda_L = \lambda_u(1 + K^2/2)/(2\gamma^2)$, where λ_u is the undulator period, γ is the Lorentz factor, and $K = eB_0\lambda_u/(2\pi m_e c) = 0.934\lambda_u[\text{cm}]B_0[\text{T}]$ is the deflection parameter of the undulator with the peak magnetic field of B_0. In addition to the resonance condition, the transverse mode and the spectral bandwidth matching between the laser and the undulator spontaneous radiation must be satisfied [9].

A scheme of femtosecond bunch-slicing is shown in Fig. 2. A small part of the main laser pulse with duration of 20 fs is split as an injection pulse to interact with a few ps long electron bunch in the undulator within a sub-picosecond time jitter. An energy of electrons overlapping with a 20 fs laser pulse will be modulated by 10 % with a $\sim 200\mu$J pulse energy. In the chicane magnets, the either low energy or high energy portion will be separated from the unmodulated part to be dumped. The sliced portion of the electron bunch with nearly the same pulse width as the injection laser pulse will be injected into a correct accelerating phase of the wakefield excited by the main laser pulse within less femtosecond time jitter.

In order to increase the energy gain beyond the diffraction limitation, it is essential to propagate a laser pulse in an underdense plasma beyond the vacuum Rayleigh length. A promising method is the relativistic self-guiding induced by relativistic quiver motion of the plasma electrons for the laser power exceeding a critical power, given by $P_c = 17(\omega^2/\omega_p^2)$ GW. Since the index of refraction, however, becomes modified by the laser pulse on the plasma frequency time scale, $\sim 1/\omega_p$, relativistic optical guiding is ineffective in preventing diffraction of ultrashort pulses, $L_L \leq \lambda_p/\gamma_L$ [10]. It is known that the relativistic self-guiding is associated with instabilities induced by ultraintense laser interactions with plasmas, such as filamentations and hose instabilities.

Optical guiding of a Gaussian laser pulse with a focal spot radius of r_0 can be made through the plasma density channel with a parabolic electron-density profile of the form $n(r) = n_0 + \Delta n_c r^2/r_0^2$, where $\Delta n \geq 0$. This parabolic plasma channel can guide a Gaussian beam with the spot radius r_0 provided that the density channel depth Δn is equal to the critical depth, $\Delta n_c = 1/(\pi r_e r_0^2)$ where $r_e = e^2/(m_e c^2)$ is the classical electron radius [11].

We have developed a stable cm-scale plasma channel produced by an imploding phase of fast Z-pinch discharge in a gas-filled capillary without wall ablation. A high current fast Z-pinch discharge generates strong azimuthal magnetic field, which contracts the plasma radially inward down to ~ 100 μm in diameter. The imploding current sheet drives the converging shock wave ahead of it, producing a concave electron density profile in the radial direction just before the stagnation phase. The concave profile is approximately parabolic to out a radius of ~ 50 μm, after which the density falls off. We have constructed the Z-pinch capillary discharge system assembled from the Marx generator and the water capacitor with four laser trigger spark gaps shown in Fig. 3. This system can generate a 10 cm long capillary discharge in a diameter of 1 mm driven by 100 kA current without time jitter.

NEW REGIME OF PONDEROMOTIVE SCATTERING IN PLASMAS

Let us consider the ultraintense ultrashort laser pulse propagating in a plasma with the electron density n_e and the plasma frequency ω_p. In the reference frame that is moving with the group velocity $v_g = c(1 - \omega_p^2/\omega^2)^{1/2} < c$ of the laser pulse along the z axis, the wavelength of the electromagnetic field becomes $\lambda\gamma^* = 2\pi\gamma^*/k$, where $\gamma^* = (1 - \beta^{*2})^{-1/2}$, $\beta^* = v_g/c$, ω is the laser frequency, and k is the laser wave number. The characteristic scale length of the field region $l^* = c\beta^*\gamma^*\tau_p$, is the laser pulse length, where τ_p is the pulse duration in the laboratory frame.

In the moving frame, since the laser ponderomotive potential becomes static, the interaction can be considered now as the elastic potential scattering. Let the initial energy and velocity of the particle γ_0 and β_0, respectively, and the initial angle of collision is θ_0. The Lorentz transformation gives the energy of the particle in the moving frame:

$$\gamma' = (1 - \beta^*\beta_0 \cos\theta_0)\gamma_0\gamma^*, \tag{6}$$

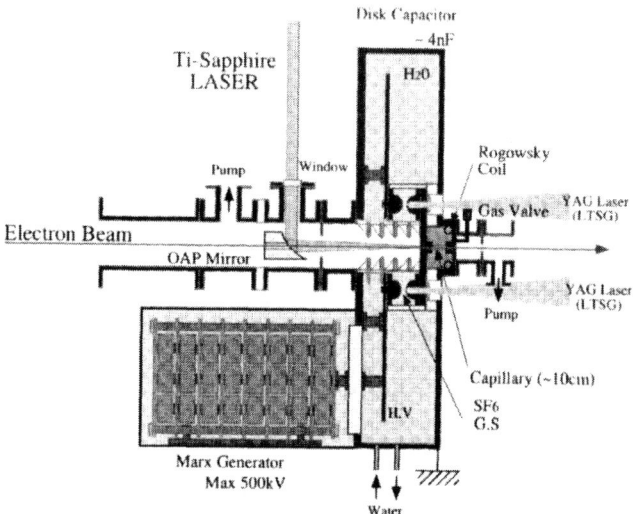

FIGURE 3. A schematic of the jitter-free fast Z-pinch capillary discharge system

where $\gamma^* = (1 - \beta^{*2})^{-1/2}$ is the characteristic Lorentz-factor of the laser pulse. According to the classical scattering theory, if this energy is less than the field peak ponderomotive potential:

$$\gamma' \leq \gamma_L = (1 + a_0^2/2)^{1/2}, \tag{7}$$

the particle has to be reflected. Here, a_0 is the normalized peak amplitude of the vector potential. As an effective Hamiltonian is time-independent in the moving frame, the final energy γ of the particle does not change after scattering (in the infinity), and, after the Lorentz transformation, we have in the laboratory frame

$$\gamma^* \gamma (1 - \beta^* \beta \cos\theta) = \gamma' \tag{8}$$

In this relation, β is the particle velocity, and θ is the final scattering angle, where $\theta = 0$ corresponds to the co-propagation case. From these equations, the following conservation law is given by

$$\gamma(1 - \beta^* \beta_z) = \gamma_0(1 - \beta^* \beta_{0z}) = \text{const.} \tag{9}$$

This result of the simple kinematics approach coincides with that of electrodynamics consideration [13].

From Eqs. 6 and 8, the maximum final energy of particle is given for scattering at zero angle $\theta = 0$ as

$$\gamma_{\pm} = [(\beta^* \pm \beta_0)^2 + \gamma_0^{-2}]\gamma_0\gamma^{*2}, \tag{10}$$

where signs correspond to the initially counter- ("+") and co-propagating ("-") laser pulse and accelerated particle. Although the energy gain is maximal for counter-propagating injection, this interaction geometry is efficient for sufficiently low particle energies (low group velocities) only, which is due to the threshold condition given by Eq. 7 on the particle energy in the moving reference frame. At high injection energies and energy gains, the co-propagating injection is preferable ($\theta_0 = \theta = 0$).

In a quantum mechanical analysis, the value of ponderomotive potential determines the probability of scattering. This probability of reflection determines the number of particles which will be accelerated. To estimate its value, one can start with Klein - Gordon equation, as the spin effects are evidently not important for the scattering. It is convenient to consider the problem in the moving reference frame, in which the potential is time-independent. We are interested in the backscattering, i.e. scattering at $\theta = 0$, and, consequently, one can use the 1D approximation. In the Lorentz gauge and paraxial ray approximation, the vector potential $A(z)$ is the transverse one, and the Klein - Gordon equation is reduced to the following equation

$$\frac{\partial^2 \psi}{\partial z^2} - \frac{1}{c^2}\frac{\partial^2 \psi}{\partial t^2} - \frac{m^2 c^2}{\hbar^2}(1 + \frac{e^2 A^2(z)}{m^2 c^4})\psi = 0. \tag{11}$$

Here and below we omit (') in the coordinates of the moving reference frame. As the potential is time-independent, the energy of the particle is conserved. One can separate the variables and seek the solution in the form

$$\psi(z,t) = R(z)\exp(-i\varepsilon t/\hbar), \tag{12}$$

where ε is the particle energy in the moving frame, $\varepsilon = m\gamma' c^2$. In fact, we assume the stationary flow of particles collides with the potential, and seek the parts of this flow which will be reflected and transmitted. The correspondent asymptotic behavior of solution must be the following

$$R(z \to \infty) = \exp(-ik_0 z) + a\exp(ik_0 z), \quad R(z \to -\infty) = b\exp(-ik_0 z), \tag{13}$$

where a and b are the reflection and transmission amplitudes, normalized to the amplitude of the incident flow. As the Klein - Gordon current,

$$j = -i\frac{e\hbar}{2mc^2}(\psi\nabla\psi^* - \psi^*\nabla\psi), \tag{14}$$

and asymptotic wavenumbers are the same for all three waves, the absolute values $|a|^2$ and $|b|^2$ are the reflected and transmitted fractions of the incident flow, respectively. The asymptotic wavenumber , $k_0 = \sqrt{\varepsilon^2 - m^2 c^4}/\hbar c$ corresponds to the field-free momentum of an electron of a given energy. The solution of the Klein-Gordon equation gives the probability of the reflection, $|a|^2 = 1$ for the condition of the particle energy in the moving reference frame,

$$\gamma' = \gamma^* \gamma_0 (1 - \beta_0 \beta^*) \leq \gamma_L \equiv (1 + a_0^2/2)^{1/2}, \tag{15}$$

where γ_L is the poderomotive energy of the laser field. The scattering occurs at the point where $\gamma' = \gamma_L$. In this regime the ponderomotive scattering off the laser pulse works as particle acceleration mechanism which we call as "Dirac Accelerator".

HIGH ENERGY ELECTRON DIRAC ACCELERATOR

The ponderomotive potential scattering results in acceleration of electrons via a point-like interaction with the strong laser fields. This acceleration length will be at most a half of the laser pulse length when the reflection condition Eq. 15 is satisfied. Let us consider the acceleration of the electron from the injection energy γ_0 to the accelerated final energyl γ via the ponderomotive interaction with the laser field in a plasma with the density n_e. The required group velocity, the corresponding Lorentz factor of the laser pulse and the ponderomotive energy of the laser field are

$$\beta^* = \frac{\gamma_0\beta_0 + \gamma\beta}{\gamma_0 + \gamma}, \quad \gamma^* = \frac{\gamma_0 + \gamma}{\sqrt{2}[1 + \gamma\gamma_0(1 - \beta\beta_0)]^{1/2}}, \quad \gamma_L = [\frac{1 + \gamma\gamma_0(1 - \beta\beta_0)}{2}]^{1/2}. \quad (16)$$

The plasma density can be determined as

$$n_e[\text{cm}^{-3}] = \frac{\pi}{r_e\lambda_L^2\gamma^{*2}} = \frac{1.115 \times 10^{21}}{\lambda_L^2[\mu\text{m}]\gamma^{*2}}, \quad (17)$$

where r_e is the classical electron radius and λ_L is the laser wavelength. The minimum required laser field is given by

$$a_0 = \sqrt{2}[\gamma^{*2}\gamma_0^2(1 - \beta^*\beta_0)^2 - 1]^{1/2} \quad (18)$$

The corresponding laser intensity is $I[\text{W/cm}^2] = 1.37 \times 10^{18}\lambda_L^{-2}[\mu\text{m}]a_0^2$. When the laser pulse is focued on the spot radius r_0, the required peak power is given by $P[\text{TW}] = 0.0215a_0^2(r_0/\lambda_L)^2$. Fig. 4 shows (a) the required laser field and intensity, and (b) the Lorentz factor corresponding to the required group velocity and the corresponding plasma density for the electron acceleration from the injection energy $E_{inj} = 150$ MeV to the GeV range of the final energy. The laser pulse with the

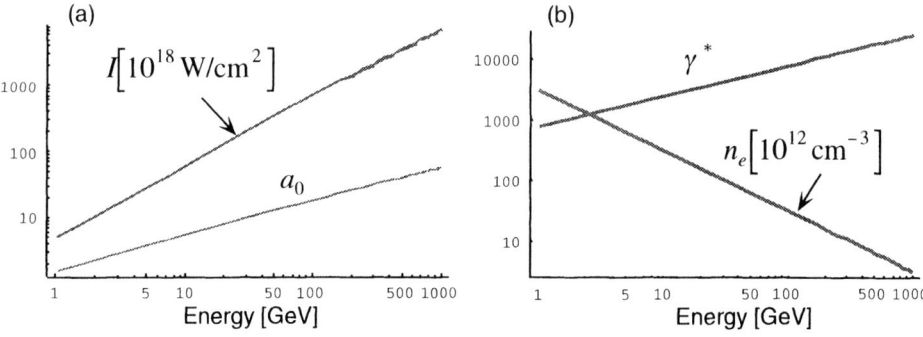

FIGURE 4. (a) The required laser field a_0 and intensity I, and (b) the Lorentz factor γ^* corresponding to the required group velocity and the corresponding plasma density for the electron acceleration from the injection energy $E_{inj} = 150$ MeV to the GeV range of the final energy.

ponderomotive energy γ_L can accelerate an electron with the initial energy γ_0 up to the maximum final energy γ_{max} given by

$$\gamma_{max} = 2\gamma_L^2\gamma_0(1 + \beta_0\beta_L) - \gamma_0 \tag{19}$$

For super high energy acceleration, these parameters are approximately given as

$$\gamma_L \approx (\gamma/4\gamma_0)^{1/2}, \ \gamma^* \approx (\gamma\gamma_0)^{1/2} \approx \gamma/2\gamma_L, \ \gamma_0 \approx \gamma/4\gamma_L^2 \approx \gamma^{*2}/\gamma. \tag{20}$$

The Dirac accelerator makes it possible to accelerate electrons up to 1 PeV ($\gamma \approx 2 \times 10^9$) energy provided with the injection energy 250 MeV ($\gamma_0 \approx 500$), the plasma density $n_e \approx 1.1 \times 10^9$ cm^{-3} ($\gamma^* \approx 10^6$), and the laser intensity 2.7×10^{24} W/cm^2 ($\gamma_L \approx 1000$, $a_0 \approx 1414$) for the wavelength $\lambda_L = 1\mu$m. This laser intensity can be produced by focusing ~ 400 PW on the spot radius $r_0 = 3\mu$m.

VISION OF SUPER HIGH ENERGY AND SUPER HIGH LUMINOSITY

The dispersion relation of relativistically strong electromagnetic waves is $\omega^2 = k^2c^2 + \omega_p^2/\gamma_L$. This gives a group velocity of the intense laser pulses: $v_g = c(1 - \omega_p^2/\gamma_L\omega^2)^{1/2} = c(1 - n_e/\gamma_L n_c)^{1/2}$, where n_c is the critical plasma density. It implies that the laser pulses can propagate overdense plasmas for $n_e < \gamma_L n_c$. This corresponds to the relativistic transparency of overdense plasmas. It means that a sufficiently intense laser pulse propagating the overdense plasma with density n_{e1}, assuming nearly solid density, accelerates initially stational plasma electrons up to $\gamma_1 = 2\gamma_1^{*2} - 1 = 2\gamma_{L1}n_c/n_{e1} - 1$ from $\gamma_0 = 1$, where the Lorentz factor for the group velocity v_g is $\gamma^*_1 = \omega\sqrt{\gamma_{L1}}/\omega_{p1} = (\gamma_L n_c/n_{e1})^{1/2}$ and the γ_{L1} corresponds to the intensity at the interaction front of the laser pulse in the overdense plasma. When the accelerated electrons are emitted to an underdense plasmas from the overdense plasma region with thickness Δ, the electron bunch length becomes $L_{b1} = \Delta/2\gamma_1$. In an underdense plasma with density n_{e2}, a group velocity of the transmitted laser pulse with γ_{L2} is accelerated to overtake the emitted electrons and to accelerate them again up to the energy, $\gamma_2 = 2\gamma_{L2}^2\gamma_1(1 + \beta_1\beta_{L2}) - \gamma_1$. The underdense plasma density should be $n_{e2} = \gamma_{L2}n_c/\gamma_2^{*2}$, where $\gamma_2^* = \gamma_{L2}\gamma_1(1 + \beta_1\beta_{L2})$. The electron bunch length in the underdense plasma becomes $L_b \sim 2\gamma_1\Delta$.

As an example, let us consider the interaction of the laser pulse with the intensity $I = 1.2 \times 10^{23}$ W/cm^2 ($a_0 = 300$) with the overdense plasma with density $n_{e1} = 1.1 \times 10^{22}$ cm^{-3} and thickness $\Delta = 30\mu$m. Assuming a half intensity $I/2$ ($\gamma_{L1} = 150$) at the interaction front, this interaction emits 15 MeV electrons with bunch length of $\sim 1\mu$m in the front of laser pulse. In the underdense plasma with density $n_{e2} = 1.6 \times 10^{15}$ cm^{-3}, the laser pulse with the poderomotive energy $\gamma_{L2} = 212$ again accelerates electrons to the energy 2.6 TeV in the length 1.7 mm.

We consider an application of the super high energy acceleration scheme based on the Dirac accelerator to particle collision with a super high luminosity. The particle collider is composed of two Dirac accelerators that produce counter-propagating electron beams. Two electron beams are self-focused on the collision point by self-pinching effect in an overdense plasma, $n_e > n_b$, where n_b is the beam electron density. In the overdense plasma, the self-focusing force is given by $F_t = -2\pi r_e m_e c^2 n_b r$ at the radius r [14]. Assuming a bi-Gaussian beam density profile with the rms beam radius σ_{br} and the rms bunch length σ_{bz}: $n(r, z) = n_b \exp[-r^2/2\sigma_{br}^2 - z^2/2\sigma_{bz}^2]$, the beam density is given by $n_b = N/[(2\pi)^{3/2}\sigma_{br}^2\sigma_{bz}]$, where N is the total number of electrons. The focusing strength is

$$K_F = \frac{F_t}{\gamma m_e c^2 r} = \frac{r_e N}{\sqrt{2\pi}\gamma\sigma_{br}^2\sigma_{bz}} = \frac{r_e N}{\sqrt{2\pi}\varepsilon_n\beta_0\sigma_{bz}}, \tag{21}$$

where ε_n is the normalized beam emittance, and $\beta_0 = \gamma\sigma_{br}^2/\varepsilon_n$ is the beta function at the plasma. Due to the plasma lens focusing, the beam size at the collision point becomes

$$\sigma^{*2} = \varepsilon\beta^* \approx \frac{\varepsilon_n f^2}{\gamma\beta_0} = 2\pi\left(\frac{\varepsilon_n\sigma_{br}\sigma_{bz}}{r_e N l}\right)^2, \tag{22}$$

where $f = 1/K_F l$ is the focal length of the plasma lens with length l. The luminosity for electron-electron beam collision per shot is given by

$$L = \frac{N^2}{4\pi\sigma^{*2}} = \frac{1}{8\pi^2}\left(\frac{r_e N^2 l}{\varepsilon_n\sigma_{br}\sigma_{bz}}\right)^2 = \frac{\gamma}{8\pi^2\varepsilon\beta_0}\left(\frac{r_e N^2 l}{\varepsilon_n\sigma_{bz}}\right)^2, \tag{23}$$

Assuming $\varepsilon_n \approx \lambda_L/\pi$, and $\beta_0 \approx \pi r_L^2/\lambda_L$ for the laser-produced beams,

$$L = \frac{\gamma}{8 r_L^2}\left(\frac{r_e N^2 l}{\lambda_L\sigma_{bz}}\right)^2. \tag{24}$$

As an example, let us consider the electron-electron collider with the center of mass energy of 5 TeV. Assuming collision of two electron bunces with the number of particles $N = 10^{11}$ and the bunch length $\sigma_{bz} \sim 0.7$ mm produced by the laser with $\lambda_L = 1\mu$m and $r_L = 3\mu$m in the overdense plasma lens with length of ~ 3 mm, the luminosity becomes $L \sim 1 \times 10^{41}$ cm^{-2}/shot. In order to make this collider, two laser pulses with peak power 20 PW and the total pulse energy of 80 kJ will be required.

CONCLUSIONS

In the sub- or the relativistic laser intensity less than 10^{20} W/cm^2, the laser wakefields generated by the ponderomotive force of ultrashort laser pulses make it possible to accelerate particles in the accelerating gradient of the order of 100 GeV/m in underdense plasmas. We have made developments of the high quality beam injector consisting of the photocathode RF gun and the 150 MeV microtron and the 2 cm plasma waveguide using the fast Z-pinch capillary discharge to achieve

GeV range electron acceleration by LWFA using the 100 TW, 20fs, 10 Hz laser at JAER-APR. In addition to these experimental achievements, we will construct the femtosecond bunch slicing stage as a part of the laser acceleration test facility to generate a femtosecond electron pulse injected into laser wakefields with femtosecond accuracy.

In the ultra-relativistic regime of the laser intensity higher than 10^{20} W/cm^2, the laser ponderomotive scattering results in particle acceleration in plasmas, which is called as Dirac accelerator based on the strong field-particle point-like interaction different from classical acceleration mechanism such as RF accelerators and LWFA. The Dirac accelerators can accelerate electrons up to PeV range energies with the laser intensity $> 10^{24}$ W/cm^2 in a point-like interaction.

The superstrong laser-matter interactions results in a superhigh energy superhigh luminosity particle collider, which is expected to explore the frontier of ultrahigh energy particle physics such as Higgs boson, Supersymmetry and Grand Unified Theory.

REFERENCES

1. T. Tajima and J. M. Dawson, Phy. Rev. Lett. **43**, 267 (1979).
2. K. Nakajima et al., Rhy. Rev. Lett. **74**, 4428 (1995).
3. H. Dewa et al., Nucl. Instr. and Meth. in Phys. Res. **A410**, 357 (1998); M. Kando et al., Jpn. J. Appl. Phys. **38**, L967 (1999).
4. P. L. Kapitza and P. A. M. Dirac, Proc. Cambridge Philos. Soc. **29**, 297 (1933).
5. K. Nakajima, Nucl. Instr. and Meth. in Phys. Res. **A410**, 514 (1998).
6. K. Yamakawa, M. Aoyama, S. Matsuoka, T. Kase, Y. Akahane, and H. Takuma, Opt. Lett. **23**, 1468 (1998).
7. M. Kando et al., Proc. of the 1999 Part. Accel. Conf. 5, 3704 (1999).
8. R. W. Schoenlein et al., Science, **287**, 2237-2240 (2000).
9. A. A. Zholents and M. S. Zolotorev, Phys. Rev. Lett. **76**, 912-915 (1996).
10. P. Sprangle et al., Phys. Rev. Lett. **69**, 2200 (1992).
11. E. Esarey et al., IEEE Trans. Plasma Sci. **24**, 252 (1996).
12. T. Hosokai et al., Optics Letters, **25**,10-12 (2000).
13. F. V. Hartemann, S. N. Fochs, G. P. LeSage, N. C. Luhmann, Jr., J. G. Woodworth, M. D. Perry, Y. J. Chen, A. K. Kerman, Phys. Rev. E, **51**, 4833 (1995).
14. J. B. Rosenzweig and Pisin Chen, Phys. Rev. D, **39**, 2039-2045 (1989).

FORUM on Superstrong Fields
and High Energy Physics

Toshi Tajima* and Gerard Mourou¶

*LLNL, Livermore, CA 94551, USA
¶ CUOS, University of Michigan, Ann Arbor, MI 48109, USA

The Livingston chart of the evolution of the accelerator energy as a function of time is just one testimony of how spectacular the accelerators and high energy physics have progressed in the last century. Equally impressive in its advance has been that of lasers, though its inception was a quarter century younger. These explosive developments and successes in the respective fields have felt the presence or progress of the other in recent years in some of specific occasions and endeavors. Nonetheless, these two fields so far have marched relatively independently from each other. We are perhaps now at a crossroad where the facilitation of the one field by the other may become mutually stimulating and even more profitable for the future progress. At the Varenna Conference on Superstrong Fields in Plasmas, we held a Forum on applications of intense lasers to high energy physics (HEP). Five panelists with a broad range of expertise from laser physics to plasma physics to high energy physics discussed this emerging cross-road of both fields in front of the Conference participants: T. Tajima, chair, G.Mourou, D.Habs, K.Nakajima, and P. Chen.

The new technology based on lasers has a potential to bring in (1) extremely high accelerating gradient and (2) very bright beams.

With the adoption of lasers, it was suggested that low emittance (in another word, very high quality and thus bright) beams of particles may be generated, which may be injected into high energy accelerators. Such bright injection beams may also help the performance of an accelerator based laser called 'Free Electron Laser', in turn, which is touted to be employed with unprecedented brightness, clarity and resolution (in ~Angstrom) as a probe of material and biological science. The high accelerating gradient has been one hallmark of laser driven acceleration (routinely exceeding by several orders of magnitude over the conventional methods, but only as a proof-of-principle fashion so far), but its control of quality of beams has lagged so far. This emerging new capability with intense lasers with its 'greenness' inherent in a new technology may perhaps be best utilized with some combination with the existing more established high energy physics technologies. This way, one can combine the strengths of two different technologies in a fashion to complement each other using their respective strength. Laser-cooled crystalline beams, which have been discussed at the Forum as a first experimentally demonstrated case, will bring in extremely

CP611, *Superstrong Fields in Plasmas:* Second Int'l. Conf., edited by M. Lontano et al.
© 2002 American Institute of Physics 0-7354-0057-1/02/$19.00

bright beams to HEP accelerators in a different method. As another application of the large gradient of laser driven accelerators, to prompt accelerate unstable particles such as pions and muons has been suggested. Such capability will help increase the quality of beams of muons and even neutrinos in the future.

The Forum discussed the new vista of the laser power leap, which is dubbed as 'en route to Zettawatt' during the conference, reminiscent of the paper 'en route to Petawatt' by Mourou a generation ago when the new technology chirped pulse amplification (CPA) was invented. With the current development of large energy lasers such as NIF (National Ignition Facility) at Livermore and LMJ(Laser Mega Joule) in France in combination of CPA and other technology will allow us an opportunity to make this leap in laser power in several orders of magnitude. Though the current intense lasers are already relativistic, these new lasers will push us into ultrarelativistic regimes. A class of new physics associated with accelerating particles to high energies, sometime to extreme high energies, has been discussed. For example, extreme high accelerating gradient of laser causes equivalent extreme high gravity, which in turn may serve to explore strong gravity effects. This includes studies to test the strong limit of Einstein's Equivalence Principle of General Theory of Relativity and even to probe the presence of (gravitationally) curved spacetime horizon that lies within a finite distance. Some of these signatures may be manifested in a phenomenon of the so-called Unruh radiation (a sister to the Hawking radiation near a black hole). There is also a potential to detect the 'leakage' of extra-dimensions of quantum gravity if the gravitational interaction 'leaks' out sufficiently far beyond the shrunken horizon.

In spite of these far-out fundamental science frontiers that extreme high field lasers may bring in, there have been also a substantial set of suggestions that may be realized in a nearer term in collaboration with the existing high energy scientists at various facilities. Mentioned are:
(a) laser electron injector with ultra-low emittance, (b) laser ion injector beam sources, (c) prototype laser-based next generation accelerators, (d) prompt acceleration of unstable particles, (e) spin-polarized positrons, (f) testing cosmic acceleration in laboratory, and (g) combination of high laser field and high energy beams such as gamma-gamma collider, and other energy boosting ideas.

In the end there was a suggestion to draw a resolution by the Forum participants to call for a greater interaction with the high energy physics community, which read: "In order to facilitate the activity of promotion of the development of near-term projects relevant to High Energy Physics Laboratories, we form an ad hoc committee to advance the liaison between HFS (high field science) and HEP communities and, if necessary, to communicate with the leaders and management of HEP labs to enhance this relation. This committee can also sensitize our understanding of the needs of lasers that are necessary to fulfill such developments. Any interested person in this Forum/Conference is invited to join this committee."

*The work was in part supported by US DOE W-7405-ENG-48 and NSF.